Formation and Trapping
of
Free Radicals

Formation and Trapping
of
Free Radicals

——— *Edited by* ———

ARNOLD M. BASS and **H. P. BROIDA**

National Bureau of Standards
Washington, D.C.

1960

ACADEMIC PRESS  *New York and London*

QD
471
.B38F
1960

CONTRIBUTORS

R. S. ALGER, *United States Naval Radiological Defense Laboratory, San Francisco, California*

C. H. BAMFORD, *Courtaulds Ltd., Maidenhead, Berks., England*

D. E. CARR, *Phillips Petroleum Company, Bartlesville, Oklahoma*

BERTRAM DONN, *Department of Physics, Wayne State University, Detroit, Michigan**

JAMES W. EDWARDS, *Research and Engineering Division, Monsanto Chemical Company, Dayton, Ohio*

J. L. JACKSON, *National Bureau of Standards, Washington, D. C.*

C. K. JEN, *Applied Physics Laboratory, The Johns Hopkins University, Silver Spring, Maryland*

A. D. JENKINS, *Courtaulds Ltd., Maidenhead, Berks., England*

F. A. MAUER, *National Bureau of Standards, United States Department of Commerce, Washington, D. C.*

H. MORAWETZ, *Polymer Research Institute, Polytechnic Institute of Brooklyn, Brooklyn, New York*

H. S. PEISER, *National Bureau of Standards, Washington, D. C.*

GEORGE C. PIMENTEL, *Department of Chemistry, University of California, Berkeley, California*

D. A. RAMSAY, *Division of Pure Physics, National Research Council, Ottawa, Canada*

F. O. RICE, *Department of Chemistry, Georgetown University, Washington, Washington, D. C.*

THOMAS M. SHAW, *General Electric Company, Microwave Laboratory, Palo Alto, California*

B. A. THRUSH, *Department of Physical Chemistry, University of Cambridge, Cambridge, England*

MAURICE W. WINDSOR, *Space Technology Laboratories, Los Angeles, California, Guest Scientist, Free Radicals Research Program, National Bureau of Standards, Washington, D. C.*

* Present address: Goddard Space Flight Center, National Aeronautics and Space Administration, Washington, D. C.

PREFACE

This book has been planned to provide in a single source an account of the present status and techniques of free radical stabilization and to indicate potential new areas of interest. Thus it will serve as a guide to new workers as well as more experienced researchers interested in special aspects of the field. It should aid the reader in the task of extracting from the literature the background, techniques, ideas, and possible fruitful areas of research.

Free radicals—defined simply as molecular fragments which normally have a very short lifetime (of the order of milliseconds), which are highly reactive, and which are generally characterized by having an unpaired electron—have a long history in the field of studies of chemical reactions. Since the earliest direct evidence of their existence was obtained at the turn of the twentieth century, the published literature has been rich with papers describing experiments designed to unravel their properties as well as with discussions of mechanisms for their production and destruction. Much of the work on the investigation of free radical reactions in the gas and liquid phases has been summarized in other volumes.

Advances in techniques for low temperature experimentation have once again focused attention on free radicals. The recent developments are in the direction of studies of stabilization of reactive species, i.e., greatly increasing lifetimes of very reactive species. Under the proper circumstances of preparation, free atoms and other low molecular weight radicals can be captured and stored for long periods of time in solidified gases at low temperature. This approach suggests many new kinds of experiments to provide information about properties of free radicals and of solidified gases. Physicists have joined the physical chemists and chemical kineticists in a very broad attack on these problems. The motivations for such a broad approach arise from such diverse interests as the roles of free radicals in chemical reactions, the physics of the interactions of free radicals with solid lattices, and practical considerations related to energy storage for propulsive purposes, construction of polymers, radiation damage, and catalytic processes. Preparation of this volume was undertaken when, in considering the status of work in this new field, the need became apparent for an integrated reference work that would be useful to those concerned with all of these aspects.

The editors wish to thank the individual contributors for their excellent cooperation in the preparation of this volume. We are especially indebted to Miss Helga Viertel for the tireless assistance she provided at all stages of the book.

A. M. Bass

H. P. Broida

November 30, 1959
Washington, D. C.

CONTENTS

INTRODUCTION

Although the knowledge of free radicals goes back more than half a century, so much work has been done lately on free radical stabilization that it is time for a book on the subject. Surges in research in an older field usually occur because of the sudden availability of new powerful research tools or because a utilitarian interest has been aroused. Free radical stabilization is no exception.

Stability of radicals at ordinary temperatures is, of course, demonstrated in irradiated polymers. However, it was. the maturing of the cryogenics technology which held out the promise of stabilization in higher and more interesting concentrations, especially for the lighter, highly active species. In addition to the ingenious extension and application of conventional analytical techniques such as emission and absorption spectroscopy, X-ray diffraction, and calorimetry, the new electron spin resonance machines have opened up dramatic new possibilities.

The widespread application of thoroughly modern research techniques has been added to the sometime frenetic support of the propellant interests. The more prosaic, but economically more important application areas of catalysis and new polymer formation have yielded, from the point of view of fashion, to that of missiles, rockets, and satellites. In the experimental missile business the total cost per firing will continue to be enormous for some time ahead. Therefore any prospects of system simplification, however remote the realization, demand intensive investigation. In this case, however, a dearth of basic information stimulated the initiation in 1956 of a three year free radicals research program, at the National Bureau of Standards, which was to be directed toward increasing fundamental knowledge. Several of the contributors to this book were associated with this program. Since this book was written near the termination of the program it is possible now to assess the great amount of basic data accumulated and yet, to all outward appearances, the goal of application to propulsion is as far away as before. The difficulties can be described more accurately, many limitations for stabilization as a function of temperature and environmental matrix have been found, and all the easy approaches have been tried.

We are left, therefore, almost where we were before the space age reached us, except that we have explored more rapidly the formation and stabilization of light species. The emphasis returns to utilization in chemical processes, to the scientific indication of radiation effects and certain biological processes, and perhaps, to the slow problem of unravelling the chemical reactions observed in astrophysical phenomena. The industry is grateful for the support of the missile czars—science ran for a short way where it formerly walked.

The authors have drawn upon their own recent researches as well as that of other leaders in their respective fields to cover the fields of formation, the analytical techniques, the measurement of properties of materials in which free radicals are present, and finally to the several paths for utilization. It seems to me that the area least known is that of the role of free radicals in biology. I would hope that this book would serve to interest more physicists and chemists in the new interdisciplinary field and also give courage to the biologist who wishes to become aware of the nonbiological and, for the most part, nondestructive, analytic techniques recently developed.

One special feature of recent research in the field is the emphasis on measurement of physical properties of material containing free radicals. Whereas previously free radicals were pursued from the standpoint of reaction kinetics and chemical intermediates, no one had accumulated sufficient quantities to study their effects *per se*. Thus gross effects such as total heat released on recombination, changes in thermal conductivity and dielectric properties, and crystal structure alterations are now measurable at the higher concentrations.

We do not wish to imply, however, that spectacular success in precision measurements has been achieved. It is only in the most favorable instances that concentrations as high as a few tenths of a molar per cent have been realized in a stable solid. The measurements of properties are in most cases barely above threshold. We do not have that happy state which exists in semiconductor physics. There, straight electrical data taken with small error can give significant information about chemical impurities present in the parts per million, or parts per billion range. One reason for the difference is that in the case of stabilized free radicals we are dealing with immobile noncharged units instead of mobile charges.

Free radicals, as discussed in this volume, are considered as highly reactive fragments of molecules, electrically neutral, but characterized by an unpaired electron. Many of the fragments described are elemental or atomic, such as hydrogen, nitrogen, or oxygen atoms. There is much emphasis in this volume on light species, partly because they are more attractive to the low temperature physicists and partly because they are the most interesting, energetically speaking.

And yet, the reader may wonder, if the field is embraced with such enthusiasm by both chemists and physicists why is the understanding so little? The answer may lie in the complexity of the phenomena observed and in the relative newness of the research tools. For each trained specialist in a given field has had to become acquainted with one or more new fields and, by the slow process of self-education or by the sometimes equally inhibitive acts of communication with a co-worker trained in a different discipline, the limitations of seemingly attractive new approaches have been discovered.

This is truly an interdisciplinary field. The spectroscopist, the crystallographer, the microwave physicist, the thermodynamicist, the radiation chemist must combine their talents with those of the organic chemist, the chemical engineer, and the cryogenicist before the technical picture can be brought in focus. Add to these the interest of the biologist and the astrophysicist and one begins to appreciate how this field really touches all of modern science.

In deference to those who seek, with diverse motivations, the "connection" between basic science and the economics of applied science, a substantial portion of this volume is devoted to free radical utilization. For some this may be the most exciting reading, for while a few examples of trapped radicals are already common knowledge, the applications are limited only by a correct grasp of the facts and by one's imagination.

The section of the book on utilization is important from at least two standpoints. One is that of establishing perspective for those who would invest blindly in the utility of a currently fashionable branch of science. The other is to present a challenge to those who must support research in universities, in industry, and in government, so that this field of universal potential application will continue to receive its share, so to speak, of the resources available. This appeal will be felt by those concerned with space and upper atmosphere research, with polymer chemistry, with the development of nuclear energy, with the quest for cancer control, as well as that vast segment of the chemical industry concerned with petroleum derivatives.

In the case of propulsion applications the research carried out so far has gone a long way to establish pessimistic upper limits, at least for systems based solely on recombination effects. The contribution to be made by the use of free radical "additives" however, has not been assessed.

For space exploitation, continued strong research is an absolute necessity if we are to understand the new environment. Although near the earth at least the species are limited to a very few, nevertheless we know that conditions exist for reasonable lifetimes. Thus the information obtained at low temperatures in the laboratory may tell us much about upper atmosphere phenomena, especially with respect to photochemical reactions. If and when the planetary atmospheres are penetrated a greater variety of species will be encountered but our sophistication in initial encounters will be enormously enhanced by whatever we shall learn in the meantime.

The free radical interpretation of radiation effects in certain solids has been so valid as to presage the success of continued understanding in the radiobiological field. Some of the analytical techniques, such as electron spin resonance, are nondestructive to tissue. If means are found to induce stabilization through modest excursions in the low temperature field, greater insight into the nature of carcinogen effects as well as radiation effects may be realized. It is vital to increase our knowledge of the possible deleterious effects of small quantities of free radicals themselves.

In the field of industrial chemistry we may expect not only new compounds to be derived from free radical research, but major changes in processes caused by the deliberate injection of controlled quantities of free radicals for controlled periods of time, thus yielding a new process parameter or dimension for the engineer. Free radical reactions may be the controlling factors in combustion, in smog, and in the improvements in conversion of fuels to electricity.

This volume will serve a further useful purpose if it provides a degree of inspiration for teachers and research students. While an impressive survey of various aspects of free radicals investigations is included it will be obvious that the chief impact is made by the material omitted. Many times only order of magnitude results are given because only that is available. Provocative questions and new lines of inquiry are suggested. In addition to the theory of the condensation process, we may add that little progress has been made on understanding the kinetics of recombination, the effects of diffusion of radicals on surfaces, the behavior of ionic free radicals, the fate of the various species created in a gaseous discharge, the details of energy absorption and redistribution in the photochemistry of solids, as well as the details of bringing theory and experiment together in the electron spin resonance and solid glow and afterglow emission spectrum observations.

To illustrate the missing material a final comment may be made about the hydroxyl radical. For a radical which is one of the most abundant in the world, albeit frequently charged, there has been little success in stabilizing it in such quantities as to permit measurements. It is a key species in biology, in combustion, and the ocean (both liquid and solid), and thus should be one of the most rewarding radicals for research.

JAMES W. MOYER

General Electric Company,
Santa Barbara, California

1. History of Radical Trapping

F. O. RICE

Department of Chemistry, Georgetown University, Washington, D.C.

The history of radical trapping goes back barely two decades and, as a serious and widespread study, it has been undertaken only in the last few years. As a matter of fact, following earlier unsuccessful attempts to isolate and study free radicals, most of the organic texts published even as late as the early 1930's contained statements to the effect that free radicals do not exist. For example: "The assumption of the existence of radicals, capable of existing alone and playing a special role in chemical reactions, has long been abandoned" (Richter, 1915). "It took a long time before it was finally recognized that the very nature of the organic radicals is inherently such as to preclude the possibility of isolating them" (Ostwald, 1896). "Negative results gradually established the doctrine that a free carbon radical was incapable of independent existence" (Porter, 1926). Since free radicals not only exist and can be trapped, but actually play a vital role in a multitude of chemical reactions, we are faced with the question: "Why is it that they are being studied only at this late date?"

In order to answer this question, we have to go back a long time and recognize that chemistry as a science made several false starts. Greek science vastly oversimplified the problem, mainly directing attention to properties which, it was supposed, could be changed by suitable manipulation of matter. This resulted, for example, in a vain search over a period of hundreds of years for a process by which baser metals could be transmuted to gold or for an elixir that would confer eternal youth. Hundreds of years later, even as late as the eighteenth century, chemistry was dominated by the phlogiston theory, originally developed to explain combustion. According to this theory, burning was caused by the escape of phlogiston from substances rich in this material. The extension of this theory throughout chemistry led to such absurd situations as occurred with the discovery of chlorine by Scheele (1774) by heating hydrochloric acid with manganese dioxide. At that time, the name for hydrochloric acid was marine acid gas since it was made by heating sea salt with oil of vitriol. Manganese dioxide was regarded as a substance lacking in phlogiston, and so chlorine obviously should be (and was) called dephlogisticated marine acid gas. The phlogiston

theory was finally discredited at the beginning of the nineteenth century as
a result of the work of Lavoisier. Only then, at long last, was the science
of chemistry started in the right direction with the enunciation of the atomic
theory by John Dalton (1808). We are thus confronted with several thousand years of almost fruitless labor, followed by a century and a half of
richest accomplishment. Curiously enough, the recognition and general
adoption of the doctrine of valency, necessary indeed for the unparalleled
rate of progress of chemistry in the second half of the nineteenth century,
ruled out all consideration of the possibility of the existence of free radicals.
Consequently, it is only in very recent decades that chemists and physicists
have joined in exploiting the vast potentialities that lie in chemical transformation involving free radicals.

Both the theoretical and the technical difficulties confronting the chemists of a century and a half ago were of fantastic proportions. Even in
physics, which as a science had a respectable antiquity as compared with
chemistry, a gas was regarded as consisting of heavy particles surrounded
by caloric or heat which gave the gas its elastic properties. In spite of this,
Avogadro had recognized in 1811 that relative molecular weights could be
determined by weighing equal volumes of gases under the same conditions
of temperature and pressure. Unfortunately, neither the importance nor
the practical value of his hypothesis were recognized for several decades,
so that it was only by the middle of the nineteenth century that atomic
weights could be determined with any degree of certainty; previously what
were thought to be atomic weights were often equivalent weights.

In the first half of the nineteenth century, chemists began to achieve an
understanding of the real meaning of chemical union. Thus, hydrogen
atoms unite not only with chlorine atoms, but also with bromine atoms
and with iodine atoms, and it seemed only natural to expect that the very
reactive chlorine atom would combine with more atoms of hydrogen than
would the less reactive iodine atom. On the other hand, the highly reactive
fluorine atom has no attraction for either the helium or the neon atom
though it combines in a 1:1 ratio with hydrogen, just as iodine does. In
the 1860's, chiefly owing to the work of Kekule (1857), the doctrine of
valency seems to have been definitely established. At approximately the
same time, and also based on Kekule's work, came the recognition of the
fact that it is the bond strength that increases or diminishes in a series of
the elements. An astonishing proportion of Kekule's work on bonds, for
example his ideas on constant valency, multiple bonds, and the consequent
formula for benzene, has been confirmed by modern studies. This achievement is all the more remarkable in view of the nineteenth century scientists'
opinion of the atom as expressed in somewhat theological terms by Newton
in "Opticks" (1730): "God, in the beginning formed matter in solid, massy,

hard, impenetrable, movable particles, so very hard as never to wear or break in pieces, no ordinary power being able to divide what God himself made one in the first creation." The great Russian scientist, Mendeleef (1889) who was of a very untheological turn of mind, all his life fought the idea that we ever could learn anything about atomic structure. To him the notion of a single kind of matter out of which all atoms were made "was taken from a remote antiquity, when it was found convenient to admit the existence of many gods—and of a unique matter." The establishment and consolidation of the doctrine of valency was finally accomplished by Mendeleef's periodic system of classification of the elements. At last it was clear that valency is a periodic property of the elements, and not just a convenient method of classifying compounds.

During the first quarter of the twentieth century, the physicists, obviously quite unaware of the concept of valence, were very busily studying free radicals produced in flames and by electrical excitation; at the same time the chemists were steadfastly denying the very possibility of the existence of free radicals. Such dissension is more comprehensible when we recall that the radicals studied by the physicist were very simple ones, such as hydrogen or nitrogen atoms or, at most, diatomic fragments. In the light of the experimental evidence on which the doctrine of valency was based, complicated radicals consisting of three, four, or more atoms could no more be expected to be stable than a bridge from which the keystone was removed.

During the middle of the nineteenth century, scarcely more than a hundred years ago, there was intensive work in many laboratories on free radicals. Thus, Frankland (1849) wrote: "The whole of the above facts taken together prove beyond doubt, that when iodide of ethyl is decomposed by zinc, at an elevated temperature, the ethyl radical is present among the gaseous products and may easily be separated in a state of perfect purity." In the same year, Kolbe (1849), who was studying the electrolysis of acetic acid, wrote: "Hence it appears that the gas evolved with carbonic acid at the positive pole is actually methyl, containing not even a trace of marsh gas, which requires the double volume of oxygen for its complete combustion and produces only an equal volume of carbonic acid."

Almost immediately, the work of Frankland and of Kolbe was questioned, especially by Laurent (1850), who was convinced that the purported radicals were actually dimers and, therefore, homologs of methane. Kolbe (1851) vigorously protested this interpretation of the experimental data, and it was not until 1864 that the matter seemed finally settled. Schorlemmer (1862), who was well versed in hydrocarbon chemistry, studied the products of the destructive distillation of coal. In 1864, he finally proved that what was thought to be methyl was really ethane. "It appears from these experiments that what I have formerly proved for the higher terms of the hydrocarbons of the series C_nH_{2n+2}—namely that no chemical difference exists between the hydrides and the radicals, holds also good for the lowest isomers of these groups." At this point, for several decades, all mention of free radicals seems to have vanished from chemical literature.

It was only toward the end of the nineteenth century that the possible importance of radicals in organic chemistry was revived by Nef (1892). He challenged the idea of the constant valency of carbon and tried to interpret the mechanism of a wide variety of organic reactions as proceeding through bivalent carbon. For example, he assumed that there were two cyanogen radicals which could form three compounds having the formula C_2N_2. Thus, he contrasted methyl cyanide $CH_3C{=}N$ with its isomer methyl isocyanide, which he wrote $CH_3N{=}C$. Although Nef's ideas proved to be largely incorrect, he doubtless influenced chemical thinking toward the possibility of free radical mechanisms and thus made somewhat less difficult the acceptance of Gomberg's subsequent discovery of triphenylmethyl and related compounds.

ʃThe first free radical of proved independent existence was accidentally prepared by Gomberg in 1900 by the action of metallic silver on triphenylchloromethane.

$$2(C_6H_5)_3C\ Cl + 2\ Ag \rightarrow 2\ AgCl + (C_6H_5)_3C{-}C(C_6H_5)_3$$

Instead of the anticipated inert colorless hydrocarbon, the reaction produced a highly reactive substance which gave colored solutions and reacted readily with a wide variety of substances, such as oxygen, nitric oxide, iodine, and metallic sodium. Gomberg proposed that the compound dissociated into two electrically neutral free radicals according to the equation:

$$(C_6H_5)_3\ C{-}C(C_6H_5)_3 \rightarrow 2(C_6H_5)_3C$$

Subsequent work not only supported the original proposal, but also showed that the degree of dissociation is very considerably enhanced when the phenyl groups in hexaphenylethane are replaced by more complex aromatic nuclei. ʅ

A complication of the above simple picture occurred when it was found that hexaphenylethane and its analogs dissociated into ions when dissolved in liquid sulfur dioxide or other ionizing solvents:

$$(C_6H_5)_3C{-}C(C_6H_5)_3 \rightarrow (C_6H_5)_3C^- + (C_6H_5)_3C^+$$

With increased understanding of atomic and molecular structure it finally became clear that the stability of these complicated aromatic species is related to the greater possibilities of resonance that the structures allow.

Doubtless Gomberg's work resulted, in the first quarter of the twentieth century, in a renewed vain search for ways to prepare aliphatic radicals such as CH_3 and its homologs; since attempts using classical chemical methods were bound to fail, there are very few references to such work. In one of the few published accounts (Salzberg and Marvel, 1928) an attempt is described to prepare an aliphatic free radical by synthesizing hexatertiary-butylethynylethane

$$[(CH_3)_3C-C\equiv C]_3\ C-C[C\equiv C-C(CH_3)_3]_3$$

with the hope that it would dissociate. However, it was found that the resulting compound had the normal molecular weight, thus proving that it did not dissociate into free radicals.

Considering that a free radical is produced whenever an ion is discharged at an electrode, it seems strange that the ordinary electrolytic cell has been so little used as a tool for the production and study of radicals. The experimental demonstration of this fact was systematically carried out by Hein and Segitz (1926), who studied solutions of sodium ethyl dissolved in zinc diethyl. Presumably, such a solution conducts the electric current through ionization of the sodium ethyl:

$$NaC_2H_5 \rightarrow Na^+ + C_2H_5^-$$

Experimentally, when a current is passed through the solution using platinum electrodes, a mixture of ethane, ethylene, and butane appears at the anode. When, however, lead, antimony, bismuth, or similar electrodes are used, one finds that the corresponding organometallic compound is formed. Thus, it is evident that the ethyl radical liberated at a lead anode forms tetraethyl lead.

Audubert and Verdier (1939) report that when solutions of hydrazoic acid are electrolyzed, flashes of light appear at the anode presumably resulting from some sort of reaction of the radical N_3 formed from N_3^-.

The actual discovery of the methyl radical was achieved by Paneth and Hofeditz (1929) by heating lead tetramethyl in a flowing system at pressures of a few tenths of a millimeter. It turned out that, at room temperatures, the methyl radical is extraordinarily reactive. Thus, it requires little or no activation energy to combine with itself to form ethane. Under the conditions of Paneth's experiment, the half-life of methyl is about 0.006 sec and hence the methyl radical cannot be preserved at room temperatures. Subsequent investigations indicated that the curious reaction of methyl with certain metals with which it forms volatile organometallic compounds is the most effective way of studying the radical directly. To accomplish this, the following conditions are necessary: (1) The gas pressure must be extremely low in order that the methyl radical can rapidly diffuse to the metallic surface. (2) The metal surface must be absolutely clean. (3) The metal must be in the form of a mirror. Thus, methyl radicals do not, for example, remove lead atoms from the bulk metal even though the surface be clean.

Paneth's work was quickly followed by the experimental demonstration (Rice *et al.*, 1932) that, on pyrolysis, ordinary organic compounds such as hydrocarbons, ketones, ethers, etc., also decompose into free radicals which can be detected by their reactions with metallic mirrors.

Although all this work was of interest to the chemist, it had no great influence either on fundamental chemical thought or practical chemical experimentation. The consensus of the time may be illustrated by the following two quotations: "The possibility of free radicals being formed as intermediate products in the course of chemical reactions whilst always admissible, is only occasionally supported by the experimental evidence" (Bennett and Chapman, 1930). "The most important use of work on free atoms and radicals appears to be in the interpretation of the mechanism of thermal and photochemical reactions" (Bowen, 1933).

The greatest difficulty in accepting free radicals as playing a dominant role in the thermal decomposition of organic compounds lay in the fact that many such thermal decompositions had been found to follow simple laws when they were studied kinetically. It was hardly to be expected that the complicated chain reactions would be found to follow such simple laws. The difficulties of the time are well presented in the following extract taken from a review (Bowen, 1933): "In all reactions of the unimolecular type, it appears that a mechanism involving a bimolecular interaction of two parts of the same molecule, rather than a simple splitting of a link, is more probable, e.g., in the decomposition of acetaldehyde, the hydrogen atom begins to link up with the methyl group before detaching itself from the carbonyl group; and while the C—H and C—CH$_3$ linkages are weakening, the carbonyl bond is simultaneously undergoing reorganization to give carbon monoxide."

These difficulties were finally met by the theoretical demonstration (Rice and Herzfeld, 1934) that, contrary to expectation, complicated free radical mechanisms can follow kinetically simple laws. However, the matter is not yet finally settled since many papers appear each year (Danby *et. al.*, 1955; see however Voevodsky, 1959) presenting evidence that thermal decompositions are composite, proceeding partly through a free radical chain and partly through a molecular transformation.

The production, trapping, and stabilization of free radicals at low temperatures in rigid media seems to have been first systematically studied by Lewis and his students in the early 1940's. Working with a wide variety of rather complex aromatic molecules, they showed that illumination with ultraviolet light could cause dissociation either (1) into two radicals or (2) into a positive and a negative ion or (3) into a positive ion and an electron. From their work describing changes in electronic absorption spectra, it appeared likely that stabilization is closely connected with the rigidity of the solvent and is relatively independent of temperature. Subsequently, spectrophotometry and electron spin resonance have been used to study free radicals stabilized in glasses.

The period 1950 to 1959 marks the beginning of a massive attack in which all the resources of the chemist and the physicist have been brought to bear on the problem. Supplementing the purely scientific interest, there were the very practical applications in a wide variety of industries dependent on chemical reactions which are unquestionably free radical in nature. In addition, free radicals are of the greatest interest in connection with the revolution now occurring in the use of fuels. This revolution consists of the replacement of the steam engine and internal combustion engine by the rocket motor—the only means of making available the energy of fuels for space travel. Since the bonds joining the atoms of stable compounds do not vary very greatly in strength, the fuel load can be kept low only by having a fuel consisting of elements of low atomic weight. If, in addition, the fuel could consist in whole or in part of free atoms or radicals the energy available per pound of fuel would be very greatly increased. These ideas are discussed in more detail in Chapter 13.

Work during the period 1930 to 1950 had made it evident that radicals combine with little or no activation energy (Steacie, 1954), so that there seems to be no possibility at all of preserving them in high concentration at ambient temperatures. Presumably at or near the absolute zero, one might hope to preserve radicals in 100 % concentration, but at any higher temperature it would be necessary to dilute the radicals with molecules that form a rigid structure. The most obvious way, therefore, to produce stabilized radicals is to freeze a substance at very low temperatures and either expose it to some suitable form of electromagnetic radiation or to bombard it with particles such as electrons, neutrons, or protons. As stated above, Lewis and his students had already used ultraviolet radiation for this purpose; but the method was now greatly extended.

Radicals may also be trapped (Rice and Freamo, 1951, 1953) by preparing them in the gas phase in a flowing system at low pressure and quickly freezing the gases leaving the reaction zone. The gases may be cooled to liquid nitrogen or even to liquid helium temperatures. Radical production in the gas phase may be most easily brought about by heat or some form of electric discharge.

The method of establishing an equilibrium or stationary state at high temperatures and rapidly cooling the system to low temperatures is very old, and it is not limited to man-made processes. Ages ago, at great depths in the earth, diamonds formed at very high pressures and temperatures were suddenly brought to the surface by movements of the underlying magma and were preserved by the sudden cooling. A similar sort of rapid cooling and consequent freezing of an equilibrium established in a condensed system is demonstrated in the freshman chemistry experiment in which plastic sulfur is formed by pouring molten sulfur into cold water.

The formation of nitric oxide from oxygen and nitrogen is another example of the formation of a free radical at high temperatures and its stabilization by sudden cooling. At room temperature the equilibrium $N_2 + O_2 \rightleftharpoons 2NO$ lies almost entirely to the left, whereas at 3000° C the equilibrium concentration of NO is greater than 99 %. By sudden cooling to room temperature, the nitric oxide can be stabilized. This process has been utilized on the commercial scale in the production of nitric oxide. However, because of practical difficulties inherent in heating gases to a very high temperature to be followed by rapid cooling, the process has been displaced by the Ostwald process of oxidizing ammonia. In this extraordinary reaction, ammonia is oxidized by air on the surface of a solid catalyst such as ferric oxide or platinum, the main reactions being:

$$4 \ NH_3 + 3 \ O_2 \rightarrow 2 \ N_2 + 6 \ H_2O$$

$$4 \ NH_3 + 5 \ O_2 \rightarrow 4 \ NO + 6 \ H_2O$$

The temperature at which oxidation begins, depends on the method of preparation of the catalyst; at low temperatures, the formation of NO is favored. This process is used on a large scale in the preparation of nitric oxide for ultimate conversion into nitric acid. Since nitric oxide, which is unquestionably a free radical, can be prepared in this way, the catalytic preparation of other free radicals seems also possible.

Krieger (1957) has published some interesting work in which butane and other substances are passed over catalysts and the free radicals identified. The radicals produced depend both on the catalyst and on the temperature. Thus it seems clear that the course of radical decompositions can be profoundly affected by catalysts.

We are now satisfied that practically all radicals can be trapped and stabilized by these various methods, so that we now have the problem of identifying them, estimating the concentrations, and studying their chemical reactivities as well as their interactions with radiation and with magnetic and electric fields. So far as chemical reactivity is concerned very little has been accomplished since, at the low temperatures necessary, matter is normally in the solid state and solids do not readily react chemically. Unfortunately, the radical materials do not dissolve in the few substances known to remain liquid to low temperatures. The chemical estimation of radical concentration can be made only in a few rather specialized individual cases. For example, HN_3 when treated by one or another of the above methods gives a blue solid (Rice and Freamo, 1951). On warming this solid to $-125°$ C, it turns white; subsequently, only ammonium azide is found on the cold finger. It is tempting to attribute the formation of the ammonium azide, at least in part, to polymerization of NH. This point could be settled by preparing and decomposing $N—N^{15}—NH$ and examining the product; ammonium azide from NH should have no N^{15}.

Another chemical method of identification is available if the active species gives off a permanent gas at the transition temperature. This has been observed in two cases. (1) the yellow product (Rice and Sherber, 1955) formed in the thermal decomposition of hydrazine which gives off almost pure nitrogen at $-178°$ C and, (2) the green product formed in the decomposition of dimethylamine, which, even at liquid nitrogen temperatures, slowly evolves almost pure hydrogen (Rice and Grelecki, 1957).

There is at least one situation in which radical chemistry at low temperatures can be readily studied. Geib and Harteck (1933) seem to have been the first to make a detailed study of the reactions of free radicals at very low temperatures, both with each other and with certain molecules. Bonhoeffer's earlier work had shown that hydrogen atoms react vigorously with both inorganic and organic substances, in the latter case usually giving a wide variety of products. Geib and Harteck (1933) showed that the over-all reaction becomes very much less complicated at low temperatures, where the only elementary steps occurring at an appreciable rate are those with a small activation energy.

We can further predict that those chemical reactions will occur at very low temperatures which involve mainly rearrangement of the electronic cloud and only very little movement of atomic centers. Presumably the most favored reactions will be either actual combination of radicals or such reactions as $O_2 + O \rightarrow O_3$ or $H + NO \rightarrow HNO$. The addition of radicals to molecules containing double or triple bonds would also be expected to occur readily at low temperatures, whereas reaction of radicals with saturated molecules would not be favored at very low temperatures. The experimental results are fully in accord with the theoretical considerations At $-190°$ C, hydrogen atoms do not react with ammonia, nitrous oxide, or methylamine whereas reaction occurs readily with oxygen, nitric oxide, sulfur dioxide, hydrocyanic acid, cyanogen, ethylene, and benzene. Similar results were obtained with oxygen atoms. In view of the high reactivity of atomic hydrogen it was surprising to find that it does not react with the blue compound (Rice and Freamo, 1953) produced by the decomposition of HN_3. Recently Klein and Scheer (1959a) have used Langmuir's method to produce hydrogen atoms and have shown that they react rapidly with solid olefins at low temperatures.

In view of the marked limitations inherent in chemical methods of analysis under the imposed experimental conditions, it was only natural that the great preponderance of thought and work has gone into the exploitation of physical methods. It must be said at the outset that the results are still highly uncertain, the reported values for the maximum free radical concentrations in different reactions ranging from a high of about 10 % to a low of a few tenths of 1 %. Obviously, gross methods such as thermal conductivity, index of refraction, and calorimetry can be expected to yield quantitative results only if the percentage of radicals present is large. Warming curves obtained from species deposited on liquid nitrogen-cooled surfaces indicate the presence of low temperature phases.

In principle, optical spectra should provide a powerful tool for the study of frozen radicals. Unfortunately, however, partly because of scattering and partly because, at room temperature, structure is often wiped out in the visible and near ultraviolet range (Rice and Ingalls, 1959), these regions of the spectrum do not add greatly to our knowledge of frozen radicals, many of which show a broad band in the visible. This usually disappears between 75° and 175° K. However, at very low temperatures, where molecular motion is greatly diminished, electronic spectra usually show structure, and, under these conditions, electronic spectra are a powerful tool for investigating free radicals. In the infrared, characteristic lines corresponding to vibrational and rotational structures appear at room temperature and scattering by the solid is much less serious than in the visible and ultraviolet. The optical spectroscopy of trapped radicals is discussed in detail in Chapter 6. By producing the free radicals in a matrix of a rare gas, Pimentel (1958) and his students have succeeded in identifying many species, since the conditions are not too different from those existing in the gas phase.

Electron spin resonance (see Chapter 7) promises to be a powerful tool for studying frozen free radicals, especially in that it gives information enabling one to distinguish between different radicals. In this capacity, it is superior to measurements of magnetic susceptibility, which do, however, distinguish between radicals and molecules. A series of X-ray and electron diffraction studies (see Chapter 9) also have been carried out with various materials containing frozen free radicals. Although the mass spectrometer, which yields a complete analysis, is our most powerful and promising tool for the study of free radicals produced in the gas phase, in general, it is not applicable, once they are frozen out.

During the years 1956 to 1959, four symposia were held devoted primarily to the study and stabilization of radicals. The first at Laval University, Quebec, September, 1956 (19 papers); the second at the National Bureau of Standards, Washington, D. C., September, 1957 (22 papers); the third under the auspices of the Faraday Society at Sheffield University, September, 1958 (21 papers); and, the fourth under the sponsorship of the National Bureau of Standards in Washington, D. C., September, 1959 (27 papers). A study of the papers given in these symposia, together with the published literature and reports during these years gives a fairly complete picture of the huge effort presently directed toward the production of free radicals and their stabilization and subsequent study.

During the course of this work, radicals have been produced by almost every conceivable means. At low temperatures, a substance may be decomposed into radicals by means of ultraviolet light or by gamma radiation. Radicals have been produced also by bombarding a deposit kept at low temperatures with charged particles, using either electrons or positive ions. Radicals may be produced also in a flowing gas system at low pressures by means of either thermal or electrical excitation and frozen out at liquid nitrogen temperatures or lower. For example, the field produced by a 2450-mc generator is a very convenient method of producing free radicals in the gas phase (see Chapter 3). In addition, there are a few special methods for producing radicals. For example, during flash photolysis and in shock tube experiments, molecules are subjected to conditions that exist only at temperatures of several thousand degrees.

A review of the study of free radicals during the period 1956 to 1959 reveals that physical methods of attack on the problem far outweighed in number and diversity the chemical methods of approach. Judging simply from the number of communications, the methods may be classified, in accordance with relative amounts of effort, in descending order as follows: (1) absorption spectra: visible, near ultraviolet, infrared, far ultraviolet; (2) mass spectrometry; (3) electron spin resonance; (4) X-ray studies at low temperatures; (5) emission spectra at low temperatures, flash photolysis. In addition to the foregoing, there is a large and diversified number of special methods among which may be listed: measurements of the dielectric constant, of magnetic susceptibility, of thermal conductivity, and of the index of refraction, as well as calorimetry.

To the chemist, of course, the transformations and reactions of radicals are of the greatest interest, and it is not surprising that a great deal of effort has gone into this study. Because the results, except in a few outstanding cases (Ruehrwein and Hashman, 1959; Klein and Scheer, 1959b), have not come up even to modest hopes, the entire chemical approach to the problem of free radicals may be unduly minimized in favor of the more powerful physical methods. Since the chemist is interested primarily in transformations of matter, there seems to be no way to avoid great disturbance of the reacting system. However, in spite of this fact, long ago the chemist had a detailed knowledge of molecular structure that only in recent years has been confirmed and extended by modern physical methods. The organic chemist knew that methane is not planar, because if it were, two chloromethanes could be made:

$$
\begin{array}{cc}
\text{H} & \text{Cl} \\
| & | \\
\text{Cl}-\text{C}-\text{Cl} \quad & \text{H}-\text{C}-\text{Cl} \\
| & | \\
\text{H} & \text{H}
\end{array}
$$

This knowledge was confirmed and extended long before X-rays by the discovery and study of optical activity. Nearly one hundred years ago Kekule realized the necessity of resonance because only one, not two, *o*-dibromo-benzenes could be prepared:

and Thiele showed that the old-fashioned formula for 1,3-butadiene

$$CH_2{=}CH{-}CH{=}CH_2$$

would have to be rewritten to explain the reactions of this compound.

It would seem, therefore, that there should be continued activity in basic research on the chemical behavior of frozen radicals to supplement the huge amount of effort put into applied work on the chemical utilization of free radicals in reactions at or near ambient temperatures. Although chemical work on frozen free radicals is severely limited by the complicated reaction mixtures, it should be remembered that there are compensations in that the multitude of competing reactions occurring at room temperature are replaced at low temperatures often by a single reaction or at most by two or three.

References

Audubert, R., and Verdier, E. T. (1939). *Compt. rend. acad. sci.* **208**, 1984. Sur l'émission du rayonnement ultraviolet par l'électrolyse de solutions d'acide azothydrique et d'azoture de sodium.

Avogadro (1811). Alembic Club Reprint, No. 4, p. 28. Williams & Wilkins, Baltimore, Maryland.

Bennett, G. M., and Chapman, A. W. (1930). *Ann. Repts. on Progr. Chem. (Chem. Soc. London),* p. 127.

Bowen, E. J. (1933). *Ann. Repts. on Progr. Chem. (Chem. Soc. London),* pp. 47, 48.

Brown, H. W. and Pimentel, G. C. (1958). *J. Chem. Phys.* **29**, 883. Photolysis of nitromethane and methyl nitrite.

Dalton, J. (1808). Alembic Club Reprint, No. 4. Williams & Wilkins, Baltimore, Maryland.

Danby, C. J., Spall, B. C., Stubbs, F. J., and Hinshelwood, C. (1955). *Proc. Roy. Soc.* **A228,** 448. The formation of CH_3D by the decomposition of n-butane in the presence of deuterium.

Fleming, S. W., and Krieger, K. A. (1957). *J. Am. Chem. Soc.* **79,** 4003. Free radicals from the heterogeneous decomposition of butane.

Frankland, E. (1849). *J. Chem. Soc.* **2,** 263. On the isolation of organic radicals.

Geib, K. H., and Harteck, P. (1933). *Ber.* **66,** 1815. Anlagerungs-Reaktionen mit H- und O-Atomen bei tiefen Temperaturen.

Gomberg, M. (1900). *J. Am. Chem. Soc.* **22,** 757. An instance of trivalent carbon: triphenylmethyl.

Hein, F., and Segitz, F. A. (1926). *Z. anorg. u. allgem. Chem.* **158,** 153. Beiträge zur Kenntnis der Natur der Metallalkyle.

Kekule, A. (1857). *Ann.* **104,** 129. Ueber die Constitution und die Metamorphosen der chemischen Verbindungen.

Klein, R., and Scheer, M. D. (1959a). *J. Am. Chem. Soc.* **80,** 1007. The addition of hydrogen atoms to solid olefins at $-195°$.

Klein, R., and Scheer, M. D. (1959b). *J. Phys. Chem.* **62,** 1011. The reaction of H atoms with solid olefins at $-195°$.

Kolbe, H. (1849). *J. Chem. Soc.* **2,** 175. Organic radicals. See Alembic Club Reprints, No. 15. The electrolysis of organic Compounds. Williams & Wilkins, Baltimore, Maryland.

Kolbe, H. (1851). *J. Chem. Soc.* **4,** 41. On the chemical constitution and nature of organic radicals.

Krieger, K. A. (1957) See Fleming and Krieger (1957).

Laurent, A. (1850). See Rice, F. O., and Rice, V.V. "The Aliphatic Free Radicals." Johns Hopkins Press, Baltimore, Maryland, 1935.

Lewis, G. N., and Lipkin, D. (1942). *J. Am. Chem. Soc.* **64,** 2801. Dissociation of organic molecules into radicals and ions; Lewis, G. N., and Bigeleisen, J. (1943). *J. Am. Chem. Soc.* **65,** 2424. Photo-oxidations in rigid media.

Mendeleef, D. J. (1889). See article by Kablukoff, *Acta Physicochim. U.R.S.S.* **1,** 9 (1934), p. 9. In memoriam D. J. Mendeleef.

Milligan, D. E., and Pimentel, G. C. (1958). *J. Chem. Phys.* **29,** 1405. Infrared spectra of diazomethane and CH_2.

Nef, J. U. (1892). *Ann.* **270,** 267. Ueber das zweiwerthige Kohlenstoffatom.

Newton, I. (1730). "Opticks," based on the 4th ed., London, 1730. Dover Publications, New York, p. 400.

Ostwald, C. W. W. (1896). Quoted by W. E. Bachman in "Organic Chemistry," H. Gilman, p. 582. Wiley, New York, 1938.

Paneth, F., and Hofeditz, W. (1929). *Ber.* **62,** 1335. Über die Darstellung von freiem Methyl.

Pimentel, G. C. (1958). See Milligan and Pimentel (1958); Brown and Pimentel (1958).

Porter, C. W. (1926). "The Carbon Compounds," p. 312. Ginn, New York.

Rice, F. O., and Freamo, M. (1951). *J. Am. Chem. Soc.* **73,** 5529. The imine radical.

Rice, F. O., and Freamo, M. (1953). *J. Am. Chem. Soc.* **75,** 548. The imine radical in the electrical discharge.

Rice, F. O., and Grelecki, C. (1957). *J. Phys. Chem.* **61,** 824. Decomposition of dimethylamine.

Rice, F. O., and Herzfeld, K. F. (1934). *J. Am. Chem. Soc.* **56,** 284. Mechanism of some chain reactions.

Rice, F. O., and Ingalls, R. (1959). *J. Am. Chem. Soc.,* **81,** 1856. Absorption spectra of some active species.

Rice, F. O., and Sherber, F. (1955). *J. Am. Chem. Soc.* **77**, 291. The hydrazino radical and tetrazane.

Rice, F. O., Johnston, W. R., and Evering, B. L. (1932). *J. Am. Chem. Soc.* **54**, 3529. Free radicals from organic compounds.

Richter, V. von (1915). "Organic Chemistry," p. 24. Keegan Paul, Trench, Trubner, London.

Ruehrwein, R. A., and Hashman, J. A. (1959). *J. Chem. Phys.* **30**, 823. Formation of ozone from atomic oxygen at low temperatures. See also Ruehrwein, R. A., Hashman, J. S., and Edwards, J. W. *J. Phys. Chem.* To be published. Chemical reaction of free radicals at low temperatures.

Salzberg, P. L., and Marvel, C. S. (1928). *J. Am. Chem. Soc.* **50**, 1737. Hexa-tertiary-butylethynylethane.

Scheele, C. W. (1774). A translation of Scheele's paper is available in Alembic Club Reprint, No. 13. Williams & Wilkins, Baltimore, Maryland.

Schorlemmer, C. (1862). *J. Chem. Soc.* **15**, 419. Hydrides of the alcohol radicals.

Schorlemmer, C. (1864). *J. Chem. Soc.* **17**, 262. On the identity of methyl and the hydride of ethyl.

Steacie, E. W. R. (1954). "Atomic and Free Radical Reactions," 2nd ed. Reinhold, New York.

Voevodsky, V. V. (1959). *Trans. Faraday Soc.* **55**, 65. Thermal decomposition of paraffin hydrocarbons.

2. Radical Formation and Trapping from the Gas Phase

B. A. THRUSH

Department of Physical Chemistry, University of Cambridge, Cambridge, England

I. Introduction

A. GENERAL

Although many of the workers in the field of trapped radicals have produced the labile species in the gas phase before trapping them, they have normally preferred electrical or thermal methods of forming free radicals; on the other hand, most of the quantitative data on free radicals and their reactions in the gas phase have been obtained from photochemical systems or chemical reactions. The production of free radicals in electric discharges is discussed in Chapter 3. This chapter is mainly concerned with other methods of producing free radicals in systems suitable for their eventual stabilization on a cold surface. The vast literature on the formation and reactions of free radicals has been excellently reviewed in the second edition of Steacie's book (1954), and reviews of more recent papers can be found in the *Annual Reports on the Progress of Chemistry* published by the Chemical Society of London. Most of the studies have been made under conditions where the proportion of free radicals would be inadequate for them to be detected in a solid matrix, even if the system could be frozen down without further reaction occurring. It is unfortunate that the techniques where radicals can be studied at relatively high concentrations, that is, flash photolysis (Norrish and Thrush, 1956) and shock waves (Greene and Toennies) are transient rather than continuous, and so unsuitable for producing radicals to be stabilized. Another surprising feature is the absence of quantitative data on the rates of free radical reactions, even in systems where the mechanism has been elucidated. Part of this difficulty arises because many reactions yield excited radicals which have different reactions from the normal species. Trotman-Dickenson (1955) has reviewed the quantitative data on free radical reactions in the gas phase fairly recently.

B. Criteria for Stabilization

The conditions used for the production of free radicals and their transfer to the surface on which they are to be stabilized is to a large extent governed by the desired state of the condensed system. Even when the trapped-radical technique is used to isolate and identify species present in electric discharges and chemically reacting systems, the criteria are essentially the same in that the active species must first be trapped and then identified.

The principal methods of studying trapped radicals are discussed in later chapters, and only those features of the solid state needed by these techniques will be considered here. These are largely features of the matrix, since the concentration of active species is quite low in all the experiments carried out to date. This matrix may consist of either undecomposed parent molecules, inert reaction products, or a specially introduced inert substance. The most obvious desiderata for the matrix are rigidity, inertness, correct volatility, good thermal conductivity, and ability to accommodate the active species (Becker and Pimentel, 1956). These properties will be considered in turn.

Diffusion in a solid is strongly temperature dependent, and this makes it important to work at the lowest possible temperature in order to obtain a really rigid matrix. The only objection to this procedure is that, at low temperatures, the form of deposition may make the matrix unsuitable for the type of measurement to be made; for instance, Becker and Pimentel (1956) state that xenon deposits as a fairly transparent matrix at 66°K, but scatters light strongly if formed below 50°K. At the present time, the best estimates one can make of the ceiling temperatures for rigidity are based on observations of changes occurring during warm-up of matrices containing active species. For nitrogen or oxygen atoms trapped in a solid formed of the parent molecule, recombination occurs between 10°K and 30°K during warm-up (Fontana, 1958; Ruehrwein and Hashman, 1959) although Fontana considers that, if the recombination is controlled by diffusion, the temperature dependence of the viscosity must be unusual. With species trapped in more solid matrices such as HN_3 or H_2O where hydrogen bonding can occur, the softening temperatures are around 120°K (Dows and Pimentel, 1955; Jones and Winkler, 1951; Livingston *et al.*, 1956). It is significant that no workers appear to have been able to obtain concentrations of trapped radicals within a factor of ten of the 10 to 14% calculated by Jackson and Montroll (1958) and by Golden (1958) for systems condensing from the gas phase, and this shows clearly that diffusion either in the solid or along the surface during condensation is an important factor in limiting the trapped radical concentration.

In addition to chemical inertness toward the stabilized radical, it is important that the matrix material should not react with it during or before the deposition process. The suitability of a matrix for a particular system depends largely on the reactivity of the species to be stabilized, and, though the inert gases and nitrogen are suitable for any substance, more caution is needed with such substrates as fluorocarbons, hydrocarbons, and carbon dioxide. In particular, hydrocarbon glasses such as ether-isopentane-alcohol, and isopentane plus methylcyclohexane (Lewis and Lipkin, 1942), albeit excellent for the study of the electronic spectra of aromatic radicals (Norman and Porter, 1955), are not suitable for stabilizing aliphatic free radicals which can abstract hydrogen from these media. In systems involving condensation of free radicals from the gas phase, chemical reactions are normally important only before or during the condensation process, since the temperatures of most matrices will be too low for the occurrence of reactions with appreciable activation energies. This is an important distinction from systems in which the radicals are formed by irradiation of the matrix, where the excess energy absorbed by the parent substance is often sufficient to overcome appreciable activation barriers.

The conditions governing volatility of the matrix material are that it should have negligible vapor pressure ($<10^{-4}$ mm) at the temperature of the collecting surface, and that it must have sufficient vapor pressure at, and somewhat below, room temperature so that it does not condense before reaching the correct surface. The latter condition is rarely important, but the former gives an upper limit to the Dewar temperature of 70°K for xenon, 35°K for argon, and 30°K for nitrogen (Becker and Pimentel, 1956).

Although good thermal conductivity is important so that the heat released on condensation can be conducted away without softening the matrix unduly, the other criteria listed are usually harder to satisfy, and softening of the deposit is usually avoided simply by having a low rate of deposition and working with the matrix well below its softening temperature. This is particularly important in spectroscopic experiments where substances are deposited on windows which have much lower thermal conductivities than copper and where the refrigerant is some distance from the point of deposition. Most workers do not give the optimum rate of deposition in their papers, but the published data indicate that these are normally around 10^{-5} moles/cm^2/sec and lie mainly within a factor of 3 of this value. For instance, both Rice and Freamo (1951) and Foner and Hudson (1958a) used flows of 3 to 4 $\times$ 10^{-6} moles/cm^2/sec to study intermediates in the decomposition of hydrazoic acid. The published data of Bass and Broida (1956) and Jen *et al.* (1959) indicate flows slightly greater than 10^{-5} moles/cm^2/sec when stabilizing nitrogen and hydrogen atoms in a matrix of their parent molecules at 4°K.

The last of these criteria, solubility in the matrix material can normally only be tested experimentally at the present time, although the inert gases are of particular value as matrices, in that they give a wide range of lattice spacings with similar physical properties. The temperature and rate of deposition are also important, because differences in surface mobility during deposition will affect the final structure. It is difficult to determine how far the maximum concentration of radicals which can be trapped in a lattice is due to limited solubility rather than to reaction due to surface migration on deposition or diffusion through the lattice. The presence of dimers of the labile species in the lattice at high concentration suggests that it is not solubility which gives an upper limit to the concentration of the labile species.

Of the criteria for particular studies, those for optical spectroscopy will be considered first. With absorption spectroscopy, it is important that the matrix should not absorb or scatter large amounts of radiation in the region to be studied; the low concentrations of active species normally present ($<1\%$) make these criteria harder to meet than might otherwise be supposed. In particular, small amounts of impurity in the matrix material can easily yield spurious bands. Most free radicals have strong electronic spectra between 2000 and 9000 A (Herzberg, 1950), many of which have been observed in absorption (Norrish and Thrush, 1956; Ramsay, 1957). This, combined with the absence of absorption by any of the substances likely to be used as matrices (except for those containing double bonds) makes this region very attractive, although considerable difficulty is normally encountered from light scattering, a problem which can be tackled only by varying the rate and temperature of deposition, or the composition of the system. The light-scattering problem is much less serious in the infrared, but this gain is largely neutralized by the extra thickness of material needed because of the relative weakness of infrared spectra. The choice of matrix materials is also rather more restricted, as fluorocarbon and hydrocarbon glasses have too many absorption bands to be useful and even relatively simple substances such as carbon dioxide, methane, and carbon tetrachloride absorb over an appreciable part of the spectrum. For this reason only the inert gases and nitrogen can normally be used for infrared studies. With emission spectroscopy, light scattering by the matrix is no problem, but it may be necessary to consider the possibility of fluorescence of the matrix induced by energy transfer from the active species.

For electron spin resonance and chemical studies, it is important that the sites in the matrix occupied by the active species should be as similar as possible, so as to simplify interpretation of the results.

II. Gas Phase Production and Reactions

A. PRODUCTION

1. Photochemical and Photosensitized

Superficially, photochemical methods for the production of free radicals appear to have many advantages in that the identity of the species formed in the primary act is generally known. Further, the decomposition can be carried out at low temperatures, where less stable radicals will persist for longer than in a discharge or thermal decomposition. Although very high transient concentrations of free radicals can be obtained by the technique of flash photolysis, the radical concentrations in most conventional photochemical arrangements is low and these would not give a high enough rate of decomposition in a flow system for detectable concentrations of free radicals to be condensed on a cold surface. The problem is therefore one of concentrating the maximum amount of photochemically active light on the tube through which the reactant gas is flowing, and ensuring that this light is absorbed by a high proportion of the reactant molecules. This clearly demands a compact source of high intrinsic brightness; extended sources are less useful unless they surround the flow tube. The two sources best answering this description are the mercury and xenon arcs, as the maximum intensities of both carbon arcs and tungsten filament lamps lie at too long a wavelength to be generally useful.

A 1-kw high pressure mercury arc gives a total output of about 5×10^{-4} einsteins per second. Below 4000 A, its spectrum contains broad lines with an underlying continuum. The lines around 3650 A are particularly intense, but there is little output between 2450 and 2600 A due to intense reversal of the 2537 A line. Medium-pressure lamps have similar efficiencies, but are less suitable for this work owing to their greater arc lengths (200 mm as against 20 mm for a 1-kw tube). For substances which do not absorb strongly any of the prominent mercury lines in this region, the xenon arc provides the most powerful available source of an ultraviolet (UV) continuum. For example, a 350-watt compact source lamp will give 2.6×10^{-5} einsteins per second total UV output with an arc length of only 6 mm. These lamps can be obtained up to 5 kw in power, giving a proportionately increased light output with an arc length of 65 mm. For both types of lamp, the UV output is a good continuum, the number of quanta per unit wavelength increasing smoothly by a factor of 8 from 2400 to 4000 A. Few workers appear to have tackled the problem of obtaining maximum photochemical decomposition in a flow system, and the work of Hicks and Melville (1954) seems to be most relevant to our considerations. They used an elliptical stainless steel reflector 16×14 cm and 5 cm high to concentrate the light from a 125-watt high pressure mercury lamp on a capillary tube 1 mm in diameter, and in this way they were able to obtain an incident light intensity of 9×10^{-6} einsteins/cm²/sec mainly around 3657 A. For a substance having a molar extinction coefficient of 50 this corresponds to a rate of decomposition of 0.1% per millisecond of irradiation time, providing the pressure is low enough to give incomplete light absorption (in this case less than atmospheric pressure for 1-cm path length). Unfortunately, the absorption coefficients of the aldehydes and ketones are only about 15 in this region, and other established sources of free radicals, such as azomethane, diazomethane, and ketene absorb even more weakly. However, other substances such as the chlorine oxides and nitrogen dioxide absorb strongly enough to make this method feasible for the production of ClO and oxygen atoms.

The problem of increasing the light absorption, which is the limiting factor above 3000 A for the systems discussed, can be tackled in two ways. The first is to use shorter wavelengths where substances absorb more strongly, and the second is to use a photosensitizer. In the first case, really strong absorption can be obtained by working in the vacuum ultraviolet, but it is difficult to obtain really intense light sources there, and though Lossing and Tanaka (1956) have detected free radicals with a mass spectrometer after photolysis in this regions, it seems unlikely that any of the sources currently available would be intense enough for the systems considered here.

Mercury resonance lamps at 2537 A merit special consideration as photochemical sources both for photolysis and photosensitized decomposition. Because of the ease with which this radiation is reversed, most workers have used relatively large lamps to obtain high outputs at this wavelength (Heidt and Boyles, 1951), but Lossing et al. (1956) have recently developed a compact source for flow studies using a mass spectrometer. Their flow tube was 7 mm in diameter and they irradiated a 36-mm length of it with a ring-shaped lamp 13 mm internal diameter and 23 mm external diameter. The lamp had three internal septa to lengthen the electrical path, and delivered approximately 1.5×10^{-6} einsteins per second at 2537 A. Using mercury at 25μ pressure (its vapor pressure at 60°C), these workers were able completely to decompose ethylene at 10μ pressure in 1 millisecond and to obtain more than 50 % decomposition of a few microns of acetone in 8 mm of helium in 2 milliseconds. These decompositions would produce high enough concentrations of free radicals for stabilization, if the presence of mercury in the condensed phase was unobjectionable. With straight photolysis, a lamp of this type would give adequate decomposition for all aromatic hydrocarbons (such as toluene) which are decomposed at this wavelength, as well as appreciable decomposition of alkyl iodides.

2. Thermal

Thermal decomposition will give a much higher rate of dissociation than photolysis, but much less is known about the absolute concentration and identity of radicals produced thermally. The high rates of decomposition obtainable appear to be the reason why workers in this field have preferred pyrolysis in a heated tube to photolysis; for instance, Rice and co-workers (Rice and Freamo, 1951; Rice and Grelecki, 1957) have partially decomposed hydrazoic acid by passing it through a quartz tube at 1000°; they used pressures around 0.1 mm with flow rates of the order of 10 meters per second. With hydrazine, Rice and Scherber (1955) used a higher pressure (0.5 mm) and a lower temperature (850°). These conditions can be expected to give the highest concentration of free radicals, that is, a short contact time and a high temperature giving very rapid decomposition, combined with a fast flow at low pressure so that very few radical reactions can occur prior to condensation.

As mentioned above, the vast literature on pyrolytic systems contains very few data on the free radical concentrations actually present in these systems, and the most relevant data are probably to be found in the work of Lossing and Tickner (1952) and Ingold and Lossing (1953). These workers pyrolyzed mercury dimethyl and di-t-butyl peroxide using a helium carrier gas and were able to measure the absolute concentration of methyl radicals produced with a mass spectrometer. With 0.014 mm mercury dimethyl in 15.5 mm helium, and a contact time of 1 millisecond at 850°C, they obtained 0.008 mm of methyl radicals in the presence of only a slightly greater total pressure of other decomposition products. Their method was to bleed off a small quantity of the hot gas through a hole about 50μ in diameter in the tip of a quartz thimble which projects into the gas stream. There seems to be no reason why this method should not be used for a large variety of systems, providing particular care is taken to minimize radical reactions during and after pyrolysis. All the workers quoted in this section used quartz tube furnaces; these have a decided advantage over heated filaments (Tsuchiya, 1957) in that it is much easier to decompose a high proportion of the reactant gas. An alternative approach is to trap the species emerging from a furnace or evaporated from a filament under conditions when there are appreciable numbers of radicals in thermal equilibrium. This method has been used by Scheer (1959) to produce a molecular beam of iodine atoms which emerged through a hole 0.5 mm in diameter in a quartz tube at 850°C in which the iodine pressure was 0.03 to 0.13 mm. The beam intensity was around 10^{-9} moles/cm^2/min. This would seem to be suitable for other atoms (particularly halogens) and for such radicals as C_2, C_3, CF, CF_2, CN, etc., which can be present with relatively high concentrations at high temperatures (Honig, 1954; Margrave and Wieland, 1953; Brewer and Templeton, 1949) but which are hard to produce photochemically or in pyrolyses where thermal equilibrium is not established.

3. Electrical

The production of free radicals in electric discharges is discussed in Chapter 3. It should be noted in passing that the Wood's type of discharge tube generally gives the most complete dissociation into atoms, and that electrodeless radio-frequency and microwave discharges are efficient and simple ways of producing the more stable radicals such as NH_2, OH, etc. Whereas the ions formed in the discharge disappear rapidly, metastable excited atoms and molecules and vibrationally excited molecules persist for relatively long times (Broida and Peyron, 1957; Fowler and Strutt, 1910; Kaufman and Kelso, 1958) and can sometimes be stabilized on a cold surface. In general, any excited species which is stable enough to persist appreciably in a solid matrix will not be deactivated easily by collisions in the gas phase and chemical reactions will be the only cause of loss during its transfer to the cold surface.

4. Chemical Reaction

In theory, it would be possible to trap the free radicals produced in a thermally or photochemically initiated branched-chain reactions using a flow system, but this approach would involve many experimental difficulties, due mainly to the irreproducibility of these reactions and the difficulty of controlling the conditions accurately enough. McNesby (1959) has tried unsuccessfully to trap radicals produced in an explosion by allowing the explosion to break a thin glass diaphragm which separated the exploding gases from the cold surface.

A more fruitful method of approach is to produce radicals by reacting atoms or free radicals produced thermally or in an electric discharge with a stable molecule to produce the desired radical. This process is distinguished from formation of new radicals by secondary reaction with molecules present in the radical source by the act of mixing after forming the first radical. This technique was first used by Geib and Harteck (1932, 1933) to trap the products of the reaction of hydrogen atoms with oxygen and with a variety of other fairly simple substances. They carried out similar studies with oxygen atoms. The reaction between hydrogen atoms and oxygen has been studied in much greater detail than have any of the others, mainly with a view to detecting the elusive HO_2, which has not so far been identified optically although it has been detected with the mass spectrometer (Foner and Hudson, 1953, 1955). The HNO radical which has been observed in solid state photolysis (Brown and Pimentel, 1958) was first observed by Dalby (1958) in flash photolysis experiments including ones in which hydrogen atoms produced by the photolysis of ammonia reacted with nitric oxide. Recently, Harvey and Brown (1959) have formed this species in the gas phase by reaction between hydrogen atoms and nitric oxide and stabilized the HNO radical by condensation.

The experimental methods of Foner and Hudson (1953, 1955) although intended for mass spectrometry are probably more suitable than those used by other workers (Geib and Harteck, 1932; McKinley and Garvin, 1955). All these workers use a coaxial mixing system in which a fine tube injects the reactant into the center of a wide tube through which the atoms from the discharge flow. Foner and Hudson place a plate containing a small leak immediately in front of the mixing zone and pump most of the products away through the outermost coaxial tube (see Lossing and Tickner, 1952). The beam of molecules passing through the small leak (in this case a 0.01 cm diameter hole in 0.0025 cm gold foil was used) can be condensed on a cold surface. By altering the size of the hole, it is possible to vary the pressure in the reaction zone without appreciably changing the rate of condensation.

B. REACTIONS

The reactivity of a free radical both in the gas phase and in the solid state is the most important factor in determining whether it is possible to stabilize it at low temperatures. Unfortunately, there is a great lack of quantitative data on free radical reactions in the gas phase, even when the mechanism of reaction is clearly established, and we have very little information as yet on the nature of reactions occurring on the surface during condensation. Emphasis has therefore been placed as far as possible on such quantitative data as are available on these reactions, even when they have been obtained with systems far removed from those with which this book is concerned. An attempt has been made to group the homogeneous reactions so that one can infer the undetermined rate constants from those which are already known. These reactions fall naturally into two groups, those removing two radicals with or without the participation of a third body, and those in which the identity of the radical is changed either by decomposition or a metathetical reaction with a molecule. In both cases atomic as well as free radical reactions will be considered.

1. Homogeneous

a. *Recombination and disproportionation.* The simplest combination reaction which can be considered is that of two atoms recombining in the presence of a third body:

$$X + X + M = X_2 + M^*$$

These reactions are kinetically third order, the rate constant being defined by

$$\frac{-d[X]}{dt} = k_1[X]^2[M]$$

To date these rate constants have been accurately measured only for bromine and iodine atoms with a variety of third bodies, the techniques used being steady photolysis (Rabinowitch and Wood, 1936 a, b) and flash photolysis (Russell and Simons, 1953; Christie *et al.*, 1955; Strong *et al.*, 1957).

Recently several groups of workers have reported preliminary measurements on the recombination of nitrogen atoms in flow systems, using titration with nitric oxide (Harteck *et al.*, 1958; Herron *et al.*, 1958) or thermal detection (Wentink *et al.*, 1958). Typical values of these rate constants at room temperature are given in Table I.

TABLE I

VALUES OF $k_1 l^{-2}$ mole^{-2} sec^{-1}

M	N	Br	I
He	8×10^8	2.7×10^9	2.4×10^9
A	3×10^9	7×10^9	6.6×10^9
N_2	1×10^{10}	7.4×10^9	9×10^9
CO_2	—	2×10^{10}	2.7×10^{10}
CH_4	—	1.3×10^{10}	1.8×10^{10}
C_6H_6	—	—	1.8×10^{11}
I_2	—	—	1.7×10^{12}

Except for the recombination process

$$I + I + I_2 = I_2 + I_2^*$$

which shows a large negative temperature coefficient due to the stability of I_3 (Bunker and Davidson, 1958), all these reactions have a negative activation energy of about -2.0 kcal, and the values of the individual rate constants for various inert gases indicate that the rate of recombination is governed by van der Waals' interaction between halogen and the third body. We can therefore confidently predict that the rates of recombination of other atoms will be similar to those for halogen atoms except where specific interactions occur. On this basis, an atomic recombination would be expected to have a half-life of about 1 sec for 0.1 mm of atoms in 1 mm of argon near room temperature, the corresponding times for nitrogen and carbon dioxide being 0.4 and 0.15 sec, respectively. For different compositions, the half-life of the atoms is inversely proportional to the initial atom pressure and to the inert gas pressure. Measurements of the decay of OH radicals in products of the discharge through water vapor (Oldenburg and Rieke, 1939) and of recombination processes in hydrogen flames (Bulewicz and Sugden, 1958; Padley and Sugden, 1958) yield values of the rate constant for the recombination, which is predominantly of H and OH to form water. Oldenburg and Rieke (1939) obtain values of 2.1 and 4.9 $\times 10^{11}$ l^2 mole^{-2} sec^{-1} at 300°C, respectively, for helium and water acting as third bodies. The flame measurements give similar rate constants around 2000°K for water, but their extrapolation to room temperature would yield a value about four times greater than the discharge method. The rates of recombination in dissociated water vapor will be fairly similar to the values quoted for atomic systems.

When polyatomic radicals recombine, the life of the collision complex is so much greater than for atoms that deactivation normally occurs during the life of the complex; the reaction is then kinetically second order and only becomes third order at low inert gas pressures. The second-order rate constants for the recombination of CH_3, CF_3, and C_2H_5 radicals are all close to 2×10^{10} l mole^{-1} sec^{-1} between 100 and 200°C, with activation energies close to zero (Shepp, 1956; Ayscough, 1956; Shepp and Kutschke, 1957). For inert gas pressures below 5 mm of acetone (and presumably 80 mm of helium) the recombination of methyl radicals becomes third order with a rate constant of about 10^{14} l^2 mole^{-2} sec^{-1} with acetone as third body (Gomer and Kistiakowsky, 1951). The rate of thermal decomposition of hydrocarbons clearly indicates that all the recombinations of hydrocarbon radicals should proceed at closely similar rates. On this basis we can estimate half-lives for radicals of this type to be about 3×10^{-5} and 8×10^{-6}

seconds for 0.1 mm radical in 1 mm of acetone (or similar gas) and for high inert gas pressures, respectively. The half-life of the radical is again inversely proportional to its initial pressure, but in this case it is only inversely proportional to the inert gas pressure if this is below 5 mm for acetone and correspondingly higher pressures for less efficient deactivators. Above this threshold the half-life is independent of inert gas pressure. It should be mentioned in passing that the reactions removing hydrocarbon radicals in the gas phase proceed at closely similar rates by two paths— combination and disproportionation. For the ethyl radical these are

$$2 \ C_2H_5 = C_4H_{10}$$

$$= C_2H_4 + C_2H_6$$

A bimolecular radical removal reaction which is known to differ widely in rate from those cited above is that for two ClO radicals: $ClO + ClO = Cl_2 + O_2$; this has a rate constant of 2×10^7 l mole^{-1} sec^{-1} over a wide range of temperatures (Edgecombe et $al.$, 1958), corresponding to a half-life of 8×10^{-3} seconds at 0.1 mm pressure of ClO.

It can be seen that bimolecular removal reactions in the gas phase are very much more important in the case of the more labile large species than for atoms, and that the pumping speeds between 500 and 2000 cm/sec normally used in flow systems for trapping radicals (see, for instance, Rice and Freamo, 1951; Foner and Hudson, 1958a) will not be adequate for such species as alkyl radicals unless the total pressure is well below 0.1 mm. Some improvement can be effected by using approximately molecular beam conditions (Jen et $al.$, 1956) or by admitting the radicals and matrix materials through separate orifices so that the flows do not combine until they meet the surface (Foner et $al.$, 1958a), thus minimizing third-order recombination.

$b.$ $Decomposition$ and $metathesis.$ Whereas the rates of the reactions discussed in the previous section are proportional to the square of the radical concentration, the reactions of radicals with stable species are first order with respect to the radical. Their half-lives are independent of the initial radical concentration and inversely proportional to the pressure of the other species participating. These reactions can be classified into three groups—addition, decomposition, and metathetical.

The first group includes addition to ethylenic compounds and to molecules with unpaired electrons such as oxygen and nitric oxide. For the methyl radical, these additions are all bimolecular at normal pressures and have frequency factors close to 10^8 l mole^{-1} sec^{-1}, (Trotman-Dickenson, 1955). The products of the reactions with oxygen and nitric oxide are not of interest in our present considerations, and the addition reactions with olefins, although yielding higher radicals, for example, $CH_3 + C_2H_4 = C_3H_7$, have activation energies of about 6 kcal; this reduces their rates by a factor of 10^4 at room temperature to make them relatively inefficient. The addition of hydrogen atoms to olefins is somewhat more rapid and provides a possible method of synthesizing alkyl radicals, as these reactions proceed with a rate constant slightly greater than 10^7 l mole^{-1} sec^{-1} at room temperature (Darwent and Roberts, 1953) giving half-lives of less than 2×10^{-3} sec at 1 mm pressure of each component. Unfortunately, the rate constant for the addition of $H + NO$ is unknown and that for $H + O_2$ is somewhat uncertain. The rate constant for the termolecular addition $O + NO + M$ is however known to be 2×10^{10} l^2 mole^{-2} sec^{-1} at room temperature (Kaufman, 1958). Very little is known about the rate of halogen atom addition to double bonds owing to the rapidity of the succeeding reaction, in which the radical formed reacts with a halogen molecule.

Thermal decomposition of a radical is essentially the reverse of the processes described above. At room temperature, these processes will not be important, as the least stable radicals which are likely to be encountered are HCO and CH_3CO, which require activation energies of at least 14 kcal to split off CO (Marcotte and Noyes, 1952).

Metathetical reactions of atoms and radicals can be represented by equations of the type

$$H + RH = H_2 + R$$

$$Cl + RH = HCl + R$$

$$CH_3 + RH = CH_4 + R$$

and most of the known rate constants refer to reactions of hydrogen or chlorine atoms or methyl radicals. For hydrocarbons, these rate constants are around 10^5, 10^7, and 10^3 l mole^{-1} sec^{-1}, respectively, at room temperature (Schiff and Steacie, 1951; Trotman-Dickenson, 1955). Except for the reaction of chlorine atoms, which could serve as an alternative source of alkyl radicals to the addition of hydrogen to an olefin, these processes are too slow to be of interest in this context. The more rapid reaction of hydrogen atoms with such species as halogens do not yield interesting products.

2. Heterogeneous

As with homogeneous reactions, the amount of quantitative information available about heterogeneous reactions is very limited. Reliable measurements appear to have been made only with hydrogen atoms (Smith, 1943) and oxygen atoms (Linnett and Marsden, 1956; Greaves and Linnett, 1958). These workers measured the fraction of atoms (γ) colliding with

TABLE II

Surface	$\gamma \times 10^5$ (H)	$\gamma \times 10^5$ (O)
Quartz	70	16
Pyrex glass	46	3[a]
Phosphoric acid	2	12
Ag_2O	—	8
B_2O_3	—	6
CaO	—	160
CuO	—	4300

[a] HF treated.

the surface which were lost at total pressures of the order of 0.1 mm. Some of their values are given in Table II.

It was found that the efficiencies for the removal of atoms by the same surface were fairly similar for oxygen and hydrogen atoms. Greaves and Linnett showed that the oxides of transition elements had the highest efficiencies and that acidic oxides normally had lower efficiencies than basic oxides, and that successive washings of Pyrex glass with hydrofluoric or nitric acids gave progressively lowered efficiencies, but that one treatment with ammonia completely destroyed this gain in performance. These figures show clearly that the surface losses of hydrogen or oxygen atoms on acid-washed Pyrex glass, quartz, or phosphoric acid-coated surfaces will be negligibly small in any conventional flow system. Of these surfaces, the phosphoric acid coating has the merit that it is probably the least easily affected by impurities in the gases.

There are no systematic data on the heterogeneous recombination of other atoms or radicals, although Herron et al. (1958) and Wentink et al. (1958) report values of γ of 1.5 and 3.0 $\times$ 10^{-5} for N atoms on Pyrex. Smith's work showed clearly that the rate of OH recombination was similar to that of H on all the surfaces studied, with the exception of KCl, where it was much higher. On an intuitive basis one would expect the surface recombination of larger radicals to proceed at a higher rate, since Greaves and Linnett's data were obtained under conditions where chemisorption did not occur and larger radicals may be expected to give stronger van der Waals' adsorption. In the absence of more extensive data for complex radicals it seems best to use rapid flow in wide tubes, or to use a molecular beam technique if the radicals have to travel far from their source to the deposition point.

III. Examples

In this section, an attempt is made to give some details of successful experiments in which the species concerned was condensed from the gas phase. In other cases, an attempt is made to suggest suitable sources of the radical, based on published data on its formation, identification, and stability in the gas phase. Atoms are considered first, owing to the preponderance of studies. The subdivision of the section on free radicals is somewhat arbitrary but is based partly on similarity of stability and method of preparation.

A. FREE ATOMS

These fall naturally into three groups; of these the most important consists of hydrogen, nitrogen, and oxygen, all of which are best obtained from electrical discharges. This method is also suitable for the halogens, which form the second group, although pyrolytic techniques may prove better. Members of the third group, which includes the metals, are probably best atomized from an arc or a heated wire.

1. Hydrogen

In his classic work Wood (1922) used a low frequency electrode discharge through hydrogen at pressures between 0.2 and 1.0 mm to obtain more than 50 % dissociation into atoms. Recent workers have favored radiofrequency (rf) and microwave discharges; these avoid the possibility of contamination by material evaporated from the electrodes, (Bass and Broida, 1956; Jen et al., 1956). Jen and associates pumped hydrogen or deuterium at 0.1 mm pressure through a 100-watt 4 Mc/sec discharge at 1500 cm/sec. A short side tube terminated in a glass slit of dimensions 10 × 0.5 mm gave a molecular beam which was condensed on a sapphire rod. They were able to show by electron spin resonance that they had trapped hydrogen atoms at liquid helium temperature. Spectroscopic studies of the solid provide no evidence of large amounts of atoms in the solid (Bass and Broida, 1958). Apart from the condensation process, the main loss of hydrogen atoms in these experiments is due to surface recombination. This can be minimized by treatment with phosphoric acid or repeated washing with hydrofluoric acid (see Section II, B) or by coating with dichlorodimethyl silane (Zletz, 1959). This coating would be adversely affected by material evaporated from the electrodes. This, plus the relative simplicity and convenience of electrodeless discharges, probably explains their general preference over the original type of Wood's tube.

Historically, Langmuir (1915) first produced hydrogen atoms by dissociation on a heated filament, and more recently this method has been used by Klein and Scheer (1958) to study reactions of hydrogen atoms with olefins condensed on a surface at 77°K.

2. Nitrogen

A great deal of work has been carried out on the products of electrical discharges through nitrogen, following the original discoveries of Lewis (1900) and Strutt, later Lord Rayleigh (1910, 1915). The presence of excited ^2D and ^2P atoms as well as ground state ^4S atoms in the discharge products has been clearly established spectroscopically by Tanaka et al. (1957) and mass-spectrometrically by Jackson and Schiff (1955). It has also been shown that the main features of the afterglow are the $^3\Pi_g - {}^3\Sigma_u{}^+$ bands of N_2 and that the upper state of this transition is formed by recombination of two ^4S nitrogen atoms (Berkowitz et al., 1956).

No author appears to have studied the optimum discharge conditions for the production of nitrogen atoms in any detail, although Harteck et al. (1958) report up to 6% of atoms in a condensed discharge. With microwave discharges, a figure of 4% has been recorded by Fontana (1958), and this worker also suggests (1959) that the free atom concentration is reduced by thorough drying of the nitrogen before passing through the discharge. Much higher concentrations of nitrogen atoms have been produced in pulsed discharges (Armstrong and Winkler, 1956), but such systems are not of interest here. Kaufman and Kelso (1958) have presented strong evidence for the presence of vibrationally excited nitrogen molecules in the products of a microwave discharge, presumably formed from metastable electronically excited states in collisions. Recently Dressler (1959) has shown by vacuum ultraviolet spectroscopy that vibrationally excited N_2 is present in discharge products. Most workers who employ microwave discharges use a Raytheon unit which delivers approximately 100 watts at 2450 Mc/sec. Typical operating pressures are around 1 mm, and Bass and Broida (1956) reported that the optimum flow rate for stabilizing excited nitrogen atoms was about 20 cc/min. For studies with electron spin resonance, Foner et al. (1958 a, b) used a matrix of undischarged nitrogen in a hundredfold excess to that which had passed through the discharge. They were able to detect N atoms in N_2, H_2, and CH_4 matrices. The maximum concentration of N atoms which has been stabilized from a discharge is probably less than 1% (Broida and Lutes, 1956; Fontana, 1959) although measurements of the heat released during warm-up suggest a higher value (Minkoff et al., 1959).

The rates of disappearance of N atoms by gas phase recombination and by wall collisions have been given in Section II, and it seems that these processes should prove to be unimportant in a well-designed system.

The electric discharge is the only practicable source of nitrogen atoms in the gas phase. There appears to be no suitable photochemical source, and the N_2 bond energy is too great for Langmuir's method to succeed.

3. Oxygen

As with nitrogen, most recent workers have preferred the microwave discharge as a source of oxygen atoms. Kaufman (1958) has found up to 20 % of oxygen atoms by titration with NO_2, when using 800 watts power at 2400 Mc/sec. By a similar method Herron and Schiff (1958) found an oxygen atom concentration of 8 % at 1 mm pressure, together with 10 to 20 % of excited O_2, probably in the $^1\Delta_g$ state. They found the ozone concentration to be less than 0.02 % by mass spectrometric analysis. It seems clear from this work that the reaction

$$O + O_3 = O_2 + O_2$$

for which the rate constant is not yet known, must proceed faster under these conditions than

$$O + O_2 + O_2 = O_3 + O_2{}^*$$

for which Kaufman (1958) has found a rate constant of $4 \times 10^8 \, l^2 \, mole^{-2} \, sec^{-1}$.

Although Peyron and Broida (1959) have observed the $^1S_0 - {}^1D_2$ transition of atomic oxygen in the afterglow of solid nitrogen, there is no spectroscopic evidence for O atoms being trapped in an oxygen lattice. Further, Jen et al. (1958) have failed to detect O atoms in hydrogen or argon matrices, perhaps because of strong line broadening caused by interaction between the orbital angular momentum of the atom and the crystal field.

The solid condensed from a discharge through oxygen does however contain large amounts of ozone; for instance Ruehrwein and Hashman (1959) found conversions of up to 78 % by condensing on a surface at 4.2°K and then warming up. Harvey and Bass (1958) were unable to detect any change in the infrared absorption of ozone during warm-up. Their work suggests that the oxygen atom concentration is less than 3 % in the solid and that the ozone is formed almost exclusively during condensation by reaction of oxygen atoms and molecules on the surface. Calorimetric measurements (Broida and Lutes, 1956; Minkoff et al., 1959) suggest oxygen atom concentrations of 1 % and over 6 %, respectively, but these techniques probably need further refinement.

As with nitrogen and hydrogen atoms, the loss of oxygen atoms by recombination in the gas phase and on walls of vessels is easily controlled. The main difficulty in this case is to prevent ozone formation during condensation; this is probably best tackled by the use of an inert matrix.

4. Halogens

The halogens have much lower bond dissociation energies than the elements so far discussed, the highest being 57 kcal for chlorine as compared with 103, 225, and 117 kcal for hydrogen, nitrogen, and oxygen, respectively. For this reason, thermal decomposition is probably the best method of producing halogen atoms, although electrical discharges could equally well be used.

The only recorded use of halogen atoms in connection with trapping species from the gas phase is the work of Scheer (1959) which has been described in Section II, A, 2. Preliminary results of this work indicate that recombination of iodine atoms occurs on a surface at 77°K and at 20° because of translation of the atoms along the surface. Introduction of a 250-fold excess of n-heptane as matrix, using a separate beam, did not appear to isolate the iodine atoms sufficiently for stabilization.

5. Metals

The possibility of trapping metals as atoms or diatomic molecules in an inert matrix is of considerable interest, and some preliminary experiments have been carried out by Edwards and Hashman (1958). They vaporized magnesium into a stream of argon. This mixture gave a red deposit which on warming formed black pyrophoric magnesium. This type of study could clearly be extended to other metals using either a heated wire or an arc to vaporize the element chosen. Recently, McCarty and Robinson (1959) have trapped mercury atoms in a cold matrix.

B. FREE RADICALS AND LABILE MOLECULES

1. Radicals Containing Carbon but no Hydrogen

a. C_2, C_3. Both these species have appreciable vapor pressures over carbon at temperatures above 2000°K and could therefore be obtained from heated carbon filaments (Honig, 1954) or from a carbon tube furnace (Phillips and Brewer, 1955).

b. CN. This species is probably best obtained from a mild electric discharge through cyanogen or by the thermal decomposition of this molecule (Kistiakowsky and Gershinowitz, 1933; White, 1940). Photochemical production of CN is likely to be insufficient owing to the low extinction coefficient of cyanogen (Ramsay, 1953). CN could also in theory be obtained from a carbon tube furnace containing nitrogen (Brewer and Templeton, 1949) where it would have an appreciable equilibrium pressure.

c. CNO, CNS. The spectra of these species have been observed in fluorescence and absorption (Holland et al., 1958) using far UV excitation and flash photolysis, respectively. The spectrum of NCO has also been observed in emission from a high frequency discharge through ethyl isocyanate, and it seems that concentrations of these species sufficient for trapping could be produced in the gas phase by the passing of a discharge through isocyanates on isothiocyanates.

d. COCl. A detailed investigation of the chlorination of carbon monoxide by Burns and Dainton (1952) shows that the reaction between a chlorine atom and carbon monoxide is 6.2 kcal exothermic and equilibrium is rapidly established. It should therefore be possible to produce appreciable concentrations of this radical by reacting chlorine atoms with carbon monoxide.

e. CF, CF$_2$, CF$_3$. Both CF and CF$_2$ have been obtained in equilibrium involving carbon-fluorine systems at high temperatures (Margrave and Wieland, 1953). An electrodeless rf discharge through fluorocarbon vapors is probably a better source (Andrews and Barrow, 1950) as CF$_2$ in particular lasts for about 1 sec under these conditions (Laird *et al.*, 1950). Much work has been done on the kinetics of the reactions of the CF$_3$ radical formed in the photolysis of hexafluoroacetone (Ayscough, 1956). Flash photolysis of this substance has, however, yielded the spectrum of CF$_2$, but not of CF$_3$ (Herzberg, 1956), suggesting that the CF$_3$ radical is relatively unstable thermally and readily disproportionates:

$$CF_3 + CF_3 = CF_4 + CF_2$$

It seems doubtful if it would be possible to produce an adequate concentration of CF$_3$ radicals in the gas phase under conditions where they would be stable, except possibly in the mercury-sensitized photolysis.

f. CS. This was apparently the first labile species to be stabilized at low temperatures, having been condensed with liquid air from the products of a silent electric discharge through CS$_2$ by Martin in 1913. More recent investigations have shown that the CS radical is remarkably stable in the gas phase (Porter and Wright, 1953), and Martin's work makes it clear that there should be no difficulty in condensing a solid which contains a large proportion of CS from the products of an electrical discharge through CS$_2$.

2. Radicals Containing Hydrogen

a. CH, CH$_2$, C$_2$H. Of these three species CH is observed strongly in discharges through hydrocarbons (Harkins, 1934) but is short-lived under most conditions. CH$_2$ has recently been observed in the vacuum ultraviolet (Herzberg and Shoosmith, 1959). C$_2$H has not been detected spectroscopically. Both species are exceptionally labile. It has recently been shown that CH$_2$ radicals react with the ketene molecules from which they are formed with a collision efficiency greater than 1 in 100 (Kistiakowsky and Sauer, 1956). Flash photolysis of diacetylene (Callomon and Ramsay, 1957) has provided no spectrum of C$_2$H, and it is probable that this radical is also very reactive. McCarty and Robinson (1959) were able to trap the CH radical at 4°K and observed its spectrum.

b. Alkyl radicals. A tremendous amount of work has been done on the kinetics of reactions involving alkyl radicals. Steacie (1954) has provided an excellent account of this literature, and the chief work of interest since then has been the determination of the rates of recombination of methyl and ethyl radicals (Ingold and Lossing, 1953; Shepp, 1956; Shepp and Kutschke, 1957).

To date, only Jen and his co-workers (1958) have trapped alkyl radicals formed in the gas phase. They produced methyl radicals in a mild electrical discharge through methane and stabilized the products at liquid helium temperature. This deposit yielded the same quartet electron spin resonance spectrum as methyl iodide photolyzed at this temperature. Higher alkyl radicals have so far been trapped only by solid state photolysis or irradiation. There would seem to be no good reason why some of the higher alkyl radicals could not be formed in the gas phase and then stabilized. It is doubtful however, whether they are stable enough to withstand the conditions in a mild electric discharge of conventional design. A modification of the very mild type of discharge tube used by Schüler and co-workers (1952) would probably be more suitable. Other profitable approaches would be pyrolysis of metal alkyl and such substances and di-t-butyl peroxide (Lossing and Tickner, 1952) or mercury-sensitized decomposition of aldehydes and ketones (Lossing, 1957). It is unlikely that normal photolysis of metal alkyls or of carbonyl compounds would give enough decomposition for detectable amounts of alkyl radicals to be stabilized. In pyrolytic experiments, the inert gas used to form the matrix could be used to advantage as a carrier for the radical source during the pyrolysis. But in photolytic experiments it would be better to combine the radical beam and the matrix as near to the surface as possible, because of the high rate of recombination of alkyl radicals. The pressures required are sure to be so low that the recombination reaction will be kinetically third order.

   c. HCO, CH$_3$CO, etc. These radicals are relatively unstable thermally, and Marcotte and Noyes (1952) have estimated the activation energy for the thermal decomposition of the first two members to be 14 and 16 kcal, respectively. For this reason, it is unlikely that appreciable concentrations of these radicals could be produced in pyrolytic reactions, and it is therefore of particular interest that Herzberg and Ramsay (1956) have observed the spectrum of HCO in flash photolysis of a wide range of aldehydic compounds. These authors discuss the stability of HCO in some detail in view of the conflicting evidence from photochemical studies (Steacie, 1954; Schoen, 1955) and conclude that the lowest observed electronic state of HCO cannot be formed from H + CO both in their electronic ground states. This is a serious obstacle to producing this radical by reacting hydrogen atoms with carbon monoxide, and it would appear that the mercury-sensitized photolysis represents the best chance of producing usable concentration of it and its homologs in the gas phase.

d. CH_3O, etc. The alkoxy radicals are much more stable than the carbonyl radicals and are readily produced in the photolysis and pyrolysis of alkyl nitrites, etc., although they have not yet been detected spectroscopically. Their stability and reactions have been discussed by Gray (1955).

e. Aromatic radicals. In view of the ease with which the spectra of many aromatic radicals may be obtained by solid state photolysis (Norman and Porter, 1955), it seems possible that these radicals could be trapped after production by gas phase photolysis (Porter and Wright, 1955), pyrolysis (Swarc, 1950), or in a specially designed electric discharge tube (Schüler and Michel, 1955). Unfortunately, most of these radicals are very short-lived in the gas phase (Porter and Wright, 1955), but it seems likely that the more stable ones such as benzyl (Swarc, 1950) and cyclopentadienyl (Thrush, 1956a) could be produced in sufficient concentration to make this method practicable.

3. Radicals Containing Nitrogen but No Carbon

a. N_3, NH, NH_2, N_2H_2, N_2H_3. This large group of radicals are best considered together, as the evidence for their formation in the gas phase and subsequent stabilization comes from studies of three closely related substances, hydrazoic acid, hydrazine, and ammonia.

In these considerations, N_3 is taken to be the symmetrical species rather than the loosely bound complex which Herzfeld and Broida (1956) and Herzfeld (1957) have considered on the basis of the emission spectrum of the solid condensed from a discharge through nitrogen. Evidence for N_3 in the gas phase comes from the flash photolysis of hydrazoic acid (Thrush, 1956b) which yields the ultraviolet absorption spectrum of a polyatomic molecule containing no hydrogen. This was interpreted as arising from the expected linear inverted $^2\Pi_g$ ground state of N_3. Other evidence is the observation by Milligan and co-workers (1956) of absorption bands at 2150, 962, and 737 cm^{-1} in the solid deposited from a glow discharge through nitrogen. Harvey and Brown (1959) have not been able to repeat these observations, and it should be noted that a $^2\Pi_g$ state of N_3 cannot arise from ground state N plus ground state N_2.

The most extensive work on stabilizing radicals produced in systems containing nitrogen and hydrogen has been carried out by Rice and his co-workers. They have studied the thermal decomposition of hydrazoic acid (Rice and Freamo, 1951; Rice and Grelecki, 1957), its decomposition in an electric discharge (Rice and Freamo, 1953), and the thermal decompositions of hydrazine (Rice and Scherber, 1955). A short description of the experimental conditions used by these workers has been given in Section II, A. The identity of the species responsible for the blue color of the solid condensed from partially decomposed hydrazoic acid has not yet been determined, and only evidence obtained before and during deposition comes within the scope of this chapter.

Foner and Hudson (1958a) flowed hydrazoic acid at 1000 cm per second and 0.05 mm pressure through a 6 Mc/sec rf discharge. Under conditions where a blue deposit was obtained a mass spectrometric investigation showed that the gases contained about 2 % of N_2H_2, the remainder being HN_3, N_2, or NH_3; no evidence of NH or other labile species was found. Di-imide(N_2H_2) was found to be remarkably stable, and in the decomposition of hydrazine it could be passed through a trap at dry-ice temperature, condensed as a yellow solid with liquid nitrogen, and then volatilized. Foner and Hudson also showed that di-imide was too stable to be responsible for the blue color. Franklin *et al.* (1958) were unable to detect any labile species in the products of a Tesla discharge through hydrazoic acid. The yellow solid which Foner and Hudson obtained in this work was different from the yellow deposit obtained by Rice and Scherber (1955) in the thermal decomposition of hydrazine, where the color was probably due to tetrazane $NH_2 \cdot NH \cdot NH \cdot NH_2$. In more recent work on the decomposition of hydrazine, using similar techniques Foner and Hudson (1958b) have detected NH_2 and N_2H_3 and N_3H_3 in the gas phase and have identified triazene (N_3H_3) and tetrazene (N_4H_4) in the species evaporating from the trapped solid during warm-up. Tetrazene was not however detected in the gas phase before condensation.

None of these observations provides any real evidence as to the nature of the blue color in hydrazoic acid deposits, and it is unlikely that it is due to NH, which has no known visible absorption spectrum. NH_2 absorbs in the visible, but should be more easily stabilized from hydrazine. Recent investigations using electron spin resonance have suggested that the color may be due to F-center formation (Gager and Rice, 1959).

The electronic spectra of both NH and NH_2 have, however, been observed in the solid state by Robinson and McCarty (1958a). These workers passed a mixture of one part of hydrazine and twenty-five parts of argon through a 15-Mc/sec electrodeless discharge and condensed the products at liquid helium temperature. They used several diffusion pumps to obtain low pressures in the flow tube and near the condensing surface, so that the radical-radical mean free path was of the order of 1 cm. The sharpness of the spectra suggested that the radicals were trapped in well-ordered argon microcrystals. Foner and co-workers (1958b) have also stabilized NH_2 in an argon matrix; their radicals were produced with an electric discharge through ammonia and detected by electron spin resonance.

b. HNO. This is a relatively stable species in the gas phase, and its spectrum has been detected in the flash photolysis of nitroethane, amyl nitrite, and ammonia with nitric oxide (Dalby, 1958). In the last case this species is undoubtedly formed by the reaction between nitric oxide and hydrogen atoms from the ammonia. For trapped radical work, direct reaction of hydrogen atoms with nitric oxide would seem to be the best source of this radical; Harvey and Brown (1959) have successfully stabilized HNO from this source at liquid helium temperature and obtained the same infrared spectrum as that Brown and Pimentel (1958) had observed by solid state photolysis. Robinson and McCarty (1958b) have observed the electronic spectrum of HNO in the deposit obtained by condensing the products of an electrodeless discharge through moist hydrazine and argon, using the apparatus mentioned in the preceding paragraph.

c. NO_2. This species can readily be isolated in a rare gas matrix at liquid helium temperatures by spraying it with a 25-fold, or more, excess of inert gas from a low pressure source at room temperature (Becker and Pimentel, 1956; Robinson *et al.*, 1957).

4. Hydrides Not Previously Given

a. OH, O_2H. A great deal of work has been done on the solids trapped from systems containing hydrogen and oxygen, either by reaction of hydrogen atoms with oxygen (Geib and Harteck, 1932, 1934; Lavin and Stewart, 1929) or by the action of an electric discharge on water vapor (Geib, 1936; Ghormley, 1957; Giguère *et al.*, 1953; Jones and Winkler, 1951; Rodebush *et al.*, 1947; Bass and Broida, 1958). None of these studies have yielded direct evidence of the stabilization of OH or HO_2 radicals in the solid state, but all the deposits have given oxygen and hydrogen peroxide on warming. It seems likely that the oxygen is held in an open hydrogen-bonded gel of water and hydrogen peroxide (Edwards and Hashman, 1959; Hogg and Spice, 1957), although some workers have considered it to be formed from OH (Rodebush *et al.*, 1947), HO_2 (Giguère 1954; Giguère and Harvey, 1956), and H_2O_3 and H_2O_4 (Ghormley, 1957).

The UV absorption spectrum of OH has been detected by Robinson and McCarty (1958b) in the solid condensed from a discharge through moist hydrazine and argon. Livingston and co-workers (1956) have observed electron spin resonance at 77°K in the products of a discharge through water vapor, and consider this may be due to about 0.2 % of OH or HO_2. It can be seen that the evidence for stabilization of these radicals from the gas phase is surprisingly little considering that these species appear to be formed quite readily in the gas phase, to judge from mass spectrometric studies (Foner and Hudson, 1953; Ingold and Bryce, 1956; Robertson, 1952).

b. Other hydrides. In view of the interest which has been focused on the hydrides of nitrogen and oxygen, it is relevant to note that there is spectroscopic evidence for the existence of the corresponding hydrides in the next row of the periodic table, PH, PH_2, SH, and S_2H having all been detected in flash photolysis experiments (Ramsay, 1956; Porter, 1950). Although these species are probably all less stable than the corresponding nitrogen and oxygen hydrides, it is possible that they could be stabilized by similar means.

5. Other Species

An examination of the books by Herzberg (1950) or Pearse and Gaydon (1950) will show that there are a large number of radicals known from spectroscopic evidence to exist in the gas phase but which have not yet been stabilized in the solid state. In addition a large number of oxides, hydrides, and halides exist in different forms in the gas phase or are stable only at high temperatures (see, for example, Brewer, 1953; Brewer and Searcy, 1956). There are too many of these species and too little is known about many of them to make a comprehensive review worth while, but one may cite the halogen monoxides and SO and S_2O as examples. Flame spectroscopy provides ample evidence for the stability of ClO, BrO, IO, at high temperatures (Gaydon, 1957); dissociation energies are known approximately and the ClO radical is readily formed and is relatively long-lived in photochemical systems (Edgecombe *et al.*, 1958). Both SO and S_2O are present in reasonable concentration in sulfur-oxygen systems at high temperatures (Dewing and Richardson, 1958; Myers and Meschi, 1956) and could be produced in sufficient proportions for stabilization, using a suitable furnace.

REFERENCES

Andrews, E. B., and Barrow, R. F. (1950). *Nature* **165**, 890. Ultra-violet band systems of CF.

Armstrong, D. A., and Winkler, C. A. (1956). *J. Phys. Chem.* **60**, 1100. Comparative production of active nitrogen from nitrogen, nitric oxide, and ammonia and from nitrogen at different discharge potentials.

Ayscough, P. B. (1956). *J. Chem. Phys.* **24**, 944. Rate of recombination of radicals II. The rate of recombination of trifluoromethyl radicals.

Bass, A. M., and Broida, H. P. (1956). *Phys. Rev.* **101**, 1740. Spectra emitted from solid nitrogen condensed at 4.2°K from a gas discharge.

Bass, A. M., and Broida, H. P. (1958). *J. Mol. Spectroscopy* **2**, 42. Absorption spectra of solids condensed at low temperatures from electric discharges.

Becker, E. D. and Pimentel, G. C. (1956). *J. Chem. Phys.* **25**, 224. Spectroscopic studies of reactive molecules by the matrix isolation method.

Berkowitz, J., Chupka, W. A., and Kistiakowsky, G. B. (1956). *J. Chem. Phys.* **25**, 457. Mass spectrometric study of the kinetics of nitrogen afterglow.

Brewer, L. (1953). *Chem. Revs.* **52**, 1. The thermodynamic properties of the oxides and the vaporization properties.

Brewer, L., and Searcy, A. W. (1956). *Ann. Rev. Phys. Chem.* **7**, 259. High temperature chemistry.

Brewer, L., and Templeton, L. K. (1949). *Univ. of Calif. (Berkeley) Rept. No.* **529**.

Broida, H. P., and Lutes, O. S. (1956). *J. Chem. Phys.* **24**, 484. Abundance of free atoms in solid nitrogen condensed at 4.2°K from a gas discharge.

Broida, H. P., and Peyron, M. (1957). *J. phys. radium* **18**, 593. Luminescence de l'azote solide (4.2°K) contenant des atomes ou radicaux libres. Effet de la dilution par l'argon.

Broida, H. P., and Peyron M. (1958). *J. phys. radium* **19**, 480. Luminescence de l'azote solide (4.2°K) contenant des atomes ou radicaux libres. Effet de traces d'oxygène, d'hydrogène et de vapeur d'eau.

Brown, H. W., and Pimentel, G. C. (1958). *J. Chem. Phys.* **29**, 883. Photolysis of nitromethane and of methyl nitrite in an argon matrix: infrared detection of nitroxyl: HNO.

Bulewicz, E. M., and Sugden, T. M. (1958). *Trans. Faraday Soc.* **54**, 1855. The recombination of hydrogen atoms and hydroxyl radicals in hydrogen flame gases.

Bunker, D. L., and Davidson, N. (1958). *J. Am. Chem. Soc.* **80**, 5085, 5090. A further study of the flash photolysis of iodine. On the interpretation of halogen atom recombination rates.

Burns, W. G., and Dainton, F. S. (1952). *Trans. Faraday Soc.* **48**, 39. The determination of the equilibrium and rate constants of the chain propagation and termination reactions in the photochemical formation of phosgene.

Callomon, J. H., and Ramsay, D. A. (1957). *Can. J. Phys.* **35**, 129. The flash photolysis of diacetylene.

Christie, M. I., Harrison, A. J., Norrish, R. G. W., and Porter, G. (1955). *Proc. Roy. Soc.* **A231**, 446. The recombination of atoms II. Causes of variation in the observed rate constant for iodine atoms.

Dalby, F. W. (1958). *Can. J. Phys.* **36**, 1336. The spectrum and structure of the HNO molecule.

Darwent, B. de B., and Roberts, R. (1953). *Discussions Faraday Soc.* **14**, 55. The reactions of hydrogen atoms with hydrocarbons.

Dewing, E. W., and Richardson, D. (1958). *Trans. Faraday Soc.* **54**, 679. The heat of formation of sulphur monoxide.

Dows, D. A., and Pimentel, G. C. (1955). *J. Chem. Phys.* **23**, 1258. Infrared spectra of gaseous and solid hydrazoic acid and deutero-hydrazoic acid: the thermodynamic properties of HN_3 .

Dressler, K. (1959). *J. Chem. Phys.* **30**, 1621. Absorption spectrum of vibrationally excited N_2 in active nitrogen.

Edgecombe, F. H. C., Norrish, R. G. W., and Thrush, B. A. (1958). *Chem. Soc. (London) Spec. Publ. No.* **9**, 121. Studies of the ClO radical by flash photolysis.

Edwards, J. W., and Hashman, J. S. (1958). Unpublished results.

Edwards, J. W., and Hashman, J. S. (1959). *J. Am. Chem. Soc.* To be published. Reaction of oxygen and atomic hydrogen.

Foner, S. N., and Hudson, R. L. (1953). *J. Chem. Phys.* **21**, 1608. Detection of the HO_2 radical by mass spectrometry.

Foner, S. N., and Hudson, R. L. (1955). *J. Chem. Phys.* **23**, 1974. OH, HO_2 , and H_2O production in the reaction of atomic hydrogen with molecular oxygen.

Foner, S. N., and Hudson, R. L. (1958a). *J. Chem. Phys.* **28**, 719. Diimide-identification and study by mass spectrometry.

Foner, S. N., and Hudson, R. L. (1958b). *J. Chem. Phys.* **29**, 442. Mass spectrometric detection of triazene and tetrazene and studies of the free radicals NH_2 and N_2H_3.

Foner, S. N., Jen, C. K., Cochran, E. L., and Bowers, V. A. (1958a). *J. Chem. Phys.* **28**, 351. Electron spin resonance of nitrogen atoms trapped at liquid helium temperature.

Foner, S. N., Cochran, E. L., Bowers, V. A., and Jen, C. K. (1958b). *Phys. Rev. Letters* **1**, 91. Electron spin resonance spectra of the free NH_2 and ND_2 radicals at 4.2°K.

Fontana, B. J. (1958). *J. Appl. Phys.* **29**, 1668. Thermometric study of the frozen products from nitrogen.

Fontana, B. J. (1959). *J. Chem. Phys.* **31**, 148. Magnetic study of the frozen products from the nitrogen microwave discharge.

Fowler, A., and Strutt, R. J. (1910). *Proc. Roy. Soc.* **A85**, 377. Spectroscopic investigations in connection with the active modification of nitrogen I. Spectrum of the afterglow.

Franklin, J. L., Herron, J. T., Bradt, P., and Dibeler, V. H. (1958). *J. Am. Chem. Soc.* **80**, 6188. Mass spectrometric study of the decomposition of hydrazoic acid by the electric discharge.

Gager, W., and Rice, F. O. (1959). *J. Chem. Phys.* **31**, 564. Paramagnetic resonance spectra of active species. Blue material from hydrazoic acid.

Gaydon, A. G. (1957). "The Spectroscopy of Flames." Chapman & Hall, London.

Geib, K. H. (1936). *J. Chem. Phys.* **4**, 391. Water vapor discharge and hydrogen peroxide formation.

Geib, K. H., and Harteck, P. (1932). *Ber.* **B65**, 1551. Eine neue Form von H_2O_2.

Geib, K. H., and Harteck, P. (1933). *Ber.* **B66**, 1815. Anlagerungs-Reaktionen mit H- und O-Atomen bei tiefen Temperaturen.

Geib, K. H., and Harteck, P. (1934). *Z. physik. Chem.* **A170**, 1. Durch H-Atome ausgelöste Oxydationsreaktionen.

Ghormley, J. A. (1957). *J. Am. Chem. Soc.* **79**, 1862. Warming curves for the condensed product of dissociated water vapor and for hydrogen peroxide glass.

Giguère, P. A. (1954). *J. Chem. Phys.* **22**, 2085. Spectroscopic evidence for stabilized HO_2 radicals.

Giguère, P. A., and Harvey, K. B. (1956). *J. Chem. Phys.* **25**, 373. On the presumed spectroscopic evidence for trapped HO_2 radicals.

Giguère, P. A., Secco, E. A., and Eaton, R. S. (1953). *Discussions Faraday Soc.* **14**, 104. Production of hydrogen and deuterium peroxides in the electrodeless discharge.

Golden, S. (1958). *J. Chem. Phys.* **29**, 61. Free radical stabilization in condensed phases.

Gomer, R., and Kistiakowsky, G. B. (1951). *J. Chem. Phys.* **19**, 85. The rate constant of ethane formation from methyl radicals.

Gray, P. (1955). Free Radicals in Combustion Processes: Thermochemistry of the Alkoxy Radicals RO. "Fifth Symposium on Combustion," p. 535. Reinhold, New York.

Greaves, J. C., and Linnett, J. W. (1958). *Trans. Faraday Soc.* **54**, 1323. The recombination of oxygen atoms at surfaces.

Greene, E. F., and Toennies, P. "Shock Waves." In press.

Harkins, W. D. (1934). *Trans. Faraday Soc.* **30**, 221. Free radicals in electrical discharges.

Harteck, P., Reeves, R. R., and Mannella, G. (1958). *J. Chem. Phys.* **29**, 608. Rate of recombination of nitrogen atoms.

Harvey, K. B., and Bass, A. M. (1958). *J. Mol. Spectroscopy* **2**, 405. Infrared absorption spectrum of oxygen discharge products and of ozone at 4°K.

Harvey, K. B., and Brown, H. W. (1959). *J. chim. phys.* **56**, 745. Etude aux infrarouges de certains solides condensés à porter de decharges en phase gazeuse.

Heidt, L. T., and Boyles, H. B. (1951). *J. Am. Chem. Soc.* **73**, 5728. Influence of several variables encountered in photochemical work upon the intensity of light of λ = 254 mμ produced by a quartz mercury vapour lamp of the low pressure type operating in water at 0 to 95°.

Herron, J. T., and Schiff, H. I. (1958). *Can. J. Chem.* **36**, 1159. A mass spectrometric study of normal oxygen and oxygen subjected to the electric discharge.

Herron, J. T., Franklin, J. L., Bradt, P., and Dibeler, V. H. (1958). *J. Chem. Phys.* **29**, 230. Kinetics of nitrogen atom recombination.

Herzberg, G. (1950). "Spectra of Diatomic Molecules," 2nd ed. Van Nostrand, New York.

Herzberg, G. (1956). "Chemical Institute of Canada Symposium on Free Radicals." Laval University, Quebec.

Herzberg, G., and Ramsay, D. A. (1956). *Proc. Roy. Soc.* **A233**, 34. The 7500 to 4500 Å absorption system of the free HCO radical.

Herzberg, G., and Shoosmith, J. (1959). *Nature* **183**, 1801. Spectrum and structure of the free methylene radical.

Herzfeld, C. M. (1957). *Phys. Rev.* **107**, 1239. Theory of forbidden transitions of nitrogen atoms trapped in solids.

Herzfeld, C. M., and Broida, H. P. (1956). *Phys. Rev.* **101**, 606. Interpretation of spectra of atoms and molecules in solid nitrogen condensed at 4.2°K.

Hicks, J. A., and Melville, H. W. (1954). *Proc. Roy. Soc.* **A226**, 314. The synthesis of block copolymers in a flow system.

Hogg, M. A. P., and Spice, J. E. (1957). *J. Chem. Soc.*, p. 3971. The nature of the low temperature transition in hydrogen peroxide prepared by discharge tube methods.

Holland, R., Style, D. W. G., Dixon, R. N., and Ramsay, D. A. (1958). *Nature* **182**, 336. Emission and absorption spectra of NCO and NCS.

Honig, R. E. (1954). *J. Chem. Phys.* **22**, 126. Mass spectrometric study of the molecular sublimation of graphite.

Ingold, K. U., and Bryce, W. A. (1956). *J. Chem. Phys.* **24**, 360. Mass spectrometric investigation of the hydrogen-oxygen and methyl-oxygen reactions.

Ingold, K. U., and Lossing, F. P. (1953). *J. Chem. Phys.* **21**, 368. The rate of combination of methyl radicals.

Jackson, D. S., and Schiff, H. I. (1955). *J. Chem. Phys.* **23**, 2333. Mass spectrometric investigation of active nitrogen.

Jackson, J. L., and Montroll, E. W. (1958). *J. Chem. Phys.* **28**, 1101. Free radical statistics.

Jen, C. K., Foner, S. N., Cochran, E. L., and Bowers, V. A. (1956). *Phys. Rev.* **104**, 846. Paramagnetic resonance of hydrogen atoms trapped at liquid He temperature.

Jen, C. K., Foner, S. N., Cochran, E. L., and Bowers, V. A. (1958). *Phys. Rev.* **112**,

1169. Electron spin resonance of atomic and molecular free radicals trapped at liquid helium temperature.

Jones, R. A., and Winkler, C. A. (1951). *Can. J. Chem.* **29,** 1010. Reactions in dissociated water vapour.

Kaufman, F. (1958). *J. Chem. Phys.* **28,** 352. Air afterglow and kinetics of some reactions of atomic oxygen.

Kaufman, F., and Kelso, J. R. (1958). *J. Chem. Phys.* **28,** 510. Vibrationally excited ground-state nitrogen in active nitrogen.

Kistiakowsky, G. B., and Gershinowitz, H. (1933). *J. Chem. Phys.* **1,** 432. The thermal dissociation of cyanogen into cyanide radicals.

Kistiakowsky, G. B., and Sauer, K. (1956). *J. Am. Chem. Soc.* **78,** 5699. The rate and mechanism of some reactions of methylene.

Klein, R., and Scheer, M. D. (1958). *J. Phys. Chem.* **62,** 1011. The reaction of H atoms with solid olefins at −195°C.

Laird, R. K., Andrews, E. B., and Barrow, R. F. (1950). *Trans. Faraday Soc.* **46,** 803. The absorption spectrum of CF_2 .

Langmuir, I. (1915). *J. Am. Chem. Soc.* **37,** 417. The dissociation of hydrogen into atoms II. Calculation of the degree of dissociation and the heat of formation.

Lavin, G. I., and Stewart, F. B. (1929). *Proc. Natl. Acad. Sci. U. S.* **15,** 829. Production of hydroxyl by the water-vapor discharge.

Lewis, G. N., and Lipkin, D. (1942). *J. Am. Chem. Soc.* **64,** 2801. Reversible photochemical processes in rigid media: the dissociation of organic molecules into radicals and ions.

Lewis, P. (1900). *Astrophys. J.* **12,** 8. Some new fluorescence and afterglow phenomena in vacuum tubes containing nitrogen.

Linnett, J. W., and Marsden, D. G. H. (1956). *Proc. Roy. Soc.* **A234,** 489, 504. The kinetics of recombination of oxygen atoms at a glass surface. The recombination of oxygen atoms at salt and oxide surfaces.

Livingston, R., Ghormley, J., and Zeldes, H. (1956). *J. Chem. Phys.* **24,** 483. Paramagnetic resonance observations on the condensed products of electric discharges through water vapor and related substances.

Lossing, F. P. (1957). *Can. J. Chem.* **35,** 305. Free radicals by mass spectrometry XII. Primary steps in the mercury photosensitized decompositions of acetone and acetaldehyde.

Lossing, F. P., and Tanaka, I. (1956). *J. Chem. Phys.* **25,** 1031. Photoionization as a source of ions for mass spectrometry.

Lossing, F. P., and Tickner, A. W. (1952). *J. Chem. Phys.* **20,** 907. Free radicals by mass spectrometry I. The measurement of methyl radical concentrations.

Lossing, F. P., Marsden, D. G. H., and Farmer, J. B. (1956). *Can. J. Chem.* **34,** 701. Free radicals by mass spectrometry XI. The mercury photosensitized decomposition of C_2-C_4 olefins.

McCarty, M., and Robinson, G. W. (1959). *J. chim. phys.* **56,** 723. Les spectres d'absorption électroniques des radicaux libres de petite dimension dans les milieux rigides constitués pardes gaz rares.

McKinley, J. D., and Garvin, D. (1955). *J. Am. Chem. Soc.* **77,** 5802. The reactions of atomic hydrogen with ozone and with oxygen.

McNesby, J. R. (1959). Unpublished work.

Marcotte, F. B., and Noyes, W. A. (1952). *J. Am. Chem. Soc.* **74,** 783. Photochemical studies XLV. The reactions of methyl and acetyl radicals with oxygen.

Margrave, J. L., and Wieland, K. (1953). *J. Chem. Phys.* **21**, 1552. Equilibria involving $CF(g)$ and $CF_2(g)$ radicals at high temperatures.

Martin, L. C. (1913). *Proc. Roy. Soc.* **A89**, 127. A band spectrum attributed to carbon monosulphide.

Milligan, D. E., Brown, H. W., and Pimentel, G. C. (1956). *J. Chem. Phys.* **25**, 1080. Infrared absorption by the N_3 radical.

Minkoff, G. J., Scherber, F. I., and Gallagher, J. S. (1959). *J. Chem. Phys.* **30**, 753. Energetic species trapped at 4.2°K from electric discharges.

Myers, R. J., and Meschi, D. J. (1956). *J. Am. Chem. Soc.* **78**, 6220. Disulphur monoxide I. Its identification as the major constituent in Schenk's "Sulphur Monoxide."

Norman, I., and Porter, G. (1955). *Proc. Roy. Soc.* **A230**, 399. Trapped atoms and radicals in rigid solvents.

Norrish, R. G. W., and Thrush, B. A. (1956). *Quart. Revs. (London)* **10**, 149. Flash photolysis and kinetic spectroscopy.

Oldenberg, O., and Rieke, F. F. (1939). *J. Chem. Phys.* **7**, 485. Kinetics of OH radicals as determined by their absorption spectrum V.

Padley, P. J., and Sugden, T. M. (1958). *Proc. Roy. Soc.* **A248**, 248. Photometric investigations of alkali metals in hydrogen flame gases IV.

Pearse, R. W. B., and Gaydon, A. G. (1950). "The Identification of Molecular Spectra." Chapman & Hall, London.

Peyron, M., and Broida, H. P. (1959). *J. Chem. Phys.* **30**, 139. Spectra emitted from solid nitrogen condensed at very low temperatures from a gas discharge.

Phillips, J. G., and Brewer, L. (1955). *Mém. soc. roy. sci. Liège, vol. hors sér.* **15**, 341. An ultraviolet continuum in the spectrum of carbon.

Porter, G. (1950). *Discussions Faraday Soc.* **9**, 60. The absorption spectroscopy of substances of short life.

Porter, G., and Wright, F. J. (1953). *Discussions Faraday Soc.* **14**, 23. Studies of free radical reactivity by the methods of flash photolysis. The photochemical reaction between chlorine and oxygen.

Porter, G., and Wright, F. J. (1955). *Trans. Faraday Soc.* **51**, 1469. Primary photochemical processes in aromatic molecules. Part 3. Absorption spectra of benzyl, anilino, phenoxy and related free radicals.

Rabinowitch, E., and Wood, W. C. (1936a). *Trans. Faraday Soc.* **32**, 907. Kinetics of recombination of bromine atoms II.

Rabinowitch, E., and Wood, W. C. (1936b). *J. Chem. Phys.* **4**, 497. Kinetics of recombination of iodine atoms.

Ramsay, D. A. (1953). *J. Chem. Phys.*, **21**, 165. Absorption spectra of free radicals in continuously irradiated photochemical systems.

Ramsay, D. A. (1956). *Nature* **178**, 374. Absorption spectra of free PH_2 and PD_2 radicals.

Ramsay, D. A. (1957). *Ann. N. Y. Acad. Sci.* **67**, 485. Electronic spectra of polyatomic free radicals.

Rice, F. O., and Freamo, M. (1951). *J. Am. Chem. Soc.* **73**, 5529. The imine radical.

Rice, F. O., and Freamo, M. (1953). *J. Am. Chem. Soc.* **75**, 548. The formation of the imine radical in the electrical discharge.

Rice, F. O., and Grelecki, C. (1957). *J. Am. Chem. Soc.* **79**, 1880. The imine radical.

Rice, F. O., and Scherber, F. (1955). *J. Am. Chem. Soc.* **77**, 291. The hydrazino radical and tetrazane.

Robertson, A. J. B. (1952). "Mass Spectrometry," p. 47. Institute of Petroleum, London.

Robinson, G. W., and McCarty, M. (1958a). *J. Chem. Phys.* **28,** 349. Electronic spectra of free radicals at 4°K—NH_2 .

Robinson, G. W., and McCarty, M. (1958b). *J. Chem. Phys.* **28,** 350. Electronic spectra of free radicals at 4°K—HNO, NH and OH.

Robinson, G. W., McCarty, M., and Keelty, M. G. (1957). *J. Chem. Phys.* **27,** 972. Electronic spectrum of monomeric nitrogen dioxide at liquid helium temperature.

Rodebush, W. H., Keizer, C. R., McKee, S., and Quagliano, J. V. (1947). *J. Am. Chem. Soc.* **69,** 538. The reactions of the hydroxyl radical.

Ruehrwein, R. A., and Hashman, J. S. (1959). *J. Chem. Phys.* **30,** 823. The formation of ozone from atomic oxygen at low temperatures.

Ruehrwein, R. A., Hashman, J. S., and Edwards, J. W. (1959). *J. Phys. Chem.* To be published. Chemical reactions of free radicals at low temperatures.

Russell, K. E., and Simons, J. (1953). *Proc. Roy. Soc.* **A217,** 271. Studies in energy transfer I. The combination of iodine atoms.

Scheer, M. D. (1959). Unpublished results.

Schiff, H. I., and Steacie, E. W. R. (1951). *Can. J. Chem.* **29,** 1. The reactions of H and D atoms with cyclic and paraffin hydrocarbons.

Schoen, L. (1955). Photodecomposition of Formaldehyde. "Fifth Symposium on Combustion," p. 786. Reinhold, New York.

Schüler, H., and Michel, A. (1955). *Z. Naturforsch.* **10a,** 459. Über den spektroskopischen Nachweis des Benzyl- und des Benzalradikals ($C_6H_5CH_2$ und C_6H_5CH).

Schüler, H., Reinebeck, L., and Koberle, R. (1952). *Z. Naturforsch.* **7a,** 421. Über das Auftreten von gemeinsamen Bruchstücken (mehratomige Radikale?) bei Benzolderivaten in der Glimmentladung. 1. Mitt. Über das "V-Spektrum" der Monoderivate des Benzols.

Shepp, A. (1956). *J. Chem. Phys.* **24,** 939. Rate of recombination of radicals I. A general sector theory: a correction to the methyl recombination rate.

Shepp, A., and Kutschke, K. O. (1957). *J. Chem. Phys.* **26,** 1020. Rate of recombination of radicals III. Rate of recombination of ethyl radicals.

Smith, W. V. (1943). *J. Chem. Phys.* **11,** 110. The surface recombination of H atoms and OH radicals.

Steacie, E. W. R. (1954). "Atomic and Free Radical Reactions." Reinhold, New York.

Strong, R. L., Chein, J. C. W., Graf, P. E., and Willard, J. E. (1957). *J. Chem. Phys.* **26,** 1287. Studies of I atom and Br atom recombination following flash photolysis of gaseous I_2 and Br_2 .

Strutt, R. J. (1910). *Proc. Roy. Soc.* **A85,** 219. A chemically active modification of nitrogen, produced by the electric discharge I.

Strutt, R. J. (1915). *Proc. Roy. Soc.* **A91,** 303. A chemically active modification of nitrogen, produced by the electric discharge VI.

Swarc, M. (1950). *Chem. Revs.* **47,** 75. The determination of bond dissociation energies by pyrolytic methods.

Tanaka, Y., Jursa, A., and Le Blanc, F. (1957). "The Threshold of Space," p. 89. Pergamon Press, London.

Thrush, B. A. (1956a). *Nature* **178,** 155. Spectrum of the cyclopentadienyl radical.

Thrush, B. A. (1956b). *Proc. Roy. Soc.* **A235,** 143. The detection of free radicals in the high intensity photolysis of hydrogen azide.

Trotman-Dickenson, A. F. (1955). "Gas Kinetics." Academic Press, New York.

Tsuchiya, T. (1957). *Tokyo Kôgyô Shikensho Hôkoku* **52,** 289. Mass spectrometric study of the thermal decomposition of gas molecules I.

Wentink, T., Sullivan, J. O., and Wray, K. L. (1958). *J. Chem. Phys.* **29,** 231. Nitrogen atom recombination at room temperature.

White, J. U. (1940). *J. Chem. Phys.* **8,** 79, 459. Spectroscopic measurements of gaseous CN.I. Dissociation in the electric discharge. II. Thermal dissociation of cyanogen.

Wood, R. W. (1922). *Proc. Roy. Soc.* **A102,** 1. Spontaneous incandescence of substances in atomic hydrogen gas.

Zletz, A. (1959). Unpublished results.

3. Techniques of Electrical Discharge for Radical Production

General Electric Company, Microwave Laboratory, Palo Alto, California

I. Introduction

The electrical discharge is universally recognized as an effective and convenient means for the production of radicals by the dissociation of molecular gases. It is the purpose of this chapter to describe and discuss some of the more important discharge techniques. An exhaustive review of the field has not been attempted. Rather, the approach has been to present the essential features of those methods which have found a reasonably widespread use for radical production in the last decade.

In an electrical discharge, the energy necessary for radical production by molecular dissociation is supplied to the gas primarily by electron collisions. Dissociation into radicals is the result of complex interactions between electrons, ions, and molecules. In the ordinary low pressure discharge, dissociation is almost entirely due to electron-molecule collisions. Collisions between atoms and molecules become important only in high pressure discharges where gas temperatures reach values of the order of 10^3 degrees or more.

The Wood's tube (Wood, 1922) is probably the best-known type of low pressure, low frequency electrical discharge which is used for radical production. Although originated as a device for investigating the Balmer series of hydrogen, the Wood's tube has been employed in numerous investigations as a source of hydrogen, oxygen, and nitrogen atoms.

As a result of the development of efficient sources of high frequency current, the Wood's tube has been largely displaced by the electrodeless radiofrequency or microwave discharge as a device for the production of radicals. Adoption of the electrodeless discharge is to be attributed to a combination of factors such as ease of control, simplicity of construction of the discharge tube, greater stability, and, probably most important, a greater purity of the discharge products because metallic electrodes in contact with the discharge gases are not required for high frequency operation. The absence of metallic electrodes may also increase the yield of radicals by elimination of recombination on electrodes.

II. Production Mechanisms

A. Dissociation by Electron-Molecule Collisions

The generally accepted picture of the dissociation process in a low pressure gaseous electrical discharge postulates a swarm of electrons moving through the gas under the influence of an applied electric field (Massey and Burhop, 1952). The majority of the electrons arise from ionizing collisions in the discharge. To sustain the discharge, the rate of electron production must balance the rate of electron loss. In the direct current (dc) discharge, electrons are swept out of the discharge by the steady field. In a high frequency discharge, however, electrons are lost primarily by diffusion to the walls of the containing vessel.

In the positive column of a dc discharge or in the plasma of a high frequency discharge, the velocity distribution of electrons is essentially Maxwellian. Thus, an electron temperature can be defined which is characteristic of the mean energy of the electrons. The electron temperature depends upon the ionization potential and pressure of the gas and is usually very large compared with the temperature of the gas. For example, in a low pressure discharge, it is doubtful if the gas temperature is very much higher than 500°K compared with an electron temperature of about 50,000°K. For high currents and pressures, on the other hand, the electron and gas temperatures tend to equalize at a value of a few thousand degrees Kelvin.

Dissociation by means of electron-molecule collisions may be the direct result of electron impact or may occur indirectly, as, for example, by the production of a metastable particle in the primary electron-molecule encounter with dissociation following a secondary encounter between the metastable and a neutral molecule. Production of a neutral atom may also follow initial formation of an ion via an electron-induced transition from the ground state of a molecule to an unstable level. In a polyatomic molecule, the dissociation processes may be very complex with many possible modes of dissociation.

Despite the wide use of the electrical discharge for dissociation, except with regard to hydrogen very little progress has been made in working out the details of the dissociation process. For hydrogen, use has been made of a theory due to Emeléus et al. (1936) and Lunt and Meek (1936). According to this theory, the number of electronic excitations leading to the dissociation of molecules into atoms is directly proportional to the number of electrons supplied to the discharge, for conditions such that the density of undissociated molecules is effectively a constant. In both the dc and the high frequency discharge plasmas for a given pressure, the electric field intensity is essentially constant and independent of discharge current. Thus, for a discharge operating at constant pressure, a linear relationship is found between the rate of radical production and the power input to the discharge at very low power levels for which the density of undissociated molecules is practically a constant. Data for the production of hydrogen atoms in dc and high frequency discharges at low power levels are in agreement with the theory (see Figs. 2 and 9).

A value of the energy efficiency, η, defined as the number of dissociations produced per unit of energy input to the plasma has been calculated for hydrogen (Lunt and Meek, 1936; Lunt et al., 1937). It is assumed that essentially all dissociation of hydrogen occurs by direct excitation of the lowest repulsive state $^1\Sigma_g^+ \to {}^3\Sigma_u^+$ and to a lesser extent via the process $^1\Sigma_g^+ \to {}^3\Sigma_g^+$ where after emission $^3\Sigma_g^+$ falls to $^3\Sigma_u^+$ and dissociates. For these processes, the calculated value of η is about 2×10^{-2} dissociations per electron volt based on a value of $0.2\ \pi a_0^2$ (a_0 = radius first Bohr orbit of H atom) for the collision cross section for the $^3\Sigma_u^+$ and $^3\Sigma_g^+$ excitations. This value of η is to be compared with experimental values of 5.8×10^{-2}, 5×10^{-2}, and 6.5×10^{-2} dissociations per electron volt for dc, 5 Mc, and 3000 Mc discharges, respectively (Poole, 1937c; Corrigan and Von Engel, 1958; Shaw, 1959). A recent theoretical calculation of the collision cross section of the $^1\Sigma_g^+ \to {}^3\Sigma_u^+$ transition has resulted in a theoretical value of η in good agreement with the experimental value (Edelstein, 1958).

B. THERMAL DISSOCIATION IN AN ELECTRICAL DISCHARGE

The simple thermal dissociation of molecules is, of course, well understood and has been widely employed, especially in research on atomic beams. For simple gases, the temperatures required are in the several thousands of degrees. Such temperatures are readily achieved in a high pressure electrical arc (Edels and Gambling, 1958) and lead to essentially complete dissociation of the arc gases. For polyatomic molecules, the completeness of dissociation achieved in an arc may constitute an objection in situations where, for example, production of the desired radical species requires removal of only one atom from the molecule.

III. Recombination

The yield of radicals obtainable from an electrical discharge represents the net result of the production processes and the losses due to recombination. For the purposes of this discussion, such losses are conveniently grouped into two classes: surface or wall phase and gas phase. The extent to which one or both of these effects is applicable to a particular discharge depends on the construction of the discharge tube (size and nature of electrode and wall materials) and on the conditions of operation (pressure, temperature, and gas flow rate). In discharges used for radical production, it appears that losses due to wall effects outweigh those due to gas phase recombination. Since wall recombination processes are not well understood, the methods used to offset it are largely empirical. These are discussed in Section IV, E insofar as they relate to the production of specific radicals.

IV. Typical Electrical Discharge Techniques

A great many different electrical discharge techniques and devices have been used for the production of free radicals. Since it is impractical to attempt to review them all, this discussion will be limited to a description of representative examples of discharges operating in three principal frequency regions: (1) low frequency, (2) radio frequency (rf), and (3) microwave.

A. ELECTRICAL DISCHARGE TUBES

Electrical discharge tubes used for radical production are generally made of glass or quartz. To a reasonable extent, these materials satisfy the requirements of mechanical strength, good electrical insulation, and high melting temperature. They fail, however, to meet the requirements for a low coefficient of recombination for many of the more common radicals. Despite the fact that the rate of recombination on glass is very much less than on metals, loss due to radical recombination on glass or quartz is frequently found to be a limiting factor and special procedures must be employed to reduce it to a tolerable level. This aspect of the problem is discussed in more detail in Section IV, F.

For a low frequency discharge tube, practically any type of glass can be used. In the Wood's tube, for example, the relatively great volume of the tube results in a comparatively low power density and a resultant low temperature of operation. In the body of the tube, the temperature may be no more than 50° above room temperature (Poole, 1937a). In any case, with tubes of this type, water cooling is easily arranged.

For radio-frequency or microwave discharge tubes, the requirements are more critical. Because electrodes in contact with the gas ordinarily are not used with tubes of this type, it is necessary for the walls of the tube to be transparent to rf or microwave energy. Thus, material of low dielectric loss is desirable to obtain high electrical efficiency and to avoid electrical heating of the walls. Although glasses of sufficiently low electrical loss to meet this requirement are available (Von Hipple, 1954), quartz is recommended because of its higher melting point. In rf and microwave discharges, heating of the tube due to recombination at the walls and dielectric losses may in some circumstances be sufficient to melt Pyrex and other similar glasses.

It is possible, of course, to cool the discharge tube to avoid excessive temperatures. Water cooling cannot be used efficiently because of the high dielectric losses in water at radio and microwave frequencies. Cooling by an air blast or a jacket of low dielectric loss liquid is satisfactory.

B. Low Frequency Discharges

The best-known example of the low frequency discharge is the Wood's tube (Wood, 1922). Figure 1 shows a typical Wood's tube and associated power supply. The principal feature of this discharge tube is the length of the discharge path. Large electrodes are separated by a tube about 1 or 2 meters long. For radical production, gas is flowed through the tube continuously. In the arrangement shown, gas enters at each electrode and the partially dissociated gas is removed at the middle of the tube. By locating the exit a long distance from the electrodes the loss of atoms due to recombination on the metal electrodes is greatly reduced. This was first discovered by Wood in work with hydrogen in which it was discovered that the atomic hydrogen spectrum was very intense in the central portion of the tube. In the regions near the electrodes, the spectrum was characteristic of molecular hydrogen.

To operate a Wood's tube of the dimensions shown in Fig. 1, a potential of about 2 kv is required at 60 cps. For hydrogen at a pressure near 0.5 mm Hg, the discharge current is about 0.1 ampere. For some conditions of operation, water-cooled electrodes may be required. Details of water-cooled electrodes are given by Weinrich and Hughes (1954). For hydrogen and oxygen, it is satisfactory to operate the Wood's tube on dc or low frequency ac. To dissociate nitrogen in a Wood's tube, however, it has been found that an intermittent type of discharge is required in which strong fields exist

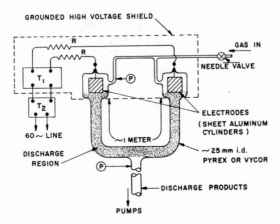

FIG. 1. Low frequency discharge tube and power supply. P, pressure gauge; R, current-limiting resistors; T_1, high voltage transformer; T_2, adjustable autotransformer.

throughout the discharge tube (Greenblatt and Winkler, 1949; Heald and Beringer, 1954). A satisfactory explanation of the behavior of nitrogen in this regard does not appear to have been given. With the exception of hydrogen, very little information is available concerning the influence of pressure, power input to the discharge, and other factors on the yields of radicals obtainable with the Wood's tube. For hydrogen, Poole (1937b, c) made a systematic study of the factors that affect the production of hydrogen atoms. Figure 2 summarizes some of Poole's results to show the effect of power and flow rate on the rate of hydrogen atom yield. Maximum efficiency of production occurred for a pressure of about 0.6 mm Hg and high rates of flow of hydrogen. More complete results are given in Poole's original articles, but much less complete data are available concerning the use of a Wood's tube for oxygen (Fryburg, 1956), nitrogen (Greenblatt and Winkler, 1949; Harteck *et al.*, 1958), and sulfur monoxide (Jones, 1950).

C. Radio-Frequency Discharges

At low frequencies, the power required to initiate and sustain a low pressure discharge does not differ significantly from that required for direct current. At radio frequencies, however, a significant difference is found in both the power required for breakdown and that required to sustain the discharge. This difference is ascribed to the fact that, at high frequencies, losses of electrons and ions to electrodes are reduced to a negligible amount relative to the losses which occur at low frequencies.

For radical production, an important advantage of the high frequency glow discharge results from the fact that the discharge can be completely

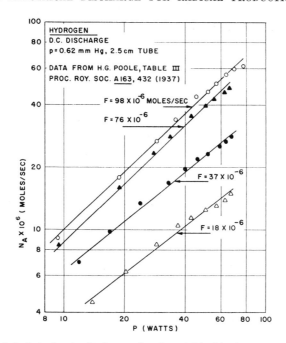

FIG. 2. Poole's data for dc discharge showing yield of hydrogen atoms as a function of hydrogen flow rate and of power dissipated in positive column of the discharge.

contained within an envelope of glass or other nonporous insulating material without the need for direct contact between the gas and metal electrodes (so-called electrodeless discharge). Since metals are usually good catalysts for effecting the recombination of radicals, the elimination of electrodes removes one of the factors which tends to limit the yield of radicals.

Coupling of high frequency power into a discharge can be accomplished with relatively simple arrangements. At radio frequencies, either inductive or capacitative types of coupling are used. Typical coupling arrangements are illustrated in Fig. 3. These sketches are intended merely to suggest some of the possible arrangements. Specific details will vary depending on the frequency, the power, and the size of the discharge tube. Radio-frequency power is obtained from conventional oscillators and amplifiers. Small commercially available radio transmitters or industrial type rf heating or diathermy equipment will serve as a convenient packaged form of rf power complete with circuits to provide a satisfactory coupling of rf power to a discharge tube.

Many examples of well-designed rf discharge systems have been described. Two which possess unique features are the design described by

Prodell and Kusch (1957) which provides for cooling the discharge tube and minimizes contact of the dissociated products with the discharge tube walls, and a design by Garvin *et al.* (1958) which includes an effective regulator for maintaining a constant discharge current.

So far as is known, no systematic studies have been made of the factors which influence the yield of radicals produced in an rf discharge. The following authors have used an rf discharge for the production of atoms of nitrogen (Barth and Kaplan, 1957), hydrogen (Jennings and Linnett, 1958), hydroxyl (Dousmanis *et al.*, 1955), iodine (Garvin *et al.*, 1958), and tritium (Prodell and Kusch, 1957). These references should be consulted for further details concerning problems encountered with particular gases.

One application of an rf discharge of which a detailed discussion is justified concerns the production of hydroxyl radicals. Considerable difficulty was encountered in finding the correct discharge conditions under which the hydroxyl radical was produced (Sanders *et al.*, 1954). Conditions for production of up to possibly 30 % hydroxyl were established on the basis of microwave spectroscopy which gave unequivocal evidence for the radicals, contrary to other types of evidence. The production of radicals was found to be critically dependent on pressure and discharge current. For constant discharge current, the yield of radicals increased rapidly with increasing pressure. The optimum pressure was of the order of 0.1 mm Hg. For optimum line intensity, the dc power supplied to the rf discharge was about 100 watts. For a dc discharge with internal electrodes, more than 500

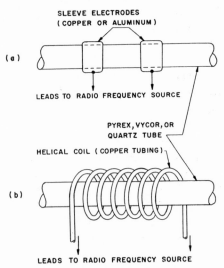

FIG. 3. Typical coupling arrangements for use with radio-frequency discharge.

watts was required to produce the same number of radicals (Dousmanis *et al.*, 1955).

D. MICROWAVE DISCHARGE

The use of microwaves for the dissociation of gases followed the development during World War II of powerful sources of microwave energy. Among the early users of this type of discharge are Nagle *et al.*, (1947) (referred to in Davis *et al.*, 1949) and Broida and Moyer (1952), who found that a microwave discharge is an efficient, stable source of hydrogen atoms. More recently, a microwave discharge has been used for the production of nitrogen atoms (Broida and Pellam, 1954; Broida and Chapman, 1958; Kaufman, 1958), oxygen atoms (Broida and Pellam, 1955; Kaufman, 1958), boron and chlorine (Holzmann and Morris, 1958), chlorine (Davis *et al.* 1949), and bromine (King and Jaccarino, 1954). Also, the dissociation of hydrogen, oxygen, water vapor, and methane has been studied by McCarthy (1954).

The main advantages of the microwave discharge are the same as for the rf electrodeless discharge, with the added feature that an efficient coupling of microwave power to a discharge can be achieved readily.

At microwave frequencies, accurate measurements of the power dissipated by the discharge are facilitated because conventional microwave coupling systems ensure that essentially all of the incident microwave energy is confined to the discharge, and radiation and reflection losses are usually small and readily determined. By contrast, in an rf discharge, an efficient coupling may be difficult to achieve.

It has frequently been asserted that the efficiency of the microwave discharge for radical production is significantly higher than that of the low frequency discharge. It appears, however, that a meaningful comparison of relative efficiency cannot be made on the basis of available information. With the possible exception of the production of hydrogen atoms, measurements suitable for determination of the efficiency of the various types of discharge are not available. For hydrogen, no significant difference in efficiency has been found when comparison was made between the power supplied to the positive column of a dc discharge (Poole, 1937c) and the power supplied to the plasma of a microwave discharge (Shaw, 1958). The validity of this comparison depends on the assumption that the methods used for measuring radical yields of the dc and microwave discharges are equally efficient.

A wide variety of microwave circuitry and components can be used to produce a microwave discharge. The two circuits described below are capable of satisfying most requirements.

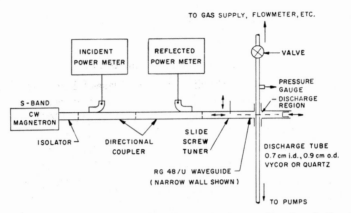

FIG. 4. Microwave apparatus used to excite discharge. Circuits similar to that shown have been constructed from commercial components and operated at frequencies in S- and X-band (3000 Mc and 9000 Mc, respectively).

The circuit shown in Fig. 4 is designed for flexibility and is capable of operation under a wide range of conditions of power input, discharge tube pressure, etc. (Shaw, 1958). Provisions are made for a variation of the power input to the discharge and for measurement of the power dissipated in the discharge. Commercial microwave tubes and waveguide components are available which are suitable for operation of discharge tubes up to continuous-wave (CW) power levels of the order of one kilowatt. Power is coupled to the discharge tube by means of the arrangement shown in Fig. 5. The coupler is fabricated from a section of conventional rectangular waveguide. Short cylindrical waveguides around the discharge tube help to prevent propagation of microwave energy along the discharge tube. An impedance match between the discharge and the microwave power source is obtained by means of a tuner and a sliding short. The power supplied to the discharge can be varied by means of a power divider (Teeter and Bushore, 1957) or a water attenuator (Alpert, 1949; Kaufman, 1958).

The circuit shown in Fig. 6 is somewhat simplified and is intended for operation at power levels up to about 100 watts. An arrangement similar to that shown has been used successfully with H_2, O_2, N_2, D_2, and water vapor (Shaw, 1958). No special means of achieving an impedance match between the discharge and the microwave power source has been provided. The isolator is capable of protecting the magnetron from the effects of a reasonably bad mismatch. A limited range of impedance matching can be secured by adjustment of the magnetron frequency and by changing the position of the discharge tube in the resonator. This resonator was designed at the National Bureau of Standards and a detailed description has been published (Broida and Chapman, 1958). Figure 7 shows a discharge tube

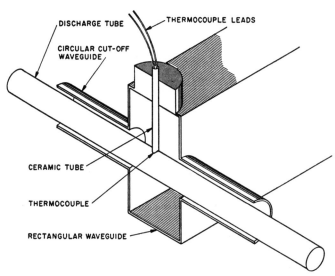

FIG. 5. Cut-away sketch showing arrangement of discharge tube in waveguide. The tube can be cooled by forcing cool air into the circular waveguides. The thermocouple can be used to monitor the temperature of the discharge tube.

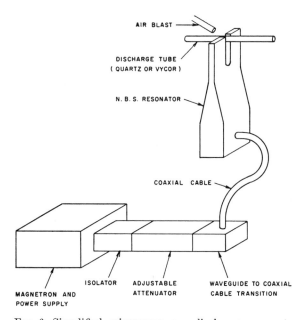

FIG. 6. Simplified microwave gas discharge apparatus

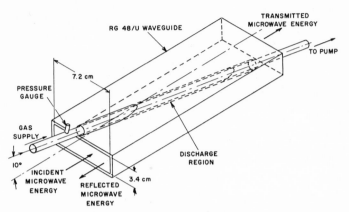

Fig. 7. Discharge tube-waveguide arrangement designed to achieve essentially complete absorption of incident microwave power.

arrangement that can be designed so that practically all incident power is absorbed in the discharge (Knol, 1951; Johnson and Deremer, 1951). An arrangement of this type is useful for operation under fixed conditions of pressure, flow rate, and power.

Systematic studies of the effects of discharge parameters on the yields of radicals produced in electrical discharges have been made in only a few instances. The principal results obtained in a recent study of the production of hydrogen atoms at microwave frequencies are summarized below (Shaw, 1958). The discharge was produced by coupling power from a microwave source to a stream of low pressure hydrogen contained in a cylindrical quartz or Vycor tube. The microwave equipment was similar to that described in Figs. 4 and 5 above. CW and pulsed microwave power was tested at frequencies of 3000 Mc and 9000 Mc at power levels of approximately 5 watts to 300 watts. No significant differences in hydrogen atom production were found for operation at the two frequencies for CW or pulsed power. The rate of production of hydrogen atoms was determined by calorimetric and electron paramagnetic resonance techniques.

The maximum yield of hydrogen atoms was obtained for a pressure of 0.5 mm Hg and was about 25 % less at 0.25 and 1.0 mm. At 0.5 mm with 100 watts of 3000-Mc power dissipated in the discharge and flow rate of 10^{-5} moles per second, the hydrogen flowing from the discharge is about 90 % dissociated. For the same pressure and power, the yield of atoms increased with increasing flow rate, but the extent of dissociation decreased to about 25 % for a flow rate of 1.5×10^{-4} moles per second.

The pressure dependence for H atom yield is shown in detail in Fig. 8. This curve is similar to that obtained by Broida et al. (1954) for the intensity of the H_α and D_α lines as a function of pressure in a 150-Mc discharge. A

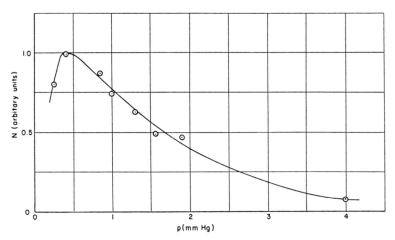

FIG. 8. Effect of hydrogen pressure on yield of hydrogen atoms from 3000 Mc CW discharge. The power dissipated in the discharge was 100 watts.

somewhat similar relation was obtained also by Poole (1937c) for a Wood's tube discharge. Poole showed a relationship of this type was consistent with the theory for dissociation developed by Emeléus et al. (1936) and Lunt and Meek (1936). Corresponding relations for other common gases apparently have not been determined.

For hydrogen in a microwave discharge at low rates of flow, the yield of atoms increases linearly with increasing power at low power and levels off at higher power levels as the degree of dissociation of the gas increases; at higher flow rates, only partial dissociation occurs even at high power levels. Typical results showing the effects of power and flow rate are shown in Fig. 9. The results shown are for a 1.75-cm i.d. discharge tube. A higher yield of atoms was obtained for a 0.9-cm i.d. tube for the same power input and flow rate. For discharge tube diameters between 0.9 and 1.75 cm i.d., the maximum power that could be utilized effectively at a flow rate of 140 $\times$ 10^{-6} moles per second was approximately 200 watts. For higher input power, the yield decreased. This effect was shown to be related to the temperature of the discharge tube in the discharge zone. In this region, a marked effect of temperature on yield was found, with maximum yield being obtained at lower temperatures. For example, for a flow of 140 $\times$ 10^{-6} moles per second of hydrogen, the yield of hydrogen atoms decreased about sixfold when the temperature of the tube was increased from 60°C to 350°C.

E. THERMAL DISSOCIATION IN AN ELECTRICAL DISCHARGE

Very little work on the production of radicals by this type of dissociation has been published. Dehmelt (1955) employed a high pressure dc arc for

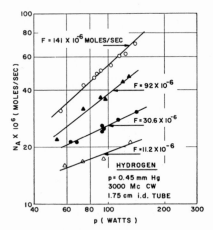

FIG. 9. Effect of power and flow rate on yield of hydrogen atoms from 3000 Mc CW discharge.

the production of hydrogen, nitrogen, and phosphorous atoms. An inert gas carrier at a pressure of 10 to 100 mm Hg was used to establish a dc arc. The gas of interest was added to the carrier gas at a pressure between 0.01 and 0.1 mm Hg. Substantial dissociation of hydrogen, nitrogen, and phosphorous was achieved, as evidenced by electron paramagnetic resonance spectra for the various atoms produced. The method should be investigated further because of its almost universal application to all types of radicals and to determine the effect of discharge parameters on radical production. Unfortunately, the method has the same distinct disadvantage common to all discharges in which electrodes are required, i.e., that metallic vapors may be sputtered or evaporated from the electrodes and may contaminate the discharge system or its products of dissociation.

McCarthy (1954) also has reported operation of a microwave discharge at pressures up to 1 atmosphere. It is not clear from the results given whether operation of a microwave discharge at very high pressure resulted in an increased yield of radicals.

F. WALL EFFECTS

The recombination coefficient for atom-atom recombination on clean glass or silica surfaces is a quantity of the order of 10^{-5}. Despite the smallness of this value compared with a value of the order of unity for common metals, the loss of radicals due to wall collisions probably constitutes the most significant loss mechanism which affects the yield of radicals obtainable in a low pressure discharge. To reduce this type of loss, various procedures for reducing the catalytic activity of the walls have been devised. Such procedures generally have resulted in an increased yield from the discharge tube, but it is not clear whether such increases are owing solely to a reduction in wall activity or to an effect on the production processes in the discharge. Some of the pertinent (and sometimes conflicting) observations on these matters are summarized in the following paragraphs as applied to hydrogen, oxygen, and nitrogen.

1. *Hydrogen*

In the course of his pioneering experiments on the dissociation of hydrogen in a low frequency discharge, Wood (1922) concluded that gas films adsorbed on the walls of the tube were an important factor in determining the yield of atomic hydrogen that could be obtained from the discharge tube. Water vapor or oxygen was believed to supply oxygen which served to reduce the ability of the walls to catalyze recombination. No atomic hydrogen was obtained with dry hydrogen in the conventional low power CW discharge. Wood found, however, that a very powerful pulsed discharge produced hydrogen atoms in dry hydrogen.

Contrary to Wood's conclusion, and as the result of a long series of experiments with an rf discharge, Finch (1949), concluded that dry hydrogen is not dissociated and that water vapor or oxygen is somehow involved in the dissociation process.

Poole (1937b) made an extensive study of wall effects in connection with the dissociation of hydrogen in a Wood's tube. He concluded that the use of water vapor as an anticatalyst in glass or silica discharge tubes is very unreliable and likely to yield erratic results. Tubes coated with metaphosphoric acid when used with wet hydrogen gave reproducible and high yields of atomic hydrogen.

In this connection it is of interest to note that Dingle and LeRoy (1950), in work with atomic hydrogen produced by thermal dissociation of dry hydrogen on a hot filament, obtained no hydrogen atoms in clean glass apparatus. A satisfactory yield was obtained, however, when the walls of the apparatus were coated with a mixture of orthophosphoric acid and phosphorus pentoxide. However, the efficiency of this substance as an anticatalyst was reduced if water vapor or oxygen were present.

More recently, Wittke and Dicke, (1956) in an effort to avoid the inevitable contamination which results from the use of wet hydrogen and acid catalysts, obtained satisfactory yields of hydrogen in a Wood's tube the walls of which were coated with Dri-film, a mixture of dimethyl dichlorosilane and methyltrichlorosilane. This material was found to provide a film that was heat stable, had a very low vapor pressure at room temperature, and was more effective than water vapor or acids in preventing recombination of hydrogen atoms at glass surfaces. In recent work (Shaw, 1958; Zletz, 1958), Dri-film was used effectively to prevent recombination of hydrogen atoms in glass, quartz, and Vycor tubes used in a microwave discharge. In Shaw's work, tank hydrogen with less than 0.5 % impurities was passed through a deoxo unit and silica gel to remove oxygen and water vapor. When the discharge tubes were cleaned with 12 % hydrofluoric acid, the yield was small and variable. After coating the tubes with Dri-film, the yield was increased and was stable. It was found, however, that when the hydrogen was subjected to further drying in a liquid nitrogen trap, the yield of hydrogen atoms was reduced from tenfold to thirtyfold. Saturation of hydrogen with water resulted in only a 25 % increase in yield over that obtained with the hydrogen that was passed through the deoxo unit and silica gel. It was noted that, upon drying the hydrogen in liquid nitrogen, the reduction in yield was complete in about 2 minutes. The speed with which the yield changed suggests that the observed effect is not due solely to recombination at the walls. As found by Wood (1922), the desorption of water from the walls of the tube is a slow effect. This observation appears to support the suggestion of Finch (1949) that water or oxygen affects the dissociation process. Further evidence that water is involved in the discharge processes is obtained from the fact that the power required to sustain a microwave discharge in wet hydrogen is only about one-third that required for hydrogen that was dried over silica gel.

2. Oxygen

The behavior of oxygen with respect to dissociation and recombination appears to be somewhat different from that of hydrogen. Finch (1949) found no difficulty in dissociating dry oxygen, but concluded that the atoms formed had a very short lifetime. Recently, Kaufman (1958), working with a microwave discharge, obtained yields of up to 18 % atomic oxygen from dry oxygen in clean quartz apparatus. The addition of a few per cent water to the oxygen supplied to the discharge increased the yield of atomic oxygen up to about 50 % of the oxygen input. The added water also increased the rate of disappearance of atomic oxygen. Since water vapor added to the gas stream beyond the discharge zone had no effect upon the disappearance of atomic oxygen, it was concluded that the accelerated decay was due to a reaction with some species produced from water vapor in the discharge. In earlier work with a low frequency Wood's tube and an rf discharge, Linnett and Marsden (1956) also concluded that water vapor added to oxygen increased the rate of production of oxygen atoms, probably by modifying the cycle of reactions involved in atom production.

3. Nitrogen

The conditions necessary to obtain a large yield of nitrogen atoms in an electrical discharge have not been well established. The water content of the nitrogen appears to be a very significant factor. In some instances phosphoric acid has been used to coat discharge tubes, but the necessity for this practice is questionable.

With dry nitrogen, not more than about 1 % atomic nitrogen is produced in a Wood's tube (Berkowitz *et al.*, 1956; Jackson and Schiff, 1955) or in a microwave discharge (Kistiakowsky and Volpi, 1957). Heald and Beringer (1954) noted difficulty in obtaining the high-peak fields required to maintain a condensed discharge in a Wood's tube operated on dry nitrogen, and consequently they obtained little evidence for the production of atoms. On the other hand, when water vapor was added to the nitrogen supply, the discharge was a dependable source of ground-state nitrogen atoms. Fontana (1958a, b) made the observation that the gaseous nitrogen atom concentration formed in a microwave discharge is increased by traces of water vapor. On the basis of thermometric measurements Fontana estimated a minimum nitrogen atom concentration of 2 to 4 % atoms arriving at a condensing surface. A nitrogen atom concentration of the same order has been obtained for nitrogen containing small amounts of oxygen with a microwave discharge by Herron *et al.* (1958a, b). Greenblatt and Winkler (1949) report yields of up to 30 % nitrogen atoms obtained with wet nitrogen in a condensed discharge. Essentially complete dissociation of nitrogen was obtained in an rf discharge by Debeau (1942), but details concerning the nature of the discharge have not been published. More recently 100 % dissociation of nitrogen by a condensed discharge has been reported by Armstrong and Winkler (1956) and Kelly and Winkler (1959).

The observations concerning the role of water vapor or oxygen in the dissociation of hydrogen, oxygen, and nitrogen and in wall recombination of atoms raises a number of questions which are pertinent to the production of atoms or radicals in an electrical discharge. A basic question concerns the manner in which water is involved in the discharge processes. Various lines of evidence seem to indicate that water has an effect on the electric fields required for dissociation. At the same time, water vapor may act as an anticatalyst for wall recombination. No satisfactory experimental evidence appears to have been obtained by which these two functions can be clearly distinguished.

V. Detection and Measurement of Radicals

Strictly speaking, the detection and measurement of radicals is not a factor in radical production. In practice, however, it is found that methods of detection and measurement frequently play an essential role in the establishment of the conditions required to produce a particular species of radical (Sanders *et al.*, 1954).

A variety of methods are commonly employed for the detection and measurement of radicals in the gas phase (Steacie, 1954). In general, such methods are based upon either (1) chemical reactivity or (2) a physical property such as the spectrum of the radical. In a great many instances, the methods used are suitable only for relative measurements of radical production, so that a comparison of the results of different investigations is difficult or impossible.

A new method which offers considerable promise for quantitative measurements of radicals is the method of electron spin resonance (ESR) (Ingram, 1958). This method utilizes the unique characteristic of a radical, i.e., that one or more of the electrons normally involved in chemical binding is unpaired. Because of the unpaired electron, the radical is paramagnetic and may be detected by the methods of electron spin resonance. The ESR method is currently being widely employed for the investigation of radicals in liquids and solids (see Chapter 7 by Jen). Recent studies of hydrogen, nitrogen, deuterium, and oxygen atoms show that the ESR method can be used effectively in the gas phase for the quantitative measurement of radicals produced in an electrical discharge (Shaw, 1958). The gas under observation is not disturbed since it is not necessary to insert into the gas stream any probes or other devices that can catalyze the recombination of radicals. Unequivocal identification of radicals is possible by making use of the characteristic spectra associated with each species of radical.

Although further refinement of the quantitative aspects of the method is necessary, it appears that the ESR method provides a better basis for the measurement of radical production than has been available in the past.

REFERENCES

Alpert, D. (1949). *Rev. Sci. Instr.* **20**, 779. A high power attenuator for microwaves.

Armstrong, D. A., and Winkler, C. A. (1956). *J. Phys. Chem.* **60**, 1100. Comparative production of active nitrogen from nitrogen, nitric oxide, and ammonia, and from nitrogen at different discharge potentials.

Barth, C. A., and Kaplan, J. (1957). *J. Chem. Phys.* **26**, 506. Oxygen bands in "air" afterglows and the night airglow.

Berkowitz, J., Chupka, W. A., and Kistiakowsky, G. B. (1956). *J. Chem. Phys.* **25**, 457. Mass spectrometric study of the kinetics of nitrogen afterglow.

Broida, H. P., and Chapman, M. W. (1958). *Anal. Chem.* **30**, 2049. Stable isotope analysis by optical spectroscopy.

Broida, H. P., and Moyer, J. W. (1952). *J. Opt. Soc. Am.* **42**, 37. Spectroscopic analysis of deuterium in hydrogen-deuterium mixtures.

Broida, H. P., and Pellam, J. R. (1954). *Phys. Rev.* **95**, 845. Phosphorescence of atoms and molecules of solid nitrogen at 4.2°K.

Broida, H. P., and Pellam, J. R. (1955). *J. Chem. Phys.* **23**, 409. A note on the preparation of solid ozone and atomic oxygen.

Broida, H. P., Morowitz, H. J., and Selgin, M. (1954). *J. Research Natl. Bur. Standards* **52**, 293. Optical spectroscopic determination of hydrogen isotopes in aqueous mixtures.

Corrigan, S. J. B., and Von Engel, A. (1958). *Proc. Roy. Soc.* **A245**, 335. Excitation and dissociation of hydrogen by an electron swarm.

Davis, L., Jr., Feld, B. T., Zabel, C. W., and Zacharias, J. R. (1949). *Phys. Rev.* **76**, 1076. The hyperfine structure and nuclear moments of the stable chlorine isotopes.

Debeau, D. E. (1942). *Phys. Rev.* **61**, 668. The nature of the reactions occurring in the production of the afterglow of active nitrogen and the effect of temperature on the phenomena.

Dehmelt, H. G. (1955). *Phys. Rev.* **99**, 527. Atomic phosphorous paramagnetic resonance experiment employing universal dissociator.

Dingle, J. R., and LeRoy, D. J. (1950). *J. Chem. Phys.* **18**, 1632. Kinetics of the reaction of atomic hydrogen with acetylene.

Dousmanis, G. C., Sanders, T. M., Jr., and Townes, C. H. (1955). *Phys. Rev.* **100**, 1735. Microwave spectra of the free radicals OH and OD.

Edels, H., and Gambling, W. A. (1958). *Proc. Roy. Soc.* **A249**, 225. Excitation temperature measurements in glow and arc discharge in hydrogen.

Edelstein, L. A. (1958). *Nature* **182**, 932. Anomalous dissociation of molecular hydrogen by electron impact.

Emeléus, K. G., Lunt, R. W., and Meek, C. A. (1936). *Proc. Roy. Soc.* **A156**, 394. Ionization, excitation, and chemical reaction in uniform electric fields. 1—The Townsend coefficient of ionization.

Finch, G. I. (1949). *Proc. Phys. Soc.* (*London*) **A62**, 465. Steam in the ring discharge.

Fontana, B. J. (1958a). Unpublished report. Properties of stabilized free nitrogen atoms in the solid state.

Fontana, B. J. (1958b). *J. Appl. Phys.* **29**, 1668. Thermometric study of the frozen products from the nitrogen microwave discharge.

Fryburg, G. C. (1956). *J. Chem. Phys.* **24**, 175. Enhanced oxidation of platinum in activated oxygen.

Garvin, H. L., Green, T. M., and Lipworth, E. (1958). *Phys. Rev.* **111**, 534. Spins of some radioactive iodine isotopes.

Greenblatt, J. H., and Winkler, C. A. (1949). *Can. J. Research* **27 B**, 721. The reaction of nitrogen atoms with ethylene.

Harteck, P., Reeves, R. R., and Mannella, G. (1958). *J. Chem. Phys.* **29**, 608. Rate of recombination of nitrogen atoms.

Heald, M. A., and Beringer, R. (1954). *Phys. Rev.* **96**, 645. Hyperfine structure of nitrogen.

Herron, J. T., Franklin, J. L., Bradt, P., and Dibeler, V. H. (1958a). Unpublished report. Mass spectroscopic studies of free radicals.

Herron, J. T., Franklin, J. L., Bradt, P., and Dibeler, V. H. (1958b). *J. Chem. Phys.* **29**, 230. Kinetics of nitrogen atom recombination.

Holzmann, R. T., and Morris, W. F. (1958). *J. Chem. Phys.* **29**, 677. Some precursors produced in the electrodeless discharge synthesis of B_2Cl_4.

Ingram, D. J. E. (1958). "Free Radicals as Studied by Electron Spin Resonance." Academic Press, New York.

Jackson, D. S., and Schiff, H. I. (1955). *J. Chem. Phys.* **23**, 2333. Mass spectrometric investigation of active nitrogen.

Jennings, K. R., and Linnett. J. W. (1958). *Nature* **182**, 597. Production of high concentration of hydrogen atoms.

Johnson, H., and Deremer, K. R. (1951). *Proc. I. R. E.* (*Inst. Radio Engrs.*) **39**, 908. Gaseous discharge super-high frequency noise sources.

Jones, A. F. (1950). *J. Chem. Phys.* **18,** 1263. Infra-red and ultraviolet spectra of sulphur monoxide.

Kaufman, F. (1958). *Proc. Roy. Soc.* **A245,** 123. The air afterglow and its use in the study of some reactions of atomic oxygen.

Kelly, R., and Winkler, C. A. (1959). *Can. J. Chem.* **37,** 62. The kinetics of the decay of nitrogen atoms as determined from chemical measurements of atom concentrations as a function of pressure.

King, J. G., and Jaccarino, V. (1954). *Phys. Rev.* **94,** 1610. Hyperfine structure and nuclear moments of the stable bromine isotopes.

Kistiakowsky, G. B., and Volpi, G. G. (1957). *J. Chem. Phys.* **27,** 1141. Reactions of nitrogen atoms—I. Oxygen and oxides of nitrogen.

Knol, K. S. (1951). *Philips Research Repts.* **6,** 288. Determination of the electron temperature in gas discharges by noise measurements.

Linnett, J. W., and Marsden, D. G. H. (1956). *Proc. Roy. Soc.* **A234,** 489. The kinetics of the recombination of oxygen atoms at a glass surface.

Lunt, R. W., and Meek, C. A. (1936). *Proc. Roy. Soc.* **A157,** 146. Ionization excitation, and chemical reaction in uniform electric fields. II—The energy balance and energy efficiencies for the principal electron processes in hydrogen.

Lunt, R. W., Meek, C. A., and Smith, E. C. W. (1937). *Proc. Roy. Soc.* **A158,** 729. Ionization, excitation and chemical reaction in uniform electric fields. III—The excitation of the continuous spectrum of hydrogen.

McCarthy, R. L. (1954). *J. Chem. Phys.* **22,** 1360. Chemical synthesis from free radicals produced in microwave fields.

Massey, H. S. W., and Burhop, E. H. S. (1952). "Electronic and Ionic Impact Phenomena." Oxford Univ. Press, London and New York.

Nagle, D. E., Julian, R. S., and Zacharias, J. R. (1947). *Phys. Rev.* **72,** 971. The hyperfine structure of atomic hydrogen and deuterium.

Poole, H. G. (1937a). *Proc. Roy. Soc.* **A163,** 404. Atomic hydrogen. I—The calorimetry of hydrogen atoms.

Poole, H. G. (1937b). *Proc. Roy. Soc.* **A163,** 415. Atomic hydrogen. II—Surface effects in the discharge tube.

Poole, H. G. (1937c). *Proc. Roy. Soc.* **A163,** 424. Atomic hydrogen. III—The energy efficiency of atom production in a glow discharge.

Prodell, A. G., and Kusch, P. (1957). *Phys. Rev.* **106,** 87. Hyperfine structure of tritium in the ground state.

Sanders, T. M., Schawlow, A. L., Dousmanis, G. C., and Townes, C. H. (1954). *J. Chem. Phys.* **22,** 245. Examination of methods for detecting OH.

Shaw, T. M. (1958). *Gen. Elec. Microwave Lab. Rept. No.* **TIS R 58ELM 115.** Unpublished. Studies of microwave gas discharges: production of free radicals in a microwave discharge.

Shaw, T. M. (1959). *J. Chem. Phys.* **30,** 1366. Dissociation of hydrogen in a microwave discharge.

Steacie, E. W. R. (1954). "Atomic and Free Radical Reactions," 2nd ed. Reinhold, New York.

Teeter, W. L., and Bushore, K. R. (1957). *IRE Trans. on MTT-4* **5,** 227. A variable-ratio microwave power divider and multiplexer.

Von Hipple, A., ed. (1954). "Dielectric Materials and Applications," pp. 309–311. Technol. Press of MIT, Cambridge, Massachusetts and Wiley, New York.

Weinrich, G., and Hughes, V. W. (1954). *Phys. Rev.* **95,** 1451. Hyperfine structure of helium-3 in the metastable triplet state.

Wittke, J. P., and Dicke, R. H. (1956). *Phys. Rev.* **103,** 620. Redetermination of the hyperfine splitting in the ground state of atomic hydrogen.

Wood, R. W. (1922). *Phil. Mag.* **44,** 538. Atomic hydrogen and the Balmer series spectrum.

Zletz, A. (1958). Unpublished report. Reaction studies of hydrogen atoms with molecular oxygen.

4. Radical Formation and Trapping in the Solid Phase

GEORGE C. PIMENTEL

Department of Chemistry,
University of California, Berkeley, California

A suspension of reactive species in a solid can be prepared by generation of the species *in situ* in the solid or by condensation of a gas mixture containing the reactive species. The former method, formation of free radicals *in situ*, is considered in this chapter. Following a historical introduction we shall consider diffusion in a solid, a process which limits the stability of the desired suspension of free radicals. The methods of *in situ* production then will be compared. The chapter is concluded with a summary of the evidence concerning radical concentrations which have been achieved both by *in situ* and by gas phase generation methods.

I. Introduction

Two methods of production of free radicals in solids, ultraviolet irradiation and electron bombardment, were applied many years ago. In 1901, Dewar irradiated unspecified organic substances as well as zinc sulfide at 21°K (the light source was not specified). Phosphorescence of the organic substances was described as "... very marked, when compared with the same effects brought about by the use of liquid air." Enhancement of phosphorescence at low temperatures is now a well-recognized behavior of many molecules which phosphoresce in forbidden triplet-singlet transitions, free radicals usually not being involved. Dewar's irradiation of zinc sulfide resulted in a more stable energy storage: zinc sulfide showed "... brilliant phosphorescence on the temperature being allowed to rise." Once again free radical formation seems to be a less likely explanation of this behavior than an alternative one, the low temperature stabilization of some sort of lattice defect.

Over two decades later Vegard began a sustained study of the luminescence of solid nitrogen at 21°K during and following bombardment with electrons and positive ions ("canal rays") (Vegard, 1924a). The light production was by no means a surprise to Vegard, rather it was sought in verification of his proposal that certain auroral emissions result from electron bombardment of frozen nitrogen dust in the upper atmosphere. In 1926, Vegard described an intense luminescence of solid nitrogen when the solid was warmed long after the cessation of electron bombardment. This "thermoluminescence" was confirmed in the work of McLennan and co-workers (1928), who shared Vegard's interest in, but doubted his explanation of, the origin of the beautiful auroral emissions. McLennan et al. (1928) report: "If the solid nitrogen were allowed to warm up slightly after the luminescence had apparently ceased, however, the bands flashed out quite strongly again for a short time, ..." "As the temperature of the solid nitrogen rose gradually, ... a point was reached where the luminescence effects disappeared totally, ... between 21°K and 62.5° and appreciably below the latter." These phenomena undoubtedly were caused by the formation and storage of free radicals (e.g., nitrogen atoms). The thermoluminescence resulted from the reactions of these chemically active species when diffusion was permitted.

After the studies of Vegard and of McLennan, more than a decade passed before Lewis and Lipkin investigated the photolysis of complex organic molecules in rigid, glassy media (1942). They enumerate the types of reaction processes which could occur under these conditions: free radical formation, ion pair formation, and ionization. Lewis and Lipkin were successful in producing stable free radical suspensions in hydrocarbon glasses at 77 and 20°K. The free radicals detected, such as triphenylmethyl and diphenylnitrogen, possess optimum characteristics for such an experiment: they are relatively stable because of "resonance stabilization," they are cumbersome molecules which would diffuse slowly, and they absorb strongly in the visible spectral region.

The excitement experienced by each of these early workers has had a rich heritage of thrills for the many workers currently performing various modifications of these glamorous experiments. Yet it cannot be said that the recent resurgence of activity was triggered or substantially aided by the sporadic pioneering work just described. The modern phase probably began in 1951 when Rice and Freamo reported the formation of a blue, paramagnetic solid upon condensation at 77°K of the products of hydrazoic acid passed through a glow discharge (1951, 1953). Then in 1954 seven separate laboratories published within a span of eight months brief reports (all were letters) of their attempts to detect free radicals suspended in some sort of solid matrix. The language used by each group suggests that they were quite unaware of other current activity in the field. The first two of these publications described paramagnetic (spin) resonance studies (hereinafter, ESR) of radiation damage in solids at low temperatures. Smaller *et al.* described studies of irradiated ice (1954) and Livingston *et al.* reported that "Paramagnetic resonance absorption lines have been observed in a variety of substances after irradiation with Co^{60} gamma rays at 77°K and are attributed to free radicals" (1954). Mador and Williams used both ultraviolet-visible and infrared methods in a variety of experiments intended to produce free radicals suspended in a solid at a low temperature (Mador, 1954; Mador and Williams, 1954). This work was undoubtedly an imaginative exploitation of the potentialities revealed in the work of Rice and Freamo. Giguère attempted the infrared detection of the HO_2 radical in the solid obtained by freezing (at 77°K) the products of H_2O passed through a glow discharge (1954). [This attempt was unsuccessful, as shown by Giguère and Harvey in later work (1956).] Broida and Pellam (1954) discovered that condensation of active nitrogen at 4°K gave a variety of spectacular luminescent phenomena which are now recognized to be virtually identical to those resulting from the electron bombardment experiments of Vegard *et al.* and McLennan *et al.*

Finally, Norman and Porter (1954) and, quite independently, Whittle *et al.* (1954), proposed what is now called the matrix isolation method. Each of these groups anticipated the wide applicability of the method and accurately foretold, in practically identical statements, the required experimental conditions.

Norman and Porter (1954): "Here we describe a general method for the preparation of . . . [free radical] species under conditions which permit their observation at high concentrations for an indefinitely long period of time." "The method involves the photolytic dissociation of a substance dissolved in a transparent rigid solvent at very low temperatures." "1. diffusion to the vicinity of another atom must be prevented . . ., 2. reaction with the solvent must not occur . . . ."

Whittle *et al.* (1954): "The matrix isolation method proposed here

TABLE I

METHODS OF PRODUCTION AND DETECTION OF FREE RADICALS *in situ*

Production
  1. Electron bombardment
      a. "cathode ray" bombardment
      b. beta ray bombardment
      c. "Tesla coil" ion bombardment
  2. Positive ion bombardment
      a. "canal ray" bombardment
      b. alpha ray bombardment
  3. Photolysis
  4. Radiolysis: gamma or X-rays
  5. Atom bombardment
Detection
  1. Visible-ultraviolet spectroscopy
  2. Infrared spectroscopy
  3. Electron spin resonance (ESR)
  4. Magnetic susceptibility
  5. Calorimetry

involves accumulation of a reactive substance under environmental conditions which prevent reaction. The intent is to trap active molecules in a solid matrix of inert material, ... If the temperature is sufficiently low, the matrix will inhibit diffusion, . . . holding the active molecules effectively immobile in a nonreactive environment."

Following these pioneering studies an increasing number of workers have been attracted to the study of free radicals suspended in a solid. A variety of methods of production and of detection are in use. Table I lists the methods which are applicable to the *in situ* production. Furthermore it has become apparent that certain processes which affect and limit the production of free radicals must be better understood. Some of these processes which will be taken up in this chapter are listed below.

a. Diffusion. Since free radicals react exothermically (possibly with zero activation energy) diffusion must be carefully controlled. Though certain aspects of self-diffusion in metals are rather well understood, relatively little is known about the diffusion of an impurity in a host crystal or in a glass.

b. Cage effect. The *in situ* production of free radicals might be inhibited by the matrix cage through two types of processes. First, any process of de-excitation could reduce the quantum yield of radical formation. If energy is efficiently drained from an excited molecule, fragmentation may become less probable or even impossible. Even if fragmentation does occur, however, the two fragments may be constrained to remain in proximity by the solid environment. If so, recombination becomes extremely likely. Either of these effects will tend to reduce quantum efficiency.

c. Local heating. The *in situ* generation of a free radical usually occurs in an energy-rich event. As this excess energy is degraded into heat the local environment is raised to a temperature above the ambient temperature. The free radical is inevitably born, then, in a locally heated region. The maximum temperature reached and its duration is of crucial importance to the maintenance of matrix rigidity.

d. Secondary reactions. Two types of secondary processes must be considered. If two radicals are formed simultaneously, they must separate or recombination may occur. This separation implies that contact with other parent molecules is likely. To the extent that this occurs, the "isolation" function of the matrix is lost. Second, the possibility of reaction with the matrix must be considered. Though the matrix may be relatively unreactive in the gas phase, there is no guarantee that reaction will not take place in a condensed phase if highly excited free radicals are involved. Photolysis frequently leaves excited molecules with energies of excitation in excess of the activation energy for hydrogen abstraction from a hydrocarbon. Hence a hydrocarbon glass must be viewed with suspicion as an inert matrix. Clearly the reaction possibilities of an ether-alcohol-isopentane glass are manifold.

II. Diffusion in Solids

Little is known about the free energy of activation for recombination of free radicals. One experimental datum has been given by Carrington and Davidson (1953) for the reaction of formation of N_2O_4 from two NO_2 radicals in the gas phase. They found $\Delta H^{\ddagger}$ to be about two kilocalories *negative*, a result rationalized in terms of a partial three-body dependence in the association and a negligible enthalpy of activation. This zero or near-zero activation enthalpy is generally expected for a radical-radical reaction, but rarely is there more evidence than the observation that it is immeasurably fast. Nevertheless the most likely assumption that can be made at this time is that free radicals will react with nearly zero activation free energy, an assumption which implies that two radicals "in contact" in a matrix will surely react. It is this likelihood which necessitates careful control of diffusion. We shall examine the available evidence concerning diffusion in solids.

A. Diffusion in Crystalline Solids

We could not attempt a general survey of the present knowledge concerning self-diffusion. Instead we shall consider some of the results presented by Nachtrieb and co-workers since their work is extensive and since references to other work can be found there. Of particular interest are the empirical correlations which have been found between the activation energy for self-diffusion in cubic closest packed metals and, first, the melting temperature, and second, the heat of fusion (Nachtrieb *et al.*, 1952).

$$\Delta H^{\ddagger} = C_T T_M \tag{1}$$

where $\Delta H^{\ddagger}$ = activation energy for self-diffusion (kcal)
        T_M = melting temperature
        C_T = a constant = 28 to 32

$$\Delta H^{\ddagger} = C_L L_f \tag{2}$$

where L_f = latent heat of fusion (kcal)
       C_L = a constant = 16.5

Nachtrieb and Handler (1954) have devised a model which explains these correlations based on creation and movement of "relaxed vacancies." This model is a modified vacancy-diffusion model since the diffusing regions considered in the model are postulated to be formed at lattice vacancies by the inward relaxation of neighboring atoms. The model leads to the result that the self-diffusion coefficient is determined by the reduced temperature, T/T_M, alone.

These results are significant because they lead to the expectation that *a given rate of diffusion will be obtained at a particular reduced temperature,* T/T_M. This rule of thumb encourages the use of empirical observations on one matrix in predicting properties of others. Application of this rule for nonmetallic matrices is presumptuous, however, since it is supported by but one example, α white phosphorus (Nachtrieb and Handler, 1955). Nevertheless evidence will be presented which shows that the result is useful in consideration of the diffusion of impurities in rapidly condensed solids.

Livingston *et al.* (1955) have detected by ESR a radical species produced by gamma ray radiolysis of 86 % H_3PO_4. They identify the species as free H atoms and measure its disappearance over some 25 to 50 minutes at 113.8°K. The melting point of 86 % H_3PO_4 is about 20°C.

Since the disappearance approximately fits a second-order law, the rate is meaningful only if the concentration is known. Livingston *et al.* (1955) reported the mole fraction of H atoms to be near 10^{-4}, but later estimates suggest that the concentration was higher (private communication). These workers indicated that further work directed toward determination of activation enthalpies is in progress. Until this appears, we can observe that T/T_M is about 0.4 for H atoms in this hydrogen-bonded lattice.

Ghosh and Whiffen (1959) have presented ESR spectra of a single crystal of glycine after gamma irradiation. A radical species identified as NH_3^+—$CHCO_2^-$ was formed in a suspension stable at room temperature. Not only is diffusion immeasurably slow at this temperature ($T/T_M \sim 0.58$), over a period of a year, but also the radicals are precisely oriented in the crystal lattice (within 5 to 10°). This remarkable alignment suggests that the radicals are held quite rigidly in the lattice through specific interactions with the lattice cage. Surely such effects must act to inhibit diffusion.

B. DIFFUSION IN RAPIDLY CONDENSED SOLIDS

Many of the experiments described in this chapter and in Chapter 2 have involved solid matrices formed by the rapid condensation of a gas. Before presenting the available information concerning diffusion in such solids, we should examine what is known about their nature. A number of X-ray studies of ice condensed at temperatures near 145°K indicate that the structure obtained is a sensitive function of the deposition conditions. Shallcross and Carpenter (1957) report, for example, that in some cases hexagonal (room temperature) ice is obtained, the form stable above about 175°K. In other similar experiments the X-ray diffraction patterns show the presence of cubic ice-hexagonal ice mixtures, and occasionally amorphous material is obtained.

There is ample calorimetric and spectroscopic evidence indicating that materials which form molecular crystals are obtained as imperfect crystals or amorphous solids if condensed at a sufficiently low temperature. De Nordwall and Staveley (1956), for example, find that for each of a number of solids prepared by quickly freezing a gas there is a characteristic temperature at which heat is released during warming. This temperature, T_D, is presumed to be due to a devitrification process. Their measurements show that the ratio T_D/T_M varies from about 0.4 to 0.7 for a dozen substances of various types. These results are shown in Table II. Similar information can be derived from infrared spectra. A number of solids formed by rapid condensation have been examined (including ammonium azide, HN_3, hydroxylamine, and hydrogen peroxide). The infrared spectra of annealed crystals of these substances often show band multiplets, the splitting being caused by crystal interactions. For the rapidly condensed solids, however, the spectra are diffuse, as is characteristic of an imperfect lattice or an amorphous solid. For some of these solids a diffusion temperature has been measured by locating the temperature at which crystal splitting develops. These are shown in Table II in terms of T_D/T_M.

Infrared data furnish another kind of information which is useful in evaluating the effect of diffusion in the matrix isolation of free radicals. Chemical changes of reactive species suspended in an inert matrix can be

TABLE II
RATIO OF THE ANNEALING TEMPERATURE, T_D, TO THE MELTING POINT, T_M

Substance	Calorimetry[a] T_D (°K)	T_D/T_M
p-Xylene	125	0.44
Ammonia	92–95	0.48
Pyridine	112–113	0.49
Methyl amine	97 (and 110)	0.54 (and .61)
Water	140–160 (and 105)	0.55 (and .38)
Methyl chloride	102–105	0.59
Thiophene	138?	0.59?
Tin tetramethyl	138–139	0.63
Silicon tetramethyl	119–123	0.69
Toluene	123–124	0.69
Methyl bromide	132–133?	0.74?
Substance	Infrared: crystal splitting T_D (°K)	T_D/T_M
NH_4N_3	150–190[b]	0.35–0.44
H_2O_2	140[c]	0.52
NH_3	113[d]	0.57
HN_3	120[e]	0.62

[a] De Nordwall and Staveley (1956).
[b] Dows and Pimentel (1955a).
[c] Giguère and Harvey (1959).
[d] Unpublished data of G. C. Pimentel.
[e] Dows and Pimentel (1955b).

traced spectroscopically as the temperature is raised. The onset of reaction indicates diffusion is no longer prevented by the matrix rigidity. Pimentel and co-workers have made such measurements for a variety of well-known molecules in several matrices. These temperatures are given in Table III, T_D now referring to the temperature at which diffusion is rapid compared to the observation time, a few minutes. A few ESR measurements of diffusion are included in Table III as well.

Both the heat release and the diffuse spectral features suggest that highly imperfect crystals or glasses are obtained on the rapid condensation of a gas if the temperature is well below half that of the melting point. The variety of substances represented in Table II shows that the phenomenon is a general one. Turning to Table III, we see that diffusion of impurity molecules in such a solid also becomes rapid at temperatures near $\frac{1}{2}$ T_M. Thus a rule of thumb similar to that for self-diffusion in solids can be applied

TABLE III
RATIO OF THE DIFFUSION TEMPERATURE, T_D, TO THE MELTING POINT, T_M

Diffusing species	Matrix	T_D (°K)	T_D/T_M	Reference
H_2O	Xe	37–45	0.25	E. D. Becker[a]
NO	CO_2	63	0.29	J. D. Baldeschwieler[a]
H, OH	H_2O	115 ± 10	0.42	Matheson and Smaller (1955).
OH(?)	$Ca(OH)_2$	~273	0.47	R. Livingston[b]
NO	N_2	30	0.48	J. D. Baldeschwieler[a]
NH_3	A	40–47	0.52	Pimentel (1958)
NH_3	CH_4	45–48	0.52	C. D. Bass[a]
HN_3	N_2	32–35	0.53	M. Van Thiel[a]
H, NH_2(?)	NH_3	100–110	0.54	Matheson and Smaller (1955)
CH_3OH	CCl_4	135	0.54	J. D. Baldeschwieler[a]
NO_2	N_2	38	0.60	J. D. Baldeschwieler[a]
H_2O	N_2	39	0.62	E. D. Becker[a]
"C"	NH_3	140	0.72	Matheson and Smaller (1955).

[a] Unpublished data from the Chemistry Department, University of California, Berkeley, California.
[b] Private communication.

to the matrix isolation method. Small molecules (i.e., containing 3 to 10 atoms) diffuse rapidly and cannot be preserved for periods of many minutes if the temperature is allowed to rise to the range 0.4 to 0.6 times the melting point of the pure matrix material.[1]

It is reasonable to expect single atoms to diffuse much more rapidly than polyatomic molecules. Some evidence corroborating this expectation is found in Fontana's careful calorimetric measurements (1958). Nitrogen gas was passed through a glow discharge and then condensed at 4°K. During slow warm-up of the resulting solid, heat release accompanied by light emission occurred at 8.9 ± 0.4°K, or at $T/T_M = 0.14$. Though the light emission is characteristic of nitrogen atoms, the heat release is assigned by Fontana to atom recombination (just as it was earlier by Broida and

[1] During deposition of the matrix the film temperature may rise well above that of the support if deposition rate is rapid. This effect is serious because of the poor thermal conductivity of most of the common matrix materials. Quantitative estimates of the effect are provided by Fontana's measurements of temperature rise during deposition of nitrogen at 4°K (1958). Depositing 31 cc (STP) of gas per minute, temperature rises of 2 to 11° were observed without the discharge and 5 to 15° with the glow discharge. Baldeschwieler and Pimentel (unpublished data) have observed diffusion requiring a film temperature rise of at least 55° during deposition of CCl_4 onto a support held at 80°K. The deposition rate in this experiment was about 2 cc (STP) of gas per minute. Of course the high heat capacity of CCl_4 (relative to N_2) accounts in part for the much larger effect in CCl_4.

Lutes, 1956). More surprising is the observation that heat release and light emission continued until the temperature reached about 35°K. Similar though more qualitative results had been reported earlier by Milligan *et al.* (1956). These authors observed initiation of visible light emission from a sample similar to that studied by Fontana just after the thermocouple indicated the temperature was rising above 4°K. This emission then persisted to a temperature near 18°K. Further warming of the solid to 35° resulted in the loss of spectral features assigned to N_3 radicals, but without visible light emission.

In summary, solids formed from a gas by rapid condensation at a temperature well below half the melting point are probably extremely imperfect crystals or amorphous. Free radicals formed in such a solid will probably be lost by diffusion (and reaction) if the temperature ever reaches or exceeds about $\frac{1}{2} T_M$ except, probably, for single atoms which might diffuse rapidly at temperatures as low as 0.1 to 0.4 T_M.

C. DIFFUSION IN GLASSES

Rapid cooling of branched hydrocarbons in the liquid or of various solutions often results in glass formation. Such a medium has long been in use for phosphorescence studies (for example, see Lewis *et al.*, 1940, 1941, and references cited there dating back as far as 1896). The use of such glasses for matrix isolation experiments was restimulated by Norman and Porter (1955). They list five possible matrices and abbreviations:

    (1) 3 MeP.P 3-methylpentane (3 parts), isopentane (2 parts)

    (2) P. MeH isopentane (3 parts), methylcyclopentane (2 parts) (also referred to as M.P.: see Porter and Strachan, 1958)

    (3) E.P.A. ether (5 parts), isopentane (5 parts), ethyl alcohol (2 parts)

    (4) pure isopentane

    (5) pure 3-methylpentane

Norman and Porter report that E.P.A. is probably the most satisfactory from the standpoint of diffusion. However, the potential reactivity of such a mixture would seem to limit its usefulness as a matrix for *in situ* radical production. Pure isopentane glass still can be deformed manually at 77°K and was not found to be particularly useful in radical-trapping experiments. Pure 3-methylpentane glass was satisfactory at 77°K except under the most intense illumination during photolysis, a difficulty shared by mixtures 3 MeP.P and P. MeH. Whereas these glasses are sufficiently rigid at 77°K to prevent diffusion of the free radicals studied by these authors (including iodine atoms, CS, ClO, and aromatic radicals), it is clear that this temperature is not far below T_D, as defined for Table III. A quantitative estimate of T_D, the temperature at which diffusion is rapid, has been given by Sowden and Davidson (1956). Using a mixture similar to 3 MeP.P (except in the proportion 1:5), these workers observed spectral evidence of chemical changes when an irradiated glass containing bromal was warmed to 90°K.

III. Production of Free Radicals in the Solid State

Methods of production of free radicals *in situ* have been collected in Table I. In this section advantages and limitations of each method are discussed briefly and then the free radicals which have been studied are considered in turn.

A. ELECTRON BOMBARDMENT

1. N, N₃ Radicals

Vegard in reporting his pioneering studies describes in ample detail his experimental apparatus for electron and positive ion bombardment (1926). Still it seems likely that a modern reference would better serve those who wish to begin experiments of this type. Field and Franklin (1957) describe the experimental technique of electron bombardment and give references to reviews concerning design of apparatus. Hörl and Marton (1958) and, more recently, Hörl (1959a) describe in detail equipment specifically adapted to the problem of electron bombardment of a refrigerated sample.

Vegard studied solid nitrogen under electron bombardment most thoroughly (1926, 1930). He found that electron energies in the range 75 to 350 volts produce band systems of excited nitrogen molecules. Vegard reported that electron energies exceeding 350 volts caused a broad emission to appear in the spectral region near the 5577 A auroral green line and this emission became the dominant spectral feature at electron energies of 700 to 750 volts. Another emission near 5230A displayed phosphorescence, the glow persisting many seconds after cessation of bombardment. Both emission groups displayed thermoluminescence, the glow reappearing on warming. These phenomena have been reproduced in great detail by Broida *et al.*, using methods described in Chapter 3 (e.g., see Bass and Broida, 1956; Broida and Peyron, 1957; and Peyron *et al.*, 1959.) From this recent work it is clear that Vegard had produced free radicals and the thermoluminescence he reported was associated with diffusion of some sort of trapped free radical. Many experiments were performed by Vegard as well as by McLennan and co-workers (1928, 1929) in which the effects of dilution of the solid with rare gases contributed information concerning the identification of the trapped active species. This early work is connected with the recent studies in which gaseous active nitrogen is condensed at 4°K by Peyron *et al.* (1959). Consequently the discussion of the identity of the species produced in the *in situ* electron bombardment experiments of Vegard, McLennan, and most recently of Hörl (1959a), is based on *all* the available data.

There has been substantial opinion that the active species trapped in solid nitrogen is atomic nitrogen. This opinion was supported by the paper of Herzfeld and Broida (1956) in which the α-lines (blue-green, centered at 5230 A) are interpreted as the ^2D–^4S transition of nitrogen atoms shifted slightly by the matrix environment. In an early report of measurements of ESR detection of this radical species, Jen[2] indicated the hyperfine splitting was practically identical with that observed in gaseous nitrogen atoms. This observation solidified the feeling that the environmental perturbations are indeed small. Four experimental facts detract from the picture. The emission of the green α-line accompanying warm-up to 9°K is attributed to an atomic transition, although the chemiluminescence accompanying recombination of two nitrogen atoms would be expected to be characteristic of an excited product molecule. A second surprising result has been reported by Broida and Peyron (1958). They found that the active species can be evaporated after deposition at 20°K provided the temperature is raised before loss of the phosphorescent glow following deposition. If the emission process which occurs at 9°K were diffusion and recombination of nitrogen atoms, this process should be immeasurably rapid at temperatures over 20°K. The third and possibly most embarrassing observation is the appearance in the thermoluminescence of the ^2P–^2D atomic transition (which has a 4-sec half-life in the gas). As clearly noted by Peyron et al. (1959), "It is implausible to expect a site within the solid to increase the lifetime so that radiation from the ^2P state would be observed during the warm-up several hours after deposition." Finally, ESR studies have recently been reported which indicate a distinct matrix effect and some unexplained resonance features. Cole et al. (1957) observe, superimposed on the triplet attributed to nitrogen atoms, a pair of weak satellites indicating that some other free radical is present. Furthermore, the hyperfine splitting of the triplet differs somewhat from the splitting of gaseous nitrogen atoms. With what seem to be more precise measurements, Foner et al. (1958) verify the finding that the hyperfine splitting in the triplet *is* larger than that of gaseous nitrogen atoms.

The proposal that a polyatomic active species may be present was made by Milligan et al. (1956). Infrared detection of a species absorbing at 2150 cm^{-1} was attributed to a molecular species, N_3 radical. Isotopic substitution definitely confirmed the presence of more than two nitrogen atoms, but the intensities were not obviously amenable to the linear, symmetric N_3 postulated by Milligan et al. In a very recent publication, Peyron et al. (1959) also decide that a triatomic molecule is responsible for the emissions

---

[2] Reported in a paper by Jen at the Free Radicals Symposium, Laval University, Quebec, September 10, 1956. In a later report, Jen and co-workers report there *is* a perturbation by the matrix (see Foner et al., 1958).

earlier attributed to free nitrogen atoms. Vibrational spacings of 2140 and 2155 cm⁻¹ are assigned to excited states and spacings of 1924 and 2302 cm⁻¹ to the ground state. Despite the nearness of these spacings to the 2150 cm⁻¹ infrared absorption, they conclude that "this cannot be the same N_3 molecule proposed by Milligan, Brown, and Pimentel." Nevertheless it seems that the least complicated interpretation of these results is that the species observed by the two laboratories are the same. If this be true, the linear symmetrical model proposed by Milligan *et al.* (1956) must be discarded (see Section III, B, 9).

2. The HN_3 "Blue Color": Tesla Coil Technique

Papazian (1957, 1958) has proposed that a Tesla coil provides a convenient laboratory tool for conducting ion bombardment of a cold solid. The technique was demonstrated by discharging a Tesla coil in proximity to solid HN_3 condensed at 77°K. After 20 seconds the HN_3 became blue opposite the tip of the coil, reproducing the visual evidence of active species which had been obtained earlier by Rice and Freamo (1951, 1953) by condensing HN_3 after pyrolysis. Presence of a magnet near the Tesla coil deflected the blue spot into two spots, showing that bombardment by both positive and negative ions is involved.

Baldeschwieler and Pimentel reproduced these experiments and examined the infrared spectrum of the resulting solid.[3] A significant decomposition of HN_3 occurs and the principal product obtained under these conditions is the final product, NH_4N_3. No evidence was found of intermediate species.

In conclusion, this method does provide a convenient tool for ion bombardment, though in the particular case cited there is no direct evidence, as yet, that the blue color is caused by trapped free radicals. Possibly the greatest disadvantages of this method are concerned with lack of control of locale and energy of bombardment.

3. Miscellaneous

Only a few other solids have been studied after ion bombardment and none as extensively as those cited above. Vegard (1924b, 1926) impinged electron beams onto solid air, oxygen, ammonia, and nitrous oxide at 20°K. Vegard and Keesom (1927) studied solid hydrogen in the same manner, but at 4°K. McLennan and Shrum (1924) studied solid carbon monoxide, ammonia, oxygen, and water. Although some of these solids displayed luminescence, evidence was not presented which we can now interpret to indicate storage of free radicals. Papazian (1957) has applied the Tesla coil method to a number of solids condensed at 77°K. Some of the solids (CH_3I, CH_3OH, CH_3NHCH_3) became colored after a few seconds, surely

---

[3] Unpublished experiments performed by J. D. Baldeschwieler.

indicating chemical reactions. Of course the color by itself can hardly be taken as convincing evidence of the presence of free radicals.

Hörl (1959a) reported at the Sheffield symposium in 1958 that he has work in progress in which various solids were bombarded with electrons at energies up to 50 kev. Hörl (1959b) reported some results of bombardment of acetone and oxygen at 4°K. In the case of acetone a thermoluminescent behavior indicates the storage of energy-rich species (see Hörl 1959b).

B. Photolysis

Of all methods available, photolysis *in situ* provides the most controlled means of producing reactive species in solids. The mode of excitation, the available energy, and the environment of the site of excitation can be known and, in principle, can be varied according to the aim of the experimenter. Higher energy quanta, X-rays and gamma rays, operate through secondary processes involving extremely high, and, necessarily, a wide range of, energies. Particle bombardment suffers from the same disadvantage. In the gas phase production of free radicals the heat released during condensation, including that due to undesired reactions, must be removed before the advantages of the matrix isolation method can be realized. Furthermore, undesired reactions may occur before condensation, an extreme difficulty in the glow discharge or microwave discharge methods.[4]

Unfortunately, there are limitations to the photolysis method, as well. The most obvious of these is that only a limited number of substances can be decomposed photolytically. Secondly, the events which follow a photolytic excitation are influenced by cage effects. As first noted by Franck and Rabinowitsch (1934), the quantum yield of a dissociation reaction tends to be much lower in condensed phases than in gaseous photolysis because the fragments are held in proximity by the environment. Another possibility is the accumulation of photolysis products in particular orientations, either through the primary photolytic event or in secondary light absorption processes.[5] Lewis and Lipkin (1942) observed such effects in the photolysis of substances suspended in a rigid hydrocarbon matrix. The effect does

---

[4] As a single example, consider the unpublished experiment conducted by Pimentel and Whittle in which acetone plus xenon was passed through a 60-cycle glow discharge and the products were condensed at 20°K. Infrared spectra revealed, among the variety of products, a substantial amount of acetylene.

[5] If the solid environment removes electronic excitation energy in the form of heat, light absorption may cause local diffusion and, hence, molecular reorientation. Such a process will gradually accumulate the absorbing molecules in orientations unfavorable for excitation. Of course this could occur either for parent molecules or absorbing products. Albrecht (1957) has called attention to this possible effect, naming it "photo-orientation."

not, in itself, prevent photolytic production of radicals, but it might confuse the interpretation of spectral results. A third possibility is that secondary reactions may occur which are unexpected on the basis of gas phase reactions. For example, free radical reactions which are slow in the gas phase because of need for a third body will not be inhibited in the solid state. Reactions may even occur involving supposedly inert matrices because of "hot radical" effects. DeMore and Davidson have reported, for example, formation of N_2O through photolysis of ozone in solid nitrogen (see DeMore, 1958). This result has been duplicated by I. Haller and Pimentel (unpublished results). DeMore and Davidson estimate the activation energy for the reaction $O + N_2 \rightarrow N_2O$ to be in the range 14.5 to 21.5 kcal but are not able to decide whether this activation energy is provided in their matrix experiment through electronic or translational excitation.

A final limitation of the photolysis method is the time necessary for preparation of the sample. The rate of photolysis is limited by the intensity of the light source. Mercury arcs are in common use: the high pressure arcs offer high intensities over a range of frequencies, whereas medium and low pressure arcs provide well defined excitation frequencies. Light scattering by the solid can reduce substantially the rate of photolysis as well.

1. Triphenylmethyl

Lewis and Lipkin (1942) produced triphenylmethyl by photolyzing hexaphenylethane in E.P.A. matrix at 90°K. A second experiment involved photolysis of tetraphenylhydrazine which was considered to lead to diphenylnitrogen. In each case the electronic absorption spectrum showed the storage of the free radical species. When the matrix was warmed about 10° above liquid air temperature, the active substance was lost and the parent material regenerated. In the case of triphenylmethyl radical, the spectrum was already well known from studies of this species in solution. These are among the earliest examples of photolytic production of free radicals *in situ* in a solid.

There seems little doubt that the success of this pioneering experiment depended in part on the unique stability of these free radicals.

2. Iodine Atoms, CS

The use of photolysis for *in situ* production of free radicals in solids was revitalized and given its present impetus by Norman and Porter (1954, 1955). Two experiments, the photolysis of iodine and of carbon disulfide in solid organic glasses, provided convincing examples of the usefulness of the technique. Iodine apparently could not be photolyzed in hydrocarbon glasses at 77°K, presumably because of diffusion during photolysis. However, in the E.P.A. glass at 77°K, a 10^{-4} M solution of iodine was completely

TABLE IV

BANDS PRODUCED BY PHOTOLYSIS OF CS_2 IN HYDROCARBON GLASS[a]

Band	Gas state: CS		Hydrocarbon glass		
	$\lambda(A)$	$\nu(cm^{-1})$	$\lambda(A)$	$\nu(cm^{-1})$	$\nu_{gas} - \nu_{glass}$ (cm^{-1})
0, 0	2575.6	38,826	2621	38,153	673
1, 0	2507.3	39,884	2553	39,170	714
2, 0	2444.8	40,903	2493	40,112	791

[a] Norman and Porter (1955).

decolorized in 1 hour by the radiation from a 1-kw high-pressure mercury lamp. The iodine color reappeared on warming, presumably confirming the trapping of iodine atoms. Carbon disulfide was photolyzed at 10^{-2} M concentration in a hydrocarbon glass (3 MeP.P) at 77°K. After only a few minutes' irradiation a strong band system appeared with a reasonable correspondence to band heads present in the spectrum of gaseous CS. These are shown in Table IV. The differences are rather large but, as noted by Norman and Porter, an exact frequency comparison is not possible because the intensity distributions in the bands are different in the two media. These workers present spectra which seem to indicate continued growth of the "CS" band heads during 2 hours of irradiation. One more spectrum is presented after another hour at 77°K without irradiation. This spectrum seems to indicate a distinct loss of CS_2 during this last waiting period, whereas no noticeable change in the "CS" bands can be discerned.[6]

Sowden and Davidson (1956) used this same experiment, the photolysis of CS_2 in a hydrocarbon matrix, to test their own technique. They were able to reproduce the results of Norman and Porter and agree that the CS molecule is readily prepared in hydrocarbon glasses by photolysis of CS_2.

Two sets of experiments detract from these, positive results. The first consisted of unsuccessful experiments conducted by Milligan and Pimentel (unpublished). These workers suspended CS_2 in matrices of solid nitrogen, argon, and 3-methylhexane. In no case did photolysis give a detectable decrease in the infrared absorptions of CS_2 (i.e., not more than about 3 % of the CS_2 could have decomposed) and no infrared absorption was detected near 1285 cm^{-1}, the ground state fundamental stretching mode of CS as

[6] These remarks about the spectral changes during the wait period are made by this author on the basis of the published spectra (Norman and Porter, 1955) and are subject to the uncertainty occasioned by the difficulties of reproduction of spectrographic data. Norman and Porter made no reference to the apparent loss of CS_2 during the wait period.

deduced from electronic spectra. The only experiments in which products were detected by these workers involved CS_2-nitrogen mixtures to which a small amount of O_2 had been added. Then an infrared absorption was observed close to the location of the asymmetric stretching mode of OCS.

The second experiment which gave negative results was performed by Sowden and Davidson (1956). They found it impossible to decolorize an iodine solution in hydrocarbon glass, even on prolonged irradiation. Similar experiments with bromine failed to give evidence of photolysis but led these authors to suspect Br_2-olefin complex formation. Sowden and Davidson remark that iodine-olefin complex formation is also well known.

3. ClO

Another of the substances photolyzed by Norman and Porter (1955) was ClO_2 in a hydrocarbon glass (P.MeH) at 77°K and at an unspecified concentration. Only about 5 min of photolysis was needed to remove completely the diffuse absorption attributed to ClO_2 and replace it by a diffuse absorption in the region in which gaseous ClO absorbs. Melting this mixture regenerated most of the ClO_2. Although the authors favor the interpretation that ClO radicals and oxygen atoms are trapped, they admit alternative explanations are possible, such as the formation of ClO_3 and ClO from the photolysis of dimeric ClO_2, Cl_2O_4, formed during the cooling process.

4. Aromatic Radicals: Benzyl, Etc.

Porter and co-workers have made a most substantial contribution in their study of the photolysis of aromatic substances in glasses at 77°K (Norman and Porter, 1955; Porter and Strachan, 1958). In general the identifications rest on detection of the same product spectrum from several dissimilar but reasonable precursors. These workers have accumulated a large enough number of examples (some forty molecules) to enable them to generalize concerning the relative probability of dissociation of different bonds.

It would not be possible to discuss here each of these aromatic free radicals individually. Rather we present Fig. 1 illustrating typical spectral data obtained by Porter et al.[7] and Table V, which includes most of the data collected by Porter and Strachan (1958). The brevity of this treatment in no sense reflects a lack of appreciation for the very significant contribution made by Porter et al. in their exploitation of the matrix method for the study of aromatic free radicals. Rather it would not seem possible to do justice to their work except by example. The benzyl radical, then, will be singled out for more detailed consideration.

Fifteen minutes of photolysis of toluene in E.P.A. at 77°K is sufficient

---

[7] This figure was kindly furnished by Professor G. Porter. It presents the same data given in Fig. 1, Porter and Strachan, 1958.

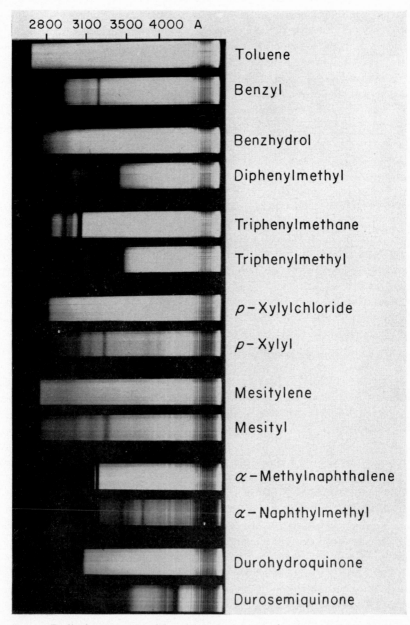

Fig. 1. Radical spectra resulting from the photolysis of aromatic molecules at −197°C. In each case the first spectrum is recorded before, and the second after, irradiation of the rigid solution for 15 min. (Porter and Strachan, 1958).

TABLE V

AROMATIC FREE RADICAL SPECTRA[a, b]

Radical	Parent molecule	λ_A	$\nu_{cm^{-1}}$	Relative intensities
$C_6H_5CH_2$	$C_6H_5CH_3$, $C_6H_5CH_2Cl$,	3187	31380	10
	$C_6H_5CH_2OH$, $C_6H_5CH_2NH_2$,	3082	32450	3
	$C_6H_5CH_2CN$, $C_6H_5CH_2COOH$,	3047	32810	3
	$C_6H_5CH_2CH_2NH_2$, $C_6H_5CH_2CH_2OH$,			
	$C_6H_5CH_2CH_2CH_2OH$.			
$C_6H_5CH \cdot CH_3$	$C_6H_5CH_2CH_3$, $C_6H_5CHOHCH_3$,	3222	31040	10
	$C_6H_5CHNH_2CH_3$, $C_6H_5CH_2CH_2OH$.	3167	31580	5
		3129	31960	6
		3083	32440	6
$C_6H_5C(CH_3)_2$	$C_6H_5CH(CH_3)_2$, $C_6H_5C(CH_3)_3$	3242	30840	—
$(C_6H_5)_2CH$	$(C_6H_5)_2CH_2$, $(C_6H_5)_2CHOH$	3355	29810	10
		3305	30260	8
		3240	30860	5
		3180	31450	3
		3122	32030	4
$(C_6H_5)_3C$	$(C_6H_5)_3CH$	3415	29290	10
		3358	29780	8
		3303	30270	2
$C_6H_5CH_2CHC_6H_5$	$(C_6H_5CH_2)_2$	3625	27580	—
$o\text{-}CH_3C_6H_4CH_2$	$o\text{-}CH_3C_6H_4CH_3$	3230	30960	10
		3170	31550	1
		3100	32260	1
$m\text{-}CH_3C_6H_4CH_2$	$m\text{-}CH_3C_6H_4CH_3$	3230	30960	10
		3100	32260	6
$p\text{-}CH_3C_6H_4CH_2$	$p\text{-}CH_3C_6H_4CH_3$	3230	30960	10
	$p\text{-}CH_3C_6H_4CH_2Cl$	3170	31550	1
		3100	32260	3
		3047	32810	2
$1,3(CH_3)_25CH_2C_6H_3$	$1,3,5(CH_3)_3C_6H_3$	3249	30770	10
		3109	32170	6
		2964	33730	2
C_6H_5CHCl	$C_6H_5CHCl_2$	3231	30950	10
		3190	31350	5
		3100	32260	4
$C_6H_5CCl_2$	$C_6H_5CCl_3$	3238	30880	10
		3108	32180	8
α-Methylenenaphthalene	α-Methylnaphthalene	3700	27030	7
		3554	28140	5
		3500	28570	7
		3424	29210	10

TABLE V—*Continued*

Radical	Parent molecule	λ_A	$\nu_{cm^{-1}}$	Relative intensities
β-Methylenenaphthalene	β-Methylnaphthalene	3840	26040	7
		3647	27420	5
		3687	27120	4
		3500	28570	8
		3424	29210	10
C_6H_5NH	$C_6H_5NH_2$	3109	32060	—
$C_6H_5NCH_3$	$C_6H_5NHCH_3$ $C_6H_5N(CH_3)_2$	3166	31590	←
$C_6H_5CH_2NC_6H_5$ or $C_6H_5CHNHC_6H_5$	$C_6H_5CH_2NHC_6H_5$	3745	26700	10
		3635	27510	8
C_6H_5O	C_6H_5OH, $C_6H_5OCH_3$ $C_6H_5OCH_2CH_3$	2870	34840	—
		Limit	—	—
p-HOC_6H_4O	p-HOC_6H_4OH	4140	24150	10
Durohydroquinone	Durosemiquinone	4220	23700	10
		4041	24740	4
		3300	30300	8

[a] All these radicals were produced in E.P.A. glass at 77°K except for the two naphthalene examples which were produced in P.MeH glass. For most experiments a 1.5-cm cell and a 1-kw high pressure mercury arc were used. The bands were usually sharp and could be estimated to ±5 A. All solutions were outgassed.

[b] Porter and Strachan (1958).

to produce the absorption shown in the second spectrum of Fig. 1, a diffuse spectrum beginning at 2839 A and extending to shorter wavelengths together with a "strong line-like band at 3187 Å and two weaker bands at 3082 and 3039 A" (Norman and Porter, 1955).[8] The diffuse absorption remains after the glass is melted and refrozen, hence is assigned to a stable product, probably a methyl-substituted hexatriene. The three sharp features, on the other hand, are lost completely when the glass is softened. These sharp lines were reproduced within 1 A through photolysis of three other benzyl compounds: benzyl chloride, benzyl alcohol, and benzylamine. Each spectral feature is shifted 137 A (1350 cm^{-1}) toward the red from a similar feature previously observed in flash photolysis studies and considered to be gaseous benzyl radical. Norman and Porter (1955) remark that "such shifts . . . [are] characteristic of spectra in rigid media."

Later work by Porter and Strachan (1958) provides confirmatory evi-

[8] In the more recent publication, the third band has been listed as 3047 A, as in Table I (Porter and Strachan, 1958).

dence on the assignment. Three possible processes are considered:

$$C_6H_5CH_2X \rightarrow C_6H_5CH_2 + X \tag{1}$$

$$\rightarrow C_6H_5CH + HX \tag{2}$$

$$\rightarrow C_6H_5 + CH_2X \tag{3}$$

To distinguish between these three, they studied a series such as $C_6H_5CH_3$, $C_6H_5CH_2Cl$, $C_6H_5CHCl_2$, and $C_6H_5CCl_3$. The radical absorptions recorded were, respectively, 3187 A, 3187 A, 3231 A, and 3238 A. If only one of the three processes listed accounts for the photolytic decomposition of these four molecules, it must be the first process. The second process is expected to give the same radical, C_6H_5CH, in three of the four cases, whereas the third process would give the same radical, C_6H_5, in every case. Only the first process provides the expectation that the first two radical products will be the same, as observed.

Finally, the same absorption spectrum was obtained by photolyzing a total of nine substances of composition $C_6H_5CH_2X$, as shown in Table V. This large number of examples provides basis for the generalization that β-bond fission is the process by which benzyl-substituted molecules decompose photolytically, giving benzyl radical irrespective of the nature of X (see also Ingram *et al.*, 1955).

5. $(CH_3)_2N$; $(CH_3)_2N_3$

Sowden and Davidson (1956) photolyzed tetramethyltetrazene, $(CH_3)_2N$ —N=N—N$(CH_3)_2$, in a hydrocarbon matrix at 77°K. They found evidence for the production and storage of some active species, possibly $(CH_3)_2N_3$, but absorption which could be attributed to $(CH_3)_2N$ was not detected in the expected region. In their discussion, Sowden and Davidson observe that identification of a free radical by the matrix method using ultraviolet-visible spectroscopy is exceedingly difficult if the free radical has not been identified previously by other methods.

6. SCl

Sowden and Davidson (1956) photolyzed both S_2Cl_2 and SCl_2 in hydrocarbon glasses at 77°K. An absorption at 2370 A appears during photolysis and disappears on warming the solid. These authors consider the assignment of this absorption to SCl to be a likely but unconfirmed hypothesis.

7. HNO

Brown and Pimentel (1958) photolyzed nitromethane in solid argon at 20°K and used infrared spectroscopic study to detect the products. These workers were able to detect and identify a variety of products and showed that fragmentation to CH_3 and NO_2 followed by isolation of these radicals in the matrix does not occur. Instead an unexpected and previously unreported isomerization to methyl nitrite was discovered, apparently followed by photolysis of the methyl nitrite. The results are consistent with the scheme:

$$CH_3NO_2 + h\nu = CH_3ONO$$

$$CH_3ONO + h\nu = H_2CO + HNO$$

$$H_2CO + h\nu = H_2 + CO$$

$$HNO + h\nu = \text{products}$$

The identification of HNO was positively confirmed through deuteration and normal coordinate calculations.

Although the nitroxyl molecule is not a free radical, its discovery by the matrix isolation method properly gives it a place here. At the time the matrix isolation work on HNO was performed, the only spectroscopic evidence for the species was an unpublished report by Dalby. He assigned to HNO certain visible range spectral features observed on flash photolysis of NH_3 in the presence of NO.[9] The matrix work presents, then, two important contrasts with other types of work. First, when infrared detection is used, almost all the products are liable to be detected. In the nitromethane study, for example, Brown and Pimentel identified among the products methyl nitrite, HNO, CH_2O, CO, CO_2, N_2O, NO, HNCO, and possibly H_2O and NH_3. Visible-ultraviolet studies could have given spectral evidence of only a few of these molecules and in an ESR study, only the radical species NO would be noticed. Surely by these techniques it would have been much more difficult to discern the details, possibly even the existence, of the complex process which actually occurs. Either of these two selective methods of study could lead to an oversimplified and perhaps naive interpretation based on the illusion that *detection* of few products implies there *are* few products. Again in contrast with a number of visible-ultraviolet and of ESR studies, this work suggests that important cage effects are operative. This topic will be discussed in detail later in this chapter.

8. CH_2

Milligan and Pimentel (1958) have photolyzed diazomethane in solid nitrogen and argon at $20°K$.[10] The reaction expected and desired, $CH_2N_2 +$

---

[9] F. W. Dalby, Symposium on Molecular Structure and Spectroscopy, Columbus, Ohio, June 13, 1957.

[10] A substantial study of the photolysis of diazomethane by the matrix isolation method has been conducted, but not yet published, by DeMore and Davidson (DeMore, 1958).

$h\nu = CH_2 + N_2$, probably occurs under these experimental conditions, but it surely cannot be the *only* reaction which occurs. Rather intense infrared spectral features were detected at 1362 and at 1114 cm^{-1} which are surely assignable to unstable species, and either could be the H—C—H bending mode of CH_2. These two features are lost if the matrix is allowed to warm momentarily to 35°K, the temperature at which diffusion takes place. This loss is accompanied by two phenomena which suggest that CH_2 may be present: first, a red glow appears during the diffusion process, persisting for 1 or 2 min, and, second, new infrared spectral features appear which have been associated with the stable products, ethylene, methane, propylene, cyclopropane, and polyethylene. The glow is presumably a chemiluminescence such as might accompany reactions of methylene

$$CH_2 + CH_2 = C_2H_4{}^*$$
or
$$CH_2 + CH_2N_2 = C_2H_4{}^* + N_2$$
or
$$CH_2 + CH_2N_2 = C_2H_4 + N_2{}^*.$$

The stable products provide even more convincing evidence since they are the same products obtained in the photolysis of gaseous CH_2N_2 where it is generally agreed that CH_2 is formed.

Yet the situation is complicated by the detection of more bands assignable to unstable species than can be justified on the CH_2 interpretation alone. Milligan and Pimentel (1958) find it necessary to postulate an additional process, the tautomerization of diazomethane. This is not a new proposal nor an unreasonable one from the standpoint of chemical bonding. Yet the necessity for the proposal of two processes detracts from the certainty of the interpretation. Again we find the special ability of the infrared detection method for revealing complex details of the photolytic processes occurring in the matrix isolation method. Although these details may inhibit a comprehensive explanation, they have the desirable feature of discouraging premature or oversimplified interpretations.

9. NH, NH₂, N₂H₂, N₃

These four species, three of them free radicals, are conveniently grouped together in a discussion of the matrix isolation studies of hydrazoic acid. Despite a large amount of work on this problem in at least three laboratories, doubt remains concerning which, if any, of these species are formed in the matrix decomposition of HN_3.

The early work of Rice and Freamo (1951, 1953) was based upon the rapid condensation at 77°K of the products of thermal or electric discharge decomposition of hydrazoic acid. The resulting solid has a fascinating blue color, a phenomenon which has as yet not received a positive explanation. This work is related to the *in situ* production of radicals by Rice and

Grelecki (1957). These workers find that the same blue color, indeed, the same spectrum in the visible region, is obtained if solid HN_3 at 77°K is irradiated with the light from a mercury discharge lamp. Rice and Grelecki (1957) propose that NH may be present when the blue color is produced by *in situ* photolysis as revealed by absorption between 3400 and 3500 A (see also Mador and Williams, 1954). Another absorption centered near 6500 A is responsible for the blue coloration. Mador and Williams (1954) assigned this broad band to NH_2 radical, but Rice and Grelecki disagree, stating ". . . the . . . [band] . . . at 6500 A must . . . be due to a highly reactive intermediate which may be diimide (N_2H_2)."[11]

Almost all of the study of the blue solid was conducted by the gas condensation method (see Chapters 2 and 3). Most of the *in situ* study of HN_3 decomposition has been performed by Pimentel and co-workers using inert matrices and the blue color was never observed. An absorption spectrum in the visible region of a sample of HN_3 photolyzed in solid nitrogen showed no trace of the band near 6500 A. The results of only one experiment of Becker *et al.* (1957) are directly comparable to the photolysis experiment performed by Rice and Grelecki (1957). Becker and associates photolyzed annealed crystalline HN_3 at 66°K and observed the infrared spectrum of the product. A noncondensable gas was produced during photolysis, and the spectrum showed the formation of a substantial amount of ammonium azide.

Turning to the inert matrix work, Becker *et al.* (1957) have made a thorough infrared study of the photolysis of hydrazoic acid, HN_3, in solid nitrogen, argon, and xenon at 20°K. In xenon, the most rigid of the three matrices, up to 40 % of the HN_3 could be decomposed but no new absorptions were detected. Figure 2 shows the spectral region near 2100 cm^{-1}: the loss of HN_3 during photolysis is plainly evident in Fig. 2, b. Upon warming the sample to 74°K, however, the infrared spectral features of the usual final product, NH_4N_3, appear, particularly the absorption near 2040 cm^{-1} (see Fig. 2, c). There seems to be but one reasonable interpretation: that NH has been produced and stored, as deduced by Becker *et al.* (1957). Presumably the single vibrational mode of NH was not detected because of a low absorption coefficient or because of spectral interference by HN_3 bands.

Results in solid nitrogen or argon are much more complicated. Even at N_2/HN_3 ratios of 500, several absorptions appear during photolysis. Two of these bands, 1290 and 1325 cm^{-1}, are apparently due to reactive species

---

[11] W. B. Gager and F. O. Rice have recently conducted ESR studies of these blue solids and detect a single asymmetric spin resonance line. While expressing doubt concerning the cause of this resonance, Gager and Rice conclude ". . . a mechanism similar to an F-center is very probable." [*J. Chem. Phys.* **31,** 564 (L) (1959).]

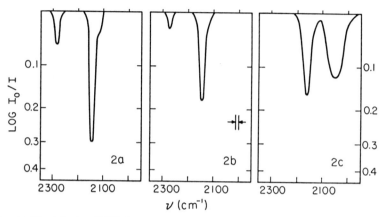

FIG. 2. Photolysis of HN_3 in xenon: region near 2100 cm^{-1}; M/A = 100. a. Before photolysis, $T = 20°K$. b. After photolysis, $T = 20°K$. c. After warming to 74°K, $T = 74°K$ (Becker et al., 1957).

and others, notably 975 cm^{-1}, are plainly due to NH_3. Becker and co-workers propose two possible reaction sequences which would lead to NH_3 through imperfect isolation of radical products in these matrices.

$$HN_3 + h\nu = NH + N_2$$

I $$NH + HN_3 = NH_2 + N_3$$

$$NH_2 + HN_3 = NH_3 + N_3$$

II $$NH + HN_3 = N_2H_2 + N_2$$

$$N_2H_2 + HN_3 = NH_3 + 2N_2$$

The two bands, 1290 and 1325 cm^{-1}, are reasonably located to be NH bending modes of either NH_2 or N_2H_2.

Van Thiel and Pimentel (1959) have continued these studies and present rather convincing evidence that either NH_2 or N_2H_2 is indeed formed. Deuterium substitution shows that surely one of the two bands in question, 1290 cm^{-1}, is a relatively pure hydrogen motion and that the species responsible for the absorption contains two equivalent hydrogen atoms. The growths of both 1290 and 1325 cm^{-1} are consistent with the interpretation that they are due to species intermediate in the formation of NH_3. Van Thiel and Pimentel conclude that the data are consistent with the assignment of 1290 cm^{-1} as the bending mode of NH_2, but they favor the alternate and equally acceptable assignment of the two absorptions, 1325 and 1290 cm^{-1}, as the bending modes of the cis- and trans-isomers of a planar diimide, N_2H_2. This species, not likely to be a radical, has not been detected previously by optical spectroscopy.

The N_3 radical has been discussed earlier in this chapter under the topic "electron bombardment." Two pieces of evidence are available from photolysis experiments. Becker et al. (1957) proposed that N_3 might be formed along with NH_2 in the photolysis of HN_3 in N_2 solid. They present spectral evidence of absorption near 2150 cm^{-1}, in agreement with that discovered by Milligan et al. (1956), but in this case obscured by overlapping absorption of HN_3.

Conflicting evidence concerning N_3 comes from infrared studies of metal azide crystals by Moore (1955). For example, a thin crystal of KN_3 was irradiated in air with the ultraviolet light from a low pressure mercury arc. A distinct growth of absorption at 1636 cm^{-1} (together with other bands) is accompanied by development of a blue coloration attributed to development of F-centers. After only 5 min of irradiation with "white" light from a 275-watt tungsten photoflood lamp, the blue color is lost and also the infrared absorption at 1636 cm^{-1} is almost eliminated. Moore considers that the formation of F-centers is accompanied by formation of azide radicals, which he refers to as "positive holes." Moore does not speculate on the structure of the N_3 radical suggested by this assignment. Apparently the possibility of a connection between this work and that of Rice and co-workers has not been explored.

10. Hydroxyl-Substituted Alkyl Radicals

Ingram and his co-workers have made fruitful use of photolytic production of free radicals and the ESR method of detection. In the earliest work, Ingram et al. (1955) duplicated some of the experiments of Norman and Porter, verifying the presence of free radicals. Almost unique, however, are their experiments in which H_2O_2 is photolyzed in dilute solution in various alcohols (Ingram and Gibson, 1957; Ingram et al., 1958; Fujimoto and Ingram, 1958). Hydrogen peroxide is dissolved in the alcohol at a mole fraction near 10^{-3}. Photolysis can be brought about with wavelengths near 3650 A, hence the experiments can be carried out in glass apparatus. Although OH radicals are sometimes detected, the usual result is a hydrogen abstraction reaction involving the matrix.

Fujimoto and Ingram (1958) summarize this work, and it is clear that the results contain interesting but complicated information. Six more or less characteristic ESR spectra are obtained from the monohydroxyl alcohols from methanol to n-hexanol (including some secondary and tertiary alcohols). In every case, the hydrogen abstraction is considered to occur at an alkyl position and the OH proton is not considered to affect the ESR spectrum. As an example, methanol plus OH gives a three-line spectrum which implies that two equivalent protons furnish hyperfine splitting, as in CH_2OH. Superimposed on this is a five-line spectrum which varies with conditions and which is assigned to a diradical formed by two CH_2OH

radicals. As a second example, both isopropyl and ethyl alcohols give complicated spectra at 77°K which change as the temperature is raised to 90°K. These temperature effects are explained in terms of onset of rotational movement.

11. Cage Effect and Quantum Yield

As mentioned earlier in this section, one of the most important questions concerning the preparation of free radicals by *in situ* photolysis is the extent to which the photolytic processes are inhibited or affected by cage effects. Strangely, the rather large amount of evidence available has led to two quite divergent schools of thought.

Norman and Porter (1955) in their pioneering work conclude that there is no significant inhibition of dissociative processes in glassy media. They state this clearly: "Although exceptions may yet be found, every substance which we have investigated which dissociates photochemically in the gas phase dissociated also in rigid media. The times of irradiation were similar . . ." Later work by Porter and Strachan (1958) strengthens this position, not only by adding examples, but by adding the only direct quantum yield measurements available. Porter and Strachan used the extinction coefficient of triphenylmethyl, as measured in toluenetriethylamine solution, to calculate the concentration and quantum yield of this radical produced by photolysis of triphenylmethane in E.P.A. glass at −197°C. The result, 0.01, is considered to represent within an order of magnitude the quantum yield of the simpler aromatic radicals, such as benzyl.

The work of Davidson and co-workers provides some basis for doubt concerning the absence of significant cage effect. In their exploratory work Sowden and Davidson (1956) found three examples, bromine, diphenylmercury, and acetophenone, which could not be decomposed by photolysis in hydrocarbon glasses at 77°K. Acetophenone, for example, shows a quantum yield between 0.1 and 1.0 in liquid solution at room temperature. At 77°K in a hydrocarbon glass, however, there was "no detectable decomposition after ten times the exposure necessary for complete decomposition at room temperature." Sowden and Davidson (1956) conclude, "photodecomposition in the rigid medium must have a negligibly small quantum yield." Practically identical conclusions were drawn concerning the photolysis of diphenylmercury.

In addition, Pimentel and co-workers have accumulated a substantial body of experience which tends to contradict the conclusion of Porter. There are a number of distinct, possibly significant, differences in the experiments conducted by Pimentel *et al.*

(1) The matrices used were solid nitrogen or solid argon, more surely providing the desired inert environment.

(2) The matrices were prepared by rapid condensation of the gas. The

resulting solids are highly scattering, hence could be crystalline rather than glassy.

(3) The temperature was 20°K.

(4) Infrared detection was used. This has the disadvantage that infrared extinction coefficients are low and the method is less sensitive than visible-ultraviolet spectroscopy. On the other hand, the decomposition of parent molecule is readily measured with reasonable accuracy and all the products present in reasonable amounts can usually be detected and measured. Thus a more detailed view of any decomposition process is obtained.

The molecules studied by Pimentel and co-workers fall into three groups. The first group includes those molecules readily photolyzed in the gas phase and difficult to photolyze in the matrix at 20°K. This group includes CS_2 in N_2, ketene in argon, ketene in nitrogen, methyl iodide in argon, HI in nitrogen, and methyl iodide in CD_4.[12] The CS_2 experiment has been described earlier, and the other systems are similar. In some cases, e.g., CH_3I in argon, a small amount of photolysis was indicated by the appearance of a small amount of product, but in none of these experiments was as much as 5% of the parent molecule decomposed during photolysis periods of 1 to 3 hours.

The second group includes molecules which are readily photolyzed under the same experimental conditions used for the first group, but the products are stable products apparently produced within the matrix cage in which photolysis occurred. This group includes azomethane, CH_3NNCH_3, in N_2 (which gave C_2H_6), CS_2 in N_2 with oxygen added (which gave OCS), cyclooctatetraene in A and in N_2 (which gave benzene and acetylene), CH_3NO_2 in A (which gave CH_3ONO), formaldehyde (which gave CO and, presumably, H_2), and CBr_3CHO in N_2 (which gave CO, $CHBr_3$, and HBr).[13] In every case it is clear that the photolysis products can be attributed to reactions within the matrix cage.

The third group includes molecules readily photolyzed which give reactive species, possibly free radicals. This group includes HN_3 in Xe, A, and N_2 (Becker et al., 1957), CH_2N_2 in A and N_2 (Milligan and Pimentel, 1958), and CH_3ONO in A (Brown and Pimentel, 1958). In each instance the photolytic decomposition gives one product with special stability, N_2 in the first two systems. Presumably both CH_2O and HNO are reasonably stable in the same cage at 20°K; note that neither is a free radical.

This pattern clearly suggests that the fragments produced by photolysis in situ do not escape from the neighborhood of the site of the primary

[12] The first three of these experiments were performed by Dr. D. E. Milligan; the others, respectively, by Dr. H. W. Brown, Dr. W. Thompson, and Mr. C. D. Bass. None of these results have been published.

[13] The first three of these experiments were performed by Dr. D. E. Milligan, the next two by Dr. H. W. Brown, and the last by Dr. G. C. Pimentel and Dr. E. D. Becker.

decomposition. If recombination is the most probable reaction, photolysis is slow or does not occur. If the fragments can react in another fashion to give stable products they may do so. If one of the fragments is sufficiently stable, the other fragment finds itself in an inert environment and can be isolated.

To clarify further the possible role of the cage, Pimentel and Rollefson (1959) studied the photolysis of gaseous nitromethane. As in the matrix studies (Brown and Pimentel, 1958), methyl nitrite is produced. However, in the gas phase, only a small fraction of the decomposed nitromethane molecules form methyl nitrite. Pimentel and Rollefson conclude that both matrix and gas phase studies can be understood in terms of a fragmentation to CH_3 and NO_2 followed by recombination, either to $CH_3NO_2^*$ or CH_3-ONO^*. In the gas phase most of the CH_3ONO^* molecules decompose to CH_3O and NO, whereas in the matrix the rapid deactivation by the solid environment prevents this decomposition, permitting accumulation of methyl nitrite.

The significance of the contradictory views concerning the cage effect leaves an interesting question yet to be answered. In the spirit of stimulating consideration of this matter, the following three factors are proposed as important contributors to the apparent discrepancy.

(1) In an inert matrix, photolysis *is* usually inhibited by the cage and only in rare events will reactive fragments escape from recombination. Only if the detection method is extremely sensitive will these small concentrations be detected. Thus infrared studies will show (correctly) that little reaction has occurred, whereas ESR or ultraviolet-visible spectroscopy may record the presence of the small amount of radicals trapped.

(2) If a radical is produced with special stability, the cage effect may be unimportant. The extreme example is, of course, triphenylmethyl, the only free radical for which the quantum yield is well established. No doubt the resonance stabilization of many of the aromatic radicals produced by Porter *et al.* figures in the success of their radical studies.

(3) If a radical is produced with special reactivity, it may react with a matrix considered to be inert. Well proven examples are provided by the reaction of oxygen atoms with a nitrogen matrix (DeMore, 1958) and the reactions of hydroxyl radicals with alcohol matrices (Fujimoto and Ingram, 1958). In many of the experiments of Porter, photolysis produces hydrogen atoms which may be lost by reaction with the matrix, particularly if E.P.A. is the glass used. This would, of course, aid in protecting the other radical fragment from recombination.

C. Gamma Ray and X-Ray Radiolysis

Gamma ray and X-ray radiolysis are properly considered together as "ionizing radiations." Most of the chemically important processes are initiated by secondary electrons, and with either radiation source, the electrons are produced with energies which are large compared to the energies of chemical reactions. The high energy implies a wide versatility: even the most stable molecules can be decomposed. The other side of this coin is that the reactions are uncontrolled, both with reference to energy and the molecular site of reaction. Nevertheless, this mode of *in situ* production of free radicals is extremely popular in teamwork with ESR detection. This current activity can be attributed principally to the sensitivity and selectivity of the detection method. Smaller and Matheson (1958) report, for example, that they can detect as few as 2×10^{12} spins of 1,1-diphenyl-2-picrylhydrazyl at 77°K. In an earlier paper (Matheson and Smaller, 1955) these same authors report a sensitivity of 10^{14} spins, a figure which may be more typical of work performed so far. These small concentrations of radicals can be selectively studied because the method blissfully ignores any reaction products which are not free radicals.

The large excess of energy dissipated per free radical produced is evident in the G-factor, the number of free radicals produced per 100 ev absorbed. Some values recorded with gamma rays are shown in Table VI. Most of

TABLE VI

G-Values Recorded by ESR Measurements

System	Temperature	G^a
HCOOH	77°K	4.9[b]
H_2O, 0.28 M H_2O_2	77°K	0.45[b]
H_2O	77°K	0.23[b]
H_2O	4°K	0.14[b]
N_2	4°K	0.2[c]
$HClO_4$	77°K	0.12[d]
NH_3	91°K	0.086[b]
H_2SO_4	77°K	0.04[d]
H_3PO_4	77°K	0.04[d]
Polyethylene		
Irradiated in vacuum	R.T.	0.00[e]
Irradiated in air	R.T.	0.3[e]
Polytetrafluoroethylene		
Irradiated in vacuum	R.T.	0.00[e]
Same, opened in air	R.T.	0.2[e]
Irradiated in air	R.T.	0.03[e]

[a] G = radicals per 100 ev absorbed (gamma radiation).

[b] Matheson and Smaller (1955).

[c] Wall *et al.* (1959b).

[d] Livingston *et al.* (1955). In a private communication Dr. R. Livingston indicated that recent work has shown higher yields.

[e] Abraham and Whiffen (1957).

the yields listed there correspond to energy dissipations of several thousand kilocalories per radical produced. This plethora of energy surely results in many undetected chemical processes, but seldom have these side reactions been thoroughly explored.

Despite the inefficiency and lack of control, many interesting results have been reported and research activity in this area is increasing rapidly. Of course one of the substantial incentives is the current interest in radiation damage to chemical systems, which is treated in more detail in Chapter 14.

1. H Atoms

Matheson and Smaller (1955) and Smaller *et al.* (1954) froze water at about $-20°C$, cooled the solid to $77°K$, and irradiated the sample with Co^{60} gamma rays. Irradiation of H_2O ice produces a doublet structure whereas D_2O ice gives a triplet structure. Records of these ESR derivative spectra are shown in Fig. 3. Matheson and Smaller (1955) observe that the doublet and triplet structures "are satisfactorily explained by assuming that they are due to the existence of a free spin (i.e., unpaired electron)

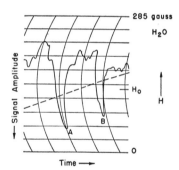

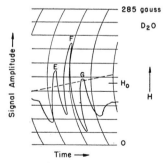

FIG. 3. Paramagnetic resonance in gamma-irradiated H_2O and D_2O ice, temperature $77°K$. ----- Field. ——— Signal amplitude during slow sweeping of magnetic field through resonance.

located near an H or a D nucleus." These authors remark that the splitting constants (85.5 Mc/sec in H_2O) are much less than the free atom values (1420 Mc/sec for H, $1s$). This result contrasts with ESR measurements of irradiated H_2SO_4 at 77°K where H atom resonances are reported with the free atom value (Livingston et al., 1955). Investigating this difference, Matheson and Smaller discovered that *both* resonances are observed when 4.4 M H_2SO_4 is irradiated at a low temperature. They explain this result in terms of crystallization into a mixture of crystals, ice, and $H_2SO_4 \cdot 4H_2O$. Their own observations on ice are discussed in terms of the effect of the strong intermolecular field. Alternative explanations, H_2O^+, H_2O^-, and electrons trapped at imperfections are considered and rejected.

Matheson and Smaller find a second doublet structure for irradiated H_2O ice at 4°K. This doublet is attributed to OH radical, a proposal which is consistent with the observation that annealing results in the loss of the features ascribed to H atom at 115 ± 10°K, whereas the alleged OH features disappear at "very slightly higher temperature."

Livingston et al. (1955) examined irradiated frozen oxygen acids: H_2SO_4, H_3PO_4, $HClO_4$, as well as boric, chromic, arsenic, and chloric acids. Only for the first three acids were ESR signals obtained, and those signals displayed hyperfine splittings within 1 % of the free atom value for hydrogen atoms. These authors are disturbed by the apparent discrepancy with the work of Matheson and Smaller, and remark that the "interpretation of the small separation . . . in ice . . . as a hyperfine separation of H modified by a polarization . . . [is] . . . difficult to reconcile with the . . . free atom value in the three acids."

Livingston et al. also find features superimposed on the resonance assigned to atomic hydrogen. Other interesting observations relate to: (1) resonances assignable to water adsorbed on the container; (2) the rates of disappearance of the radicals in each acid as the temperature was raised; and (3) the storage of the radicals in some samples for several months at 77°K without change.

ESR spectra produced by gamma irradiation of ammonia (Matheson and Smaller, 1955) and of methane (Smaller and Matheson, 1958) have also been attributed to hydrogen atoms. If these identifications are correct, the splitting in ammonia is over twice as big as in ice, but still far from the free hydrogen value. In methane, the splitting is described as "the normal H doublet of 506 gauss."

Wall et al. (1959a) report the ESR spectra of gamma-irradiated methane, hydrogen, and deuterium at 4°K. For methane they report a doublet (with a separation of about 500 gauss) which is superimposed upon a quartet of smaller spacing. A brief warming to an unknown temperature (followed by recooling to 4°K) caused the doublet to disappear and the quartet to sharpen. The doublet is assigned to H atoms, the quartet to CH_3.

The spectra of irradiated solid hydrogen and deuterium are curious in character. Hydrogen showed a triplet: the outer features are attributed to H atoms and the center line to a spurious resonance. Surprisingly, all three components have the appearance of absorptions rather than of derivatives. Deuterium shows what might be a quintet, the outermost features again resembling absorptions. Again the interpretation is complicated: the outer features are assigned to H atom from H_2 impurity (though the intensity exceeds that expected), and the center triplet is assigned to deuterium atoms together with spurious resonance at the center line. The features assigned to H atoms decayed after 19 hours at 4°K.

Additional information concerning the environmental effect on the hyperfine splitting comes from the work of Jen and co-workers (1956). These results are not based upon *in situ* radical production, but rather on rapidly frozen gases passing through a glow discharge. In solid hydrogen the hyperfine splitting constant is within 0.2 % of the free hydrogen value.

It is generally agreed that hydrogen atoms have been detected in solid matrices. Yet the very large differences among the reported hyperfine splittings, together with the spurious ESR signals such as that from adsorbed hydrogen (Livingston *et al.*, 1955), leave doubts and interesting questions.

2. N Atoms

The matrix isolation of nitrogen atoms has been discussed earlier in this chapter (see Wall *et al.*, 1959a, b). This case offers information concerning the environmental effect on ESR spectra. Foner *et al.* (1958) measured the hyperfine splitting of a triplet resonance attributed to nitrogen atoms suspended in three different matrices. The results, the hyperfine coupling constants, are shown in Table VII, including that for free nitrogen atoms. These significant deviations from the free nitrogen atom coupling constant suggest that quantitative interpretations of hyperfine splittings of more

TABLE VII

HYPERFINE COUPLING CONSTANTS, A, OF NITROGEN ATOMS IN SOLID MATRICES[a,b]

Matrix	A (Mc)	Deviation from free value (%)
N (free)	10.45	—
N in H_2	11.45	9.6
N in N_2	12.08	15.6
N in CH_4	13.54	29.5

[a] $T = 4$°K.
[b] Data from Foner *et al.* (1958).

complicated free radicals will be of doubtful value until these environmental effects are better understood.

3. OH

When Matheson and Smaller (1955) irradiated ice at 4°K (rather than 77°K), they detected a second doublet feature superimposed on the doublet assigned to hydrogen atoms. The hyperfine splitting of the new pair is about one-third that of hydrogen. This ESR signal is assigned to hydroxyl radical, OH. The effect of deuteration is consistent with the requirement that a triplet be formed though the central peak is obscured by that of hydrogen atom. Support is found in studies of H_2O_2-H_2O solutions for which the OH features are enhanced (provided an increasing hyperfine splitting with increasing H_2O_2 concentration is accepted as an environmental effect on a single species). In the H_2O_2 solutions the resonances are identical at 4°K and 77°K. The presence of the OH doublet in ice at 4°K is attributed to a quenching of the orbital angular momentum of the $^2\Pi_{1/2}$ state of OH. Its loss, then, at 77°K is assigned to a short relaxation time "owing to orbital coupling."

Matheson and Smaller (1955) turned to formic acid as a second possible source of OH (and of H) radicals. A doublet was observed which might be OH although the splitting is 50 % larger than that of the doublet in ice. On annealing, a strong single peak begins to grow at −103°C, a thermoluminescent glow occurs at −99°C, the peak becomes a maximum at −94°C and then it disappears, leaving the two side peaks until a temperature of −56 to −40°C is reached. It seems that doubt must be retained concerning the storage of a small radical such as OH in a formic acid matrix at −56°C $(T_D/T_M \sim 0.8)$.

Livingston and co-workers reported at the 1958 symposium at Sheffield, England (September, 1958) that gamma irradiation of $Ca(OH)_2$ at a low temperature produces a substantial ESR signal.[14] In this case an ultraviolet absorption spectrum provided distinct evidence of the presence of OH radical. Since the ESR signal is a singlet rather than a doublet, it cannot be assigned to OH. Hence these workers attribute the signal to an electron removed from OH^-. The question of why OH itself was not observed is not yet clear.

4. CH₃

Mador (1954) first reported experiments in which methyl radicals might have been formed *in situ* by X-radiation. Irradiation of lead tetramethyl at 4°K was followed by infrared spectroscopic study which showed methane formation. Only ambiguous evidence, such as color changes and physical

---

[14] Private communication from Dr. R. Livingston.

changes of the sample during warm-up, was presented in favor of the stabilization of methyl radicals.

Using ESR detection, Gordy and McCormick (1956) studied the X-irradiation of zinc dimethyl at 77°K. A quartet was observed, superimposed upon a weaker triplet. After 10 days of storage (presumably always at 77°K), the quartet had disappeared, leaving the triplet. Gordy and McCormick conclude that "it seems probable that the quartet with a total spread of 70 to 80 gauss is . . . the 'fingerprint' of the methyl free radical." The triplet is assigned to an ion radical, $(ZnCH_2)^+$. These workers interpret the hyperfine splitting of CH_3 in terms of a nonplanar but rapidly inverting structure. Smaller and Matheson (1958) used gamma irradiation of methane and also of dimethylmercury to obtain similar quartet spectra. These workers agree with the earlier workers that ". . . the quartet is almost certainly methyl" (see also Wall et al., 1959a). Cole and co-workers (1958) reproduced this quartet resonance by X-irradiation (50 kv) of methyl iodide at 77°K and measured the corresponding resonance in C^{13}-labeled CH_3I. Since the hyperfine splitting assigned to $C^{12}H_3$ had been observed to be about the same in each of four matrix environments (zinc dimethyl, methane, dimethylmercury, and methyl iodide), these workers were emboldened to make a quantitative structural interpretation based on the $C^{13}H_3$ datum. Estimating spin densities in part from Hartree wave functions and in part from a "semiempirical" hyperfine splitting constant, these workers obtain, appropriately, "semiquantitative" agreement with experiment assuming a planar model reorienting rapidly and isotropically. The same conclusion was reached independently by Karplus (1959), but without benefit of the $C^{13}H_3$ datum. Cole et al. (1958) express conviction that not only a pyramidal but also "a non-planar tunnelling radical . . . is excluded." Whether this confident conclusion is warranted is difficult to assess since Gordy and McCormick came to the opposite conclusion with similar data and since the effect of the matrix on hyperfine splittings remains uncertain.

5. C_2H_5

Gordy and McCormick (1956) irradiated mercury diethyl at 77°K with 40-kv X-rays. These workers detected an ESR pattern of six equally spaced components with relative intensities expected for five protons equally coupled to the electron spin. Expressing surprise that all five protons would be equivalent, Gordy and McCormick assign the signal to C_2H_5 radical.

Smaller and Matheson (1958) irradiated ethane at 77°K with gamma rays and detected a more complex spectrum which they assign to ethyl radicals. The features appear to be a quartet of triplets, a pattern appropriate to an odd electron interacting with three hydrogens and to a lesser extent with two equivalent hydrogens. This would make the electron density greatest at the hydrogen nuclei farthest from the presumed site of the free valence. This peculiar electron distribution is rationalized by Smaller and Matheson in terms of hyperconjugation.

The ESR signal observed by Smaller and Matheson has a somewhat smaller spread than that detected by the earlier workers ($\sim$80 gauss compared to $\sim$130 gauss) and, of course, has a more complex pattern. Smaller and Matheson propose that both spectra could be ethyl radical, the environmental effect of $Hg(C_2H_5)_2$ broadening the C_2H_5 spectrum into a sextet. They seek additional support for their own identification through gamma irradiation of other substances. The same features were obtained from solid ethyl chloride and from ethylene, though the spectrum was of low intensity in the case of ethylene. These workers conclude "the most likely radical which would be a common intermediate for all three compounds is the ethyl radical."

6. Allyl

Smaller and Matheson (1958) irradiated propylene with gamma rays at 77°K and obtained a five-line pattern spread only about one-half as much as the spectrum observed either from propane or n-butane. They assign the spectrum to allyl radical in which four equivalent protons are responsible for the hyperfine splitting. Smaller and Matheson remark: "The 5 lines are beautiful confirmation of the usual picture of the allyl radical as a superposition of the structures ... $CH_2\!\!=\!\!CH\!-\!\dot{C}H_2$ [and] $\dot{C}H_2\!-\!CH\!\!=\!\!CH_2$. In this radical the middle hydrogen causes no resolvable splitting of the lines."

7. Other Radicals

Several other radicals produced by gamma or X-radiation and detected by ESR methods have been tentatively identified. These are listed in Table VIII, and only those cases which involve unique aspects will be mentioned briefly here.

Studies of radiation damage to high polymers are of particular interest, and a comprehensive coverage of the work on free radicals in polymers has not been attempted here. With reference to synthetic polymers, the work of Whiffen et al. is of particular interest (e.g., Abraham and Whiffen, 1957; Abraham et al., 1957, 1958). Free radicals produced in natural polymers, particularly proteins, have been studied by Gordy and co-workers (e.g., Gordy et al., 1955a, b; McCormick and Gordy, 1958).

The study of the anisotropy of the ESR pattern produced by gamma irradiation of a single crystal of glycine by Ghosh and Whiffen (1959) is of particular interest. The orientation of the radical in the lattice can, of course, be deduced but, probably more important, the anisotropy may con-

TABLE VIII
POSTULATED FREE RADICALS: X- OR GAMMA IRRADIATION OF PURE SOLIDS

Radical	Substance irradiated	γ- or X- Rays	Temperature
NH_2	NH_3[a]	γ	77°K
$CH_3\dot{C}HCH_3$	Propane[a]	γ	77°K
$CH_3CH_2\dot{C}HCH_3$	n-Butane[a]	γ	77°K
$-CH_2-\dot{C}H-CH_2-$	n-Alkanes[a]	γ	77°K
$-CH_2-\dot{C}-(CH_2-)_2$	Polythene[b]	γ	R.T.
$-CF_2-\dot{C}-(CF_2-)_2$	Teflon[b, c, d]	γ	R.T.
$\dot{C}H_2{}^+$	CH_3OH, $NaOCH_3$, CH_3CONH_2[e]	X	77°K
$\dot{C}_2H_4{}^+$	C_2H_5OH[e]	X	77°K
	$C_2H_5CONH_2$[e]	X	R.T.
	Alanine[f]	X	R.T.
$\dot{C}H_2OH$	CH_3OH[a, g]	γ	77°K
$CH_3\dot{C}HOH$	C_2H_5OH[a]	γ	77°K
$\dot{C}_2H_6{}^+$	Valine[f]	X	R.T.
$\overset{+}{N}H_3\dot{C}HCO_2{}^-$	Glycine[h]	γ	R.T.
$CH_3CO\dot{O}H^+$	Acetic acid[i]	X	77°K

[a] Smaller and Matheson (1958).
[b] Abraham and Whiffen (1957).
[c] Ard et al. (1955).
[d] Schneider (1955).
[e] Luck and Gordy (1956).
[f] Gordy et al. (1955a); see other work of Gordy et al. on irradiation of proteins.
[g] See also Anderson and Alger (1956), Alger et al. (1956), R. Livingston et al. (private communication).
[h] Gibson et al. (1957).
[i] Gordy et al. (1955b).

tain detailed information about electron distribution within the molecule. Surely these studies by Ghosh and Whiffen provide a model which we may hope will be exploited.

IV. Concentrations of Free Radicals

One of the motivations which has stimulated development of the matrix isolation method has been the desire to produce for leisurely spectroscopic study a sample containing a readily detectable concentration of free radicals. Questions which arise quite naturally concern the concentrations which can be and which have been achieved. There are conceivable uses of the latent high energy content of the solid that would result if a high concentration of radicals could be obtained. A second practical basis for interest in high concentrations is the limitation placed on spectroscopic study of thick but dilute suspensions by the light scattering of the solid. At the concentrations needed for infrared analysis, for example, secondary reactions of the reactive species do provide a limit on the absolute number of radicals that can be accumulated and studied.

Aside from practical issues, there are important fundamental implications associated with the ultimate radical concentrations which can be achieved. Broadly stated, these questions relate, first, to the nature of the primary process by which the radical is formed and trapped and, secondly, to the nature of the inhibition of reactions under matrix conditions. To amplify the second point, it is conceivable that physical separation (isolation) of the reactive species by a chemically insulating layer of molecules is *necessary* to prevent reaction of two free radicals *at any temperature* (i.e., activation free energy for reaction may be zero). It is also possible that some free radicals are "protected" by a small activation free energy which, at sufficiently low temperature, will prevent reaction at a measurable rate. It is amply clear that both points of view have been represented in the opinions and interpretations of workers in this field.

Becker and Pimentel (1956) may have been first to provide evidence concerning concentration limitations in matrix methods. Their results are expressed in terms of the ratio of moles of matrix divided by moles of reactive species, M/R. Using a variety of stable but suitable test molecules in matrices of rare gases, nitrogen, CO_2, CCl_4, and cyclohexane, these workers concluded: "Experiments indicated that M/A^{15} less than 100 cannot generally be tolerated." Though it is true that their work did not involve *in situ* production of radicals, the experiments of Becker and Pimentel probably *do* suggest some of the limiting factors in the matrix isolation method. Two of these factors are the diffusional properties of the matrix and the relative size of the matrix molecule and the species to be isolated. These properties undoubtedly explain the fact that xenon was found to be much more effective as an isolating matrix than argon under otherwise identical conditions (Becker and Pimentel, 1956).

With these points in mind and assuming no thermodynamic inhibition of reaction between two free radicals in nearest-neighbor contact, theoretical arguments can be framed concerning the maximum possible concentration of free radicals in a matrix. In the most naive approach, we can picture the closest packing of reactive spheres and matrix spheres of identical size. To isolate completely a given reactive species, all of its twelve nearest neighbors would have to be matrix species. On the other hand, each matrix "cage" wall could be shared by two reactive species, one on each side. Hence we obtain a maximum concentration of about 16 % with a perfectly ordered arrangement. This is surely an upper limit, and it applies if the reactive

---

[15] In their earliest work, the ratio moles matrix:moles active species was designated M/A. This abbreviation, M/A, was later abandoned in favor of M/R since argon was in frequent use as a matrix.

species is the same size or smaller than a vacant lattice site in the matrix. It is clear this naive model could be extended to hypothetical situations in which the reactive species occupies two, or three, or more sites.

A more elegant attack might be based upon an assumed random distribution and take into account other important factors such as the mode of production of the matrix suspension. Jackson and Montroll (1958) attempted an attack on this problem and concluded the trapped radicals could not exceed 10 to 14 %, depending upon the packing. Golden (1958) made a still more detailed statistical analysis, and for his "densely packed bulk stabilized" model he confirmed the above limiting estimates.

A factor which has not been considered is the influence of the heat release accompanying an occasional reaction. Such an event would perhaps initiate local diffusion and, if the concentration is sufficient, cause a second such event. If this process propagates itself, the entire sample would "burn." [This type of behavior presumably accounts for the "blue-flash" phenomena reported by Bass and Broida (1956) in N atom - N_2 matrices.] Unfortunately, this phenomenon, which would place much more restriction on radical concentration than is implied by the statistical packing, involves properties of the matrix which are not easily estimated. It is particularly difficult to predict the thermal conductivity of the solid obtained by rapid condensation of a gas. Yet it can be anticipated that the maximum concentrations which can be obtained will be far below the theoretical estimates because of such "chain" reactions.[16, 17]

Fortunately a fairly large number of estimates of radical concentrations have been made. The most reliable of these are undoubtedly those based upon ESR measurements. Despite difficulties in calorimetry, such as thermal conductivity by effused gases and heat released by annealing, this method has provided useful results. The systems for which concentrations have been estimated are given in Table IX, preference being given to the largest concentration figure reported by each set of authors. For comparison, some values are included based on rapid condensation from the gas phase.

V. Summary

The matrix method has emerged as a new general method for the study of reactive species. There are a variety of methods of preparing and examin-

[16] Dr. J. L. Jackson has attempted an analysis of the "critical radical concentration" above which spontaneous ignition will occur (private communication). (See Chapter 10.)

[17] Dr. B. J. Fontana has made magnetic susceptibility measurements of nitrogen atoms suspended in solid nitrogen. He reports the following definitive results. Stable suspensions are obtained at concentrations below 0.04 mole per cent nitrogen atoms. If the nitrogen atom concentration rises to the range 0.2 to 0.5 mole per cent the sample becomes unstable [J. Chem. Phys. **31**, 148 (1959).]

TABLE IX

CONCENTRATIONS OF FREE RADICALS REPORTED

Radical	Matrix	Mole per cent radicals	Method of production and estimate[a]	Reference
O	O_2	4–20	Gas, cal	Minkoff et al. (1959).
		<3	Gas, IR	Harvey and Bass (1958)
		~1	Gas, cal	Broida and Lutes (1956)
OH	$Ca(OH)_2$	0.6	γ, ESR	R. Livingston[b]
N	N_2	4	Gas, cal	Minkoff et al. (1959)
		0.2	Gas, cal	Broida and Lutes (1956)
		0.03	γ, ESR	Wall et al. (1959b)
		>0.03	Gas, cal	Fontana (1958)
		0.01–0.04	Gas, MS	Fontana[c]
OH(?)	HCOOH	0.2	γ, ESR	Matheson and Smaller (1955)
CH_3	CH_4	0.14	γ, ESR	Wall et al. (1959a)
H	CH_4	0.1	γ, ESR	Wall et al. (1959a)
N	NH_3	0.1	Gas, ESR	Cole and Harding (1958)
H	$HClO_4$—H_2O	0.1	γ, ESR	Livingston et al. (1955)
H	H_2O	0.01	γ, ESR	Matheson and Smaller (1955)
H, NH_2(?)	NH_3	0.01	γ, ESR	Matheson and Smaller (1955)
ROH	Alcohols	~0.01	UV, ESR	D. Ingram[b]
H	H_2	0.0006	γ, ESR	Wall et al. (1959a)

[a] Abbreviations: gas = rapid condensation of gaseous radicals; γ = gamma ray in situ production; UV = photolytic in situ production; IR = infrared analysis; cal = calorimetry; MS = magnetic susceptibility.

[b] Private communication.

[c] Fontana, B. J. (1959). J. Chem. Phys. **31**, 148.

ing matrix suspensions of free radicals, and we shall conclude this chapter with a brief comparison of their advantages and disadvantages.

Photolysis in situ offers the most controlled means of producing free radicals in the solid state. There are limitations such as the possible inhibition of fragmentation by the matrix and the restriction to photosensitive precursors. There is every indication, however, that a complete understanding of the cage effect will permit design of experiments which are not impeded by this matrix effect. The restriction to photosensitive materials will probably be even less a problem. Already the way to remove this boundary has been shown: secondary reactions producing the desired species can be initiated with radicals easily produced by photolysis. The work of Fujimoto and Ingram (1958) using OH radicals from H_2O_2 in a reactive matrix provides a model.

Radiolysis, either by gamma rays or X-rays, has greater versatility in that energy limitations on the possible chemical processes are no longer present. Ironically, this may well be the principal interference to its general use. The large amount of energy dissipated per reaction of interest implies an absence of selectivity which would obstruct its use with any but the most discriminating detection method. ESR detection is suitably discriminating, yet many of the ESR studies have been interpreted on the basis of superimposed spectra.

Electron bombardment, like radiolysis, is not sufficiently discriminating to provide controlled reactions. No doubt its use for production of free radicals will be more or less confined to the simplest chemical systems.

The potentialities and limitations of atom bombardment are virtually unexplored. Klein and Scheer (1958a,b) have presented interesting work on the reaction of hydrogen atoms with solid olefins at −195°C. By a kinetic argument they propose that free alkyl radicals may have been accumulated. The technique surely deserves continued exploration. Glow discharges provide convenient sources of hydrogen, nitrogen, and oxygen atoms.

The rapid freezing of free radicals from the gas phase will surely have increasing application. The possibilities of forming desired species by mixing reactive gas streams just before freezing have not yet been tapped. Furthermore, the use of molecular beam evaporation from an oven gives the possibility of matrix study of such interesting molecules as inorganic molecules normally available only at high temperatures. The main problem in these methods is the control of the sample temperature during deposition. However, when this problem is given due consideration, it does not restrict unduly the usefulness of the method.

The several detection methods are most useful when used in combination. The ESR technique is the most selective and most sensitive for singling out free radicals. Further, the method has a unique potentiality for concentration determination since there is no issue of absorption coefficient. On the other hand, the centering of all ESR spectra around the same value of g causes a troublesome superposition of spectra. It is not an advantage, either, that spectra are recorded in terms of the derivative.

The detection of free radicals by spectroscopic study of electronic transitions is also selective and only a little less sensitive than the ESR method. It is unfortunate that highly scattering matrices make ultraviolet absorption spectroscopy rather difficult. The several glassy matrices which have been used solve this light-scattering problem but substitute reactivity which may be troublesome.

Infrared methods are most informative of the details of the processes accompanying free radical production. This versatility gives the most complete picture of the chemistry and the spectra are a substantial aid in identifying unknown species. Light scattering is far less serious, too, in this long-wavelength region. The limitations are significant, however. The most serious, probably, is the low absorption coefficient characteristic of this spectral region. This implies that much larger numbers of free radicals are needed for detection and study.

Whichever technique is preferred or most suited for a given problem, all workers in this field will agree that the *in situ* production of free radicals for matrix isolation has been established as a lucrative method of study. Surely much interesting work can be anticipated merely in exploitation of its demonstrated potentialities. Furthermore, experience has exposed some unexpected problems which demand attention on their own merits (see, for example, Pimentel, 1958a, b). Some of these are listed below.

(1) The effect of the matrix on hyperfine coupling constants in ESR spectra is not at all negligible and remains to be explained.

(2) The effect of the matrix on lifetimes of forbidden atomic electronic transitions is not clear.

(3) The importance and the nature of the inhibition of dissociative processes by the matrix cage is in dispute.

(4) The ESR technique has added a new dimension to the study of radiation-induced chemical reactions. Of course these are but matrix phenomena involving *in situ* production of energetic and reactive species.

(5) The kinetics of free radical reactions may well be examined by matrix methods, taking advantage of low temperatures to control reaction rates which are "instantaneous" under normal conditions.

(6) It is to be anticipated that unexpected and unfamiliar molecules will be detected for the first time in matrix studies.

These and many other problems contribute to the excitement of work in this field and account for the rapid development of this active scientific frontier.

References

Abraham, R. J., and Whiffen, D. H. (1957). *Trans. Faraday Soc.* **54,** 1291. Electron spin resonance spectra of some γ-irradiated polymers.

Abraham, R. J., Ovenall, D. W., and Whiffen, D. H. (1957). *Arch. sci. (Geneva)* **10,** Spec. No., pp. 84–85. Electron-spin resonance of free radicals in gamma-irradiated polymers. (*Chem. Abstr.* **52,** 1757, 1958).

Abraham, R. J., Melville, H. W., Ovenall, D. W., and Whiffen, D. H. (1958). *Trans. Faraday Soc.* **54,** 1133. Electron spin resonance spectra of free radicals in irradiated polymethyl methacrylate and related compounds.

Albrecht, A. C. (1957). *J. Chem. Phys.* **27,** 1413. Photo-orientation. (*Chem. Abstr.* **52,** 5073, 1958).

Alger, R. S., Drahmann, J. B., and Anderson, T. H. (1956). *Bull. Am. Phys. Soc.* [II] **1,** 379. Saturation and yield of free radicals by irradiation.

Anderson, T. H., and Alger, R. S. (1956). *Bull. Am. Phys. Soc.* [II] **1,** 379. The color, luminescence, and paramagnetism induced in some organic compounds by irradiation.

Ard, W. B., Shields, H., and Gordy, W. (1955). *J. Chem. Phys.* **23**, 1727. Paramagnetic resonance of x-irradiated teflon; effects of absorbed oxygen. (*Chem. Abstr.* **50**, 58, 1956).

Bass, A. M., and Broida, H. P. (1956). *Phys. Rev.* **101**, 1740. Spectra emitted from solid nitrogen condensed at 4.2°K from a gas discharge.

Becker, E. D., and Pimentel, G. C. (1956). *J. Chem. Phys.* **25**, 224. Spectroscopic studies of reactive molecules by the matrix isolation method. (*Chem. Abstr.* **50** 15244, 1956).

Becker, E. D., Pimentel, G. C., and Van Thiel, M. (1957). *J. Chem. Phys.* **26**, 145. Matrix isolation studies: infrared spectra of intermediate species in the photolysis of hydrazoic acid.

Broida, H. P., and Lutes, O. S. (1956). *J. Chem. Phys.* **24**, 484. Abundance of free atoms in solid nitrogen condensed at 4.2°K from a gas discharge.

Broida, H. P., and Pellam, J. R. (1954). *Phys. Rev.* **95**, 845. Phosphorescence of atoms and molecules of solid nitrogen at 4.2°K. (*Chem. Abstr.* **48**, 11932, 1954).

Broida, H. P., and Peyron, M. (1957). *J. phys. radium* **18**, 593. Luminescence de l'azote solide (4.2°K) contenant des atomes ou radicaux libres. Effet de la dilution par l'argon.

Broida, H. P., and Peyron, M. (1958). *J. Chem. Phys.* **28**, 725. Evaporation of active species trapped in a solid condensed from "discharged" nitrogen.

Brown, H. W., and Pimentel, G. C. (1958). *J. Chem. Phys.* **29**, 883. Photolysis of nitromethane and of methyl nitrite in an argon matrix; infrared detection of nitroxyl, HNO.

Carrington, T., and Davidson, N. (1953). *J. Phys. Chem.* **57**, 418. Shock waves in chemical kinetics: the rate of dissociation of N_2O_4. (*Chem. Abstr.* **47**, 7300, 1953).

Cole, T., and Harding, J. T. (1958). *J. Chem. Phys.* **28**, 993. Electron spin resonance spectrum of the electrical discharge products of ammonia frozen at 4.2°K.

Cole, T., Harding, J. T., Pellam, J. R., and Yost, D. M. (1957). *J. Chem. Phys.* **27**, 593. Electron paramagnetic resonance spectrum of solid nitrogen afterglow at 4.2°K.

Cole, T., Pritchard, H. O., Davidson, N. R., and McConnell, H. M. (1958). *Mol. Phys.* **1**, 406. Structure of the methyl radical.

DeMore, W. B. (1958). Ph.D. Thesis. California Institute of Technology, Pasadena, California.

De Nordwall, H. J., and Staveley, L. A. K. (1956). *Trans. Faraday Soc.* **52**, 1207. The formation and crystallization of simple organic and inorganic glasses. (*Chem. Abstr.* **51**, 5493, 1957).

Dewar, J. (1901). *Proc. Roy. Soc.* **68**, 360; *Rev. Am. Chem. Research* **7**, 150. The nadir of temperature and allied problems.

Dows, D. A., and Pimentel, G. C. (1955a). *J. Chem. Phys.* **23**, 1475. Infrared spectrum of solid ammonium azide: a vibrational assignment.

Dows, D. A., and Pimentel, G. C. (1955b). *J. Chem. Phys.* **23**, 1258. Infrared spectra of gaseous and solid hydrazoic acid and deutero-hydrazoic acid: the thermodynamic properties of HN_3.

Field, F. H., and Franklin, J. L. (1957). "Electron Impact Phenomena and the Properties of Gaseous Ions." Academic Press, New York.

Foner, S. N., Jen, C. K., Cochran, E. L., and Bowers, V. A. (1958). *J. Chem. Phys.* **28**, 351. Electron spin resonance of nitrogen atoms trapped at liquid helium temperature.

Fontana, B. J. (1958). *J. Appl. Phys.* **29**, 668. Thermometric study of the frozen products from the nitrogen microwave discharge.

Franck, J., and Rabinowitch, E. (1934). *Trans. Faraday Soc.* **30**, 120. Free radicals and the photochemistry of solutions. (*Chem. Abstr.* **28**, 2269[7], 1934).

Fujimoto, M., and Ingram, D. J. E. (1958). *Trans. Faraday Soc.* **54**, 1. Electron resonance studies of the change in free radical spectra of solid alcohols with variation of temperature and time of UV radiation.

Ghosh, D. K., and Whiffen, D. H. (1959). *Mol. Phys.* (in press). Electron spin resonance spectrum of a γ-irradiated single crystal of glycine.

Gibson, J. F., Ingram, D. J. E., Symons, M. C. R., and Townsend, M. G. (1957). *Trans. Faraday Soc.* **53**, 914. Electron resonance studies of different radical species formed in rigid solutions of hydrogen peroxide after ultraviolet irradiation. (*Chem. Abstr.* **52**, 6948, 1958).

Giguère, P. A. (1954). *J. Chem. Phys.* **22**, 2085. Spectroscopic evidence for stabilized HO_2 radicals. (*Chem. Abstr.* **49**, 3712, 1956).

Giguère, P. A., and Harvey, K. B. (1956). *J. Chem. Phys.* **25**, 373. On the presumed spectroscopic evidence for trapped HO_2 radicals.

Giguère, P. A., and Harvey, K. B. (1959). *J. Mol. Spectroscopy* **3**, 36. An infrared study of hydrogen bonding in solid H_2O_2 and H_2O—H_2O_2 mixtures.

Golden, S. (1958). *J. Chem. Phys.* **29**, 61. Free radical stabilization in condensed phases.

Gordy, W., and McCormick, C. G. (1956). *J. Am. Chem. Soc.* **78**, 3243. Microwave investigations of radiation effects in solids: methyl and ethyl compounds of Sn, Zn, and Hg. (*Chem. Abstr.* **50**, 16394, 1956).

Gordy, W., Ard, W. B., and Shields, H. (1955a). *Proc. Natl. Acad. Sci. U. S.* **41**, 983. Microwave spectroscopy of biological substances. I. Paramagnetic resonance in x-irradiated amino acids and proteins. (*Chem. Abstr.* **50**, 11816, 1956).

Gordy, W., Ard, W. B., and Shields, H. (1955b). *Proc. Natl. Acad. Sci. U. S.* **41**, 996. Microwave spectroscopy of biological substances II. Paramagnetic resonance in x-irradiated carboxylic acids and hydroxy acids. (*Chem. Abstr.* **50**, 11816, 1956).

Harvey, K. B., and Bass, A. M. (1958). *J. Mol. Spectroscopy* **2**, 405. Infrared absorption of oxygen discharge products and ozone at 4°K.

Herzfeld, C. M., and Broida, H. P. (1956). *Phys. Rev.* **101**, 606. Interpretation of spectra of atoms and molecules in solid nitrogen condensed at 4.2°K.

Hörl, E. (1959a). *J. Mol. Spectroscopy.* **3**, 425. Light emission from solid nitrogen during and after electron bombardment.

Hörl, E. (1959b). *J. Chem. Phys.* **31**, 564. Electron bombardment of solid acetone.

Hörl, E., and Marton, L. (1958). *Rev. Sci. Instr.* **29**, 859. Cryostat for electron bombardment and electron diffraction work.

Ingram, D. J. E., and Gibson, J. F. (1957). *Arch. sci.* (*Geneva*) **10**, Spec. No., pp. 81–83. Paramagnetic resonance of free radicals produced by ultraviolet radiation. (*Chem. Abstr.* **52**, 1757, 1958).

Ingram, D. J. E., Hodgson, W. G., Parker, C. A., and Rees, W. T. (1955). *Nature* **176**, 1227. Detection of labile photochemical free radicals by paramagnetic resonance. (*Chem. Abstr.* **50**, 11107, 1956).

Ingram, D. J. E., Fujimoto, M., and Gibson, J. F. (1958). *Arch. sci.* (*Geneva*) **11**, Spec. No., pp. 170–176. Etude de radicaux libres à basse température.

Jackson, J. L., and Montroll, E. W. (1958). *J. Chem. Phys.* **28**, 1101. Free radical statistics.

Jen, C. K., Foner, S. N., Cochran, E. L., and Bowers, V. A. (1956). *Phys. Rev.* **104**,

846. Paramagnetic resonance of hydrogen atoms trapped at liquid He temperature.

Karplus, M. (1959). *J. Chem. Phys.* **30**, 15. Interpretation of the electron spin resonance spectrum of the methyl radical.

Klein, R., and Scheer, M. D. (1958a). *J. Am. Chem. Soc.* **80**, 1007. The addition of hydrogen atoms to solid olefins at −195°C.

Klein, R., and Scheer, M. D. (1958b). *J. Phys. Chem.* **62**, 1011. The reaction of H atoms with solid olefins at −195°C.

Lewis, G. N., and Lipkin, D. (1942). *J. Am. Chem. Soc.* **64**, 2801. Reversible photochemical processes in rigid media: the dissociation of organic molecules into radicals and ions. (*Chem. Abstr.* **37**, 833⁸, 1943).

Lewis, G. N., Magel, T. T., and Lipkin, D. (1940). *J. Am. Chem. Soc.* **62**, 2973. Absorption and re-emission of light by cis- and trans-stilbenes and the efficiency of their photochemical isomerization. (*Chem. Abstr.* **35**, 31⁵, 1941).

Lewis, G. N., Lipkin, D., and Magel, T. T. (1941). *J. Am. Chem. Soc.* **63**, 3005. Reversible photochemical processes in rigid media. A study of the phosphorescent state.

Livingston, R., Zeldes, H., and Taylor, E. H. (1954). *Phys. Rev.* **94**, 725. Atomic hydrogen hyperfine structure in irradiated acids. (*Chem. Abstr.* **48**, 8041, 1954).

Livingston, R., Zeldes, H., and Taylor, E. H. (1955). *Discussions Faraday Soc.* **19**, 166. Paramagnetic resonance studies of atomic hydrogen produced by ionizing radiation. (*Chem. Abstr.* **50**, 9865, 1956).

Luck, C. F., and Gordy, W. (1956). *J. Am. Chem. Soc.* **78**, 3240. Effects of X-irradiation upon some organic substances in the solid state: simple alcohols, amines, amides, and mercaptans. (*Chem. Abstr.* **50**, 15245, 1956).

McCormick, G., and Gordy, W. (1958). *J. Phys. Chem.* **62**, 783. Electron spin resonance studies of radiation damage to proteins.

McLennan, J. C., and Shrum, G. M. (1924). *Proc. Roy. Soc.* **A106**, 138. On the luminescence of nitrogen, argon, and other condensed gases at very low temperatures.

McLennan, J. C., Ireton, H. J. C., and Samson, E. W. (1928). *Proc. Roy. Soc.* **A120**, 303. On the luminescence of solid N under cathode ray bombardment.

McLennan, J. C., Samson, E. W., and Ireton, H. J. C. (1929). *Trans. Roy. Soc. Can.* **23**, 25. Phosphorescence of solid A irradiated with cathode rays. (*Chem. Abstr.* **24**, 785, 1930).

Mador, I. L. (1954). *J. Chem. Phys.* **22**, 1617. The stabilization of the methyl radical. (*Chem. Abstr.* **49**, 561, 1955).

Mador, I. L., and Williams, M. C. (1954). *J. Chem. Phys.* **22**, 1627. Stabilization of free radicals from the decomposition of hydrazoic acid. (*Chem. Abstr.* **49**, 42, 1955).

Matheson, M. S., and Smaller, B. (1955). *J. Chem. Phys.* **23**, 521. Paramagnetic species in γ-irradiated ice. (*Chem. Abstr.* **49**, 10054, 1955).

Milligan, D. E., and Pimentel, G. C. (1958). *J. Chem. Phys.* **29**, 1405. Matrix isolation studies: possible infrared spectra of isomeric forms of diazomethane and of methylene, CH_2.

Milligan, D. E., Brown, H. W., and Pimentel, G. C. (1956). *J. Chem. Phys.* **25**, 1080. Infrared absorption by the N_3 radical.

Minkoff, G. J., Scherber, F. I., and Gallagher, J. S. (1959). *J. Chem. Phys.* **30**, 753. Energetic species trapped at 4.2°K from gaseous discharges.

Moore, P. W. J. (1955). Ministry of Supply, Waltham Abbey, Essex, England (unclassified), Explosives Research and Development Establishment Rept. **23/R/55,** Infrared spectra of pure and partially decomposed metallic azide crystals.

Nachtrieb, N. H., and Handler, G. S. (1954). *Acta Met.* **2,** 797. A relaxed vacancy model for diffusion in crystalline metals. (*Chem. Abstr.* **49,** 2968, 1955).

Nachtrieb, N. H., and Handler, G. S. (1955). *J. Chem. Phys.* **23,** 1187. Self-diffusion in α white phosphorous. (*Chem. Abstr.* **49,** 13747, 1955).

Nachtrieb, N. H., Weil, J. A., Catalano, E., and Lawson, A. W. (1952). *J. Chem. Phys.* **20,** 1189. Self-diffusion in solid sodium. II. The effect of pressure. (*Chem. Abstr.* **47,** 9706, 1953).

Norman, I., and Porter, G. (1954). *Nature* **174,** 508. Trapped atoms and radicals in a glass "cage". (*Chem. Abstr.* **49,** 2876, 1955).

Norman, I., and Porter, G. (1955). *Proc. Roy. Soc.* **A230,** 399. Trapped atoms and radicals in rigid solvents. (*Chem. Abstr.* **49,** 14490, 1955).

Papazian, H. A. (1957). *J. Chem. Phys.* **27,** 813. Technique to produce free radicals in the solid state. (*Chem. Abstr.* **52,** 1712, 1958).

Papazian, H. A. (1958). *J. Chem. Phys.* **29,** 448. Free radical formation in solids by ion bombardment.

Peyron, M., Hörl, E. M., Brown, H. W., and Broida, H. P. (1959). *J. Chem. Phys.* **30,** 1304. Spectroscopic evidence for triatomic nitrogen in solids at very low temperatures.

Pimentel, G. C. (1958a). *J. Am. Chem. Soc.* **80,** 62. Reaction kinetics by the matrix isolation method: diffusion in argon; cis-trans isomerization of nitrous acid. (*Chem. Abstr.* **52,** 4296, 1958).

Pimentel, G. C. (1958b). *Spectrochim. Acta* **12,** 94. The promise and problems of the matrix isolation method for spectroscopic studies.

Pimentel, G. C., and Rollefson, G. (1959). (Unpublished work.) Formation of methyl nitrite in the photolysis of gaseous nitromethane.

Porter, G., and Strachan, E. (1958). *Trans. Faraday Soc.* **54,** 1595. Primary photochemical processes in aromatic molecules. Part 4. Side-chain photolysis in rigid media.

Rice, F. O., and Freamo, M. (1951). *J. Am. Chem. Soc.* **73,** 5529. The imine radical. (*Chem. Abstr.* **46,** 6026, 1952).

Rice, F. O., and Freamo, M. (1953). *J. Am. Chem. Soc.* **75,** 548. The formation of the imine radical in the electrical discharge. (*Chem. Abstr.* **47,** 5291, 1953).

Rice, F. O., and Grelecki, C. (1957). *J. Am. Chem. Soc.* **79,** 1880. The imine radical.

Schneider, E. E. (1955). *J. Chem. Phys.* **23,** 978. Paramagnetic resonance of X-rayed Teflon. (*Chem. Abstr.* **50,** 9869, 1956).

Shallcross, F. V., and Carpenter, G. B. (1957). *J. Chem. Phys.* **26,** 782. X-Ray diffraction study of the cubic phase of ice. (*Chem. Abstr.* **51,** 11804, 1957).

Smaller, B., and Matheson, M. S. (1958). *J. Chem. Phys.* **28,** 1169. Paramagnetic species produced by gamma-irradiation of organic compounds.

Smaller, B., Matheson, M. S., and Yasaitis, E. L. (1954). *Phys. Rev.* **94,** 202. Paramagnetic resonance in irradiated ice. (*Chem. Abstr.* **48,** 7457, 1954).

Sowden, R. G., and Davidson, N. (1956). *J. Am. Chem. Soc.* **78,** 1291. Photochemical studies with rigid hydrocarbon solvents at low temperatures. (*Chem. Abstr.* **50,** 9883, 1956).

Van Thiel, M., and Pimentel, G. C. (1959). *J. Chem. Phys.* (in press). Matrix isolation studies: infrared spectra of intermediate species in the photolysis of hydrazoic acid, II.

Vegard, L. (1924a). *Nature* **113,** 716. The auroral spectrum and the upper atmosphere. (*Chem. Abstr.* **18,** 2287, 1924).

Vegard, L. (1924b). *Leiden Comm. No.* **175;** The luminescence from solidified gases down to the temperature of liquid hydrogen and its application to cosmic phenomena.

Vegard, L. (1926). *Ann. phys.* **79,** 377. The luminescence of solidified gases and its relation to cosmic processes. (*Chem. Abstr.* **20,** 2283, 1926).

Vegard, L. (1930). *Ann. phys.* **6,** 487. The spectra of solidified gases and their theoretical atomic meaning. (*Chem. Abstr.* **25,** 29, 1931).

Vegard, L., and Keesom, W. H. (1927). *Leiden Comm. No.* **186,** On the luminescence produced by bombarding solidified gases with electric rays at the temperature of liquid helium.

Wall, L. A., Brown, D., and Florin, R. E. (1959a). *J. Phys. Chem.* **63,** 1762. Atoms and free radicals by γ-irradiation at 4.2°K.

Wall, L. A., Brown, D. W., and Florin, R. E. (1959b). *J. Chem. Phys.* **30,** 602. Electron spin resonance spectra from gamma-irradiated solid nitrogen.

Whittle, E., Dows, D. A., and Pimentel, G. C. (1954). *J. Chem. Phys.* **22,** 1943. Matrix isolation method for the experimental study of unstable species. (*Chem. Abstr.* **49,** 2158, 1955).

5. Low Temperature Equipment and Techniques

F. A. MAUER

*National Bureau of Standards, United States Department of Commerce,
Washington, D. C.*

I. Introduction

A. METHOD OF TRAPPING RADICALS

A brief review of the methods available for producing and trapping free radicals will serve to introduce the low temperature apparatus and techniques used in this type of research.

Broida (1957) groups the methods into two categories: (1) those in which the radicals are frozen into a solid after preparation at high temperature; and (2) those in which the radicals are produced directly in the solid at low temperature. In methods of the first type, formation of radicals in the gas phase may be brought about by electric discharge, high temperature, or chemical reaction. Stabilization is accomplished by freezing the products on a cold surface to immobilize the active species and prevent recombinations. In methods of the second type radicals are produced in the cold solid directly by irradiation with gamma rays, X-rays, or ultraviolet radiation; or by bombardment with electrons, neutrons, or protons.

Although the methods differ as to the way in which free radicals are produced they all require low temperatures for preserving the unstable chemical species. The temperature depends on the species and matrix to be investigated: some are stable at room temperature, others only at the temperature of liquid helium. In general, the temperature required is considerably lower than the melting point of the matrix in which the radicals are to be frozen. For the simplest systems (resulting from discharges through water vapor, oxygen, nitrogen, and hydrogen) temperatures of 4.2°K and lower are required.

Maintenance of these low temperatures and their use in preserving an unstable species under conditions which permit determinations of its physical and chemical properties are essential to much of the research on free radicals. Vacuum techniques for handling volatile samples, for providing thermal isolation in apparatus containing refrigerants, and for reducing the pressure of gas in the discharge tube are a necessary adjunct. In addition, determinations of temperature and pressure are nearly always required.

B. Apparatus

The apparatus used in a typical system for trapping and studying a solid containing free radicals is shown diagrammatically in Fig. 1. It can take many forms and incorporate any or all of the components commonly used in the vacuum manipulation of volatile compounds (Sanderson, 1948). In addition, it will ordinarily include (1) a cryostat specially equipped for making observations and measurements on solids condensed on a cold surface, and (2) a microwave cavity or some other device for producing free radicals either in the gas stream or in the solid condensed on the cold surface. The system shown in Fig. 1 is by no means the least complex, having several components that are not essential in many types of work. A discussion of the function of various components in this system will aid in understanding design requirements and techniques to be considered later.

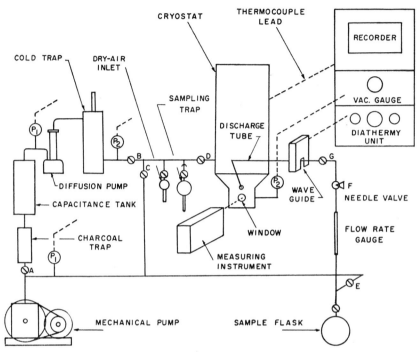

FIG. 1. A typical system for depositing and studying a solid containing free radicals.

1. Cryostat

This component will be considered in detail in Section II. Essential features are (1) a tank containing a refrigerant such as liquid nitrogen, (2) a vacuum space around this tank to decrease the rate of heat transfer to the refrigerant, (3) a cold surface on which vapors can be condensed, and (4) provisions for observing the solid and measuring its properties.

2. Measuring Instrument

Most of the observations on solids containing free radicals are made from outside the Dewar (cryostat), and the information must be transmitted as an electrical analog or as some form of electromagnetic radiation. The eye has long been an important though poorly calibrated receptor. All types of spectroscopy offer ways of making observations, as do the methods of X-ray and electron diffraction. Magnetic properties can be observed through the walls of the Dewar or by means of suitable coils inside, and thermal effects can be fairly readily converted to electrical signals for measurement by external instrumentation.

3. Vacuum System

A number of components are required to produce the low pressures and contamination-free atmospheres required in free radical research. A mechanical fore pump is used for initial evacuation of the system and to provide the low discharge pressure required for normal diffusion pump operation. The mechanical pump is occasionally used for rough-pumping a portion of the system before that portion is connected to the high vacuum line. During rough-pumping of large volumes the pressure at the mechanical pump inlet becomes too high to sustain its normal function of backing the diffusion pump. For this reason, a vacuum capacitance tank is sometimes provided to increase the time during which the mechanical pump can be used for other purposes without interfering with normal diffusion pump operation. A diffusion pump is needed if initial pressures lower than 10^{-3} mm Hg are required. The rate at which mechanical pumps are able to remove residual gas drops rapidly with decreasing pressure, and a "diffusion" or "condensation" pump is needed to maintain adequate pumping speed. A Pirani or thermocouple gauge (P_1) in the line between diffusion pump and fore pump is occasionally useful for determining whether the pressure at the diffusion pump discharge is being kept below the limit for normal operation. A charcoal or copper foil trap is needed only for research in which contamination by hydrocarbons must be avoided. A cold trap is used in nearly all systems to prevent diffusion of volatile components of pump oil into the vacuum system and to protect the pumps from contamination by vapors introduced into the vacuum chamber. A cold-cathode ionization gauge (P_2) in the line between the cold trap and the Dewar shows when the pressure in the pumping system has been reduced below that in the Dewar and is also useful in finding leaks by the technique of isolating various sections and observing the pressure rise. Valves in the fore-pump line permit the pump to be isolated from the diffusion pump while it is used to rough-pump the Dewar or to evacuate the sample train after a new flask has been attached. A sampling trap in the Dewar pumping line can be used to collect for analysis samples of the gases evolved during warm-up of the Dewar. It can also be used to remove volatile material from the system without contaminating the pump oil and polluting the atmosphere. The dry-air inlet permits bringing the pressure in the pumping line up to 1 atmosphere for ease in removing the sampling trap or in opening the cryostat. This inlet is equipped with a tube filled with Drierite or other suitable drying agent. Even better, a source of dry nitrogen may be connected.

4. Sample Deposition Train

This portion of the apparatus varies considerably in individual installations, and the arrangement shown is one of the simplest. The gas sample, having been prepared elsewhere, is brought to the apparatus in a 1-liter sample flask at a pressure of 1 atmosphere. The flow of gas into the discharge tube and thence into the Dewar is regulated with a micrometer-type needle valve, and the rate of deposition is measured with a flow gauge. The discharge tube that passes through the waveguide should be of fused silica or Vycor, as Pyrex requires cooling when exposed to the electrical energy.

5. Microwave-Discharge Equipment

Either a tunable closed resonant-cavity type of waveguide or a simple, cup-shaped director may be used for concentrating the microwave radiation on the discharge tube. A typical source of microwave power for the discharge is a Raytheon 125-watt diathermy unit operating in the 2400- to 2500-Mc band. The choice of suitable components and experimental conditions for the production of free radicals by discharge techniques has been considered in Chapter 3.

6. Vacuum Gauge and Temperature Recorder

A cold-cathode ionization gauge has been found suitable for measuring pressure in free radical research because it operates over a considerable range and is nearly immune to destruction by exposure to corrosive gases. Temperature is usually measured with a thermocouple, and a potentiometer is required for determining the thermocouple electromotive force (emf). A recording potentiometer can be used, and if a multipoint instrument is available, the output of the vacuum gauge can be recorded on the same chart with the temperature.

This survey of components is intended only as an introduction to a typical system for carrying out low temperature research on solids containing free radicals. Individual components and techniques will be considered in more detail in the sections to follow.

II. Design of Research Cryostats

An apparatus within which a constant low temperature may be maintained for long periods is termed a cryostat. In practice, the device is usually a modified Dewar flask containing a liquid refrigerant boiling under constant pressure. Double walls, with the surfaces made highly reflective and the space between evacuated, are used to reduce the flow of heat into the inner flask and increase the period during which the temperature remains constant.

Vacuum-jacketed vessels for liquid hydrogen and liquid helium are usually mounted in larger containers of liquid nitrogen to lower the temperature of the outer vacuum wall and further reduce the flow of heat into the primary refrigerant. The resulting "double Dewars" are very efficient for storing low-boiling liquids, the evaporation rate for helium being as low as 1% per day in good storage Dewars.

A. General Design Considerations

1. Types of Research Cryostats

The problem in designing a research cryostat is to make provisions for keeping the specimen in thermal contact with the primary refrigerant while observing or measuring some property at low temperatures. Two types of cryostats are widely used in free radical research.

The Pyrex glass Dewar described by Schoen et al. (1958) and used in early spectroscopic studies of condensed discharge products is shown in Fig. 2 as an example of the first type. The specimen is introduced into the vacuum space and allowed to condense on a finger, T, at the bottom of the liquid helium vessel. By the substitution of various specimen deposition surfaces and windows, this cryostat can be modified for many different types of work. It is popular in research that depends on visual observations, spectroscopy, X-ray and electron diffraction, electron spin resonance, and other methods that require the specimen to be exposed so that it can be observed from outside the cryostat by means of some kind of radiation or field.

A schematic diagram of apparatus described by Fontana (1958) is shown in Fig. 3 as an example of the second type of cryostat. It consists of a thin-walled container surrounded by a liquid refrigerant. Gaseous samples can be brought into the well thus formed through a vacuum-jacketed tube and allowed to condense on the inner surfaces. This type of cryostat is not well adapted to visual observations or spectroscopy because of the several surfaces and liquids that are interposed between the specimen and the instrument. It is preferred for calorimetry and for measurements that are made remotely by electrical and magnetic methods. The uniformity of temperature throughout the specimen chamber is an advantage and permits observations to be made in the presence of vapors that would cause excessive heat transfer to the specimen surface in the "cold-finger" type of cryostat.

2. Types of Construction

Research cryostats can be made from either glass or metal, and although the two types of construction have little in common, a cryostat to perform a given function can usually be designed for fabrication by either method. The choice between glass and metal construction is likely to depend on the availability of skilled craftsmen and shop facilities.

The fabrication of large glass Dewars requires the services of a skilled glass blower equipped with a glass-blowing lathe designed to handle tub-

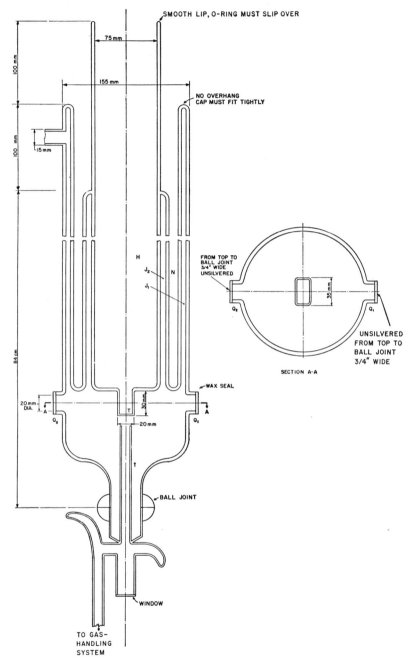

FIG. 2. A "cold finger" glass Dewar described by Schoen *et al.* (1958) and used in early spectroscopic studies of condensed discharge products. The liquid helium reservoir, H; liquid nitrogen reservoir, N; finger, T; quartz windows, Q_1 and Q_2; and vacuum jackets, J_1 and J_2, are designated.

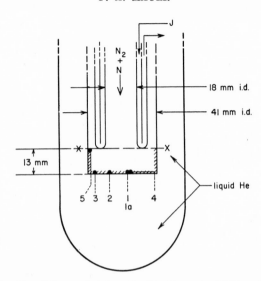

FIG. 3. The "cold well" cryostat described by Fontana (1958) and used in thermo-metric studies. Thermocouple positions, *1–5*; heater, *H*; and warm helium gas, *J*; are indicated. Line *XX* is the maximum level at which solid deposits.

ing up to 6 inches in diameter. If these are available, the fabrication of De-wars from Pyrex glass offers distinct advantages of economy and simplicity of construction as well as almost unlimited flexibility of design. Visibility of liquid refrigerants makes such Dewars convenient to use and contributes to their popularity. A number of types of Pyrex glass research cryostats are available commercially.[1]

Machine shop facilities are more generally available, and the decision to use metal construction is often based on this consideration. Furthermore, factors such as immunity from breakage, better temperature uniformity, and the ease with which certain mechanical features may be incorporated, often favor metal construction. At least one American firm is marketing a line of metal research cryostats.[2]

3. Heat Leakage into Dewar Vessels

Wexler (1951) and Sydoriak and Sommers (1951) independently studied the factors involved in the design of storage containers for liquid helium. Wexler developed a metal Dewar having a loss by evaporation of only 1% per day, and his design has been adopted by manufacturers of storage con-tainers. Sydoriak and Sommers, working with glass Dewars, arrived at simi-lar conclusions about the principles of good design and achieved a minimum loss of 5% per day.

[1] H. S. Martin & Son, Evanston, Illinois.
[2] Hofman Laboratories, Inc., Hillside, New Jersey.

Helium evaporation rates encountered in cryostats used for free radical research are often fifteen or more times as great as these. The dominant heat leak is that caused by thermal radiation through openings in the nitrogen-cooled radiation shield, and it can be shown that, compared to this, the factors that determine the ultimate heat leak into storage Dewars are relatively unimportant. Nevertheless, these must be considered.

Wexler studied the flow of heat into a spherical liquid helium container suspended in a vacuum space by means of a thin-walled neck tube. The outer walls and the top of the neck tube were cooled by liquid nitrogen. Of the various causes of heat flow into such a vacuum-insulated container those that depend on residual gas in the vacuum space can be disregarded; vapor pressures of all substances with the exception of hydrogen and helium are negligible at liquid helium temperature. If there are no supporting members across the vacuum space, the only factors of importance are radiation across the vacuum space, conduction down the neck tube, and radiation through the neck aperture. The latter can be effectively eliminated by installing a suitable baffle or by blackening the inside of the neck in the zone cooled by liquid nitrogen.

a. Conduction. Conduction down the neck tube is greatly affected by heat exchange with the effluent gas, and in a well-designed Dewar the heat capacity of helium gas rather than the small latent heat of evaporation of the liquid is used in removing heat that flows down the neck. Wexler carried out calculations to show that in the absence of other sources of heat, the loss due to conduction down the neck is reduced by a factor of 10 if perfect heat exchange with the effluent gas is achieved. In practice, this condition is approached and neck conduction losses are very small.

As heat transfer by radiation increases, the evaporation rate does not increase proportionately because the cold gas passing up through the neck removes heat that would otherwise reach the liquid. Neck conduction losses decrease rapidly and become negligible when normal boil-off supplies sufficient cold gas to remove all of the heat that flows down the neck. However, the important point brought out by this analysis is that, if the neck tube is too heavy, conduction losses may remain high, so that reducing losses due to radiation only results in an increase in those due to conduction.

Heat exchange with the cold gas resulting from normal radiation losses was found to be sufficient to eliminate conduction losses in Wexler's storage Dewar, which had a neck made of hard-drawn Inconel tubing $\frac{5}{8}$ inch o.d. by 8 inches long with a wall thickness of 0.0105 inch. Tubing of this size is more than adequate for suspending the helium container in research cryostats, and since the helium evaporation rate in most of these exceeds that in Wexler's storage Dewar by a factor of from 4 to 15, heat exchange with the effluent gas entirely eliminates heat conduction down the neck tube. In fact, excess gas usually escapes at a temperature far below that of the room and a larger neck opening may be practical if steps are taken to use fully the cooling effect of this gas. If pumping on the helium reservoir to achieve temperatures below 2°K is planned, a large neck diameter is an advantage and, since conduction losses are greatly reduced below 4°K, no increase in evaporation loss is likely to be encountered. Schoen *et al.* (1958) report that when a Dewar similar to that shown in Fig. 2 was used to achieve temperatures of around 1.2°K by pumping on the helium reservoir, the loss rate, which was 6 cc per minute at 4.2°K, was reduced to 2.8 cc per minute. This reduction is believed to be due to the sharp decrease in the thermal conductivity of the glass below 4°K (Powell and Blanpied, 1954).

It should be noted that conductive paths where there is no heat exchange with the effluent gas have a very serious effect on evaporation losses. Wexler estimated that the evaporation rate caused by conduction down the neck tube he used would have been 770 cc per day if there had been no heat exchange. If supports are essential, the use of Pyrex spheres or multiple-interface supports in the form of stacks of thin metallic plates should be considered (Mikesell and Scott, 1956).

b. Radiation. The effect of radiation across the vacuum space can be estimated from the simplified equation given by Wexler (1951),

$$\frac{dH}{dA} = e\sigma T_0^{\ 4} \tag{1}$$

where dH/dA is the radiant energy per unit area absorbed by a cold body of emissivity $e \ll 1$ in a black-body enclosure at temperature T_0. (It is assumed that the temperature of the cold body, T, satisfies the condition $T^4 \ll T_0^4$.) The Stefan-Boltzmann constant σ has the value 5.72×10^{-12} watt cm^{-2} deg^{-4}. If the enclosure has a reflective inner surface with an emissivity equal to that of the body, the equation becomes

$$\frac{dH}{dA} = \frac{e}{2}\,\sigma T_0^{\ 4} \tag{2}$$

These equations can be used to estimate the effect of heat transfer by radiation under various conditions. It is convenient to express the heat transfer directly in terms of the volume of liquid helium evaporated. Taking 4.95 cal/gm and 0.125 gm/cm^3 for the heat of vaporization and density of liquid helium (Table II), Eq. (1) becomes

$$\frac{dV}{dA} = 1.3 \times 10^{-10}\, eT_0^{\ 4}\ \text{cc/min/cm}^2 \tag{3}$$

This equation is used to calculate evaporation loss rates under various conditions. The results are summarized in Table I.

TABLE I

HELIUM EVAPORATION RATE FOR HEAT TRANSFER BY
RADIATION UNDER VARIOUS CONDITIONS[a]

T_0 (°K)	e_i	e_0	dV/dA (cc/min/cm^2)
300	1.	1.	1.
	0.02	1.	0.02
	0.02	0.02	0.01
77	1.	1.	5×10^{-3}
	0.02	1.	9×10^{-5}
	0.02	0.02	5×10^{-5}
20	1.	1.	2×10^{-5}
	0.02	1.	4×10^{-7}
	0.02	0.02	2×10^{-7}

[a] T_0 is the temperature of surrounding surfaces, e_i is the emissivity of the surface of the helium container, e_0 is the emissivity of surrounding surfaces, and dV/dA is the evaporation rate per unit area of exposed surface.

The table shows that under the most unfavorable conditions (radiation from a black body at room temperature to a black body at liquid helium temperature) the estimated evaporation rate is 1 cc/min/cm^2. This case is of interest because it approximates the condition often encountered in research Dewars where radiation from glass or other low-emissivity surfaces at room temperature falls on an opening in the nitrogen-cooled radiation shield. Nearly all of the radiation entering the opening is absorbed and the resulting evaporation rate is about 1 cc/min for each square centimeter of opening. Schoen et al. (1958) have shown that the actual performance of glass research Dewars having various openings in the liquid nitrogen-cooled shield is predicted rather accurately on this basis. Radiation through such openings is evidently the dominant factor determining evaporation losses.

If the black-body radiation from room temperature surfaces falls on a highly reflective surface ($e = 0.02$), the evaporation rate is only 2% as great, and if the room temperature surfaces are of equally low emissivity, the evaporation rate is diminished by another factor of 2. Thus, at a given temperature the transfer of heat by radiation is reduced by a factor of 100 by the use of suitable reflecting surfaces. Lowering the temperature of surrounding surfaces from 300° to 77° K and from 300° to 20° K results in reduction of the evaporation rate by a factor of 200 in the first case and 50,000 in the second. For a research Dewar with a copper helium vessel having a capacity of 1.5 liters and a surface area of 750 cm^2 inside a high-reflectivity copper radiation shield cooled by liquid nitrogen, the evaporation rate (if there were no openings in the radiation shield) would be about

50 cc per day. The fact that research Dewars often have evaporation rates that exceed this by a factor of 100 proves that the effect of other types of heat leaks is so overwhelming that unless these are eliminated very little can be gained by using better reflecting materials or additional radiation shields. For example, a glass viewing port at room temperature with an area of 1 cm² radiating to a highly absorbing specimen at 4°K may contribute 1 cc per minute to the evaporation rate. This is about thirty times the normal radiation-loss rate in a similar Dewar without openings.

c. Gaseous heat transfer. Although heat leaks due to conduction and convection in residual gases in the vacuum space can be disregarded in the case of helium storage containers, these factors are of considerable importance in research Dewars used in studying gaseous samples.

The conductivity of gases is negligible at pressures below 2×10^{-5} mm Hg, but rises rapidly above 2×10^{-4}. Above about 10^{-1} mm Hg there is no change with pressure.

The conductivity of different gases at the same pressure varies by a factor of nearly 10, low molecular weight gases having high conductivity. However, Fulk *et al.* (1957) found that residual oxygen and nitrogen in a vacuum-insulated vessel may transport more heat from a copper wall at 300°K to a silver wall at 76°K than would hydrogen or helium at equal pressure. The low effective conductivity of helium and hydrogen is attributed to the low accommodation coefficient for these gases at 76°K.

B. Detailed Design

The problem of designing a research cryostat is best approached by considering the design of a number of cryostats which have been used successfully. It is often practical to adapt one of these with only minor modification or, at least, to pick up valuable ideas.

1. Typical Research Cryostats

a. Glass cryostats. Schoen *et al.* (1958) have prepared an excellent review of glass Dewars used for optical studies. One of these was used earlier as an example of a simple cold-finger cryostat. Another is shown in Fig. 4. It was designed for studies, by absorption spectroscopy, of short-lived intermediates produced in a high frequency electric discharge and illustrates most of the features that have been incorporated in glass Dewars.

A drawing of this Dewar is shown in Fig. 5. One of the important features of this design is a movable and removable liquid helium reservoir with an infrared-transparent cold surface affixed to it. Standard-taper and ball-and-socket ground joints are used to achieve flexibility and interchangeability of accessories.

The central helium finger, *H*, is insulated by a separate vacuum jacket,

FIG. 4. Apparatus for absorption spectroscopy at low temperatures showing (1) liquid helium Dewar, (2) infrared source, (3) infrared monochromator, (4) sample flask, (5) calibrated leak, (6) discharge tube, (7) diathermy unit, (8) diathermy director, and (9) Tesla coil. A drawing of the Dewar is shown in Fig. 5.

V_1, which terminates in the ball section of the large, 102/75, ball-and-socket joint. This joint remains at room temperature and permits rotation of the helium Dewar around a vertical axis during experimentation. The glass bellows permits expansion and contraction during temperature

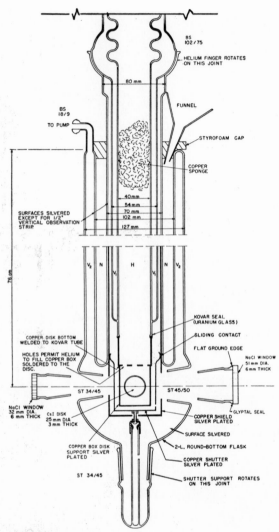

FIG. 5. Drawing of the glass Dewar used for absorption spectroscopy (Schoen *et al.*, 1958). The helium reservoir, H; nitrogen reservoir, N; and vacuum spaces, V_1 V_2; are indicated.

changes. A stopcock is provided to permit periodic evacuation of the vacuum jacket to remove helium that diffuses through the glass.

The tip of the helium finger terminates in a Kovar[3] metal cup to which has been soldered a narrow, thin-walled copper box with a 1-inch diameter tube through its center. The seal, which involves a graded sequence of

[3] Stupakoff Ceramic and Mfg. Co., Latrobe, Pennsylvania.

Pyrex-to-uranium-to-Nonex-to-Kovar sealing glass, has withstood the thermal shock of cooling to 4°K many times. Commercial glass-to-metal seals utilizing beryllium or Nonex glass alone have not proved satisfactory in this application. The copper box is connected to the helium reservoir by several holes in the Kovar cup. The tube through the box is internally threaded and equipped with copper washers and screw rings so that disks of various window materials can be installed. The window can be turned normal to any port for deposition or observations.

The base of the Dewar is fashioned from a 2-liter round-bottom flask with female sections of standard taper joints set in at 90-degree intervals. This arrangement permits the use of various plug-in windows, observation ports, wire leads, and gas carriers. This not only results in considerable flexibility, but permits removal of hygroscopic windows for storage in dessicators.

A copper shield, silver plated and highly polished, is cemented to the walls of the nitrogen reservoir by means of a metal conductive lacquer.[4] The large openings for the infrared beam are responsible for a major heat leak, and provisions have been made for covering them by means of a silver-plated copper shutter operated by rotating the standard taper joint at the bottom of the Dewar.

Additional measures to reduce the heat leak include (1) silvering the inside of the outer vacuum wall and the outside of the nitrogen container, except for vertical, diametrically situated observation strips $\frac{1}{2}$ inch wide, and (2) inserting a copper sponge in the neck of the finger just above the surface of the liquid helium.

A removable cap (not shown in Fig. 5) makes the helium reservoir a closed system from which air can be excluded to prevent frosting of the inner surfaces.

About 500 cc of helium is needed to lower the temperature of the finger to 4°K, after this a charge of 800 cc is added. The evaporation rates (determined principally by the size of openings in the nitrogen-cooled shield) are 22 cc per minute without copper sponge or shutter, 16 cc per minute with sponge but without shutter, and 4 cc per minute with both sponge and shutter. Measurements can ordinarily be continued for 45 to 60 minutes without refilling.

One feature not incorporated in this design is the use of a standard taper joint at the lower end of the helium vessel to permit the use of interchangeable specimen mountings. This feature is incorporated in another of the glass Dewars described by Schoen *et al.* A 34/45 female joint is mounted in the bottom of the helium reservoir with the large end up so that the inserts carrying the various specimen mounts can be removed through the

---

[4] Dag Dispersion No. 235, Michigan Colloids Co., Port Huron, Michigan.

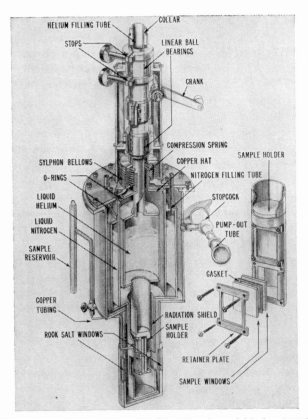

FIG. 6. The metal optical cell described by Duerig and Mador (1952).

top of the Dewar by using a hook. The ground-glass joint in contact with liquid helium is sealed with a soap-glycerine mixture (Hudson, 1958).

b. Metal cryostats. As an example of a metal cryostat, the optical cell of Duerig and Mador (1952) is shown in Fig. 6. It is a portable unit slightly over 6 inches in diameter and 34 inches tall. Heat leakage is very low: a liter of helium lasts about 20 hours and 1.5 liters of nitrogen lasts about 12 hours. A novel feature is the Sylphon bellows mounting of the helium vessel which permits the whole assembly, including sample holders, to be moved as much as 2 inches in a vertical direction. This permits either of two samples or a blank to be aligned with the beryllium X-ray windows for irradiation or the rock-salt windows for spectral absorption measurements.

The low heat leak characteristics of this cryostat result from (1) the use of a long, thin-walled neck tube of low-conductivity alloy as the sole supporting member for the helium vessel, (2) reduction of openings in the radiation shield to the minimum required, (3) the use of a baffle in the helium filling tube to exclude radiation, and (4) the use of copper plating on vacuum walls to reduce the emissivity of these surfaces. Suspension of the nitrogen tank on two small-diameter filling tubes and two supporting rods decreases the flow of heat from the top plate and results in a saving in liquid nitrogen as well as the elimination of condensation during humid weather.

A novel liquid helium research cryostat recently marketed by Hofman Laboratories, Inc. does not require the use of any other refrigerant for shielding the liquid helium vessel. With respect to the helium loss rate, which is 1.4 cc per minute, the performance of this cryostat is superior to that of most liquid nitrogen-shielded cryostats. This excellent performance is achieved by surrounding the helium vessel with thermal radiation barriers one or more of which are cooled by heat exchange with cold helium gas from normal boil off.

Another novel device described by Littauer (1958) and originally used in making liquid hydrogen and deuterium targets for high energy accelerators appears to offer some features that could be applied with advantage in free radical research. The target cup is mounted on top of a vertical vacuum-insulated tube that extends into the storage Dewar. The helium gas used to pressurize the storage Dewar is connected to a bubbler filled with silicone oil so that the pressure exceeds 1 atmosphere by an amount equal to the hydrostatic pressure of oil in the bubbler. This is easily regulated to cause the refrigerant to rise to any desired level. A cryostat patterned after this device could be quickly filled and emptied for convenience in experiments requiring frequent warm-up and quench cycles. The supply of liquid could be as great as 75 liters and storage losses would presumably be low. It is quite possible that by depositing the specimen on a long, vertical rod with the top end attached to a liquid nitrogen reservoir and the bottom end dipped in liquid helium (the height of which would be adjustable), it would be possible to vary the temperature of the specimen over fairly wide limits by shifting the temperature gradient in the rod. The device described by Janes et al. (1956) makes use of a stream of cold helium gas to liquefy hydrogen in the target. In this apparatus, temperature control can be effected by controlling the flow of cold gas.

2. Special Fittings

The examples given above have introduced a number of devices used to adapt cryostats for particular types of research. Because of the importance of this problem, a few fittings fulfilling needs that arise frequently are described below.

a. Demountable flanges. Designing a cryostat in such a way that it can be readily disassembled for cleaning, repair, or modification generally increases its utility. Glass apparatus is made demountable by the use of standard-taper and ball-and-socket ground-glass joints as described in the examples above. Additional devices have been described in books on high vacuum technique (e. g., Yarwood, 1955).

Demountable flanges in metal systems are usually made vacuum tight by the use of rubber or neoprene O-ring gaskets. The use of these is too well known to require discussion here. Recommended dimensions for O-ring grooves are given in the manufacturers' data sheets.

In order to achieve a vacuum better than 10^{-7} mm in metal systems, the walls must be baked out at 300°C or higher to eliminate adsorbed contaminants. This cannot be done if rubber O-rings are used. Moreover, certain components of the rubber or neoprene in O-rings have an appreciable vapor pressure and, in very high vacuum systems, this may place a limit on the ultimate pressure attainable (Young, 1958, 1959). For these reasons metal gaskets are used where very low pressures and extremely clean atmospheres are required. Methods of obtaining a good seal have been given by Van Heerden (1956), Higatsberger and Erbe (1956), Foote and Harrington (1957), and Spees et al. (1957).

b. *Windows.* Removable windows fitted into standard-taper ground-glass joints have been described by Schoen et al. (1958). The windows are usually cemented into the glass with baked Glyptal varnish. A method of sealing windows that permits expansion during bake-out or contraction during cooling has been described by Schwarz (1955). Chasmer et al. (1951) described sapphire-to-glass seals used in long-wavelength vacuum photocells. Mueller (1956) used an O-ring seal around a rectangular opening in a curved surface to seal beryllium windows into an X-ray camera. The same problem was solved by Black et al. (1958) by using clamping rings with baked Glyptal varnish as a sealant. The need for a transparent window in a vacuum chamber was eliminated by Louckes (1957) through the use of a flexible curtain held against an O-ring in such a way that the curtain could move across the window opening with the detector.

c. *Sample-deposition surfaces.* If the optical properties of the surface on which the sample is to be deposited are not important, copper is usually selected because of its high conductivity. Powell et al. (1957) showed that the conductivity varies considerably with purity. The purest sample had a conductivity coefficient at 4°K that exceeded that of the best commercial grade (coalesced-annealed) by a factor of 10. However, the treatment received during fabrication is often the factor that most affects the conductivity.

For research requiring transparency in the visible region, Pyrex glass makes a convenient specimen substrate. It is also useful (because of its low thermal conductivity) in apparatus designed to permit measurements of specimen temperature (Fontana, 1958).

The various surfaces used in absorption spectroscopy pose rather difficult mounting problems, especially if the window is to be in contact with liquid refrigerants. Schoen *et al.* (1958) describe a quartz liquid helium finger joined to a Pyrex flask by means of graded seals. In a similar apparatus for use in infrared spectroscopy the difficult problem of sealing cesium iodide to glass or metal was avoided by mounting the disk-shaped window between screw rings and washers in an inside-threaded copper tube passing through the liquid helium finger.

A permanent vacuum seal for joining special materials to glass or metal tubes having different thermal expansion properties has been designed by Schwarz (1955). Waters (1958) sealed a window directly into his liquid air container and used it successfully at 80°K.

Infrared windows for a helium cryostat were made by Warschauer and Paul (1956) from polyethylene film. This was stuck to the edges of a hole in the helium container with a thin film of silicone high vacuum grease and clamped with a heavy metal frame. It is claimed that the polyethylene film will withstand a pressure differential of 1 atmosphere in contact with liquid helium without leaking.

d. Electrical leads. Prefabricated glass-to-metal seals are used to provide electrical leads in both glass and metal vacuum systems. However, the need for making glass-to-metal seals in the laboratory still arises occasionally. The scientific basis for glass-to-metal sealing techniques was established by Hull and Burger (1934) and Scott (1935). Detailed descriptions of modern materials and techniques can be found in the M. I. T. "Tube Laboratory Manual" edited by Rosebury (1956). Strong (1938) has described the procedure for sealing platinum, tungsten, copper, Kovar, and Fernico[5] into glass.

Platinum can be sealed to soft glass only, and this restriction, plus the high cost has limited the use of these seals.

Tungsten wires of small diameter (<0.060 inch) are often sealed through Pyrex directly. Larger wires must first be sealed into a sleeve of Nonex glass which is, in turn, sealed to Pyrex.

Copper may be sealed to Pyrex by techniques developed by Housekeeper (1923). The metal has a large coefficient of thermal expansion, and success in sealing it to glass depends on the use of a design which permits a thin, ductile section to deform.

Kovar and a number of other alloys have been developed to match the expansion characteristics of glasses. Pieces 4 inches or more in diameter may be sealed to appropriate glasses without difficulty.

Hull and Burger (1934) and Scott (1935) have cautioned that phase transformations in both Kovar and Fernico make these alloys unsuitable for sealing to glass for use at temperatures below −40°C. Schoen *et al.*

[5] General Electric Co., Schenectady, New York.

(1958) have reported successful use of a graded Pyrex-to-uranium-to-Nonex-to-Kovar seal that is frequently cooled to liquid helium temperature.

Electrical leads for metal Dewars may be of the Kovar metal-glass-metal type, but in a few applications requiring currents in excess of 15 A or voltages in excess of 1 kv there may be a need for special terminals. Strong (1938) has described two basic types that may be adapted to fill special requirements.

Thermocouple leads, where they pass through the wall of a Dewar, should not be cut and joined to another metal because of the danger of thermal emfs caused by temperature gradients between the inner and outer junctions. In fact any lead in a measuring circuit where small parasitic voltages cannot be tolerated should be passed through a hole and sealed with wax. If a number of leads are involved, the method of Noggle et al. (1957) for casting the leads in a plastic flange may prove convenient.

e. *Mechanical linkages.* The need for transmitting force or motion through the walls of a vacuum chamber or cryostat occasionally arises, and a suitable mechanical linkage must be provided. Two types have been mentioned in the description of research cryostats: the standard-taper glass joint and the Sylphon bellows. Additional examples are given by Yarwood (1955).

Standard-taper joints or stopcocks, if well lubricated, provide a sealed shaft that can be rotated slowly without causing leaks. The rotary motion is used directly or converted to linear motion.

Sylphon bellows such as those used in Fig. 6 are especially suitable for transmitting linear motion because no lubricating materials that might contaminate the atmosphere are required.

Shafts made vacuum tight by the use of O-ring seals can be made to slide as well as rotate and are sometimes used to transmit motion in two dimensions.

Ball-and-socket joints, which permit limited rotation about two axes as well as unlimited rotation about a third, can be used to provide motion in three dimensions. A screw converts rotation about the third axis to translation parallel to it. A system for transmitting three-dimensional motion in cylindrical coordinates through a vacuum seal has been described by Smotrich and Feldman (1956).

A vacuum-tight seal for a shaft that rotates continuously at high speed was developed by Taylor (1956) for use with a rotating-anode X-ray tube and may find use in trapped-radical research.

Force or displacement inside a cryostat is often provided by using magnetic coupling through the vacuum wall or by installing an electromagnet, a solenoid, a small motor, or a milliammeter movement. Bourdon gauge movements, remotely operated by fluid pressure, have been used to provide very small but accurately reproducible displacements.

These are but a few of the devices that have been described in the literature, but they are believed to represent the basic types.

C. Construction Methods and Materials

1. Glass Dewars

The material used in making glass Dewars, at least in America, is ordinarily Pyrex. The working temperature required is high but, because of its low thermal expansion and high strength, Pyrex is more easily annealed than soft glass and is better able to withstand thermal shock. One undesirable characteristic is its high permeability for warm helium gas and the resulting need for frequent re-evacuation.

Annealing of glass Dewars is hardly more critical than the annealing of other large glass apparatus. On cooling from the working temperature, nine-tenths of the contraction occurs above room temperature, and the additional stresses that develop on cooling to liquid helium temperature are probably not severe.

The process of silvering internal surfaces is an art that few glass blowers have mastered. Anyone planning to fabricate a glass Dewar is likely to have to undertake this phase of the work himself. The Brashear process which is used has been described by Strong (1938). Application of the process to the silvering of Dewar flasks is discussed by Scott et al. (1931).

Much of the difficulty encountered in the silvering process is caused by failure to clean the surfaces adequately before the silvering solution is applied. Part of the trouble may be due to failure to maintain proper cleanliness during fabrication and annealing. Fingerprints and certain other types of dirt are almost impossible to remove if they are burned in.

The silvering is done upon completion of construction and annealing. The surface is prepared by using a potassium dichromate-sulfuric acid cleaning solution, followed by nitric acid. It is then slightly etched with hydrofluoric acid. After a thorough rinsing, the Dewar is mounted in a level position on its side and filled with the silvering solution to a level just below the half-way point. As soon as the solution is spent, the Dewar is drained, rinsed, and rotated 180°. The process is repeated, using a fresh solution. In this way, a strip is left unsilvered to permit observation of the contents of the Dewar.

The Dewar must be thoroughly cleaned and dried before fittings such as windows, the specimen deposition surface, and thermocouples are installed. Temperatures as high as 550°C may be used for outgassing. Finally, leak tests must be carried out to determine whether the Dewar is ready for use.

2. Metal Dewars

a. Materials. Copper and brass are the metals most used in the construction of cryostats, but Monel, Inconel, and stainless steel are used in applications requiring low thermal conductivity.

Because of the need for bake-out at temperatures of 300°C and higher if a vacuum of 10^{-7} mm Hg is to be attained, welded or hard-soldered joints are preferred.

Copper is easily silver soldered to produce joints that are leak free even at low temperatures. For this reason, it is preferred for the liquid helium and liquid nitrogen containers. These are made by butting tubing of the required diameter against oversized end plates which are turned down after soldering. For a Dewar having a helium capacity of 1 liter, these cylindrical containers should have a wall thickness of at least $\frac{1}{16}$ inch with $\frac{1}{8}$ inch-thick end plates. The critical collapsing pressure of cylinders can be calculated using formulas given by Strong (1938).

Because of its high thermal conductivity and low emissivity, copper is ideal for thermal radiation shields. A nitrogen-cooled shield 0.040 inch thick surrounded by copper-plated surfaces at room temperature will maintain a temperature that differs by only a few tenths of a degree from the temperature of the nitrogen reservoir.

Copper, silver, or gold is used for plating interior surfaces to reduce heat transfer by radiation. In the past, these surfaces were polished, but Blackman *et al.* (1948) have reported that a freshly reduced, smooth copper surface is more efficient in diminishing heat transfer than one that is highly polished. In fact, they found that diligent polishing to produce a mirror-like surface resulted in a twofold increase in heat transfer by radiation. The infrared absorption of oxide films and low electrical conductivity of amorphous surface layers are offered as explanations.

Yellow brass is used for the outer vacuum wall and many fittings. For a cylindrical chamber 6 inches in diameter a wall thickness of $\frac{1}{8}$ inch is suggested. This may be increased to $\frac{1}{4}$ inch for larger sections. Care is required in silver-soldering brass, as excessive heating may cause loss of zinc and result in porosity.

The thermal emissivity of brass is more than twice that of copper, and a slight reduction in heat transfer can be made by copper-plating the inside of the brass chamber.

Several alloys are in common use for the filling tube that supports the liquid helium container. The essential requirement is low thermal conductivity at low temperatures. According to thermal conductivity data by

Powell and Blanpied (1954), Monel, hard-drawn Inconel, and 18-8 stainless steel are all about equally suitable. The tube should have an inside diameter of $\frac{5}{8}$ inch so that a commercially available vacuum-insulated transfer siphon ($\frac{1}{2}$ inch o.d.) may be used. A wall thickness of 0.010 to 0.015 inch is adequate. The length of the tube should be at least 8 inches if the top end is attached to the liquid nitrogen reservoir, or 16 inches if it is supported at room temperature. A baffle cooled by the effluent helium gas should be provided to remove heat entering the neck aperture by radiation. Use of similar filling tubes to suspend the liquid nitrogen reservoir will result in lower losses and will reduce the condensation of moisture during humid weather.

Any of the alloys suggested (Monel, Inconel, and 18-8 stainless steel) may be joined to copper or brass by silver soldering, but unless the solder flows properly on the first application, trouble in obtaining a leak-free joint may be encountered, particularly in the case of the stainless alloy. Austenitic stainless steels (such as 18-8) tend to precipitate carbides at grain boundaries when heated to the temperature required for silver soldering (about 750°C), and the operation must not be prolonged if porosity is to be avoided. Approved methods of soldering and welding are found in the catalog of the Superior Tube Co. (1957).[6]

Probably because of carbide precipitation, stainless alloys have sometimes been found to give trouble when used in making containers for low temperature liquids. Riddiford and Coe (1951), for example, found that cold traps made of welded 0.018 inch-thick stainless steel leaked badly when immersed in liquid air. Attempts to repair the leaks by soldering greatly increased the low-temperature leak rate.

b. Assembly. Before actual construction is started, the sequence of operations should be planned. It may develop that the design worked out on paper is not feasible because certain joints that have to be soldered or welded after partial assembly become inaccessible.

Wherever possible, components should be leak-tested before being built into the main assembly. This may necessitate the construction of special test manifolds, but considering the difficulties that can be caused by a leak in an inaccessible region, it is worth the trouble. If a mass-spectrometer type of leak detector is available, the apparatus should be checked piece by piece, helium gas under pressure being used wherever possible. Subassemblies should be tested at low temperatures by dipping them in liquid nitrogen.

If no other facilities are available, vessels may be pressurized to about 30 psi and submerged in water.

Before final assembly, all surfaces should be cleaned and dried to rid

[6] Superior Tube Co. (1957) Catalogue Section 21, Norristown, Pennsylvania.

them of grease and water. Even after this is done, outgassing will continue for some time and may be mistaken for leakage. If the apparatus cannot be heated, pumping must be continued for three weeks or more before ultimate vacuum is obtained. Filling the nitrogen and helium tanks with hot water and applying mild heat to the outer wall speeds the process of outgassing considerably. If it is necessary to have a very clean atmosphere and low pressure, before the helium is introduced it may be necessary to bake the cryostat at 300°C or higher under high vacuum.

III. Vacuum Equipment and Techniques

The vacuum system used with a cryostat need not be elaborate. Its complexity is determined largely by the extent to which contamination from residual gases in the Dewar and from pump oil must be eliminated.

A. Pumping Systems

Many types of low temperature chemical research can be carried out with a vacuum system consisting only of a good mechanical pump connected to the Dewar by a rubber vacuum hose. A pressure of 10^{-2} mm is attained by pumping, after which the Dewar valve is closed. Upon transfer of liquid nitrogen to the outer Dewar vessel, the pressure drops by about two orders of magnitude because of condensation of the principal constituents of the residual gas: water vapor and hydrocarbons (from pump oil). The final pressure is sufficiently low for most purposes.

In some types of work, such as absorption spectroscopy, condensation on the specimen surface causes serious trouble, and a diffusion pump with a good cold trap must be used to eliminate most of the residual gases.

A suitable pumping system may consist of a 2-inch water-cooled fractionating diffusion pump with a speed of 60 liters per second at 10^{-3} mm backed by a rotary pump with a speed of at least 6 liters per second at 100 μ. With a suitable liquid nitrogen cold trap this system should reach an ultimate pressure of 10^{-6} mm in a leak-free and well out-gassed system.

B. Residual Gases

The nature of the residual gases in a high vacuum system and methods of eliminating them have been studied by Blears (1951) and by Coe and Riddiford (1955).

Blears used a mass spectrograph and analyzed the residual gases during the initial evacuation, during bake-out, and at ultimate pressure. The stainless steel mass spectrometer tube was exposed to air for 3 hours, then evacuated. The air was removed comparatively easily. Thus, the partial pressure of oxygen was less than 10^{-7} mm after 20 minutes; less than 10^{-8} mm and still decreasing after 1 hour. Air accounted for less than 0.5 % of the material in the gas phase. About 70% of the ions were from water. Most of the remainder were from hydrocarbon vapors. The largest single line in the mass spectrum (apart from the lines of water) was due to CO, the concentration of which was 3% that of water.

The rate of efflux of these gases was less by a factor of 4000 than the calculated rate. Thus, though fast pumps can maintain slightly lower pressures than slow ones, it is the forces binding molecules to the surface which determine the rate of removal of matter from a vacuum system.

Analysis of residual gases after long pumping showed that the water content diminished but the total hydrocarbon content did not. Baking the spectrometer tube at 100°C while pumping through a trap cooled to −78°C helped to remove hydrocarbons, but the vapor pressure of water condensed on the cold trap became the limiting factor. Baking at the highest possible temperature was found to be the most effective procedure for diminishing the residual gas content.

The troublesome hydrocarbon contamination was found to be caused by back diffusion from the rotary pump during the initial stages of pumping. Since the diffusion pump must be turned off to prevent oxidation and the cold trap left unfilled to prevent condensation of water vapor, the use of an active-charcoal trap in the fore-pump line is suggested as a means of eliminating backstreaming of pump oil.

Pumping at any time without the protection of a cold trap is likely to result in contamination of the apparatus with hydrocarbons, and the use of a cold trap that will hold liquid nitrogen for more than a day (Kunzler, 1956) or an automatic cold-trap filling device is required if pumping is to be continued for a long period of time. As an alternative an active-charcoal trap may be used between the diffusion pump and the cryostat (Strong, 1938).

A more recent means of preventing the backstreaming of pump vapors is the copper-foil trap, which is supposedly effective even without cooling. Burns (1957) has described a convenient form of this trap with which he claims to have attained a pressure of 10^{-8} mm using a small metal pump and Octoil.

C. Vacuum Gauges

An excellent review of gauges used for measuring low pressures has been prepared by Steckelmacher (1951). Similar information is also found in the catalogs of suppliers of vacuum equipment.

In vacuum systems that contain only low vapor pressure solids, the vacuum gauge is used almost exclusively to indicate the effectiveness of the pumping procedure and to reveal leaks. An occasional glance will tell the operator if all is well.

In low temperature chemical research, the presence of condensed gases in the cryostat can result in large and rapid pressure changes that have to be accurately measured and correlated with other data such as specimen-block temperature. The operator is occupied with other measurements and the use of gauges requiring repeated manipulations cannot be undertaken. Continuous recording of both temperature and pressure will often catch subtle or transient effects that might otherwise be overlooked.

Pressures encountered are often so low that a good ionization gauge is needed in order to get any measurements. In the important annealing range, on the other hand, the pressure may rise above the upper range limit of ionization gauges (0.5 mm), and at least one widely used instrument has an awkward gap from 0.03 to 0.07 mm in its range. An excellent feed-back-controlled Pirani gauge that is not harmed by exposure to atmosphere is available and can be used to cover the range 10^{-3} to 100 mm on two scales with automatic switching. Recorder taps permit the instrument to be used with a 10-mv recorder, and a pressure record may be obtained on the same chart with the temperature.

IV. Gas-Handling Equipment

For the procedures described in this chapter, the gas-handling equipment is extremely simple. More versatile systems are described by Sanderson (1948), as well as in many original papers.

In order to limit the scope of this chapter it is assumed that the sample of gas to be condensed in the research cryostat has been prepared in an independent system and brought to the low temperature apparatus in a flask that can be readily joined to it by a standard-taper joint.

The volume of gas deposited in the cryostat is ordinarily of the order of 50 cc (STP). The use of a 1-liter flask filled to a pressure of 1 atmosphere permits this amount of gas to be delivered without a large change of pressure.

Some measure of the quantity of gas delivered is needed, and this can be made most directly by the use of a burette, a displacement fluid such as butyl phthalate or mercury, and a liquid leveling device. The volume of gas is measured at a pressure of 1 atmosphere, and a fine needle valve between the burette and the cryostat is used to control the rate of deposition. An alternative method of determining the amount of gas delivered is to measure the pressure change in a flask of known volume.

In the use of a microwave discharge to produce free radicals in the gas phase, it is the flow rate that is most directly related to pressure in the discharge tube and gas is usually metered into the system by using a flow rate meter as indicated in Fig. 1. This device commonly consists of a ball in a vertical tapered tube with a scale engraved on the outside. Flow of gas causes the ball to rise, increasing the open area between it and the walls. The range required is 3 to 10 cc (STP) per minute, and in this range the accuracy is considerably poorer than that attained by displacement or pressure methods.

Flow into the cryostat is commonly limited and adjusted by a very fine "micrometer" needle valve. This is installed between the flow rate meter and the cryostat so that the flow rate is measured at atmospheric pressure.

V. Liquid Refrigerants

The liquid refrigerants used in low temperature research include oxygen, nitrogen, hydrogen, and helium. Some of the properties of these are given in Table II.

The boiling point of oxygen is only a little higher than that of nitrogen, and because of the hazard involved, the liquid is seldom used. Data are included in Table II mainly because of the importance of the boiling point of oxygen in thermometry.

Compact and efficient plants for liquefying all of the above gases are available, but in many areas, arrangements can be made for purchasing the liquids. Where this is possible, small laboratories are spared the trouble and expense of operating a liquefaction plant.

Distribution of low-boiling liquids by commercial firms has been made possible by the development of highly efficient Dewar flasks for storing and transporting these liquids. The loss rate for liquid nitrogen and oxygen is 2.5 to 3% per day, and transportation over considerable distances is practical. For large laboratories, 300-gallon storage containers that can be filled from a tank truck are available.

The method of supplying low temperature liquids to laboratories at the National Bureau of Standards (NBS), Washington, D. C., has been described by Stober (1959). Liquid nitrogen and oxygen are delivered to individual laboratories by a commercial supplier in 15-, 25-, or 50-liter containers. Helium is liquefied at the NBS, but not in sufficient quantity to meet the demands of the accelerated free radicals research program. Additional supplies of liquid helium are shipped each week from the NBS Cryogenic Engineering Laboratory in Boulder, Colorado. The liquid is carried by a commercial air line in 50-liter nitrogen-shielded containers. Losses during the 1700-mile flight are 1 to 2 liters and are not excessive even when shipments are delayed by unfavorable weather conditions. The total cost of liquid helium supplied by air from Boulder is approximately the same as the cost of that liquefied in Washington.

Liquid hydrogen is less readily distributed because of the explosion hazard. Nevertheless, high purity para-hydrogen is supplied by The Linde Company of Tonawanda, New York. Deliveries are made within a 500-

TABLE II

Physical Properties of Common Liquid Refrigerants

Property	He	H₂ (99.97% para)	N₂	O₂
Normal boiling point				
T_b, °K	4.215[a]	20.27₃[h]	77.34[i]	90.190[m]
Lambda point				
T_λ, °K	2.17[b]	—		
p_λ, mm Hg	38.0[b]	—		
Triple point				
T_t, °K	—	13.81₃[h]	63.24[i]	54.363[m]
p_t, mm Hg	—	52.8[h]	96.2[i]	1.14[m]
Heat of vaporization at T_b				
L_v, cal/gm	4.95[c]	105.8[h]	48.0[c]	51.0[n]
L_v, cal/cc	0.619[c,d]	7.48[h]	38.7[c,j]	58.1[n]
Density, gm/cc				
Solid	0.198 at 2°K and 37 atm.[e]	0.0865 at 13.8°K[h]	1.033 at 20°K[k]	1.38 at 50°K[o]
Liquid at T_b	0.125[d]	0.0707[h]	0.807[j]	1.14[n]
Gas at 273°K and 760 mm Hg	1.78×10^{-4}	8.99×10^{-5} [h]	1.25×10^{-3} [l]	1.46×10^{-3} [l]
Specific heat, cal/gm/°K				
C_s, Liquid at T_b and 760 mm Hg	1.08[f]	2.27[h]		
C_p, Gas at 273°K and 760 mm Hg	1.24[g]	3.39[h]		
Vapor pressure, $\log_{10} (p_{mm})$				
Solid		$(4.62438 + 0.03635\,T - 47.0172\,T^{-1})$[h]	$(7.65894 - 359.093\,T^{-1})$[l]	$(7.86224 - 0.0049832\,T - 408.740\,T^{-1})$[i]
Liquid	See reference[a]	$(4.64392 + 0.02093\,T - 44.3450\,T^{-1})$[h]	$(7.781845 - 0.0062649\,T - 341.619\,T^{-1})$[i]	

[a] Brickwedde (1958). [b] Atkins (1959). [c] Wexler (1951). [d] Kerr (1957). [e] Swenson (1950). [f] Hill and Lounasmaa (1957). [g] Rossini et al. (1952). [h] Woolley et al. (1948). [i] Linder (1950). [j] Chelton and Mann (1956). [k] Bolz et al. (1959). [l] Hilsenrath et al. (1955). [m] Hoge (1950). [n] Scott (1950). [o] Bolz et al. (1959).

mile radius. Special containers that do not require nitrogen shilding are used for shipping the liquid via company trucks.

At the NBS, a central supply of liquid helium and hydrogen is maintained. These liquids are delivered to individual laboratories and transferred to research cryostats by trained technicians.

VI. Temperature-measuring Equipment and Techniques

A. THERMOCOUPLES

The device most commonly used for measuring temperature in chemical research is the thermocouple. Being small and simple in construction it is readily incorporated in almost any part of the apparatus. No heat is dissipated in the specimen chamber, and heat leaks along leads can be made negligible by using long, fine wires. Temperature sensing can be localized to permit measurements of thermal gradients, and the rapid response of small junctions can be utilized in observing transient effects.

1. Choice of Thermocouples for Low Temperature Measurements

The thermoelectric power $(\partial E/\partial T)$ of thermocouples approaches zero at absolute zero and in most cases the sensitivity in the important 4° to 20°K range is inadequate. The best sensitivity in this range is obtained with a gold + 2.1 atomic per cent cobalt alloy, as reported by Keesom and Matthijs (1935). Wire of this composition and of suitable size for most applications is available commercially[7] and is widely used. The silver + 0.37 atomic per cent gold alloy originally used for the second wire by Keesom and Matthijs was simply the normal in a series of measurements on various alloys. Since this alloy does not contribute to the desirable properties of the thermocouple, it is usually replaced with copper. A low temperature calibration table for the gold + 2.1 atomic per cent cobalt versus copper thermocouple has been published by Fuschillo (1957) and by Bunch and Powell (1958). A revised table supplied by the latter is given in condensed form in Table III.

Users of the gold-cobalt versus copper (or silver alloy) thermocouple warn that substantial variations in calibration may be encountered. Bunch and Powell (1958) report variations in thermoelectric force of different melts of the gold + 2.1 atomic per cent cobalt element as high as 7%. Thus, the need for spot calibration by the user to obtain corrections to the calibration table is indicated.

Becker (1955) reported the results of tests carried out to check the initial homogeneity of the wire. Using a temperature gradient of 200°C, regions were found where this gradient produced parasitic voltages of about

---

[7] Sigmund Cohn Corp., Mount Vernon, New York.

TABLE III

GOLD-2.1 ATOMIC PER CENT COBALT VERSUS COPPER THERMOCOUPLE[a]

°K	0	1	2	3	4	5	6	7	8	9
					Microvolts					
0	0.00	0.52	2.04	4.56	8.04	12.44	17.76	23.96	31.01	38.89
10	47.57	57.03	67.23	78.16	89.78	102.07	115.0	128.6	142.9	157.7
20	173.1	189.1	205.8	223.1	241.0	259.5	278.6	298.2	318.3	338.9
30	360.1	381.6	403.7	426.2	449.2	472.6	496.4	520.7	545.3	570.3
40	595.8	621.5	647.7	674.2	701.1	728.3	755.8	783.6	811.8	840.2
50	869.0	898.0	927.3	956.9	986.8	1016.9	1047.2	1077.9	1108.7	1139.8
60	1171.1	1202.7	1234.4	1266.4	1298.5	1330.9	1363.4	1396.2	1429.1	1462.2
70	1495.4	1528.8	1562.4	1596.2	1630.1	1664.1	1698.3	1732.6	1767.0	1801.6
80	1836.3	1871.1	1906.1	1941.2	1976.4	2011.7	2047.2	2082.8	2118.5	2154.4
90	2190.3	2226.4	2262.6	2298.9	2335.3	2371.8	2408.4	2445.1	2481.9	2518.8
100	2555.8	2592.9	2630.1	2667.3	2704.7	2742.1	2779.6	2817.2	2854.9	2892.6
110	2930.5	2968.4	3006.4	3044.4	3082.5	3120.7	3158.9	3197.3	3235.6	3274.1
120	3312.6	3351.1	3389.7	3428.4	3467.1	3505.9	3544.7	3583.6	3622.6	3661.5
130	3700.6	3739.6	3778.8	3817.9	3857.1	3896.4	3935.7	3975.0	4014.4	4053.8
140	4093.2	4132.7	4172.3	4211.8	4251.4	4291.0	4330.7	4370.4	4410.1	4449.8
150	4489.6	4529.4	4569.2	4609.1	4649.0	4688.9	4728.8	4768.7	4808.7	4848.7
160	4888.7	4928.8	4968.8	5008.9	5049.0	5089.1	5129.3	5169.4	5209.6	5249.8
170	5290.1	5330.3	5370.6	5410.9	5451.2	5491.5	5531.8	5572.2	5612.5	5652.9
180	5693.3	5733.8	5774.2	5814.6	5855.1	5895.6	5936.1	5976.5	6017.1	6057.6
190	6098.1	6138.7	6179.2	6219.8	6260.3	6300.9	6341.5	6382.1	6422.7	6463.3
200	6504.0	6544.6	6585.2	6625.9	6666.5	6707.2	6747.9	6788.5	6829.2	6869.9
210	6910.6	6951.3	6991.9	7032.6	7073.3	7114.0	7154.8	7195.5	7236.2	7276.9
220	7317.6	7358.3	7399.0	7439.7	7480.5	7521.2	7561.9	7602.6	7643.4	7684.1
230	7724.8	7765.5	7806.2	7847.0	7887.7	7928.4	7969.1	8009.8	8050.5	8091.2
240	8131.9	8172.6	8213.3	8254.0	8294.7	8335.4	8376.1	8416.8	8457.5	8498.1
250	8538.8	8579.5	8620.2	8660.8	8701.5	8742.1	8782.8	8823.4	8864.0	8904.7
260	8945.3	8985.9	9026.5	9067.1	9107.7	9148.3	9188.9	9229.5	9270.0	9310.6
270	9351.1	9391.7	9432.2	9472.8	9513.3	9553.8	9594.3	9634.8	9675.3	9715.8
280	9756.3	9796.8	9837.2	9877.7	9918.1	9958.6	9999.0	10039.0	10079.0	10120.0
290	10160.0	10201.0	10241.0	10281.0	10322.0	10362.0	10402.0	10443.0	10483.0	10523.0
300	10564.0									

[a] Adapted from Bunch and Powell (1959).

7 μv. Stretching the wire and repeating the procedure showed that errors of about 100 μv can occur if inhomogeneities caused by cold-working lie in a region where there is a temperature gradient. The conclusion was that the gold-cobalt element should be used only to measure small temperature differences with respect to a reference temperature. It was suggested that by calibrating during each experiment at the temperatures of liquid helium, liquid nitrogen, and solid carbon dioxide, absolute temperatures in the range 77° to 300°K could be measured to ±0.5°K and relative temperatures, to ±0.2°K. In the 4° to 20°K range it was estimated that absolute temperatures would be accurate to ±1°K and relative temperatures to ±0.5°K.

Perhaps these early experiences do not reflect recent improvements in the quality of wire. Current reports indicate that the gold-cobalt versus copper thermocouple is used in many applications with little difficulty. It has a high sensitivity and its reproducibility is satisfactory for most engineering applications.

Because of reports of poor reproducibility with the gold-cobalt element, some workers prefer to use constantan versus copper thermocouples. A calibration table for temperatures down to 0°K has been made available by Bunch and Powell (1959) and is reproduced in abbreviated form in Table IV. The sensitivity of the constantan versus copper thermocouple near 4°K is 1.365 μv/°K, or about one-third that for gold-cobalt versus

TABLE IV

CONSTANTAN VERSUS COPPER THERMOCOUPLES[a]

°K	0	1	2	3	4	5	6	7	8	9
					Microvolts					
0	0.00	0.18	0.70	1.56	2.76	4.29	6.14	8.31	10.79	13.56
10	16.64	20.01	23.66	27.58	31.78	36.25	40.97	45.96	51.22	56.73
20	62.48	68.48	74.76	81.30	88.11	95.18	102.5	110.0	117.8	125.8
30	134.1	142.5	151.2	160.1	169.3	178.6	188.2	197.9	207.9	218.1
40	228.5	239.0	249.8	260.8	271.9	283.3	294.8	306.5	318.4	330.5
50	342.8	355.2	367.8	380.6	393.6	406.7	420.0	433.4	447.0	460.8
60	474.7	488.8	503.1	517.5	532.0	546.7	561.6	576.6	591.7	607.0
70	622.4	638.0	653.6	669.5	685.4	701.6	717.8	734.3	750.8	767.5
80	784.4	801.4	818.5	835.8	853.2	870.7	888.4	906.2	924.2	942.3
90	960.5	978.8	997.3	1015.9	1034.6	1053.5	1072.5	1091.6	1110.9	1130.3
100	1149.8	1169.5	1189.2	1209.1	1229.2	1249.3	1269.6	1290.0	1310.5	1331.2
110	1351.9	1372.8	1393.9	1415.0	1436.3	1457.7	1479.2	1500.8	1522.6	1544.5
120	1566.5	1588.6	1610.8	1633.2	1655.7	1678.3	1701.0	1723.8	1746.8	1769.9
130	1793.1	1816.4	1839.9	1863.5	1887.1	1911.0	1934.9	1958.9	1983.1	2007.4
140	2031.8	2056.3	2081.0	2105.7	2130.6	2155.6	2180.7	2206.0	2231.3	2256.8
150	2282.4	2308.1	2334.0	2359.9	2386.0	2412.2	2438.5	2465.0	2491.5	2518.2
160	2545.0	2572.0	2599.0	2626.2	2653.5	2680.9	2708.4	2736.1	2763.8	2791.7
170	2819.8	2847.9	2876.2	2904.6	2933.1	2961.7	2990.5	3019.3	3048.4	3077.5
180	3106.7	3136.1	3165.6	3195.2	3224.9	3254.7	3284.5	3314.6	3344.7	3374.9
190	3405.2	3435.6	3466.1	3496.8	3527.5	3558.3	3589.3	3620.3	3651.5	3682.7
200	3714.1	3745.5	3777.1	3808.7	3840.5	3872.4	3904.4	3936.4	3968.6	4000.9
210	4033.3	4065.7	4098.3	4131.0	4163.8	4196.7	4229.7	4262.8	4296.0	4329.3
220	4362.7	4396.2	4429.8	4463.5	4497.3	4531.2	4565.2	4599.3	4633.5	4667.8
230	4702.2	4736.7	4771.4	4806.1	4840.9	4875.8	4910.8	4945.9	4981.2	5016.5
240	5051.9	5087.4	5123.0	5158.8	5194.6	5230.5	5266.5	5302.7	5338.9	5375.2
250	5411.7	5448.2	5484.8	5521.6	5558.4	5595.3	5632.4	5669.5	5706.7	5744.1
260	5781.5	5819.1	5856.7	5894.5	5932.3	5970.3	6008.3	6046.5	6084.7	6123.1
270	6161.5	6200.1	6238.7	6277.5	6316.4	6355.3	6394.4	6433.6	6472.8	6512.2
280	6551.7	6591.3	6631.0	6670.7	6710.6	6750.6	6790.7	6830.9	6871.2	6911.6
290	6952.1	6992.7	7033.4	7074.2	7115.2	7156.2	7197.3	7238.5	7279.9	7321.3
300	7362.9									

[a] Adapted from Bunch and Powell (1959).

copper. This low sensitivity must be offset by the accuracy and sensitivity of the potentiometric measurements. It appears that the stability of the thermocouple is still the limiting factor, however.

Iron versus constantan and Chromel P versus Alumel thermocouples are used to a very limited extent, but only down to liquid nitrogen temperature. Reference tables have been prepared by Shenker et al. (1955).

2. Preparation of Thermocouples

The thermocouple wire ordinarily used in low temperature work is made fine (0.003 to 0.005 inch diameter) to reduce heat leaks and increase speed of response. It must be handled with care to avoid kinking or other injuries which may result in cold working and localized inhomogeneities.

Junctions are usually made by fusing or soldering the two wires together. For most applications it is not desirable to have a large junction or one that is made of twisted wires. Welding, soldering, and twisting all tend to induce changes in composition and structure, and unless all of the junction between the unmodified wires is at the same temperature, errors in calibration will arise. Thermocouple wires to be welded or soldered should be made to cross at a point, and all welding or soldering should be done on the outer ends of the wire, working back only as far as the intersection, without heating or cold-working the wires beyond it. The wires can be lightly clamped between sheets of aluminum foil in a vise so that only about $\frac{1}{2}$ inch of the ends extends above the jaws. The ends should be bent to make them touch lightly at a point, forming an "X." On applying a torch with a very fine flame to the tip of one wire and then the other it will be found that one melts more readily than the other. By applying the torch first to the very tip of the less easily melted wire a bead is formed and followed down until it reaches the junction. Then the second wire is heated until the two beads of molten metal fuse. Success depends on starting with only a little wire (about $\frac{1}{8}$ inch) extending beyond the intersection. Unless the whole operation is carried out quickly, oxidation may progress to the point where good fusion of the dissimilar metals in the bead is not achieved. If solder is used, it should not extend beyond the point where the wires first meet.

These precautions become especially important when the gold-cobalt alloy is used. Bunch and Powell (1959) caution that the material must not be held at 100°C or higher, or the cobalt will diffuse out of the gold matrix and the alloy will lose its thermoelectric sensitivity. Becker (1955) has reported that cold working of the alloy in a region where temperature gradients occur can cause serious parasitic voltages. In applications where the temperature of a block is to be measured, the problem of forming a good junction without damaging the wire can be circumvented by crossing the bare wires and clamping the junction with a screw and washer. If it is desirable to isolate the thermocouple electrically, mica washers may be used.

3. Installation of Thermocouples

Good thermal contact between thermocouple and specimen is important if the response is to be rapid and the measurements accurate, and some workers will say categorically that metal-to-metal contact achieved by spot welding or soldering is essential. In most cases of interest to the users of this book, however, the specimen will be a nonmetallic film on a comparatively massive cold surface. If the surface is of metal and the thermocouple junction is joined to it by a low-thermal-resistance path, only the temperature of the block will be measured. Heat effects in the sample will have very little effect. If the surface on which the specimen is deposited is glass, the low conductivity of the surface will make measurement of specimen temperature somewhat easier, but in any case attempts to measure specimen temperature will usually result in the introduction of high thermal resistance in the path between the thermocouple junction and the specimen block. As a result of this condition heat flow along the leads and energy transferred by conduction or radiation influence the temperature of the junction so that reproducible measurements of temperature cannot be obtained. A partial solution lies in proper "heat stationing" and shielding of the thermocouple leads.

"Heat stationing" refers to the technique of placing the thermocouple leads in very good thermal contact with something that is at nearly the same temperature as the specimen. In this way, heat transfer along the wires is greatly reduced. The technique is not required (except in calorimetry) if the thermocouple junction is soldered to a massive block of high conductivity metal, but even in this case, it may eliminate parasitic voltages resulting from inhomogeneities in the leads near the junction by greatly reducing thermal gradients along the wire. In attempts to measure film temperatures, heat stationing is essential, but another problem arises which makes additional precautions necessary.

Contrary to the usual practice in calorimetry and solid state physics research, observations of deposits containing free radicals are often made with an appreciable vapor pressure in the specimen chamber. Any thermocouple that is not in good contact with a thermal reservoir tends to function as a heat-conductivity pressure gauge, and readings become pressure sensitive. For this reason, heat stationing, to be effective must be very close to the junction. One way of mounting a thermocouple so that it gives some indication of the temperature of a specimen on a metal block is to stretch a butt-welded thermocouple across the block with a narrow space or a thin layer of insulating varnish separating it from the block. The leads are anchored thermally but not electrically at the edges of the block, about 1 to 2 cm from the junction, and are wound several times around the block just outside the observation zone. The leads wound on the block can be covered with aluminum foil to shield them from incident heat energy. The temperature of this thermocouple is somewhat pressure sensitive and differs from that of the block thermocouple. However, the readings are believed to be at least roughly indicative of the temperature of a film on the surface of the block. It is often desirable to record temperature readings from both types of thermocouples.

In bringing the thermocouple leads out of the vacuum system, it is not good practice to cut them and join the ends to a dissimilar metal in a vacuum lead-through insulator. There are likely to be thermal gradients in this region, and it is better to pass the thermocouple leads themselves through the vacuum wall and to seal the openings with wax. The insulation must be stripped from the leads before the wax is applied and it will be easier to get a good seal if single-stranded wire is used.

4. Reference Junctions

A reference junction is required so that an external junction between the dissimilar wires occurs in a known and constant temperature environment. Crushed ice in a small Dewar flask is usually used to establish a temperature of 0°C at the point of contact of each thermocouple lead to the copper potentiometer lead. However, the sensitivity of the ice junction is from ten to thirty times that of a junction at temperatures near 4°K and care is required if the ice junction is not to become a source of errors. In particular, unless some steps are taken to eliminate stratification, water at 4°C will collect at the bottom and the ice will rise to the top of the Dewar. Because of the higher sensitivity of the ice junction even the very small gradients introduced in this way cannot be tolerated if temperatures are to be correct to 0.5°K.

Two schemes have been proposed to eliminate errors that arise because of the greater sensitivity of the reference junction. Dauphinee et al. (1953) suggested the use of a pure copper versus copper + 0.066 weight per cent tin thermocouple which has greatly reduced sensitivity above 60°K. Unfortunately, wire of the required composition is not commercially available and other workers have failed to obtain the desirable properties reported by Dauphinee and associates.

The other method of reducing the effect of variations in reference junction temperature is to use a refrigerant which boils at a low temperature to cool the reference junction. Larson and Mayer (1952) modified a recording potentiometer for accurate measurements down to 10°K using a copper versus constantan thermocouple. They incorporated a reference junction immersed in liquid nitrogen and automatically compensated for variations in the boiling point that result from oxygen contamination and fluctuations in barometric pressure.

Use of either the liquid nitrogen or liquid helium reservoir as an internal reference junction is an attractive possibility because it would make practical the use of only a short length of the sensitive gold-cobalt wire with sturdier copper wire leading out of the Dewar. However, the temperature of boiling nitrogen may vary by +4°K because of contamination with oxygen and by ±0.1°K because of fluctuations in barometric pressure. If a liquid-hydrogen or liquid-helium bath were used to obtain a reference temperature the effect of contamination and of day-to-day fluctuations in barometric pressure would become negligible. Unfortunately these liquids as a rule are no longer present in the Dewar during the important warm-up period. Use of the storage Dewar for cooling reference junctions is a possible, though less attractive, alternative.

5. Measuring and Recording the emf of Thermocouples

The choice of a potentiometer for measuring the emf of thermocouples involves a compromise between accuracy, speed, and flexibility on the one hand versus cost and complexity on the other. For many applications, portable, manually balanced instruments are satisfactory, but if accurate measurements are to be made (particularly with the copper versus constantan thermocouple) a high-quality instrument with a sensitive galvanometer is essential. Continuous recording of temperature throughout the experiment is very desirable but usually involves a sacrifice of accuracy. Larson and Mayer (1952) described modifications to a commercial recording potentiometer by which they were able to achieve a stability of 0.3 μv and a sensitivity of 0.1 μv. Using this instrument and a copper versus constantan thermocouple, temperature was recorded at 10°K with a claimed accuracy of ±0.1°K. The modifications were too extreme to be undertaken in most laboratories, but the paper is of interest because it points out some of the problems that must be solved before high accuracy can be attained.

One of these problems is the relatively large and unchanging portion of the thermocouple emf that arises when the reference junction is kept at 0°C. If a recorder capable of recording the whole output of the thermocouple is used, the important temperature range 4° to 77°K covers only one-fifth of the recorder scale and chart paper shrinkage introduces errors of ±3°K. This difficulty can be solved by using a recorder with a narrow span and variable zero suppression. For use with the gold-cobalt versus copper thermocouple a recorder having a span of 2.5 mv and zero suppressions of 0, 2, 4, 6, and 8 mv can be used to cover the entire range from 4° to 273°K without an appreciable sacrifice in accuracy. Alternatively, a chopper-type dc amplifier can be used in conjunction with a strip chart recorder to obtain deflection sensitivities as great as 0.2 inch per microvolt (corresponding to 0.8 inch per degree for the gold-cobalt thermocouple at 4°K) and a wide selection of ranges. This instrumentation does not solve the problem of the large, constant portion of the thermocouple emf. Either a reference junction temperature very close to the specimen temperature must be used, or a stable and accurately calibrated bucking voltage must be provided. Some dc amplifiers do have a built-in bucking-voltage supply. One of the less obvious advantages of using a dc preamplifier is that it permits the use of a less specialized recorder that is more suitable for the simultaneous recording of other signals such as the output of a vacuum gauge. The lower cost of such a recorder partially offsets the expense of the dc amplifier.

B. Gas Thermometers

Although the constant-volume helium gas thermometer is the generally accepted standard for temperature measurements throughout most of the low temperature region, and even up to the gold point (1063°C), the instrument finds very limited use in free radicals research. The bulb, which is the sensing element, is large compared to the dimensions of the usual sample and is difficult to bring into thermal equilibrium with it. The time required for observations is rather a nuisance and prevents observation of many interesting effects. Furthermore, unless the instrument is used with great care the absolute accuracy of low temperature measurements is not likely to exceed the accuracy attained much more readily by means of thermocouples. The device is commonly used for the low temperature calibration of other devices and in calorimetry, but its use will not be considered here. For a thorough exposition of gas thermometry, the reader is referred to the work of Beattie (1955).

C. Resistance Thermometers

1. Platinum Resistance Thermometers

The platinum resistance thermometer is capable of high precision at temperatures as low as 12°K, and a group of six of these calibrated against a gas thermometer by Hoge and Brickwedde (1939) serve as the basis for National Bureau of Standards calibrations in the range 12 to 90°K. Agreement with the thermodynamic Celsius scale is estimated at ±0.02°.

The size consideration, which limits the use of the gas thermometer in low temperature chemistry, also applies to the resistance thermometer. Nevertheless, resistance thermometry is often used in calorimetry. The resistance element may be wound directly on the outside of the calorimeter can to reduce bulk and improve the speed of response. Very precise measurements of temperature averaged over the area covered by the winding are obtained in this way. Although the standard platinum thermometer loses sensitivity below about 12°K, gold wire and alloys such as phosphor bronze and constantan are used for calorimetric work at liquid helium temperature. However, it is felt that the applications of resistance thermometry in trapped-radical research are too limited to justify including a discussion of the techniques in this chapter. The reader is referred to the paper by Stimson (1955) on precision resistance thermometry and to the references at the end of that paper.

2. Indium Resistance Thermometers

An indium metal resistance thermometer which retains its sensitivity down to 3.41°K has been reported by White and Woods (1957). The low-temperature sensitivity of this metallic conductor is the result of the high purity of the indium (99.999%) and the low Debye temperature. No application of this device to low temperature chemical research has been reported.

3. Carbon Resistance Thermometers

Very high sensitivity below 20°K is achieved by the use of carbon resistors of a type commonly used in electronic circuits. These must be individually calibrated, preferably during every run, but this disadvantage is offset by the remarkable sensitivity available.

Clement and Quinell (1952) tested the thermometric properties of a number of commercially available resistors and found that Allen-Bradley[8] resistors of the 1-watt size were superior in sensitivity and reproducibility. Thermometric properties vary somewhat with nominal size and, under conditions of the test, the 56-ohm resistors were judged best. These have resistances in the range of 500 to 2500 ohms at 4.2°K.

To reduce size and heat capacity and improve thermal contact, the insulating cover is ground off and the core of the resistor is coated with Glyptal. The resistor is mounted where it is in good thermal contact with the specimen, shielded from thermal radiation, and kept out of contact with liquid helium.

Resistance measurements can be made with an equal-arm Wheatstone bridge and a three-lead connection to cancel the effect of lead resistance. It is important to limit the power dissipated in the resistor to about 10^{-6} watts and to keep the measuring voltage constant.

A semiempirical equation with three constants

$$\log R + K/\log R = A + B/T$$

[8] Allen-Bradley Company, Milwaukee, Wisconsin.

is used to relate resistance and temperature. It is necessary to determine K, A, and B experimentally, after which temperatures can be calculated from measured values of resistance within $\pm 0.5\%$ from $20°$ to $2°$K. The sensitivity increases by a factor of 100 with decreasing temperature in this range, and with apparatus capable of measuring resistance to 1 part in 5000 it is possible to detect changes in temperature smaller than $0.01°$ at $20°$K and smaller than $0.0001°$ at $2°$K.

A bridge that is self-contained except for the oscilloscope used to display the output has been designed especially for carbon resistance thermometry (Blake et al., 1958). With the dissipation of only 2×10^{-8} watts power in the resistor it is capable of measuring changes in resistance of 0.1 ohm, corresponding to 4×10^{-6} °K at $2°$K.

4. Semiconductor (Non-ohmic) Resistance Thermometers

High sensitivity and reasonably good reproducibility has been reported for various thermometers making use of the special properties of semiconductors. Friedberg (1955) has reviewed the literature on the theory of semiconductor and carbon resistance thermometry. More recently, Kunzler et al. (1957) gave a detailed description of the preparation and performance of a germanium resistance thermometer suitable for calorimetry. The results are encouraging, but work in this area is, at present, limited to a few laboratories where facilities for semiconductor research are available.

D. Vapor Pressure Thermometers

Any pure substance having a measurable vapor pressure in a given range can be used for determining temperature in that range. The vapor pressure is independent of the amount of solid or liquid present and of the configuration of the container. The pressure measured anywhere in a closed system containing one volatile component is the equilibrium vapor pressure of that component at the temperature of the coldest part of the system. For these reasons, it is sometimes possible to use the vapor pressure of the sample itself as an indication of temperature although the vapor pressure of most solids has not been accurately measured and probably depends on the crystallinity. The method is practical only if the sample is pure, or if all the components except one have negligible vapor pressure. In general, in the low-temperature range each liquid is usable in only a very narrow temperature interval. The lower the boiling point, the more sensitive the thermometer and the more limited the temperature interval in which it can be used.

Empirical data on the vapor pressure as a function of temperature are needed for the determination of temperature by this method. Such data for the five gases: helium, oxygen, nitrogen, neon, and hydrogen (normal and 20.4°K equilibrium) have been compiled by Linder (1950). In addition, vapor pressure versus temperature tables for seven compounds are given by Sanderson (1948). This author also includes a collection of melting points, boiling points, and values of vapor pressure at a few temperatures for 398 pure compounds.

The vapor pressure of helium provides a useful secondary standard of temperature in the approximate range 1° to 5°K and has received special attention. In an effort to provide a common temperature scale for the many laboratories doing work in this important interval, an "agreed" vapor pressure-temperature relation was arrived at by an informal international committee and compiled by Van Dijk and Shoenberg (1949). Evidence available by 1955 indicated the need for revision of the 1948-agreed scale, and in that year, two new p-T relations were proposed. Magnetic thermometer calibrations in the region 1.3 to 4.2°K were used by Ambler and Hudson (1956) to examine the internal consistency of two provisional helium vapor-pressure scales of temperature. The differences were finally reconciled and in October, 1958, Van Dijk, Durieux, Clement, and Logan released the "1958 ^{4}He Scale" which is expected to be adopted at the next meeting of the international committee (Brickwedde, 1958). The 1958 scale is believed to agree with the thermodynamic scale to within ±2 millidegrees.

E. CALIBRATION OF TEMPERATURE-MEASURING EQUIPMENT

In cases where it is not practical to compare a given thermometer with standards such as the helium gas thermometer or a calibrated platinum resistance thermometer, it is necessary to rely on calibration against fixed points that can be duplicated in the laboratory.

The International Temperature Scale of 1948 offers only two points below room temperature, the freezing point of water (defined as 0.0100°C below the triple point) and the boiling point of oxygen.

The ice point can be realized with accuracy sufficient for general laboratory purposes by preparing an ice-water slush using distilled water and immersing the element to be calibrated in a bath of this slush contained in a Dewar flask. Occasional stirring is required to prevent stratification since water at 4°C has a higher density than the water in equilibrium with ice at the top of the bath.

The oxygen and nitrogen boiling points are difficult to realize with accuracy because of the mutual solubility of these gases and the difficulty in purifying them. The commercially available liquids are especially subject to contamination and should never be used for calibration purposes. Errors in the boiling point of nitrogen as great as 4°K can arise because of contamination with oxygen.

Scott (1941) described apparatus and procedures for observing the boiling point of oxygen. He recommended special preparation of oxygen to obtain a sample of sufficient purity. However, research grades of oxygen and nitrogen of very high purity can now be obtained. Nitrogen is more readily purified and may be preferred for calibration purposes even though its boiling point is not one of those used in defining the International Temperature Scale. A closed system should be used to avoid contamination by the air. The pure gas is liquefied in the bulb of a vapor pressure thermometer and used to measure the temperature of a bath of commercial liquid oxygen or nitrogen boiling at atmospheric pressure. A slow stream of the same gas from a cylinder is passed through the bath to agitate it and prevent superheating. Equations for vapor pressure versus temperature for oxygen and nitrogen are given by Linder (1950):

$$\log_{10}p(\text{mm}) = 7.86224 - 0.0049832\ T - 408.740\ T^{-1} \quad \text{(oxygen)}$$

and

$$\log_{10}p(\text{mm}) = 7.781845 - 0.0062649\ T - 341.619\ T^{-1} \quad \text{(nitrogen)}$$

The temperature of the bath can be calculated from these equations.

A convenient intermediate point is the sublimation temperature of solid carbon dioxide (194.65°K), which can be calculated from the vapor pressure by means of the equation

$$\log_{10}p = 9.81137 - 1349\ T^{-1}$$

if certain precautions are taken to insure that the CO_2 sublimes in equilibrium with its own vapor at barometric pressure. Scott (1941) reported a gradual increase in the temperature of a thermometer buried in dry ice from an initial reading of $-87°C$ to a final value of $-78.51°C$ after 10 hours. This behavior is not greatly improved by the use of a low-freezing-liquid bath. Scott recommends burying the device to be calibrated in crushed CO_2 in a Dewar and using a small electric heater at the bottom of the Dewar (at least 5 cm from the thermometer) to increase the rate of sublimation so as to sweep air out of the flask. An initial input of 30 watts established equilibrium after 12 minutes, at which time the power input was reduced to 7 watts.

Fixed points depending on the properties of hydrogen and helium are somewhat more reliable than those based on the properties of oxygen and nitrogen. However, it is necessary to cope with the variation in ortho-para ratio of hydrogen. Hydrogen molecules exist in two forms: ortho-hydrogen, in which the magnetic moments of the two nuclei are parallel, and para-hydrogen in which they are antiparallel. According to Woolley et al. (1948) normal hydrogen (25 % para-hydrogen and 75 % ortho-hydrogen) has a boiling point of 20.39°K and a triple point of 13.95_7°K. Its vapor pressure is given by the equation

$$\log_{10} p(\text{mm}) = 4.66687 + 0.020537\ T - 44.9569\ T^{-1}$$

When this liquid is stored, it converts spontaneously to "equilibrium hydrogen" (99.97% para-hydrogen and 0.21% ortho-hydrogen) having a boiling point of $20.27_3°$K and a triple point of $13.81_3°$K. Its vapor pressure can be calculated from the equation

$$\log_{10} p(\text{mm}) = 4.64392 + 0.02093\ T - 44.3450\ T^{-1}$$

To prevent storage losses of hydrogen caused by energy released during the ortho-para conversion, catalysts are used at the liquefaction plant to convert the liquid almost entirely to para-hydrogen. For liquid hydrogen that has been in storage, the data for "equilibrium hydrogen" should be used.

Finally, the boiling point of helium is a reliable fixed point that is even relatively insensitive to variations in barometric pressure. It is easily determined for any barometric pressure by using the "agreed" scale of 1948 (Linder, 1950). At 760 mm Hg, the boiling point of helium is 4.211°K and it decreases about 0.0014°K per millimeter with decreasing barometric pressure.

The above points should be sufficient for most low temperature calibration purposes, but many more are given by Hoge (1941).

VII. Operating Techniques

A. Pumping and Outgassing

Procedures for evacuating a cryostat before filling it with liquid refrigerants vary considerably, and it is possible to outline only one routine here.

The initial evacuation presents some special problems. Normal performance for the system has not been established and it is difficult to distinguish between the effects of leakage and outgassing. The latter is likely to be very pronounced in a new apparatus and may continue for several weeks.

The effect of outgassing is readily recognized by isolating the cryostat from the pumping line and watching the change in pressure after adding liquid nitrogen to the outer Dewar. Nearly all of the residual gas in a tight system that has been pumped for 30 minutes or more is condensable at liquid nitrogen temperature. The pressure should drop below 10^{-5} mm and remain low if there is no leak. Both oxygen and nitrogen have high vapor pressures at 77°K, and any leakage is indicated by a steady rise in pressure.

Leaks must be repaired before the cryostat is filled with refrigerants. Outgassing can be tolerated if an impurity content of 0.5% is permissible.

However, it is much better to clean up the system by heating it to 100°C or, preferably, 300°C while pumping through a cold trap.

Once leakage and outgassing have been eliminated, the pumping routine with the system shown in Fig. 1 is completed in three steps: (1) Using the mechanical pump, the entire system is evacuated through the charcoal trap by closing valve c and opening a, b, d, g, and e. (2) After the pressure in the cryostat has been reduced below 10^{-2} mm, the cold trap is filled. The pressure should drop below 10^{-3} mm within 10 minutes. (3) Finally, the diffusion pump is turned on to reduce the pressure to 10^{-6} mm. A period of about 30 minutes normally is required.

Before filling the cryostat, valves d and f are closed. If, in the course of an experiment it becomes necessary to pump on the cryostat without contaminating the cold trap, diffusion pump, and charcoal trap, these can be bypassed by closing valves a and b and pumping through c. However, there is no cold trap or charcoal trap in this line to prevent the diffusion of pump oil into the cryostat on the one hand, or the contamination of pump oil by gasses in the cryostat on the other. This line is intended for use in pumping noncondensable and noncorrosive gases out of the cryostat. If pumping is not prolonged beyond the time required to reduce the pressure to 10^{-2} mm, diffusion of pump oil is not likely to be a problem. Installation of an auxiliary cold trap between the pump and valve c is recommended if condensable, corrosive, or noxious gases are to be handled. This trap should be made demountable to facilitate cleaning.

If degassing of the charcoal trap is required, it may be done by closing valves a, d, and e and pumping through b and c while heating the charcoal. The cold trap should be filled to avoid contaminating the mechanical pump oil. Condensation in the diffusion pump is not likely to occur if pumping speed is sufficient to keep the pressure below about 17 mm.

B. Transfer of Liquid Refrigerants

With the cryostat evacuated and valves to the pumping system and sample-deposition train closed, the liquid refrigerants may be introduced.

Liquid nitrogen, the refrigerant commonly used in the outer Dewar vessel, is easily handled. It may be kept in small glass Dewar flasks (1.6 or 4.5 liters) and poured when needed, using a small metal funnel. If quantities in excess of 1 liter are to be transferred, it is usually more convenient to use a siphon tube that passes through an air-tight cap and reaches the bottom of the storage vessel, which may have a capacity of 50 liters or more. A piece of rubber tubing connects the siphon to the receiving vessel. Air at a pressure not exceeding 6 psi is commonly used to force the liquid through the siphon, but if contamination with liquid oxygen is to be avoided, nitrogen gas may be used instead. This can be obtained by introducing heat into the storage flask to vaporize some of the liquid nitrogen.

After the outer Dewar has been filled, liquid nitrogen may be used to precool the inner vessel in preparation for filling with liquid helium or hydrogen. However, all nitrogen must be removed before a lower-boiling liquid is transferred to avoid freezing the siphon in the cryostat.

The heat of vaporization of liquid helium is very small, and special techniques are required for handling it. Liquid hydrogen is handled in the same way, but additional precautions are necessary because of the explosive nature of hydrogen-air mixtures (Hernandez *et al.*, 1957).

The technique for siphoning liquid helium while recovering all of the gas has been described by Broom and Rose-Innes (1956). The procedure used in America, where the gas is not recovered, is somewhat different. Stober (1959) has described the method used at the National Bureau of Standards.

Liquid helium is brought to the cryostat in a 50-liter storage Dewar or a glass transport flask of 2- or 3-liter capacity. The latter is easier to use if strips have been left unsilvered to permit observation of the level of the liquid and the position of the transfer siphon. Glass transport vessels are often graduated to facilitate measuring the volume of liquid transferred. However, the double transfer of liquid from storage Dewar to transport flask to research cryostat is somewhat wasteful, requiring an average of 250 cc extra per transfer of liquid helium.

The low heat of vaporization of helium (5 cal per gram) makes it necessary to use a vacuum-insulated transfer siphon in handling the liquid. Jacobs and Richards (1957) have described a suitable siphon and have given references to five other papers on the subject. The siphon is made up of two tubes held concentric by Teflon spacers. The space between is evacuated to reduce heat transfer to the inner tube. Thin walled tubing of German silver or other low-conductivity metal is used to reduce the heat introduced into the helium Dewar when the siphon is inserted. Suitable dimensions for the inner and outer tubes respectively are $\frac{1}{2}$ inch o.d. with a 0.015 inch-thick wall and $\frac{3}{16}$ inch o.d. with a 0.010-inch wall. The siphon is usually made in the shape of an inverted "U" the vertical arms of which fit into the neck tubes of the storage Dewar and research cryostat. These vertical arms slide through rubber sleeves that slip over the neck tubes of the two vessels to exclude air. The sleeve on the storage Dewar is fitted with a side arm that can be used to pressurize the vessel. The receiving vessel must be vented.

Before the siphon is inserted, the neck opening of the storage Dewar is cleared of any frozen air by a quick thrust with a $\frac{1}{2}$-inch wooden dowel. This helps to eliminate the dangerous situation that may develop if the siphon freezes in the neck tube and becomes plugged.

The level of liquid in the Dewar should be determined before the transfer and the siphon marked to indicate the point where contact with the liquid occurs during insertion. A dip stick similar to that described by

Gaffney and Clement (1955) can be used. It consists of a $\frac{1}{16}$-inch o.d. Monel tube 3 to 4 feet long with one end enlarged to $\frac{3}{8}$ inch i.d. This end is covered tightly with the index finger while lowering the other slowly into the helium Dewar. After a few seconds, a strong vibration is set up which changes in frequency and amplitude when the lower end of the tube touches the liquid. The effect may be obscured if the tube becomes plugged with frozen air.

Success in controlling the transfer of liquid helium and avoiding waste depends on observing three precautions: (1) The gas that is evolved when the siphon is first introduced must be made to pass through the siphon, cooling it and driving out condensable vapors. (2) The siphon must not be allowed to make contact with the liquid until it has been thoroughly cooled. (3) The transfer must not be allowed to begin prematurely.

The rubber sleeves on both vertical arms of the siphon are slid all the way down so that they can be slipped over the neck tubes of the two vessels without inserting the siphon into the storage Dewar more than a few inches. With the side tube on the storage vessel sleeve closed and the receiving vessel vented, the siphon is lowered a few inches at a time until the lower end extends through the neck tube and an inch or two into the vessel. A considerable volume of gas will be discharged at this stage because of the heat introduced. The cold helium gas is made to pass through the siphon, cooling it and sweeping air out of the system. The heat of vaporization of helium is only about 1 % of the heat required to raise its temperature to 300°K, and it is wasteful to allow the siphon to touch liquid helium before it has been thoroughly cooled by the gas.

The gas coming through the siphon at first is quite warm and may cause a considerable temperature rise. If this cannot be tolerated, the siphon should not be inserted in the research cryostat until it has been cooled. Collection of frost during cooling outside the cryostat can be prevented by slipping a polyethylene bag over the end of the siphon and securing it with a rubber band. The bag is pierced when the siphon is to be inserted.

Before the siphon is allowed to touch the liquid helium, pressure in the two vessels must be equalized to avoid having the transfer begin spontaneously. This is done by venting both vessels until the boil-off that results from inserting the siphon has subsided.

With the siphon in place, the transfer is made by pressurizing the storage vessel with helium gas from a cylinder. A pressure of 0.5 psi is sufficient to transfer liquid helium at a rate of about 1 liter per minute.

Some operators prefer to connect the storage and receiving vessels with a piece of rubber tubing before lowering the siphon into the liquid. Pressure is allowed to build up in both vessels, and transfer is brought about by venting the receiving vessel while maintaining pressure in the storage container.

The first liquid to come over boils away immediately and large volumes of gas leave the cryostat without reaching thermal equilibrium. This waste can be reduced to some extent by beginning the transfer slowly and increasing the rate when liquid begins to collect. It is not economical to prolong the transfer unduly, however, because heat leakage to the liquid in the transfer siphon is an important factor.

A volume of 300 to 500 cc of liquid helium is required to cool most systems enough for liquid helium to collect. The completion of cooling is marked by a noticeable subsidence in the flow of gas from the vent.

In metal cryostats it is difficult to judge the amount of liquid transferred. Liquid-level indicators that depend on the difference in heat dissipation from a wire in liquid and in gas have been described by Wexler and Corak (1951) and by Maimoni (1956). However, the change in resistance of a carbon resistor on being covered by liquid helium is the basis for most indicators. Some power must be dissipated in the resistor so that its temperature depends on the rate of heat loss. In one version, developed by Stober and Bass (1956) this is accomplished by using a crude ohmmeter circuit consisting of a 1.5-volt battery in series with a 100-mamp meter and a 15 ohm-, $\frac{1}{10}$-watt resistor. The current, which fluctuates considerably while the resistor is in the turbulent gas, falls to a constant and predictable value when it is immersed in the liquid. The resistor is mounted on a sliding rod and can be set at any desired level. The point on the meter scale corresponding to the reading with the resistor immersed in liquid helium is marked. The helium transfer is continued until the meter reading drops to this mark.

Pressure in the storage vessel must be released before the siphon is withdrawn. The level of the liquid remaining is measured to determine the amount of liquid helium used.

The neck tubes of the storage container and research cryostat are closed after removing the siphon to prevent the accumulation of frozen air and water vapor. A piece of rubber tubing with a longitudinal slit is used as a valve to permit the escape of helium gas.

C. DEPOSITION OF SAMPLES

When the cryostat has been filled with liquid refrigerants the sample is introduced as a gas and allowed to condense on the cold surface of the specimen block. A 1-liter flask containing the sample at a pressure of 1 atmosphere is joined to the deposition train shown in Fig. 1. Air is removed from the tubing by pumping and the stopcock on the flask is opened to fill the train with gas as far as the micrometer needle valve. The sample is admitted to the cryostat by opening the needle valve and the rate of deposition is adjusted with the aid of the flow-rate gauge. If no more than

10 % of the sample is used, the decrease in pressure does not seriously affect the flow rate or calibration of the gauge.

A similar procedure is used for introducing samples that are liquid at room temperature. The liquid, which is introduced into an evacuated sample flask, may be heated during deposition to increase its vapor pressure.

During deposition, the flow must be limited to avoid raising the pressure and causing excessive transfer of heat to the liquid helium. However, the helium-cooled surface is so effective as a pump that flow rates of 3 cc per minute (STP) may be used without causing the pressure to rise above 10^{-4} mm. Thus, no additional heat leak is encountered and it is necessary to allow only for cooling and condensing of the gas introduced.

Recent experiments by Foner *et al.* (1959) have shown that 60 % of the atoms in a stream of argon directed at a helium-cooled surface condense on first impact. The behavior of other gases deposited at temperatures far below their melting points probably does not differ greatly from that of argon. In analyzing a particular system, it may be assumed, as a first approximation, that 60 % of the molecules condense where they impinge on the cold surface and the remainder distribute themselves fairly uniformly over all exposed helium-cooled surfaces. On this basis, crude calculations of the thickness of the films deposited can be made.

Methods of producing free radicals in the solid as well as in the gas phase have been discussed in preceding chapters and will not be treated here. However, an example of one method that is in fairly general use is given to introduce the apparatus required.

Broida and Pellam (1954) have described the technique for producing free radicals in the gas phase and depositing them on a helium-cooled surface. A system such as that shown in Fig. 1 may be used.

The gas which is to be dissociated is passed through a silica tube in the rf field of a waveguide or cup-shaped director. Power is provided by a 125-watt microwave-diathermy unit operating in the 2400- to 2500-Mc band. With the power on, the flow of gas is increased relative to the rate of condensation. When the flow is sufficient to maintain a pressure of 0.1 to 1.0 mm in the silica tube, a Tesla coil can be used to initiate the discharge. The flow of gas through the discharge is continued until a deposit of the required thickness has accumulated. The time required varies from about 15 to 60 minutes.

D. Measuring Techniques

With the sample deposited on the helium-cooled surface, measurements can be made by methods that often do not differ greatly from those used at room temperature. Space does not permit a discussion of techniques used in the various methods such as absorption spectroscopy, X-ray diffraction, magnetic susceptibility, and calorimetry. Some of these are covered in other chapters of this book.

One situation that is encountered in studies of frozen gases arises because of the rapid changes that occur during warm-up. The sample often changes appreciably while a measurement is being made and may be lost because of evaporation a short time after the helium supply is exhausted. For these reasons careful advanced planning of experiments is required to avoid wasted time. As many of the measurements as possible should be made by recording instruments.

E. Collection of Evolved Gases for Analysis

The sampling trap shown in Fig. 1 is used to collect small quantities of evolved gases for analysis by mass spectrometry or other methods. The trap has an elongated tip which can be immersed in liquid nitrogen or hydrogen to cause condensation of gases diffusing into the evacuated trap when the stopcock is opened. The bulb has a sufficient volume to permit the trap to be isolated and warmed to room temperature without an excessive rise in pressure. For many purposes, a sufficient sample is obtained without cooling the trap to promote condensation.

The same trap can be used to condense samples that must be recovered or must not be allowed to pass through the pumping system.

F. Removal of Samples and Disposal of Toxic Materials

Many samples can be removed from the cryostat simply by pumping during warm-up. However, condensable materials that might contaminate the pump oil, corrosive samples that would be harmful to equipment, and toxic substances that cannot be allowed to escape into the laboratory all require special handling. The safest procedure is to evacuate the sampling trap and cool it with liquid nitrogen or liquid hydrogen so that volatile substances can be collected during warm-up by closing all valves to the pumping system and opening those between the trap and the cryostat.

The diffusion of vapors into the sampling trap is slow because of the small opening through the stopcock. In circumstances not requiring the most elaborate precautions, the sample may be allowed to collect in the cold trap if this is designed in such a way that it can be readily disassembled for cleaning. This procedure has several obvious disadvantages, but removal of vapors is more rapid (especially in cases where part of the gas in the system is noncondensable and remains to interfere with diffusion). If the cold trap has a sufficient liquid nitrogen capacity (or is automatically filled) it is very convenient to allow the sample to collect there while pumping is continued during the night.

VIII. Closing Remarks

Before concluding this chapter some apology must be made for the failure to include many important techniques of low temperature physics and chemistry. Some omissions must certainly be charged to oversight, but others were made deliberately to permit more adequate coverage of basic techniques. The selection of topics has been based on the author's experience in entering the field of low temperature research less than three years ago. It has been his intention to cover most adequately the problems that had to be solved before any progress could be made. The point of view throughout has been that of a beginner in the field. The chapter is intended to provide quick solutions to many problems that arise in the early phases of low temperature research. For more refined and specialized techniques the reader is asked to refer to original articles, many of which are listed in the bibliography, and to the recent books by Scott (1959) and by White (1959).

REFERENCES

Ambler, E., and Hudson, R. P. (1956). *J. Research Natl. Bur. Standards* **57,** 23. An examination of the 1955 helium vapor-pressure scales of temperature.

Atkins, K. R. (1959). "Liquid Helium." Cambridge Univ. Press, London and New York.

Beattie, J. A. (1955). *In* "Temperature—Its Measurement and Control in Science and Industry" (H. C. Wolfe, ed.), Vol. II, p. 63. Reinhold, New York; Chapman & Hall, London. Gas thermometry

Becker, J. H. (1955). Private communication.

Black, I. A., Bolz, L. H., Brooks, F. P., Mauer, F. A., and Peiser, H. S. (1958). *J. Research Natl. Bur. Standards* **61,** 367. A liquid helium cold cell for use with an X-ray diffractometer.

Blackman, M., Egerton, A., and Truter, E. V. (1948). *Proc. Roy. Soc.* **A194,** 147. Heat transfer by radiation to surfaces at low temperatures.

Blake, C., Chase, C. E., and Maxwell, E. (1958). *Rev. Sci. Instr.* **29,** 715. Resistance thermometer bridge for measurement of temperatures in the liquid helium range.

Blears, J. (1951). *J. Sci. Instr.* **28,** Suppl. No. 1, p. 36. Application of the mass spectrometer to high vacuum problems.

Bolz, L. H., Boyd, M. E., Mauer, F. A., and Peiser, H. S. (1959a). *Acta Cryst.* **12,** 247. A re-examination of the crystal structures of α and β nitrogen.

Bolz, L. H., Mauer, F. A., and Peiser, H. S. (1959b). Density of solid oxygen from an X-ray measurement of the lattice constant. Unpublished results.

Brickwedde, F. (1958). *Physica* **24,** S128. The "1958" He scale of temperature.

Broida, H. P. (1957). *Ann. N. Y. Acad. Sci.* **67,** 530. Stabilization of free radicals at low temperatures.

Broida, H. P., and Pellam, J. R. (1954). *Phys. Rev.* **95,** 845. Phosphorescence of atoms and molecules of solid nitrogen at 4.2°K.

Broom, R. F., and Rose-Innes, A. C. (1956). *J. Sci. Instr.* **33,** 420. Simple equipment for a small low temperature laboratory.

Bunch, M. D., and Powell, R. L. (1958). *Proc. Cryogenic Eng. Conf., Natl. Bur. Standards, Boulder, Colorado, 1957,* p. 269. Calibration of thermocouples at low temperatures.

Bunch, M. D., and Powell, R. L. (1959). Private communication. Calibration of thermocuples for use at low temperatures.

Burns, J. (1957). *Rev. Sci. Instr.* **28,** 469. Copper foil trap of improved form.

Chasmer, R. P., Craston, J. L., Isaacs, G., and Young, A. S. (1951). *J. Sci. Instr.* **28,** 206. A method of sealing sapphire to glass and its application to infrared photocells.

Chelton, D. B., and Mann, D. B. (1956). *Radiation Lab.,* **UCRL-3421.** Cryogenic data book. Univ. of Calif. (Berkeley).

Clement, J. R., and Quinell, E. H. (1952). *Rev. Sci. Instr.* **23,** 213. The low temperature characteristics of carbon-composition thermometers.

Coe, R. F., and Riddiford, L. (1955). *J. Sci. Instr.* **32,** 207. The final vacua of oil diffusion pumps.

Dauphinee, T. M., MacDonald, D. K. C., and Pearson, W. B. (1953). *J. Sci. Instr.* **30,** 399. The use of thermocouples for measuring temperatures below 70°K to ±0.5°K.

Duerig, W. H., and Mador, I. L. (1952). *Rev. Sci. Instr.* **23,** 421. An optical cell for use with liquid helium.

Foner, S. N., Mauer, F. A., and Bolz, L. H. (1959). *J. Chem. Phys.,* **31,** 546. Argon deposit on a 4.2°K surface.

Fontana, B. J. (1958). *J. Appl. Phys.* **29,** 1668. Thermometric study of the frozen products from the nitrogen microwave discharge.

Foote, T., and Harrington, D. B. (1957). *Rev. Sci. Instr.* **28,** 585. Demountable metal vacuum seal.

Friedberg, S. A. (1955). *In* "Temperature—Its Measurement and Control in Science and Industry" (H. C. Wolfe, ed.), Vol. II, p. 359. Reinhold, New York; Chapman & Hall, London. Semiconductors as thermometers.

Fulk, M. M., Devereux, R. J., and Schrodt, J. E. (1957). *Proc. Cryogenics Eng. Conf., Natl. Bur. Standards, Boulder, Colorado, 1956,* p. 163. Heat transport through powders.

Fuschillo, N. (1957). *J. Phys. Chem.* **61,** 644. A low temperature scale from 4 to 300°K in terms of a gold-cobalt versus copper thermocouple.

Gaffney, J., and Clement, J. R. (1955). *Rev. Sci. Instr.* **26,** 620. Liquid helium level finder.

Hernandez, H. P., Mark, J. W., and Watt, R. D. (1957). *Rev. Sci. Instr.* **28,** 528. Designing for safety in hydrogen bubble chambers.

Higatsberger, M. J., and Erbe, W. W. (1956). *Rev. Sci. Instr.* **27,** 110. Improved metal-to-metal vacuum seals.

Hill, R. W., and Lounasmaa, O. V. (1957). *Phil. Mag.* [8] **2,** 143. The specific heat of liquid helium.

Hilsenrath, J., Beckett, C. W., Benedict, W. S., Fano, L., Hoge, H. J., Masi, J. F., Nuttall, R. L., Touloukian, Y. S., and Woolley, H. W. (1955). *Natl. Bur. Standards (U. S.) Circ.* **564.** Tables of thermal properties of gases.

Hoge, H. J. (1941). *In* "Temperature—Its Measurement and Control in Science and Industry" Vol. I, p. 141. Reinhold, New York. A practical temperature scale below the oxygen point and a survey of fixed points in this range.

Hoge, H. J. (1950). *J. Research Natl. Bur. Standards* **44,** 321. Vapor pressure and fixed points of oxygen and heat capacity in the critical region.

Hoge, H. J., and Brickwedde, F. G. (1939). *J. Research Natl. Bur. Standards* **22,** 351. Establishment of a temperature scale for the calibration of thermometers between 14° and 83°K.

Housekeeper, W. G. (1923). *J. Am. Inst. Elec. Engrs.* **42,** 954. The art of sealing base metals through glass.

Hudson, R. P. (1958). Private communication.

Hull, A. W., and Burger, E. E. (1934). *Physica* **5**, 384. Glass to metal seals.

Jacobs, R. B., and Richards, R. J. (1957). *Rev. Sci. Instr.* **28**, 291. Vacuum insulated transfer tubes.

Janes, G. S., Hyman, L. G., and Strumski, C. J. (1956). *Rev. Sci. Instr.* **27**, 527. Helium-cooled liquid hydrogen-deuterium target.

Keesom, W. H., and Matthijs, C. J. (1935). *Physica* **2**, 623. Measurements on thermoelectric forces of some alloys at temperatures from 2.5 to 17.5°K.

Kerr, E. C. (1957). *J. Chem. Phys.* **26**, 511. Density of liquid He⁴.

Kunzler, J. E. (1956). *Rev. Sci. Instr.* **27**, 879. Liquid nitrogen vacuum trap containing a constant cold zone.

Kunzler, J. E., Geballe, T. H., and Hull, G. W. (1957). *Rev. Sci. Instr.* **28**, 96. Germanium resistance thermometers suitable for low temperature calorimetry.

Larson, E. V., and Mayer, R. (1952). *Rev. Sci. Instr.* **23**, 692. Recording and indicating instruments for temperature measurements down to 10°K using copper-constantan thermocouples.

Linder, C. T. (1950). *Westinghouse Research Labs., Research Rept.* **R-94433-2-A**. The measurement of low temperatures.

Littauer, R. (1958). *Rev. Sci. Instr.* **29**, 178. Pressure-fed liquid hydrogen target.

Louckes, F. I., Jr. (1957). *Rev. Sci. Instr.* **28**, 468. Unique sliding seal for a vacuum chamber.

Maimoni, A. (1956). *Rev. Sci. Instr.* **27**, 1024. Hot wire liquid-level indicator.

Mikesell, R. P., and Scott, R. B. (1956). *J. Research Natl. Bur. Standards* **57**, 371. Heat conduction through insulating supports in very low temperature equipment.

Mueller, M. H. (1956). *Rev. Sci. Instr.* **27**, 411. Vacuum seal for thin metal windows.

Noggle, T. S., Blewitt, T. H., Coltman, R. R., and Klabunde, C. E. (1957). *Rev. Sci. Instr.* **28**, 464. Thermal-EMF-free vacuum seal for electrical lead wires.

Powell, R. L., and Blanpied, W. A. (1954). *Natl. Bur. Standards (U. S.) Circ.* **556**, 24. Thermal conductivity of metals and alloys at low temperatures.

Powell, R. L., Rogers, W. M., and Roder, H. M. (1957). *Proc. Cryogenic Eng. Conf., Natl. Bur. Standards, Boulder, Colorado, 1956*, p. 166. Thermal conductivities of copper and copper alloys.

Riddiford, L., and Coe, R. F. (1951). *J. Sci. Instr.* **28**, 352. Leaks in vacuum liquid air traps.

Rosebury, F. (1956). "Tube Laboratory Manual." Research Lab. of Electronics, M. I. T., Cambridge, Massachusetts.

Rossini, F. D., Wagman, D. D., Evans, W. H., Levine, S., and Jaffe, I. (1952). *Natl. Bur. Standards (U. S.) Circ.* **500**. Selected values of chemical thermodynamic properties.

Sanderson, R. T. (1948). "Vacuum Manipulation of Volatile Compounds." Wiley, New York; Chapman & Hall, London.

Schoen, L. J., Kuentzel, L. E., and Broida, H. P. (1958). *Rev. Sci. Instr.* **29**, 633. Glass dewars for optical studies at low temperatures.

Schwarz, E. (1955). *J. Sci. Instr.* **32**, 445. A permanent vacuum seal.

Scott, H. (1935). *J. Franklin Inst.* **220**, 733. Recent developments in metals sealing into glass.

Scott, R. B. (1941). *In* "Temperature—Its Measurement and Control in Science and Industry" Vol. I, p. 206. Reinhold, New York. The calibration of thermocouples at low temperatures.

Scott, R. B. (1959). "Cryogenic Engineering." Van Nostrand, Princeton, New Jersey, and Toronto, Canada.

Scott, R. B., Cook, J. W., and Brickwedde, F. G. (1931). *J. Research Natl. Bur. Standards* **7**, 935. Silvering and evacuating Pyrex Dewar flasks.

Shenker, H., Lauritzen, J. I., Jr., Corruccini, R. J., and Lonberger, S. T. (1955). *Natl. Bur. Standards (U. S.) Circ.* **561**. Reference tables for thermocouples.

Smotrich, H., and Feldman, L. (1956). *Rev. Sci. Instr.* **27**, 970. Three directional motion vacuum sealed joint.

Spees, A. H., Reynolds, C. A., Boxer, A., and Pearson, G. (1957). *Rev. Sci. Instr.* **28**, 1090. Vacuum gasket at low temperatures.

Steckelmacher, W. (1951). *J. Sci. Instr.* **28**, Suppl. No. 1, p. 10. Review of vacuum gauges.

Stimson, H. F. (1955). *In* "Temperature—Its Measurement and Control in Science and Industry" (H. C. Wolfe, ed.), Vol. II, p. 141. Reinhold, New York; Chapman & Hall, London. Precision resistance thermometry and fixed points.

Stober, A. K. (1959). *Natl. Bur. Standards (U. S.), Tech. News Bull.* **43**(8), 146. A method of supplying low-temperature liquids to a large-scale, accelerated research program.

Stober, A. K., and Bass, A. M. (1956). Unpublished results.

Strong, J., (1938). "Procedures in Experimental Physics." Prentice-Hall, New York.

Swenson, C. A. (1950). *Phys. Rev.* **79**, 626. The liquid-solid transformation in helium near absolute zero.

Sydoriak, S. G., and Sommers, H. S., Jr. (1951). *Rev. Sci. Instr.* **22**, 915. Low evaporation rate storage vessel for liquid helium.

Taylor, A. (1956). *Rev. Sci. Instr.* **27**, 757. Improved demountable crystallographic rotating anode X-ray tube.

Van Dijk, H., and Shoenberg, D. (1949). *Nature* **164**, 151. Tables of vapor pressure of liquid helium.

Van Heerden, P. J. (1956). *Rev. Sci. Instr.* **27**, 410. Metal gaskets for demountable vacuum systems.

Warschauer, D. M., and Paul, W. (1956). *Rev. Sci. Instr.* **27**, 419. Infrared windows for helium cryostats.

Waters, J. L. (1958). *Rev. Sci. Instr.* **29**, 1053. Low temperature optical window seal used at 80°K.

Wexler, A. (1951). *J. Appl. Phys.* **22**, 1463. Evaporation rate of liquid helium.

Wexler, A., and Corak, W. S. (1951). *Rev. Sci. Instr.* **22**, 941. Measurement and control of the level of low-boiling liquids.

White, G. K. (1959). "Experimental Techniques in Low Temperature Physics." Oxford Univ. Press, London and New York.

White, G. K., and Woods, S. B. (1957). *Rev. Sci. Instr.* **28**, 638. Indium resistance thermometer; 4°–300°K.

Woolley, H. W., Scott, R. B., and Brickwedde, F. G. (1948). *J. Research Natl. Bur. Standards* **41**, 379. Compilation of thermal properties of hydrogen in its various isotopic and ortho-para modifications.

Yarwood, J. (1955). "High Vacuum Technique." Wiley, New York; Chapman & Hall, London.

Young, J. R. (1958). *Rev. Sci. Instr.* **29**, 795. Vacuum limitations of rubber O-ring joints.

Young, J. R. (1959). *Rev. Sci. Instr.* **30**, 291. Cleaning techniques for rubber O-rings used in vacuum systems.

6. Optical Spectroscopy of Trapped Radicals

D. A. RAMSAY

Division of Pure Physics, National Resarch Council, Ottawa, Canada

I. Introduction

Investigations of the optical spectra of transient species trapped in solid media fall into two categories. The first class deals with the emission and absorption spectra of metastable excited states of stable molecules, notably the phosphorescences involving the lowest excited triplet states of aromatic molecules. The second class deals with the spectra of chemically reactive molecular fragments which normally have only a transient existence in the gas phase at room temperature, but which may be trapped for relatively long periods in rigid media. The publications pertaining to the first class of spectra are numerous and have been adequately summarized in review articles by Kasha (1947) and Kasha and McGlynn (1956). The present chapter therefore contains only a brief historical sketch of the development of this field of research. Most of the papers dealing with the second class of spectra have appeared within the last five years and have not previously been reviewed. An attempt will therefore be made in this chapter to provide a comprehensive summary of work in this particular field.

II. Historical Background

A. Phosphorescent State Studies

As early as 1888, Wiedemann and Dewar noted that complex molecules emit phosphorescence when irradiated at low temperatures. Although many similar observations were reported by later workers and numerous attempts were made to interpret the phosphorescence in terms of complex solid state interactions, the mechanism of the phosphorescence remained obscure until the early 1940's, when Lewis *et al.* (1941) and Terenin (1943) suggested that the phosphorescence may be due to a triplet→singlet emission of the complex molecule. Confirmation of this view was afforded in a series of classic papers by Lewis and co-workers. This work has stimulated considerable activity in this field in recent years. In 1944 Lewis and Kasha gave spectroscopic data on the phosphorescence emission of 89 molecules in rigid glasses at liquid nitrogen temperatures. They provided the general identification that these emission spectra were due to triplet→singlet transitions from the lowest excited triplet states of these molecules. In 1945 Lewis and Calvin reported qualitative measurements of the paramagnetism of molecules in the excited state. Subsequently Lewis *et al.* (1949) showed that a quantitative determination of the paramagnetism is in agreement with the assignment of the excited state to a triplet state. In 1955 Evans reported studies of the variation of paramagnetic susceptibility with time upon extinction of the exciting radiation. He showed that the decay was exponential with a time constant equal to that for the decay of the phosphorescence.

"Absorption spectra of the phosphorescent state," now recognized as triplet←triplet absorption spectra, were recorded in the early researches of Lewis and co-workers and have been studied in detail by McClure (1951) and others. Using flash photolysis techniques Porter and Windsor (1953, 1954) were able to study triplet←triplet absorption spectra even in fluid solvents at room temperature.

The complete mechanism for the excitation of phosphorescence involves three stages:

(1) Absorption of the exciting radiation by the parent molecule, raising it to an excited singlet state,

(2) Internal conversion of the energy to the lowest excited triplet state of the molecule, and

(3) Phosphorescent emission from the lowest excited triplet state to the singlet ground state.

The factors governing the internal conversion of energy in complex molecules have been summarized by Kasha and McGlynn (1956), and a discussion of the function of the rigid medium has been given by Kasha (1947).

B. SPECTRA OF TRAPPED RADICALS

The earliest experiments on the spectra of radicals trapped in the solid state appear to be those of Vegard and of McLennan and Shrum in 1924 on the spectra emitted by solid nitrogen and other condensed gases when bombarded with electrons and positive ions ("canal" rays) at liquid hydrogen temperatures. These experiments were prompted by the suggestion of Vegard that the auroral spectrum, and in particular the green auroral line at 5577 A, is due to the irradiation of small condensed particles of solid nitrogen in the upper atmosphere by electrons. Although this theory was not corroborated by the experiments of McLennan and co-workers (1924–1929), who showed that the auroral line was due to atomic oxygen, the luminescence observed in solid nitrogen under the above conditions of irradiation retained a very considerable interest and was the subject of numerous investigations by Vegard and McLennan and their respective co-workers. Vegard's contributions have been summarized in three principal papers (Vegard, 1930b,c; Vegard and Stensholt, 1935). More recently Broida and co-workers at the National Bureau of Standards, Washington, have carried out extensive reinvestigations of the luminescence of solid nitrogen at low temperatures. This work is described in more detail in Section IV, B.

In 1930 Lavin and Bates observed an intense green glow when the products of a discharge through streaming ammonia were cooled with liquid nitrogen. These observations were confirmed by Lunt and Mills (1935) and the possibility that the glow was due to NH_2 radicals was discussed. In 1933 Harteck prepared a yellow solid by reacting hydrogen atoms with nitric oxide at liquid air temperature and postulated that the yellow color was due to HNO or to a polymer $(HNO)_n$.

Most of the work on the spectra of trapped radicals however has been carried out since 1950, and indeed most of the publications have appeared since 1954. Rice and Freamo (1951) reported observations on a blue solid obtained when hydrazoic acid at low pressures is passed through a heated quartz tube and the decomposition products are rapidly condensed on a liquid nitrogen-cooled finger. This particular observation has aroused the active interest of many subsequent investigators. Several other colored solids, possibly containing trapped radicals, have subsequently been reported by Rice and co-workers.

The year 1954 brought significant developments from three different groups of workers. A large program of work on the emission and absorption spectra of free radicals trapped in solids at low temperatures was initiated at the National Bureau of Standards, Washington, following some interesting observations by Broida and Pellam on the phosphorescence of atoms and molecules of solid nitrogen at 4.2°K. At the University of California the matrix isolation method for the study of the spectra of free radicals in inert matrices, e.g., solid Xe, N_2, at low temperatures was proposed by Whittle, Dows, and Pimentel and has been followed by several publications, mainly pertaining to the infrared region of the spectrum. In England, Norman and Porter obtained the electronic absorption spectra of several free radicals trapped in rigid glasses at liquid nitrogen temperatures. The radicals were produced *in situ* by ultraviolet irradiation. Some similar, but not entirely confirmatory, results were reported by Sowden and Davidson from the California Institute of Technology in 1956. In the last year or two, Robinson and McCarty have developed a technique, approximating to a molecular beam technique, for preparing deposits containing free radicals on a cold finger at 4.2°K and have obtained spectra of several simple free radicals. Discussion of the various experimental techniques employed and of the spectra observed is given below.

III. Experimental Methods

The experimental techniques used for preparing solid films of trapped radicals suitable for spectroscopic investigation fall into two classes: (1) preparation of the radicals in the gas phase by the decomposition of parent molecules, followed by rapid freezing, and (2) preparation of the radicals *in situ* by the decomposition of molecules in the solid state.

In the first class, pyrolysis and electrical discharge techniques have been extensively used and in one or two cases the use of the photochemical method has been reported. In the second class, a variety of techniques has been employed. Irradiation of the solid with ultraviolet radiation has been most widely used, though irradiation with X-rays and gamma rays has also been reported. In addition, bombardment of the solid with electrons and positive ions has been employed, and recently electrical discharges have been applied directly to the solid.

In all methods, the physical and chemical properties of the solid material in which the radicals are trapped are of prime importance. Physically the solid should be sufficiently rigid to prevent diffusion of the radicals, and chemically the medium should be inert. In many experiments the parent molecules are diluted with a considerable excess of an inert gas, e.g., Xe, A, N_2, before production of the radicals. Pimentel and co-workers generally use a dilution of $\sim$500 to 1 for most of their experiments, but Robinson and McCarty have found a dilution of 30 to 1 adequate for their purposes.

The Franck-Rabinowitsch "cage effect" (1934) is of considerable importance when the radicals are produced *in situ* in the solid. Unless the fragments produced by some primary decomposition process have sufficient energy to diffuse away from the center of production, they will remain in juxtaposition in the solid with a consequent high probability of recombination. The probability of recombination will thus depend on the excess energy with which the fragments are formed and on the thermal properties and viscosity of the medium. The probability will depend strongly on the energy of the incident quanta and on temperature.

The importance of the "cage" effect is well illustrated by some experiments reported by Sowden and Davidson (1956). These authors found that photolysis of a 4.7×10^{-5} M solution of acetophenone in 5:1 isopentane-3-methylpentane at room temperature produced benzene with a quantum yield of the order of 0.1 to 1.0. Irradiation of the same solution at 77°K effected no detectable decomposition after ten times the exposure necessary for complete decomposition at room temperature. Similar results were obtained in the ultraviolet irradiation of a 5×10^{-5} M solution of diphenyl-mercury at room temperature and at 77°K.

A diagram showing the metal Dewar and collection system used by Bass and Broida (1956) in their studies of the spectra emitted by solid nitrogen condensed at 4.2°K from a gas discharge is given in Fig. 1. Nitrogen gas at low pressure is subjected to an electrodeless discharge maintained by a 125-watt, 2450 Mc/sec oscillator. The products are condensed on a copper block in good thermal contact with a liquid helium reservoir. The cold surface condenses all gases except helium and thus acts as an effective high speed pump. To prevent solidification of the discharge products at temperatures above 4.2°K, the gas is carried to the cold surface via a passageway which is maintained near room temperature by passing warm helium gas through the compound walls surrounding the channel E. Emission from the condensed products is observed through the quartz windows. It should be mentioned that glass Dewars may also be used for optical investigations at low temperatures. A number of such Dewars have been described by Schoen *et al.* (1958).

In the experiments of Robinson and McCarty (1959), parent compounds are diluted with argon in the ratio of approximately 1:30, and the resulting mixtures subjected to a 100-watt, 2450 Mc/sec discharge at a pressure of

~0.1 mm Hg. The products then pass through a 1-mm diameter hole and spray onto a quartz plate, cooled with liquid helium and situated 3.5 cm from the leak. The liquid helium-cooled surface ensures rapid pumping, and in addition two oil diffusion pumps are installed on either side of the cold plate to remove randomly directed molecules. This arrangement allows most molecules to travel from the leak to the cold plate without collision and provides conditions which are well suited for the study of highly reactive species. A photograph of Robinson and McCarty's apparatus is shown in Fig. 2.

A simple apparatus which permits the ultraviolet irradiation of frozen

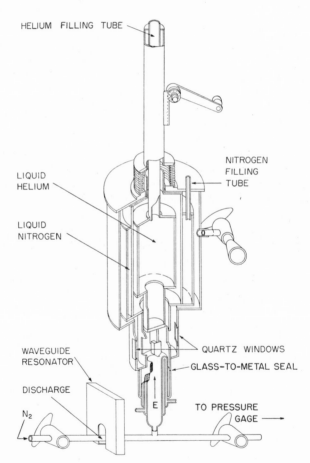

FIG. 1. Metal Dewar and collection system used by Bass and Broida (1956) to study the spectra emitted from solid nitrogen condensed at 4.2°K from a gas discharge. Reproduced by kind permission of the authors and of *Physical Review*.

FIG. 2. Photograph of the apparatus used by Robinson and McCarty (1959) in their investigations of the absorption spectra of free radicals condensed at 4.2°K from a gas discharge. A, Discharge tube; B, discharge tube exhaust; C, leak; D, one of the two external quartz viewing windows; E, diffusion pumps; F, liquid nitrogen-cooled baffles; G, auxiliary leak; H, liquid helium Dewar. Reproduced by kind permission of the authors and of the *Journal of Chemical Physics*.

solids and the study of their absorption spectra at 77°K was described by Norman and Porter (1955) and is shown in Fig. 3. A quartz cell fits closely between the two inner windows of a quartz Dewar and contains the sample to be investigated. The cell is clamped rigidly to the flask and the whole apparatus mounted so that it can be conveniently rotated by 90°. In one position the sample can be irradiated by means of a high pressure mercury lamp, and in another position the absorption spectrum of the sample can be studied. No difficulty is experienced with gas bubbles from the liquid nitrogen, providing the cell fits closely between the inner windows of the Dewar flask.

Most of the features in solid state spectra are broad compared to the corresponding gas phase features, and hence the use of spectrographs of high resolving power is not essential in solid state studies. The sharpest lines which have been reported in the work presently under review have a width of a few cm^{-1} (Robinson and McCarty, 1959; Peyron *et al.*, 1959). Spectrographs capable of a resolution of 1 cm^{-1} should therefore be adequate for most studies.

IV. Luminescences in Solids

Numerous publications have appeared on the luminescence emitted when solid nitrogen is bombarded with electrons or positive ions or when

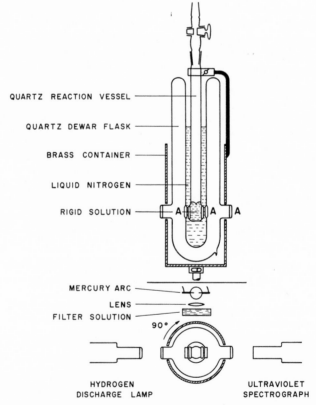

FIG. 3. Apparatus used by Norman and Porter (1955) for the ultraviolet irradiation of rigid solutions and the study of their absorption spectra. Reproduced by kind permission of the authors and of the Royal Society of London.

the products of an electrical discharge through nitrogen are condensed at low temperatures. Nitrogen gives an exceptionally intense luminescence and a considerable afterglow whereas the solid systems O_2, CO, N_2O, and NH_3 merely give a faint luminescence with no afterglow. The high luminous power of solid nitrogen is restricted to the α modification stable below 35.5°K (Vegard, 1930b). An account of the spectral investigations which have been carried out for H_2, N_2, O_2, and NH_3 is given below.

A. HYDROGEN

Vegard and Keesom (1927) observed a greenish-blue luminescence when solid hydrogen, condensed on a copper rod at 4.2°K, was bombarded with fast cathode rays. The luminescence persisted for about a minute after the cathode rays were stopped. The spectrum showed only a continuum. The same authors noted that if "canal" rays were used for the bombardment, a strong luminescence resulted, the spectrum of which contained twenty-six lines. The five most prominent lines belonged to the Balmer series of hydrogen, thus indicating that hydrogen atoms were present.

Bass and Broida (1958) observed that when the products of an electrical discharge through hydrogen were condensed to 4.2°K, an extremely feeble green glow could be seen in the gas surrounding the cold collecting surface. The glow was not emitted from the cold surface itself. The spectrum, taken with a $f/0.6$ spectrograph, shows the H_α, H_β, and H_γ lines, probably due to scatter from the discharge, and a continuum superimposed on which broad bands were seen at 4740, 4650, 4560, and 4500 A. No assignments of these bands were given.

B. NITROGEN

Extensive investigations of the luminescences produced in solid nitrogen have been carried out by several workers and a multitude of spectra have been observed. The complexity of the results is well summarized by Peyron and Broida (1959a), who state that "the number of different spectra emitted by solids condensed from gas discharges containing nitrogen is so large that a simple and cohesive presentation is not possible."

The earliest investigations were carried out independently in 1924 by Vegard and by McLennan and Shrum. These workers observed an intense luminescence when solid nitrogen at liquid hydrogen temperature was bombarded with electrons. The spectrum of the luminescence was found to contain two strong features in the green, called N_1 and N_2 by Vegard, as well as several bands in the blue and ultraviolet. N_1 is a broad band extending from about 5525 to 5660 A with a principal maximum at 5555 A and subsidiary maxima at 5616 and 5659 A. N_2 is a comparatively sharp line with a wavelength of 5230 A. In subsequent experiments Vegard (1925) observed two other features in the red, which he called N_3 and N_4. The former consists of a broad band with maxima at 6365 and 6577 A, and the latter is a comparatively sharp line at 5945 A.

Vegard carried out experiments in which nitrogen was diluted with inert gas and found that with increasing dilution the broad band N_1 becomes narrower and the position of the principal maximum changes. At very large dilutions the two subsidiary maxima disappear and the principal maximum assumes the character of a sharp line with a wavelength of 5577.38 A (see summary by Vegard, 1930b). Since this value agrees within experimental error with the wavelength of the strong auroral line at 5577.35 A, Vegard concluded that the green auroral line was due to the bombardment of finely divided particles of solid nitrogen in the upper atmosphere with electrons. McLennan and Shrum (1925), however, showed that a sharp line at 5577.35 $\pm$ 0.15 A is observed in emission from discharges through oxygen, and subsequent work by McLennan et al. (1927a) confirmed that this line is due to atomic oxygen. The line was identified with the $^1S \rightarrow {}^1D$ forbidden transition of O I (McLennan, 1928). The observation by Vegard of the N_1 feature near 5577 A in the luminescence from solid nitrogen was due to a small amount of oxygen impurity in the nitrogen.

Vegard (1924b) noted that when the bombardment of the solid nitrogen was stopped, an afterglow was observed which persisted for several minutes. When the sample was gradually warmed, an intense light emission was observed and the spectrum showed the 5230-A line very strongly, together with the N_1 band and other bands in the blue and ultraviolet. When the temperature reached 35.5°K, a bright flash was seen and a phase change observed. No phosphorescence was observed above 35.5°K. Similar observations on the phosphorescence spectrum were carried out by McLennan and co-workers (1924, 1927b, 1928).

Systematic investigations of the luminescence and phosphorescence spectra excited under a wide variety of experimental conditions were carried out by Vegard and co-workers. Experiments were carried out at liquid hydrogen and liquid helium temperatures, and the luminescences were excited both with cathode ray and with "canal" ray bombardment. The variation of the spectra with cathode ray velocity was studied over a range of velocities corresponding to potentials varying from 100 to 300,000 volts. The effect of dilution of the nitrogen with inert gases was also studied. In addition to the four features described above (N_1, N_2, N_3, and N_4), numerous band series were observed, some of which were identified with molecular transitions of nitrogen and nitric oxide (see Table I). An important new band system (the ϵ-system) was found and was later observed in the gas phase by Kaplan (1934). This system, now known as the Vegard-Kaplan system, was assigned on the basis of the

TABLE I

Summary of Spectra Emitted during Deposition of
Discharge Products Containing Nitrogen[a]

NBS designation	Veg-ard's desig-nation	Most intense feature			Wavelength region	Number of lines or bands	Tentative assignment		Shift in solid (cm⁻¹)	Width (cm⁻¹)
		Inten-sity	λ(A)	ν(cm⁻¹)			Emit-ter	Transition		
α Group										
α	N₂	1000	5229.8	19116	5212–5241	6	N	$^2D - {}^4S(0, 0)$	−107	10
α'	N₄	150	5945.7	16814	5932–5970	5	N	$^2D - {}^4S(0, 1)$		10
α″	—	10	4699.9	21271	4674–4726	9	N	$^2D - {}^4S(1, 0)$		7
α‴	—	5	5865.3	17044	5836–5895	7	N	$^2D - {}^4S(1, 2)$		≦7
δ Group										
δ	—	10000	10475	9546	10396–10495	6	N	$^2P - {}^2D(0, 0)$	−67	10
δ'	—	600	13790	7250	13330–13790	2	N	$^2P - {}^2D(0, 1)$		75
δ″	—	20	8555.1	11686	8471–8584	8	N	$^2P - {}^2D(1, 0)$		≦7
μ Group										
μ	—	10	3479.9	28728	3480–3485	2	N	$^2P - {}^4S(0, 0)$	−112	≦6
β Group										
β	N₁	2500[b]	5549	18016	5550–5657	3	O	$^1S - {}^1D(0, 0)$	+90	100
β'	N₃	100[b]	6367	15701	6367–6500	3	O	$^1S - {}^1D(0, 1)$		100
β″	—	25[b]	4940	20237	4940–5025	3	O	$^1S - {}^1D(1, 0)$		80
Bands										
VK	ε	250[c]	2946.3	33931.0	2300–4450	30	N₂	$A^3\Sigma - X^1\Sigma$	120–140	—
A	C	100[b]	4490.0	22265.5	3572–6390	10	O₂	$A^3\Sigma - X^1\Sigma$		—
M	α	125[c]	4207.4	23761.0	3085–5296	9	NO	$^4\Pi - X^2\Pi$		—
B	B	60[b]	4306	23217	3000–4700	7	NO₂?			—
D	E	10[b]	3392.6	29467.5	2613–3803	11	NO	$(\beta)B^2\Pi - X^2\Pi$	95–165	—
E	G	40[d]	2470.0	40473.6	2361–3457	20	NO	$(\gamma)A^2\Sigma^+ - X^2\Pi$	0	—
F	—	40[d]	3359.2	29760.5	3359–3370	2	NH	$A^3\Pi_i - X^3\Sigma^-$	0	—
G	—	20[d]	6612	15119	6300–6770	8	N₂	$B^3\Pi - A^3\Sigma$	15	—

[a] This table is based largely on the table given by Peyron and Broida (1959a).
[b] Intensity estimate in presence of 0.1% oxygen.
[c] Intensity estimate in a 5% nitrogen 95% argon mixture.
[d] Intensity in special conditions (Bass and Broida, 1956).

vibrational analysis of the bands to the $A^3\Sigma_u^+ - X^1\Sigma_g^+$ forbidden transition of molecular nitrogen.

Under high dispersion, Vegard noted that the features N_1, N_2, N_3, and N_4 and some of the molecular band systems are split into several components with fairly close spacings. McLennan et al. (1927b) similarly resolved the N_2 feature into eight components with roughly equal spacings between 5204 and 5240 A. The latter authors attributed the "fine" structure to rotational effects, but Vegard (1930d) rejected this conclusion since the spacings are too large, and moreover the multiplicity of the lines is more pronounced at liquid helium than at liquid hydrogen temperature. Vegard also dismissed the possibility that the splitting is due to the multiplicity of the electronic terms and ascribed the effect to vibrations of the crystal lattice.

Vegard (1930b) noted that the features N_2 and N_4 and the features N_1 and N_3 were separated by approximately 2300 cm⁻¹, an interval which lies close to the vibration frequency of the nitrogen molecule in the gas phase. He therefore postulated that the luminescence from solid nitrogen at very low temperatures may involve a process in which an electronic transition of a nitrogen atom is accompanied by a simultaneous vibrational transition of a nitrogen molecule. This idea is discussed in more detail below.

After the researches of Vegard and McLennan and their respective co-workers, no further investigations were carried out until 1954 when Broida and Pellam observed a brilliant green glow with occasional bright flashes when the products of a discharge through nitrogen were condensed at liquid helium temperature. The glow persists for several minutes with decreasing intensity after the discharge is stopped and reappears on warming to about 10°K. On further warming the green glow decreases and at about 25°K a less intense blue-green glow is observed. This glow persists until the temperature reaches about 35°K when a bright flash is seen. No further phosphorescence is then observed.

The spectra of these glows have been studied by several workers (Bass and Broida, 1956; Broida and Peyron, 1957, 1958; Peyron and Broida, 1959a; Peyron et al., 1959) and the general characteristics found to be very similar to those observed by Vegard and McLennan and their co-workers. The most prominent features and their assignments are summarized in Table I. The National Bureau of Standards workers use Greek symbols to denote atomic transitions and ordinary capital letters to denote molecular transitions. For comparison purposes the notation used by Vegard is also given. Some typical spectra are reproduced in Fig. 4.

The strongest atomic features in the visible region are the α- and β-features which have been assigned to the $^2D - {}^4S$ transition of atomic nitrogen and to the $^1S - {}^1D$ transition of atomic oxygen, respectively. The former is prominent in the spectrum of the afterglow while the latter is absent. Broida and Peyron (1958) showed that the intensity of the β-feature decreases as the amount of oxygen impurity in the nitrogen is reduced, thus providing confirmation for the assignment of this feature. The $^2P - {}^2D$ and $^2P - {}^4S$ transitions of atomic nitrogen have been identified by Peyron et al. (1959) with the δ- and μ-features observed near 10475 and 3480 A, respectively.

A series of red-degraded bands (the A-bands) extending from 3572 to 6390 A was observed by Bass and Broida (1956) and was assigned by Herzfeld and Broida (1956) to a $^5\Sigma_g^+ - A\,^3\Sigma_u^+$ transition of molecular nitrogen. Subsequent investigations by Peyron and Broida (1959a) however rendered the assignment untenable since no isotopic shifts were observed with N^{15}. In addition, Peyron and Broida (1959b) have shown by the use of O^{18} that the bands are due to molecular oxygen and may be assigned to a $^3\Sigma - {}^1\Sigma$ transition.

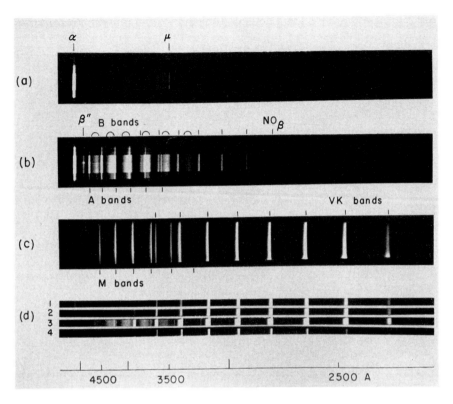

FIG. 4. Emission spectra from solid nitrogen condensed at 4.2°K from gas discharges under various conditions: (a) pure nitrogen, (b) nitrogen with 0.1% oxygen, (c) 5% nitrogen with 95% argon, (d) series of spectra taken during warm-up of solid after deposition from a mixture of 10% nitrogen and 90% argon. Reproduced from Peyron and Broida (1959a) by kind permission of the authors and of the *Journal of Chemical Physics*.

The Vegard-Kaplan bands of nitrogen ($A^3\Sigma_u^+ - X^1\Sigma_g^+$) have been observed in the glow emitted by pure nitrogen, but the intensity of the bands is considerably enhanced when the nitrogen is diluted with a rare gas. Broida and Peyron (1957) found that the bands were particularly intense when the nitrogen is diluted with 80 to 95 % argon. The bands are believed to arise from the recombination of ^4S nitrogen atoms in the solid, forming molecular nitrogen in the $^3\Sigma$ state which then radiates to the ground state.

In the blue and ultraviolet, Bass and Broida (1956) observed two series of bands which increase in intensity with increasing oxygen impurity and which appear to correspond to the beta ($B^2\Pi - X^2\Pi$) and gamma ($A^2\Sigma^+ -$

$X^2\Pi$) systems of NO. The bands of the gamma system exhibit a curious phenomenon in that the vibrational and rotational intensity distributions in the solid spectrum indicate a higher "temperature" than in the spectrum of the gas afterglow.

Bass and Broida (1956) also observed some diffuse bands (the B-bands) in the region 3000 to 4700 A which they tentatively assigned to NO_2. These bands were observed in the warm-up glow by Broida and Peyron (1957), who later showed that the intensity of the bands was sensitive to small amounts of oxygen impurity (Broida and Peyron, 1958). The presence of one or more oxygen atoms in the carrier of the bands was demonstrated by Peyron and Broida (1959b) who showed that the bands are shifted to shorter wavelengths when O^{18} is used. A band near 3360 A was assigned by Bass and Broida (1956) to the $A^3\Pi_i - X^3\Sigma^-$ transition of NH. Confirmation of this assignment was afforded by Broida and Peyron (1958), who showed that the intensity of the band was sensitive to small amounts of hydrogen impurity. Some bands of the first positive system of nitrogen ($B^3\Pi_g - A^3\Sigma_u^+$) were also reported by Bass and Broida (1956), but these observations have not been confirmed by subsequent workers (e.g., Peyron and Broida, 1959a).

A series of bands between 3080 and 5300 A (the M-bands) was observed by Broida and Peyron (1957) in the glow from the condensed products of a discharge through nitrogen diluted with argon. The bands were also observed in the warm-up glow. These authors later showed that the bands appear with mixtures of nitrogen and argon with less than 30 % nitrogen and have their maximum intensity with about 2 % nitrogen (Peyron and Broida, 1959a). Later it was found (Peyron and Broida, 1959b) that the bands appear with 1 % oxygen in argon and are not affected by small amounts of added nitrogen. Experiments with N^{15} and O^{18} show that the molecular emitter contains nitrogen and oxygen. The observations are consistent with a $^4\Pi - {}^2\Pi$ transition in NO.

Further investigations of the luminescences produced in solid nitrogen under various excitation conditions have been carried out at the National Bureau of Standards. Hörl (1959) repeated the early experiments of Vegard and McLennan with controlled electron energies and studied the effects of varying the electron energy on the spectrum excited. He also measured the lifetimes of the various phosphorescence bands. Schoen and Rebbert (1959) studied the spectral characteristics of the glows produced by applying an electrical discharge to solid nitrogen, while Wall et al. (1959) carried out similar experiments on the irradiation of solid nitrogen with gamma rays from a Co^{60} source. In all cases the spectral characteristics were similar to those described above.

Herzfeld (1957) developed a theory to explain the detailed structure of the α lines. He assumed that the $^2D - ^4S$ transition of atomic nitrogen is perturbed by neighboring molecules in the solid. Since the lines in the solid are relatively sharp, he assumed that the nitrogen atoms occupy a definite spatial relationship relative to the nitrogen molecules. The treatment envisages a nitrogen atom as being loosely bound to a neighboring molecule, and so placed that a line from the atom to the midpoint of the N—N bond is 2.5 A and makes an angle of 50° with the N—N bond. Although the treatment gives a self-consistent explanation of the gross shift and the fine splittings of the lines observed in the solid, and of the reduction of the half-life of the transition from $\sim$20 hours in the gas phase to about 15 seconds in the solid, the theory does not explain certain other features. Thus the treatment can only explain the existence of five lines in the solid whereas more have now been observed. Moreover, the theory does not account for the change in the fine splitting observed with N^{15} (Peyron $et\ al.$, 1959) or for the variation in the lifetimes observed for the individual lines (see Chapter 10 for the details of this theory). It should be pointed out that the loose complex is not the same as the N_3 radical proposed by Milligan $et\ al.$ (1956) to explain certain infrared observations, or by Thrush (1956) to explain the bands observed near 2600 A in the flash photolysis of HN_3. The N_3 radical is a 15-electron molecule and should have a linear ground state and vibration frequencies similar to those for the isoelectronic molecules CO_2^+ and NCO.

The idea of a loose complex is, however, a very useful one and has been used to explain some of the spectral features observed in the glow emitted by solid nitrogen. Broida and Peyron (1958) found that when the α and β features were overexposed, weaker features α' and β' were also observed at approximately 2300 cm^{-1} to lower frequencies. With still longer exposures further weak features α'' and β'' were observed at approximately 2200 cm^{-1} to higher frequencies. These authors suggested that the weak satellite lines may be due to double transitions involving an electronic transition of a nitrogen or oxygen atom and a simultaneous vibrational transition of a neighboring nitrogen molecule. A similar interpretation of the α' and β' features (N_4 and N_3 using Vegard's notation) had been given earlier by Vegard (1930d). The investigations of satellite lines were later extended by Peyron $et\ al.$ (1959), who observed the $^2P - ^2D$ transition of atomic nitrogen near 10470 A (δ feature), and found vibrational satellites to lower and higher frequencies, δ' and δ'' respectively. The $^2P - ^4S$ transition near 3480 A was also observed (μ feature) but the intensity was insufficient to permit the observation of satellite lines. An energy level diagram showing the transitions observed for the system N–N_2 is given in Fig. 5. The interpretation of the α''' feature observed by Peyron $et\ al.$ is perhaps questionable.

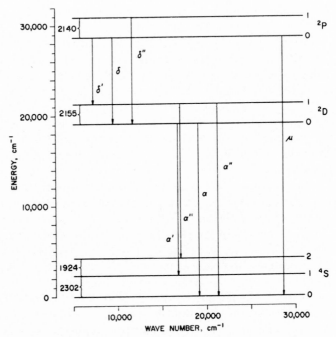

F IG. 5. Energy level diagram showing the observed transitions for the N—N₂ system. Reproduced from Peyron *et al.* (1959) by kind permission of the authors and of the *Journal of Chemical Physics*.

Confirmation of the above ideas is afforded by experiments carried out with N_2^{15} (Peyron *et al.*, 1959) (see Fig. 6). No shift of the strongest lines in the α, β, or δ features was observed, but the vibrational satellites were found to be slightly closer to the principal lines. Furthermore the displacements were roughly equal to those expected on the basis of a vibrational isotope shift. In addition it should be mentioned that the fine-structure splittings of the various groups were slightly smaller with N_2^{15}, and moreover the lifetimes of the various transitions were also modified. Supporting evidence that the satellite lines are due to simultaneous transitions is afforded by experiments carried out with a 50–50 mixture of N_2^{14} and N_2^{15}. The principal lines were found to be accompanied by three satellite lines on each side of the main line, the satellites having intensities in the ratio 1:2:1 and appearing at the wavelengths expected for N_2^{14}, $N^{14}N^{15}$, and N_2^{15}. The double transition therefore appears to depend on the interaction of a nitrogen (or oxygen) atom with a *single* neighboring nitrogen molecule, rather than with all the nitrogen molecules surrounding a given nitrogen atom.

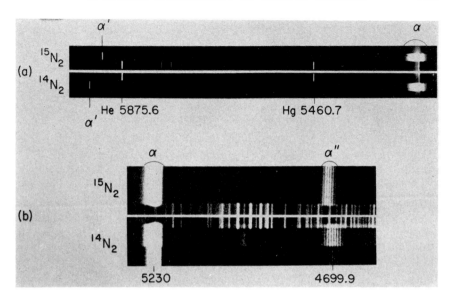

FIG. 6. Isotope shifts of the α' and α'' features with N_2^{15}. Reproduced from Peyron et al. (1959) by kind permission of the authors and of the *Journal of Chemical Physics*.

Most of the spectral features in the glows emitted by solid nitrogen at low temperatures can thus be explained in terms of electronic transitions of atoms and molecules and double transitions of the type discussed above. There are still several outstanding problems, however, and these have been discussed by Peyron and Broida (1959a). The most important are the interpretation of the fine structures of the atomic lines and the mechanisms whereby the various spectral features are excited.

C. OXYGEN

Only very faint luminescences have been excited in pure solid oxygen at low temperatures, and no spectra have been reported. If the oxygen is diluted with rare gases, however, stronger luminescences are observed and spectra may be obtained. Thus Peyron (1959) has observed the A-bands when the products of a discharge through argon containing 1% oxygen are condensed to 4.2°K. Schoen and Broida (1959) have shown that the same bands can be excited if an electrical discharge is applied to the same mixture in the solid.

Bass and Broida (1958) have investigated the absorption spectrum of the solid condensed at 4.2°K from a discharge through oxygen. Most of the bands in the region 2200 A to 3.5 microns were assigned to oxygen or ozone, but a strong band at 3496 cm^{-1} could not be identified. This band disappears on warming and does not reappear on cooling. Further investigations between 2 and 25μ were carried out by Harvey and Bass (1958). From observations of the ozone absorption as a function of temperature an upper limit of 3 % was inferred for the oxygen atom concentration in the solid.

D. Ammonia

Lavin and Bates (1930) observed an intense greenish-blue glow when the products of a discharge through ammonia were condensed with liquid nitrogen. These observations were confirmed by Lunt and Mills (1935), who also obtained the glow with N_2–H_2 mixtures. They found that the spectrum consisted of a broad continuum extending from 5300 to 4400 A, with a maximum at 5000 A. They postulated that the emission is due to NH_2 radicals. Similar spectral results were obtained by Bass and Broida (1958).

V. Absorption Spectra of Trapped Radicals

Most of the assignments of absorption spectra of free radicals in the solid state have been carried out by comparison with well-known gas phase spectra. Data on diatomic free radicals in the gas phase have been summarized by Herzberg (1950), and information on polyatomic free radicals has been recently reviewed by Ramsay (1959). The compilations of Pearse and Gaydon (1950) and of Barrow *et al.* (1952) are very useful in the identification of molecular spectra. Caution must be exercised in the assignment of solid state spectra on the basis of the approximate agreement of wavelength data with that in the gas phase, since matrix shifts are usually encountered and the nature of these shifts is not fully understood. Where possible the assignments should be confirmed by isotopic substitutions. A summary of the spectra which have been obtained is given in Table II.

A. Atoms

Little work has been carried out on the absorption spectra of atoms in the solid state, since the first resonance lines of most of the simple atoms lie in the vacuum ultraviolet region. McCarty and Robinson (1959b) have carried out some observations on the 2537-A line of mercury in absorption in a study of the matrix shifts of this line in different matrices.

B. Diatomic Radicals

1. NH

In 1951, Rice and Freamo reported a blue paramagnetic solid obtained when hydrazoic acid (HN_3) at low pressures is passed through a heated quartz tube and the decomposition products are rapidly condensed on a

TABLE II

ABSORPTION SPECTRA OF FREE RADICALS IN THE SOLID STATE

Radical	Spectral region	Method of production[a]	References
NH	3500 A (broad)	Thermal or electrical decomposition of HN_3	Rice and Freamo (1951. 1953) Mador and Williams (1954) Rice and Grelecki (1957a)
NH	3380 A (sharp)	Discharge through NH_3, N_2H_4, or N and H in any form with rare gas	Robinson and McCarty (1958b,c,d)
		Photolysis of NH_3 with vacuum ultraviolet radiation	Schnepp and Dressler (1959)
NH	3080 cm^{-1}	Photolysis of HN_3	Becker et al. (1957)
OH	3110 A	H_2O_2-argon discharge	Robinson and McCarty (1958b,c,d)
CH	4310 A and 3919 A	CH_4-argon discharge	Robinson and McCarty (1958d)
PH	3400 A	PH_3-argon discharge	McCarty and Robinson (1959b)
C_2	4700–5650 A	CH_4, C_2H_2, or isopentane discharge with argon	Robinson and McCarty (1958d) McCarty and Robinson (1959b)
CN	3500–3900 A 7880 A	Discharges in compounds containing C and N atoms	McCarty and Robinson (1959b)
CS	2490–2620 A	Photolysis of CS_2	Norman and Porter (1955)
ClO	2550–3000 A	Photolysis of ClO_2	Norman and Porter (1955)
S_2	Purple solid	Rapid freezing of sulfur vapor	Rice and Sparrow (1953)
SCl	2370 A	Photolysis of S_2Cl_2	Sowden and Davidson (1956)
NH_2	3700–7900 A	Discharge through NH_3, N_2H_4, or N and H in any form with rare gas	Robinson and McCarty (1958a,c,d; 1959)
NH_2	1290 cm^{-1}	Photolysis of HN_3	Becker et al. (1957)
PH_2	3600–5500 A	PH_3-argon discharge	Robinson and McCarty (1959)
CH_2	1362 or 1114 cm^{-1}	Photolysis of CH_2N_2	Milligan and Pimentel (1958)
HCO	5000–6500 A	CH_4—O_2-argon discharge	Robinson and McCarty (1958d)
		C_2H_2—O-krypton flame	McCarty and Robinson (1959b)

TABLE II—*Continued*

Radical	Spectral region	Method of production[a]	References
HNO	5900–7620 A	Discharges containing H, N, and O in any form	Robinson and McCarty (1958b,c,d)
HNO	3592, 1570, and 1110 cm^{-1}	Photolysis of CH_3NO_2	Brown and Pimentel (1958)
		H_2 discharge—NO	Harvey and Brown (1959)
N_3(?)	2150 cm^{-1}	N_2 discharge	Milligan et al. (1956)
		Photolysis of HN_3	Becker et al. (1957)
NCO	4400 A	CH_4—O_2—N_2-argon discharge	Robinson and McCarty (1958d)
NO_3	5000–7000 A	Photolysis of $O_3 + NO_2$ NO_2—O_2—N_2 discharge	Davidson and DeM_re (1958)
$(CH_3)_2N$	Violet solid	Pyrolysis of tetramethyltetrazene	Rice and Grelecki (1957b)
CH_3CHOH	5170 A (broad)	Photolysis of H_2O_2 in presence of ethyl alcohol	Symons and Townsend (1956)
$C_6H_5CH_2$	3187 A (strong)	Photolysis of $C_6H_5CH_2X$	Norman and Porter (1955)
	3082 and 3047 A (weak)		Porter and Strachan (1958)
C_6H_5NH	3109 A	Photolysis of $C_6H_5NH_2$	Norman and Porter (1955) Porter and Strachan (1958)
C_6H_5O[b]	2870 A (limit)	Photolysis of C_6H_5OX	Norman and Porter (1955) Porter and Strachan (1958)

[a] The discharge and pyrolysis techniques are carried out in the gas phase and are followed by rapid freezing of the products. The photolysis techniques refer to the solid state.

[b] For further spectra of aromatic free radicals see Norman and Porter (1955), Porter and Strachan (1958).

surface cooled with liquid nitrogen. They found that the blue color and the paramagnetism persist while the solid is maintained at liquid nitrogen temperature, but disappear suddenly and irreversibly when the sample is allowed to warm up to 148°K. At the transition point a white solid is formed and was identified as ammonium azide, NH_4N_3. These authors proposed that hydrazoic acid decomposes according to the equation

$$HN_3 \rightarrow NH + N_2$$

and that the blue color and the paramagnetism of the solid are due to condensed NH radicals or to some polymer of NH. They later found (1953) that the blue solid could be more conveniently prepared by freezing the products of an electrical discharge through hydrazoic acid.

Mador and Williams (1954) carried out spectroscopic investigations of the blue solid prepared by the latter method both at 4°K and 77°K. In the infrared region they found that all the absorption bands could be attributed to undecomposed hydrazoic acid or to ammonium azide. In the visible and ultraviolet regions, they observed two broad absorption bands with maxima near 3500 and 6500 A. These bands disappeared irreversibly when the solid was warmed to 148°K and were tentatively ascribed to small amounts of NH and NH_2, respectively, in the blue solid.

Dows et al. (1955) carried out independent investigations of the infrared spectrum of the blue solid and obtained results differing from those of Mador and Williams. At 90°K, Dows et al. found that all the absorption bands in the region 3500 to 750 cm^{-1} could be ascribed to HN_3, NH_4N_3, and NH_3 with the exception of two bands at 3320 and 1090 cm^{-1}. As the solid was warmed slowly, two new bands at 3230 and 860 cm^{-1} appeared between 90°K and 120°K without any observable decrease in the intensities of the other bands in the spectrum. On further warming, the first pair of bands at 3320 and 1090 cm^{-1} disappeared at about 175°K, and the second pair disappeared above 190°K. Since these bands did not reappear on cooling they were assigned to intermediate species, designated A and B, respectively. Moreover, since the spectrum of B appears without observable decrease of other bands in the spectrum, it was inferred that B is produced from a third unobserved intermediate C. Each of the intermediates A and B produces ammonium azide without evidence of other products, hence each is assumed to have the same stoichiometric composition as ammonium azide $(NH)_x$. The intermediate C was assumed to be the NH radical. This would have a single infrared frequency which could easily be masked by the strong NH_3 and NH_4N_3 absorptions in the same region. The intermediate B was interpreted as N_2H_2, diimide, and the intermediate A as $(NH)_x$ where x is greater than 2 and may be equal to 4. Becker et al. (1957) investigated the infrared spectra of intermediate species formed by the photolysis of hydrazoic acid in inert matrices at 20°K and tentatively assigned a band at 3080 cm^{-1} to NH. This frequency is in reasonable agreement with the gas phase value $\Delta G_{\frac{1}{2}}'' = 3125.6$ cm^{-1} recently determined by Dixon (1959).

From material balance studies, Rice and Grelecki (1957a) also deduced that the substance responsible for the blue color has the same stoichiometric composition as ammonium azide. These authors showed that the blue solids produced by thermal, electrical, and photochemical decompositions of equivalent amounts of hydrazoic acid all had comparable absorption bands at 6500 A and comparable ultraviolet absorption bands of ammonium azide. Only the two samples prepared by electrical and photochemical decomposition, however, showed an absorption band at 3500 A. These authors concluded that the band at 6500 A was not due to NH_2 as suggested by Mador and Williams.

Recently, Foner and Hudson (1958a, b) identified N_2H_2 in mass spectrometer studies of the decomposition products of hydrazoic acid and hydrazine subjected to mild electrodeless discharges. They were unable, however, to detect any NH radicals. The N_2H_2 was found to have a much longer lifetime in the gas phase than the substance responsible for the blue color, and moreover on condensation a yellow solid was formed. N_2H_2 is thus not responsible for the blue solid observed by Rice and Freamo.

A clue as to the possible origin of the blue color is afforded by the electron spin resonance absorption measurements of Gager and Rice (1959). These authors found a large number of unpaired spins with $g = 2.003$ in blue solids prepared by the thermal, electrical, and photolytic techniques and suggested that the blue color may be associated with F-center formation. This interpretation appears to provide a satisfactory explanation of the 6500-A band, although it is possible that it may be related to a polymer $(NH)_x$ with x greater than 2, to NH_2 in quantities too small to be detected in the infrared and mass balance experiments, or yet to some other source.

A sharp absorption band near 3380 A which almost certainly is due to NH in the solid has been reported by McCarty and Robinson, 1959a, (see also Robinson and McCarty, 1958b, c, d). These authors observed the band in the absorption spectrum of the deposit condensed at 4.2°K from discharges through ammonia, hydrazine, or nearly any mixture of hydrogen and nitrogen containing molecules in the presence of a large excess of inert gas. Four lines were observed in absorption and four more in emission (see Fig. 7). The latter lie to the red of the absorption lines and were attributed to resonance fluorescence excited by the light source used to study the absorption spectrum. For a given matrix, the lines were found to be displaced by approximately a constant wave-number shift from the lines in the gas phase spectrum originating in the lowest rotational levels, thus permitting rotational assignments to be made for the lines in the solid. The matrix shift, however, was found to be considerably different for different matrices. Confirmation of the assignment of the spectrum to NH was afforded by the observation of isotope shifts when H was replaced by D. In addition to the strong (0-0) band of the $A^3\Pi \leftarrow X^3\Sigma^-$ transition, McCarty and Robinson observed the much weaker (1-0) band for both NH and ND. Moreover in the case of NH in an argon matrix these authors also observed an extremely sharp absorption line near 3280 A which they assigned to the (0-0) band of the $c^1\Pi \leftarrow a^1\Delta$ transition, (see Fig. 7). This

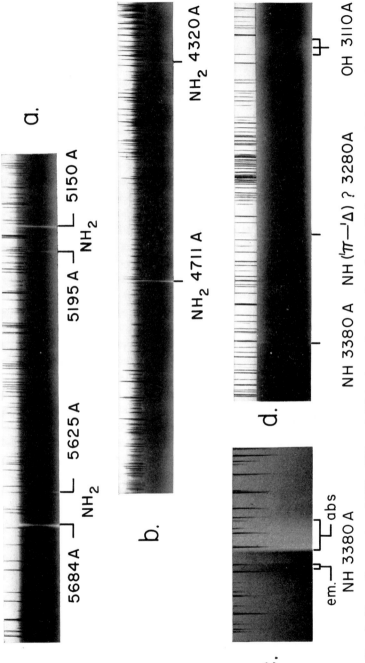

Fig. 7. Absorption spectra of NH_2, NH, and OH in the products condensed at 4.2°K from an argon-hydrazine discharge containing a trace of water and oxygen. Deposit times: (a) 10 hours, (b) 10 hours, (c) 2 hours, (d) 30 min. In (d) the resonance fluorescence excited by the light source used for the absorption measurements is seen. Reproduced from Robinson and McCarty (1958c) by kind permission of the authors and of the *Canadian Journal of Physics*.

line disappears immediately the discharge is turned off, whereas the 3380-A band persists. These observations provide strong evidence that the $^3\Sigma^-$ and not the $^1\Delta$ state is the ground state of NH.

Recently Schnepp and Dressler (1959) have produced NH radicals by the photolysis of NH_3 in solid argon, using vacuum ultraviolet radiation. They observed the absorption feature at 3380 A in agreement with the wavelength reported by McCarty and Robinson (1959a).

2. OH

Robinson and McCarty (1958b,c,d) have reported a strong absorption feature at 3110 A in the products trapped at 4.2°K from H_2O_2-argon and H_2O-argon discharges, and other discharges containing traces of oxygen and hydrogen in any form. The intensity of the band decreases as the oxygen-containing species are removed from the discharge. They assign the band to the (0-0) transition of the $A^2\Sigma^+ \leftarrow X^2\Pi_{3/2}$ system of OH, which is shifted by 395 cm^{-1} to lower frequencies compared with the gas phase spectrum. Two shaded absorption features separated by 71 $\pm$ 5 cm^{-1} were observed and assigned to $Q(1)$ and $R(1)$ transitions which are separated by 68 cm^{-1} in the gas phase spectrum. The absence of a $P(1)$ line however casts some doubt on the above interpretation.

3. CH

Absorption bands of CH have been found by Robinson and McCarty (1958d) in the products condensed at liquid helium temperatures from low pressure discharges in CH_4-argon mixtures. The bands were fairly broad compared with the sharp absorption bands of NH and OH discussed above. The $A^2\Delta \leftarrow X^2\Pi$ transition at 4310 A and the $B^2\Sigma^- \leftarrow X^2\Pi$ transition at 3919 A were both observed, the former being shifted by 50 cm^{-1} to the red and the latter by 210 cm^{-1} to the red from the corresponding gas phase transitions. The third system of CH near 3143 A was not observed because of scattering.

4. PH

The (0-0) band of the 3410-A system has been observed by McCarty and Robinson (1959b) in the products condensed from an argon-phosphine discharge. The band is exceedingly broad and no wavelength measurements have been attempted.

5. C_2

The (0-0) and (1-0) bands of the C_2 Swan system ($A^3\Pi_g \leftarrow X^3\Pi_u$) were observed at 4.2°K by McCarty and Robinson (1959b) in the products condensed from discharges using CH_4, C_2H_2 or isopentane diluted with argon or xenon. The bands were observed with undiminished intensity for as long as the solid was maintained at liquid helium temperature, i.e., a few hours. This observation provided strong support for the idea that the $^3\Pi_u$ state is the ground state of the C_2 molecule. It was later shown by Ballik and Ramsay (1959) by means of rotational perturbations, that the $^1\Sigma_g^+$ state lies $\sim$600 cm^{-1} below the $^3\Pi_u$ state in the gas phase and is indeed the true ground state of the C_2 molecule. It should be pointed out, however, that the separation between the $^3\Pi_u$ and $^1\Sigma_g^+$ states in the solid will undoubtedly be different from that in the gas phase, and it is not impossible that the $^3\Pi_u$ state is the ground state in the solid. The experiments of Robinson and McCarty are not conclusive since if the $^3\Pi_u$ state does lie a few hundred cm^{-1} above the $^1\Sigma_g^+$ state in the solid, the radiative lifetime of the C_2 molecule in the $^3\Pi_u$ state could easily be longer than the duration of their experiments.

In addition to the absorption bands, the (0-0), (0-1), (1-0), (1-1), and (1-2) bands have been observed in resonance fluorescence. One surprising result of the investigations is that the vibration frequencies in the solid are found to be considerably different from those observed in the gas phase. Thus in xenon the values of $\Delta G_{\frac{1}{2}}''$ and $\Delta G_{\frac{1}{2}}'$ are 1724 cm^{-1} and 1941 cm^{-1} compared to the gas phase values of 1618 cm^{-1} and 1754 cm^{-1}, respectively. These discrepancies might cast some doubt on the assignment of the solid state bands. However the absence of any isotope shift with CD_4, and the observation of the bands in fluorescence lend support to the assignment.

6. CN

The (0-0) and (1-0) bands of the violet system of CN ($B^2\Sigma^+ \leftarrow X^2\Sigma^+$) and the (2-0) band of the red system ($A^2\Pi \leftarrow X^2\Sigma^+$) have been observed by McCarty and Robinson (1959b) in the condensed products of discharges through compounds containing C and N atoms. The violet bands have a small matrix shift ($\sim$16 cm^{-1}) and show considerable fine structure, suggesting that the CN radicals are rotating in the solid.

7. CS

Norman and Porter (1954, 1955) observed the (0-0), (1-0), and (2-0) bands of CS following the ultraviolet irradiation of carbon disulfide in a rigid glass at 77°K. The bands are shifted by 45 to 48 A to the red compared with the corresponding gas phase bands and disappear when the glass is melted.

8. ClO

Norman and Porter (1954, 1955) reported that when chlorine dioxide in a hydrocarbon glass at 77°K is subjected to ultraviolet irradiation, the bands due to ClO_2 rapidly decrease in intensity and are replaced by a diffuse spectrum in the region 2550 to 3000 A. If the photolysis be continued the diffuse spectrum also disappears, indicating that the product itself is photochemically dissociated. No further absorption, however, appears. The diffuse spectrum was tentatively assigned to ClO on account of its general similarity to the gas phase spectrum. The possibility of an alternative assignment to ClO_3 was also discussed. The spectrum disappears when the glass is melted and most of the original absorption by chlorine dioxide reappears.

9. S_2

Rice and Sparrow (1953) found that when sulfur vapor at about 500°C and at a pressure of 0.1 to 1.0 mm is rapidly condensed to liquid nitrogen temperature, a purple solid is obtained. Since the vapor under the above conditions consists almost entirely of S_2, they suggested that the purple color in the solid may be associated with S_2 molecules. Rice and Ditter (1953) subsequently obtained a green form of sulfur at liquid nitrogen temperature and suggested that the green color may be associated with S_8 molecules.

10. SCl

Sowden and Davidson (1956) observed that when a dilute solution of S_2Cl_2 in a rigid hydrocarbon glass at liquid nitrogen temperature is subjected to ultraviolet irradiation, the parent absorption at 2640 A disappears and a weak band at 2370 A is formed. On melting, the new band disappears and the 2640-A band is partially regenerated. These authors tentatively suggested that the band at 2370 A may be due to SCl. Unfortunately no bands of SCl are known in the gas phase for comparison.

C. TRIATOMIC RADICALS

1. NH_2

a. Electronic spectrum. Robinson and McCarty (1958a,c,d; 1959) have observed extensive absorption due to NH_2 radicals in the products condensed at 4.2°K from discharges through ammonia, hydrazine, or almost any mixture of molecules containing hydrogen and nitrogen in the presence of a large excess of inert gas. About 60 to 70 lines have been observed in the region 3700 to 7900 A, the lines being very sharp and having a width of only a few cm^{-1} (see Fig. 7).

The spectrum shows an interesting correlation with the absorption spectrum in the gas phase in that alternate bands are "missing" in the solid state spectrum. This phenomenon is in agreement with the interpretation of the gas phase spectrum (Ramsay, 1957; Dressler and Ramsay, 1959) and arises from the linear equilibrium configuration of the excited state. The gas phase spectrum consists of a long progression of the bending vibration in the excited state $(0, v_2', 0) \leftarrow (0, 0, 0)$, where $v_2' = 3, 4, 5, \cdots 18$. The bands consist of Σ, Δ, $\Gamma \cdots$ vibronic subbands for v_2' odd and Π, $\Phi \cdots$ vibronic subbands with v_2' even. The only lines which can arise from the 0_0 rotational level of the nonlinear ground state are the $^R R(0)$ lines in each of the Π vibronic subbands. Since at $4.2°K$ $\sim 99\%$ of the NH_2 radicals are in the 0_0 rotational level, the low temperature spectrum should consist of single lines corresponding to the Π-bands in the gas phase spectrum. This indeed is exactly what is observed by Robinson and McCarty, apart from small matrix shifts. In very thick deposits or when the sample is warmed slightly, weak lines are observed approximately midway between the strong lines. These weak lines correspond to transitions from the 1_{-1} rotational level of the ground state to the Σ-vibronic levels in the upper state. Other weak lines are also observed from the 1_0 and 1_{+1} levels in the ground state to the Δ-vibronic levels in the excited state. It should be pointed out that the intensities of these weak lines are considerably greater than would be expected on the basis of thermal equilibrium at $4.2°K$. The excess population of the low-lying rotational levels is probably caused by absorption of radiation from the light source used in the absorption experiment, followed by fluorescence to rotationally excited levels. Such levels may have an appreciable metastability with respect to the ground state.

In addition to the principal progression, four bands of a subsidiary progression $(1, v_2', 0) \leftarrow (0, 0, 0)$ with $v_2' = 6, 7, 8, 9$ have been identified in the gas phase spectrum. In the solid state two lines of medium intensity are observed at 5625 and 5195 A corresponding to the $(1, 6, 0) \leftarrow (0, 0, 0)$ and $(1, 8, 0) \leftarrow (0, 0, 0)$ bands respectively (see Fig. 7). From the gas phase analysis it was shown that the vibrational levels $(0, v_2', 0)$ and $(1, v_2' - 4, 0)$ in the excited state are in Fermi resonance and a Fermi interaction constant of 72 ± 3 cm^{-1} was deduced. The subsidiary progression derives some, if not all, of its intensity by "borrowing" intensity from the bands of the principal progression. By measuring the relative intensities of the resonating pairs

$$\left. \begin{cases} (1, 6, 0) & \leftarrow (0, 0, 0) \\ (0, 10, 0) & \leftarrow (0, 0, 0) \end{cases} \right\} \text{ and } \left. \begin{cases} (1, 8, 0) & \leftarrow (0, 0, 0) \\ (0, 12, 0) & \leftarrow (0, 0, 0) \end{cases} \right\}$$

and assuming that all the intensity of the subsidiary progression is obtained from the resonance, Robinson and McCarty (1958d, 1959) found that $W_{ni} = 75 \pm 2$ cm^{-1}. The agreement between this value and the gas phase value shows that the proposed mechanism for the origin of the intensity of the subsidiary progression is substantially correct.

Many of the weaker NH_2 lines in the spectrum of the solid remain unexplained, especially in the region of 4300 A and in the extreme red region, 7000 to 8000 A, where considerable difficulty was also encountered in interpreting the gas phase spectrum.

b. Vibrational spectrum. Becker *et al.* (1957) investigated the photolysis of hydrazoic acid (HN_3) suspended in xenon, argon, and nitrogen matrices at 20°K. By investigating the variation of the infrared spectrum of the photolyzed products as a function of temperature they were able to identify various intermediate species. They tentatively assigned a line at 1290 cm^{-1} to the bending vibration of NH_2. The corresponding frequency in the gas phase is not yet known.

2. PH₂

An absorption spectrum of PH_2 at 4.2°K has been observed by McCarty and Robinson (1959b) in the products condensed from a phosphine-argon discharge. The spectrum lies in the same region as the gas phase spectrum observed by Ramsay (1956). Eleven bands were observed in the gas phase, of which five have been observed in the solid. It is interesting to note that all the bands of the gas phase spectrum are present in the solid in contrast to the observations for NH_2. This result suggests that PH_2 is nonlinear in its excited state, a conclusion which should be of assistance in the analysis of the gas phase spectrum.

3. CH₂

Milligan and Pimentel (1958) have investigated the infrared spectrum of the photolysis products of diazomethane in argon and nitrogen matrices at 20°K. They claim that one of the absorption bands observed at 1362 and 1114 cm^{-1} could correspond to the bending vibration of CH_2.

4. HCO

Four bands of HCO at 6486, 5915, 5439, and 5042 A have been observed by Robinson and McCarty (1958d) at liquid helium temperature in the products condensed from a CH_4–O_2–argon discharge. Similar observations were made by McCarty and Robinson (1959b) by condensing the products of an acetylene-atomic oxygen flame diluted with krypton. The bands correspond to alternate bands in the gas phase progression $(0, v_2', 0) \leftarrow (0, 0, 0)$ observed by Herzberg and Ramsay (1955). In the excited state the radical has a linear equilibrium configuration, cf NH_2, whereas in the ground state it is bent. The bands in the gas phase should consist alternately of Σ, Δ, $\Gamma \cdots$ and Π, $\Phi \cdots$ vibronic subbands. Actually only the Σ-subbands are found to be discrete. The Π-vibronic subbands are completely diffuse because of predissociation in the excited state, and no definite assignment has been given of the higher subbands. Since only the II-vibronic levels in the excited state combine with the lowest rotational level in the ground state, the spectrum in the solid at 4.2°K should correspond principally to the II-vibronic subbands, and on account of the predissociation the bands should be broad. These conclusions are confirmed by the solid state spectrum.

5. HNO

a. Electronic spectrum. Robinson and McCarty (1958b,c,d) have reported that the strong (0-0) band of HNO near 7600 A is exceedingly common in the condensed products from discharges containing hydrogen, nitrogen, and oxygen in any form. Intense HNO absorption, accompanied by strong NO_2 bands, was obtained by these authors from the products of a N_2H_4–O_2–H_2O–argon discharge condensed at 4.2°K. In addition to the three bands observed by Dalby (1958) in the gas phase, three other much weaker bands were observed in the solid and fitted readily into the vibrational scheme suggested by Dalby.

b. Infrared spectrum. Brown and Pimentel (1958) observed the infrared spectrum of HNO in the products of photolysis of nitromethane in an argon matrix at 20°K. The first stage of the reaction involves isomerization of the nitromethane to methyl nitrite and is followed by photodecomposition producing HNO and formaldehyde. Separate experiments were carried out with CH_3NO_2 and CD_3NO_2 and the following frequencies identified:

$$HNO: \quad (3300), \ 1570, \ 1110 \ cm^{-1}$$

$$DNO: \quad (2481), \ 1560, \ \ 822 \ cm^{-1}$$

These assignments have been corroborated by Harvey and Brown (1959) by studying the infrared spectrum of the products formed by reacting H (or D) atoms with NO and condensing the products with liquid helium. Under these conditions the spectrum consisted principally of the three bands of HNO (or DNO) and the NH and ND stretching frequencies were more accurately located at 3592 cm^{-1} and 2680 cm^{-1}, respectively.

6. HO_2

Giguère (1954) reported a sharp absorption band at 1305 cm^{-1} in the glassy deposit obtained by freezing the products of an electrical discharge through water vapor. The band was tentatively assigned to the HO_2 radical, but subsequent work by Giguère and Harvey (1956) revealed that the band was due to N_2O_3 impurity.

7. N_3

Milligan et al. (1956) studied the infrared absorption spectra of the products condensed at 4.2°K from a glow discharge in nitrogen. They observed a band at 2150 cm^{-1} which disappeared on warming to 35°K and which showed an isotope shift when a sample enriched in N^{15} was used. They assigned this band to the antisymmetric stretching vibration of the N$_3$ radical. Some subsequent work by Harvey and Brown (1959), however, failed to confirm this result. Evidence that the same band is present in the absorption spectrum of the products of photolysis of dilute suspensions of hydrazoic acid in argon and nitrogen at 20°K was presented by Becker *et al.* (1957).

8. NCO

Robinson and McCarty (1958d) observed an absorption band near 4400 A in the condensed products of a methane-argon discharge containing nitrogen and oxygen impurities. The band was tentatively identified as the strong $(0, 0, 0) \leftarrow (0, 0, 0)$ band of the $A^2\Sigma^+ \leftarrow X^2\Pi$ transition of NCO. The band is displaced by 70 cm^{-1} to higher frequencies from the gas phase band observed by Holland *et al.* (1958).

D. SOME LARGER RADICALS

1. CH$_3$

All attempts to obtain the optical absorption spectrum of CH$_3$ in the solid have so far been unsuccessful. An absorption band of CH$_3$ in the gas phase near 2140 A has been found by Herzberg and Shoosmith (1956) together with a Rydberg series in the vacuum ultraviolet. These absorption features however have not yet been reproduced in the solid state.

Mador (1954) carried out some experiments in an attempt to stabilize the methyl radical at liquid helium temperature. He produced the radicals *in situ* by ultraviolet irradiation of methyl iodide and by X-irradiation of lead tetramethyl. The products of pyrolysis of lead tetramethyl were also rapidly condensed. In all the experiments the infrared spectra of the products were investigated, but all the new absorption bands formed could be attributed to CH$_4$, C$_2$H$_6$, C$_2$H$_4$, C$_3$H$_8$, or CH$_2$I$_2$.

Brown (1958) investigated the photolysis of methyl iodide at 4.2°K and 20°K in an argon matrix, the ratio of argon to methyl iodide being of the order of 500:1. The only additional bands, however, that he was able to identify after photolysis could be assigned to CH$_4$ and C$_2$H$_6$.

It is interesting to note that in the photolysis studies of Norman and Porter (1955) in rigid glasses at 77°K, no new bands associated with the ethyl radical were found when ethyl iodide was investigated.

2. NO$_3$

Davidson and DeMore (1958) have produced NO$_3$ radicals in the solid by three methods: (1) the photolysis of O$_3$ in the presence of NO$_2$ in an argon or nitrogen matrix at 20°K; (2) the vapor phase photolysis of N_2O_5 and rapid condensation of the products in the presence of a matrix; and (3) trapping the products of an electrical discharge through a mixture of NO_2 and O_2 with a nitrogen matrix. In all the experiments they found absorption bands of NO_3 in the region 6000 to 7000 A. The spectrum in the solid lies in a similar region to that observed in the gas phase by Jones and Wulf (1937).

3. $(CH_3)_2N$

Rice and Grelecki (1957b) obtained a paramagnetic violet solid when tetramethyltetrazene, $(CH_3)_2$—N—N=N—N—$(CH_3)_2$, was passed over a tungsten filament at 1300°C and the products were rapidly condensed on a liquid nitrogen-cooled surface. The violet material changes suddenly and irreversibly at -160°C to a white solid which contains substantial quantities of tetramethylhydrazine, $(CH_3)_2$—N—N—$(CH_3)_2$. Rice and Grelecki suggested therefore that the violet color may be ascribed to the dimethylamino radical, $(CH_3)_2N$. The violet material was also prepared by irradiating tetramethyltetrazene vapor with ultraviolet light and rapidly condensing the products with liquid nitrogen. It is interesting to note that Sowden and Davidson (1956) investigated the photolysis of tetramethyltetrazene in hydrocarbon glasses at 77°K. Although the parent absorption band at 2730 A disappeared rapidly on photolysis and reappeared on warming, no violet color was reported and no new spectral features were observed.

4. CH_3CHOH

Symons and Townsend (1956) observed that paramagnetic violet glasses were obtained by the ultraviolet irradiation of a variety of photochemically active compounds, e.g., H_2O_2, dissolved in rigid glasses at liquid nitrogen temperature and containing ethyl alcohol as a major component. The violet color and the paramagnetism disappear when the glass softens. The spectrum of the glasses reveals a broad absorption band with a maximum at 5170 ± 20 A. The authors postulated that the violet color was caused by CH_3CHOH radicals and supported their conclusions by the results obtained from paramagnetic resonance studies. The mechanism of production of the radicals involves first the formation of a highly reactive species by photolysis, followed by reaction with the surrounding medium to form new radicals which are then trapped.

5. Others

Rice and Grelecki (1957c) have reported a green solid obtained by passing an electrical discharge through dimethylamine vapor, $(CH_3)_2NH$, and rapidly condensing the products on a liquid nitrogen-cooled surface. The green color could not be produced by thermal or photochemical techniques. The color disappeared on standing with evolution of hydrogen and small amounts of methane, and a white solid was formed. No tetramethylhydrazine was found in the residue. The authors tentatively suggested that the green color may be ascribed to CH_3NHCH_2 radicals.

A yellow solid was obtained by Rice and Scherber (1955) by rapidly freezing at $-195°C$ the products of thermal decomposition of hydrazine. The solid however was not paramagnetic and hence could not contain the radical NH_2NH. On warming to $-178°C$ the substance decomposed suddenly with evolution of nitrogen. The authors assigned the yellow solid to tetrazane, $NH_2NHNHNH_2$, and proposed that the decomposition followed the reaction

$$NH_2NHNHNH_2 \rightarrow 2NH_3 + N_2$$

The yellow color could not be produced by ultraviolet irradiation of hydrazine condensed on a cold surface.

A clear colorless deposit was obtained in a similar manner from methyl azide by Rice and Grelecki (1957d). On warming, a transition to a white opaque solid was observed. No absorption spectra were reported however.

Rice (1956) also reports a yellow solid obtained by Michaelsen (1955) from CH_3S—SCH_3 and a red solid obtained from $(CH_3)_3CS$—$SC(CH_3)_3$. These colors may be associated with CH_3S and $(CH_3)_3CS$ radicals.

The possibility that the colors on Jupiter are due to frozen free radicals of the type mentioned above has been discussed by Rice (1956).

E. AROMATIC RADICALS

The first systematic investigations of the absorption spectra of aromatic free radicals in the solid state were those of Lewis and his collaborators. These workers studied the effects of ultraviolet irradiation of various aromatic compounds dissolved in rigid glasses at liquid nitrogen temperature. Similar experiments have been carried out more recently by Linschitz, Porter, and their respective co-workers. Several types of primary dissociation process have been identified.

1. Photodissociation in an Aromatic Side Chain

The first examples were afforded by Lewis and Lipkin (1941, 1942), who obtained the absorption spectra of triphenylmethyl and diphenylnitrogen by the photolysis of hexaphenylethane and tetraphenylhydrazine, respectively.

Norman and Porter (1955) observed the spectra of the benzyl ($C_6H_5CH_2$), anilino (C_6H_5NH), and phenoxy (C_6H_5O) radicals following the photolysis of toluene, aniline, phenol, and other aromatic compounds, and also obtained the spectra of simple derivatives of these radicals, e.g., $C_6H_5 \cdot CH \cdot C_6H_5$ and $o\text{-}CH_3 \cdot C_6H_4 \cdot NH$. Further studies of these spectra have been carried out by Porter and Strachan (1958) and by Chilton et al. (1958), who used gamma irradiation as well as ultraviolet irradiation. The spectra are characterized by narrow linelike structures in which one line is very much stronger than the rest of the spectrum. The spectra show no decrease in intensity while the glass remains rigid but disappear irreversibly when the glass melts.

The assignments of the spectra are based primarily on photochemical arguments. As an example, the assignment of the spectrum of the benzyl radical will be considered. A sharp band is observed at 3185 A when compounds of the type $C_6H_5CH_2X$ are irradiated. From energetic considerations the most likely primary processes are

$$C_6H_5CH_2X + h\nu \rightarrow C_6H_5CH_2 + X \tag{1}$$

$$C_6H_5CH_2X + h\nu \rightarrow C_6H_5CH + HX \tag{2}$$

$$C_6H_5CH_2X + h\nu \rightarrow C_6H_5 + CH_2X \tag{3}$$

A clear distinction between these processes is afforded by a study of the series $C_6H_5CH_3$, $C_6H_5CH_2Cl$, $C_6H_5CHCl_2$, and $C_6H_5CCl_3$. It is found that toluene and benzyl chloride give the same spectrum whereas benzal chloride and benzotrichloride give two similar spectra at slightly different wavelengths. These observations are consistent with process (1) and the assignment of the three spectra to $C_6H_5CH_2$, C_6H_5CHCl, and $C_6H_5CCl_2$. In a similar manner the spectra of other aromatic radicals have been assigned.

2. Ring Fission

This process has been invoked by several workers (Lewis and Lipkin, 1942; Lewis and Bigeleisen, 1943; Gibson et al., 1953; Norman and Porter, 1955) to explain the formation of hexatriene during the photolysis of benzene in rigid glasses. The primary process of ring fission is followed by abstraction of two hydrogen atoms from the rigid solvent. Spectra very similar to that of hexatriene were reported by Porter and Strachan (1958) in the permanent products of photolysis of toluene and related molecules in rigid glasses, and were tentatively ascribed to substituted hexatrienes.

3. Electron Ejection

This process was first clearly demonstrated by Lewis and Lipkin (1942) and Lewis and Bigeleisen (1943) in the case of various dyes, aromatic amines, and phenols. In irradiated glassy solutions, absorption bands were formed which were identical with those of organic ions prepared by chemical oxidation. For this reason, the process was called by Lewis "photo-oxidation." For example when tri-p-tolylamine was illuminated at 77°K in EPA (ether, isopentane, and alcohol) glass, a blue color was obtained. The absorption curve showed a maximum at 6560 A and a shoulder in the region of 5700 A. The curve was very similar to that obtained for the (p-CH$_3$C$_6$H$_4$)$_3$N$^+$ ion produced by chemical oxidation of tri-p-tolylamine.

A new spectrum in the region of 4288 A was observed by Norman and Porter (1955) during the low temperature photolysis of aniline and was assigned by these workers to the aniline radical cation C$_6$H$_5$NH$_2$$^+$. Porter and Strachan (1958) reported that they have observed spectra of several radical cations after photolysis of many aromatic amines. They note that the process of electron ejection does not occur readily in nonpolar glasses and it is normally encountered only in basic molecules.

The presence of solvated electrons in "photo-oxidized" glasses was established by Linschitz *et al.* (1954a), who compared the spectral changes in the glasses with those in similar media containing dissolved alkali metals. On warming the glasses, delayed luminescences were frequently observed and were attributed to triplet-singlet emission from triplet states formed by recombination of radical ions and electrons.

4. Others

The process of photodissociation into a positive and a negative ion was postulated by Lewis and Lipkin (1942). For example, these authors postulated that ultraviolet irradiation of (C$_6$H$_5$)$_2$N·NH$_2$ produces photoionization according to the reaction

$$(C_6H_5)_2N \cdot NH_2 \rightarrow (C_6H_5)_2N^+ + NH_2^-$$

The process however does not appear to be a common one in rigid media.

The process of "photoreduction" has been discussed by Linschitz *et al.* (1954b). They observed that irradiation of quinones in rigid media produces the same radical spectra that are obtained by the photo-oxidation of the corresponding hydroquinones. Similar studies have been carried out in fluid solvents by the flash photolysis technique by Bridge and Porter (1958).

Examples of photo-orientation of free radicals in rigid media are found in the work of Lewis and Bigeleisen (1943). These authors prepared free radicals by photo-oxidation of aromatic molecules with polarized light, and established the polarization of the absorption bands of the photo-product relative to that of the original molecule.

VI. Discussion

The spectra of molecules in the solid state provide a direct source of information concerning the energy levels of the molecules in the solid. The spectra occasionally give information concerning molecular structure, but the main interest lies in the possibility of studying the interactions of molecules with the surrounding medium. The spectra also provide a means of studying the phenomenon of diffusion in solids and of following solid state reactions. The various items of interest will be considered briefly.

A. SPECTROSCOPIC CONSTANTS

The electronic spectra of molecules in the solid state provide information concerning the electronic energy levels of the molecules in the solid and the vibrational frequencies of the molecules in the various electronic states. Infrared spectra provide data on the vibrational frequencies of the molecules in their ground states. Neither type of spectrum gives much information concerning the rotational constants of the molecules, since rotation in the solid is in general severely restricted and only in a few spectra have rotational structures of bands been observed (see Section VI, E). It should be pointed out that the spectroscopic constants derived from solid state data are in general slightly different from those of the free molecule. Moreover the values of the constants vary slightly from one medium to another.

B. STRUCTURAL DETERMINATIONS

Little information concerning molecular structure can in general be deduced from solid state spectra. The most accurate method of determining the geometrical structures of molecules is by means of rotational analysis of gas phase spectra. This method is not generally applicable to solid state studies, since only in exceptional cases do molecules rotate freely in the solid. Some information concerning molecular structure may be deduced from a consideration of the magnitudes of the vibrational frequencies observed in the solid or, in the case of symmetrical molecules, from the selection rules governing the appearance or nonappearance of certain vibrational bands in the spectrum. Only for molecules where the corresponding information is not available in the gas phase, do the solid state spectra provide additional information in structural determination.

As an example of the use of solid state data we may consider the structure of the HNO molecule. Rotational analysis of the gas phase spectrum has so far been unable to distinguish unambiguously between the structures HNO and HON, although further work with isotopic species should be able to resolve this problem. The ground state vibrational frequencies in the solid are 3592, 1570, and 1110 cm^{-1} and isotopic studies reveal that the 1570 cm^{-1} frequency is associated with the NO stretching vibration (see Section V, C, 5). The magnitude of this frequency suggests an $N{=}O$ double bond rather than an $N{-}O$ single bond and hence favors the structure $H{-}N{=}O$. Such arguments, however, are suggestive rather than conclusive.

C. f-VALUES AND CONCENTRATIONS

Solid state spectroscopy at low temperatures offers two advantages compared with gas phase spectroscopy for the determination of the f-values of transitions of free radicals. First, the spectrum is greatly simplified at low temperatures since the bands arise predominantly from the lowest vibrational (and rotational) level. Second, the radicals may be trapped for long periods, thus enabling their concentration to be measured by some other method, e.g., electron spin resonance, magnetic susceptibility. Although several experimental difficulties may be encountered, e.g., deposition of uniform films and measurement of film thickness, it appears that low temperature studies provide a potentially useful method of obtaining f-values of transitions of free radicals. To date, however, little attention has been devoted to this problem.

D. MATRIX SHIFTS

Shifts between gas phase and solid state spectra provide a source of information concerning the interaction of molecules with the surrounding medium. So far, however, no adequate theory has been advanced to account for the considerable body of experimental data which has become available. Most shifts amount to less than 1 % of the total transition energy and are generally to lower frequencies in the solid. For electronic spectra the shifts depend primarily on the electronic states involved and only to a lesser extent on the vibrational levels excited. The shifts also vary considerably from one matrix to another.

In order to predict the matrix shifts it is necessary to know the difference in energy of the matrix-radical interaction between the ground and excited state. McCarty and Robinson (1959a,b) assumed that the interaction energy contains dispersion terms and in some cases terms related to electric dipole-induced dipole interactions. For the former these authors used a Lennard-Jones 6-12 potential

$$V = 4\epsilon \left[\left(\frac{\sigma}{r} \right)^{12} - \left(\frac{\sigma}{r} \right)^{6} \right] \tag{1}$$

where ϵ is the well depth, r is the nearest-neighbor distance, and σ is an effective size. They assumed the usual combination laws for mixtures viz.

$$\epsilon_{12} = \sqrt{\epsilon_1 \epsilon_2} \tag{2}$$

$$\sigma_{12} = \tfrac{1}{2}(\sigma_1 + \sigma_2) \tag{3}$$

and assumed that σ_{radical} remains the same in different electronic states. By neglecting contributions due to dipole effects, McCarty and Robinson determined effective values for the parameters $(\sqrt{\epsilon'} - \sqrt{\epsilon''})_{\text{radical}}$ and σ_{radical} from the matrix shifts observed in two different rare gas matrices. They then calculated the matrix shifts to be expected for a third matrix. For NH and NH_2 the calculated values were found to be in reasonable agreement with experiment. This theory appears to provide an interesting starting point in an understanding of the origin of matrix shifts.

E. ROTATION IN SOLIDS

Some spectroscopic evidence for the free or hindered rotation of molecules in solids has appeared in the last few years. We shall not discuss this field in detail here but will consider two examples which appear to provide evidence for nearly free rotation in an inert gas matrix. The first involves NH_2 in an argon matrix at liquid helium temperature. In addition to the strong lines arising from the lowest rotational level, Robinson and McCarty (1959) observed many weak lines in the electronic absorption spectrum of NH_2 and assigned some of these lines to transitions arising from the 1_{-1}, 1_0, 1_{+1}, and 2_{-2} rotational levels of the ground state. These assignments were based on the gas phase assignments, the matrix shifts of the weak lines being very similar to those found for the strong lines in neighboring regions of the spectrum. This latter fact provides evidence that the rotational levels of the ground state are relatively unperturbed and indicates free or nearly free rotation of the NH_2 molecules in the solid. It is interesting to note that McConnell (1958) has recently advanced an explanation for the observed hyperfine structure of the electron spin resonance spectrum of NH_2 in an argon matrix at $4.2°K$ on the assumption that NH_2 radicals are freely rotating in the matrix.

The second example involves the NH radical in an argon matrix at liquid helium temperature. McCarty and Robinson (1959b) have observed four lines of the $A^3\Pi \leftarrow X^3\Sigma^-$ transition of NH in absorption and four in resonance fluorescence. The lines have been assigned to individual rotational transitions and the matrix shifts found to lie between 185 and 210 cm^{-1}. The approximate constancy of the shifts provides evidence that the NH radicals are freely rotating, or very nearly so, in the solid matrix. The crystal structures of the inert gases have a very high degree of symmetry, and furthermore the gases are chemically inert, hence it may be anticipated that molecules which are small enough to be accommodated in the lattice will show considerable freedom of rotational motion.

ACKNOWLEDGMENTS

The author wishes to thank Dr. A. M. Bass, Dr. H. P. Broida, Professor G. W. Robinson, and Mr. M. McCarty, Jr., for kindly reading this manuscript and for offering many valuable comments and criticisms.

REFERENCES

Ballik, E. A., and Ramsay, D. A. (1959). *J. Chem. Phys.* **31**, 1128. Ground state of the C₂ molecule.

Barrow, R. F., Caunt, A. D., Downie, A. R., Herman, R., Huldt, E., McKellar, A., Miescher, E., Rosen, B., and Wieland, K. (1952). "Atlas des Longueurs D'Onde Caractéristiques des Bandes d'Émission et d'Absorption des Molécules Diatomiques." Hermann et Cie, Paris-Vᵉ.

Bass, A. M., and Broida, H. P. (1956). *Phys. Rev.* **101**, 1740. Spectra emitted from solid nitrogen condensed at 4.2°K from a gas discharge.

Bass, A. M., and Broida, H. P. (1958). *J. Mol. Spectroscopy* **2**, 42. Absorption spectra of solids condensed at low temperatures from electric discharges.

Becker, E. D., Pimentel, G. C., and Van Thiel, M. (1957). *J. Chem. Phys.* **26**, 145. Matrix isolation studies: infrared spectra of intermediate species in the photolysis of hydrazoic acid.

Bridge, N. K., and Porter, G. (1958). *Proc. Roy. Soc.* **A244**, 259. Primary photoprocesses in quinones and dyes. I. Spectroscopic detection of intermediates.

Broida, H. P., and Pellam, J. R. (1954). *Phys. Rev.* **95**, 845. Phosphorescence of atoms and molecules of solid nitrogen at 4.2°K.

Broida, H. P., and Peyron, M. (1957). *J. phys. radium* **18**, 593. Luminescence de l'azote solide (4.2°K) contenant des atomes ou radicaux libres. Effet de la dilution par l'argon.

Broida, H. P., and Peyron, M. (1958). *J. phys. radium* **19**, 480. Luminescence de l'azote solide (4.2°K) contenant des atomes ou radicaux libres. Effet de traces d'oxygène, d'hydrogène et de vapeur d'eau.

Brown, H. W. (1958). "Informal Discussion on Free Radical Stabilisation." Faraday Society, Sheffield, England. Photolysis of methyl iodide in an argon matrix.

Brown, H. W., and Pimentel, G. C. (1958). *J. Chem. Phys.* **29**, 883. Photolysis of nitromethane and of methyl nitrite in an argon matrix: infrared detection of nitroxyl, HNO.

Chilton, H. T. J., Porter, G., and Strachan, E. E. (1958). "Informal Discussion on Free Radical Stabilisation." Faraday Society, Sheffield, England. Spectroscopic studies of trapped aromatic radicals produced by ultraviolet and gamma irradiation.

Dalby, F. W. (1958). *Can. J. Phys.* **36**, 1336. The spectrum and structure of the HNO molecule.

Davidson, N., and DeMore, W. B. (1958). Private communication.

Dewar, J. (1888). *Proc. Roy. Inst. Gt. Brit.* **12**, 557. Phosphorescence and ozone.

Dixon, R. N. (1959). *Can. J. Phys.* **37**, 1171.

Dows, D. A., Pimentel, G. C., and Whittle, E. (1955). *J. Chem. Phys.* **23**, 1606. Infrared spectra of intermediate species in the formation of ammonium azide from hydrazoic acid.

Dressler, K., and Ramsay, D. A. (1959). *Phil. Trans. Roy. Soc. London.* **A251**, 553. The electronic absorption spectra of NH₂ and ND₂.

Evans, D. F. (1955). *Nature* **176**, 777. Photomagnetism of triplet states of organic molecules.

Foner, S. N., and Hudson, R. L. (1958a). *J. Chem. Phys.* **28**, 719. Diimide—identification and study by mass spectrometry.

Foner, S. N., and Hudson, R. L. (1958b). *J. Chem. Phys.* **29**, 442. Mass spectrometric detection of triazene and tetrazene and studies of the free radicals NH₂ and N₂H₃.

Franck, J., and Rabinowitsch, E. (1934). *Trans. Faraday Soc.* **30,** 120. Some remarks about free radicals and the photochemistry of solutions.

Gager, W., and Rice, F. O. (1959). *J. Chem. Phys.* **31,** 564. Paramagnetic resonance spectra of active species. Blue material from hydrazoil acid.

Gibson, G. E., Blake, N., and Kalm, M. (1953). *J. Chem. Phys.* **21,** 1000. The photochemical decomposition of benzene.

Giguère, P. A. (1954). *J. Chem. Phys.* **22,** 2085L. Spectroscopic evidence for stabilised HO_2 radicals.

Giguère, P. A., and Harvey, K. B. (1956). *J. Chem. Phys.* **25,** 373L. On the presumed spectroscopic evidence for trapped HO_2 radicals.

Harteck, P. (1933). *Ber.* **66B,** 423. The preparation of HNO or $(HNO)_n$.

Harvey, K. B., and Bass, A. M. (1958). *J. Mol. Spectroscopy* **2,** 405. Infrared absorption of oxygen discharge products and ozone at 4°K.

Harvey, K. B., and Brown, H. W. (1959). *J. chim. phys.* **56,** 745. Etude aux infrarouges de certains solides condensés a porter de décharges en phase gazeuse.

Herzberg, G. (1950). "Molecular Spectra and Molecular Structure, I. Spectra of Diatomic Molecules," 2nd ed. Van Nostrand, New York.

Herzberg, G., and Ramsay, D. A. (1955). *Proc. Roy. Soc.* **A233,** 34. The 7500 to 4500 A absorption system of the free HCO radical.

Herzberg, G., and Shoosmith, J. (1956). *Can. J. Phys.* **34,** 523. Absorption spectrum of free CH_3 and CD_3 radicals.

Herzfeld, C. M. (1957). *Phys. Rev.* **107,** 1239. Theory of the forbidden transitions of nitrogen atoms trapped in solids.

Herzfeld, C. M., and Broida, H. P. (1956). *Phys. Rev.* **101,** 606. Interpretation of spectra of atoms and molecules in solid nitrogen condensed at 4.2°K.

Holland, R., Style, D. W. G., Dixon, R. N., and Ramsay, D. A. (1958). *Nature* **182,** 336. Emission and absorption spectra of NCO and NCS.

Hörl, E. (1959). *J. Mol. Spectroscopy.* **3,** 425. Light emission from solid nitrogen during and after electron bombardment.

Jones, E. J., and Wulf, O. R. (1937). *J. Chem. Phys.* **5,** 873. The absorption coefficient of nitrogen pentoxide in the ultraviolet and the visible absorption spectrum of NO_3 .

Kaplan, J. (1934). *Phys. Rev.* **45,** 675, 898. New band system in nitrogen.

Kasha, M. (1947). *Chem. Revs.* **41,** 401. Phosphorescence and the role of the triplet state in the electronic excitation of complex molecules.

Kasha, M., and McGlynn, S. P. (1956). *Ann. Rev. Phys. Chem.* **7,** 403. Molecular electronic spectroscopy.

Lavin, G. I., and Bates, J. R. (1930). *Proc. Natl. Acad. Sci. U. S.* **16,** 804. The ammonia discharge tube.

Lewis, G. N., and Bigeleisen, J. (1943). *J. Am. Chem. Soc.* **65,** 520. The orientation of molecules produced photochemically in rigid solvents.

Lewis, G. N., and Calvin, M. (1945). *J. Am. Chem. Soc.* **67,** 1232. Paramagnetism of the phosphorescent state.

Lewis, G. N., and Kasha, M. (1944). *J. Am. Chem. Soc.* **66,** 2100. Phosphorescence and the triplet state.

Lewis, G. N., and Lipkin, D. (1941). *J. Am. Chem. Soc.* **63,** 3232. The dissociation of tetraphenylhydrazine and its derivatives.

Lewis, G. N., and Lipkin, D. (1942). *J. Am. Chem. Soc.* **64,** 2801. Reversible photochemical processes in rigid media: the dissociation of organic molecules into radicals and ions.

Lewis, G. N., Lipkin, D., and Magel, T. T. (1941). *J. Am. Chem. Soc.* **63**, 3005. Reversible photochemical processes in rigid media: a study of the phosphorescent state.

Lewis, G. N., Calvin, M., and Kasha, M. (1949). *J. Chem. Phys.* **17**, 804. Photomagnetism. Determination of the paramagnetic susceptibility of a dye in its phosphorescent state.

Linschitz, H., Berry, M. G., and Schweitzer, D. (1954a). *J. Am. Chem. Soc.* **76**, 5833. The identification of solvated electrons and radicals in rigid solutions of photooxidised organic molecules; recombination luminescence in organic phosphors.

Linschitz, H., Rennert, J., and Korn, T. M. (1954b). *J. Am. Chem. Soc.* **76**, 5839. Symmetrical semiquinone formation by reversible photo-oxidation and photoreduction.

Lunt, R. W., and Mills, J. E. (1935). *Trans. Faraday Soc.* **31**, 786. The blue glow on surfaces at $-180°C$ attributed to NH or NH_2 molecules.

McCarty, M., Jr., and Robinson, G. W. (1959a). *J. Am. Chem. Soc.* **81**, 4472. Imine and imine-d radicals trapped in argon, krypton and xenon matrices at $4.2°K$.

McCarty, M., Jr., and Robinson, G. W. (1959b). *J. chim. phys.* **56**, 723. Electronic absorption spectra of small free radicals in rare gas matrices.

McClure, D. S. (1951). *J. Chem. Phys.* **19**, 670. Excited triplet states of some polyatomic molecules.

McConnell, H. M. (1958). *J. Chem. Phys.* **29**, 1422L. Free rotation in solids at $4.2°K$.

McLennan, J. C. (1925). *Nature* **115**, 46. On the luminescence of solid nitrogen and argon.

McLennan, J. C. (1928). *Proc. Roy. Soc.* **A120**, 327. The aurora and its spectrum.

McLennan, J. C., and Shrum, G. M. (1924). *Proc. Roy. Soc.* **A106**, 138. On the luminescence of nitrogen, argon, and other condensed gases at very low temperatures.

McLennan, J. C., and Shrum, G. M. (1925). *Proc. Roy. Soc.* **A108**, 501. On the origin of the auroral green line 5577 Å, and other spectra associated with the aurora borealis.

McLennan, J. C., McLeod, J. H., and McQuarrie, W. C. (1927a). *Proc. Roy. Soc.* **A114**, 1. An investigation into the nature and occurrence of the auroral green line λ 5577 Å.

McLennan, J. C., Ireton, H. J. C., and Thomson, K. (1927b). *Proc. Roy. Soc.* **A116**, 1. The luminescence of solid nitrogen under cathode ray bombardment.

McLennan, J. C., Ireton, H. J. C., and Samson, E. W. (1928). *Proc. Roy. Soc.* **A120**, 303. On the luminescence of solid nitrogen under cathode ray bombardment.

McLennan, J. C., Samson, E. W., and Ireton, H. J. C. (1929). *Trans. Roy. Soc. Can.* **23**, 25. On the phosphorescence of solid argon irradiated with cathode rays.

Mador, I. L. (1954). *J. Chem. Phys.* **22**, 1617L. The stabilization of the methyl radical.

Mador, I. L., and Williams, M. C. (1954). *J. Chem. Phys.* **22**, 1627. Stabilization of free radicals from the decomposition of hydrazoic acid.

Michaelsen, J. D. (1955). Dissertation. Catholic University, Washington, D. C.

Milligan, D. E., and Pimentel, G. C. (1958). *J. Chem. Phys.* **29**, 1405. Matrix isolation studies: possible infrared spectra of isomeric forms of diazomethane and of methylene, CH_2.

Milligan, D. E., Brown, H. W., and Pimentel, G. C. (1956). *J. Chem. Phys.* **25**, 1080L. Infrared absorption by the N_3 radical.

Norman, I., and Porter, G. (1954). *Nature* **174**, 508L. Trapped atoms and radicals in a glass "Cage".

Norman, I., and Porter, G. (1955). *Proc. Roy. Soc.* **A230**, 399. Trapped atoms and radicals in rigid solvents.

Pearse, R. W. B., and Gaydon, A. G. (1950). "The Identification of Molecular Spectra," 2nd ed. Chapman & Hall, London.

Peyron, M. (1959). Private communication.

Peyron, M., and Broida, H. P. (1959a). *J. Chem. Phys.* **30**, 139. Spectra emitted from solid nitrogen condensed at very low temperatures from a gas discharge.

Peyron M., and Broida, H. P. (1959b). *J. Chem. Phys.* (in press). Emission spectra of N_2, O_2, and NO_2 molecules trapped in solid matrices.

Peyron, M., Hörl, E., Brown, H. W., and Broida, H. P. (1959). *J. Chem. Phys.* **30**, 1304. Spectroscopic evidence for triatomic nitrogen in solids at very low temperatures.

Porter, G., and Strachan, E. (1958). *Trans. Faraday Soc.* **54**, 1595. Primary photochemical processes in aromatic molecules. Part 4—Side chain photolysis in rigid media.

Porter, G., and Windsor, M. W. (1953). *J. Chem. Phys.* **21**, 2088L. Triplet states in solution.

Porter, G., and Windsor, M. W. (1954). *Discussions Faraday Soc.* **17**, 178. Studies of the triplet state in fluid solvents.

Ramsay, D. A. (1956). *Nature* **178**, 374. Absorption spectra of free PH_2 and PD_2 radicals.

Ramsay, D. A. (1957). *Mém. soc. roy. sci. Liège* **18**, 471. The analysis of the α bands of ammonia.

Ramsay, D. A. (1959). *Advances in Spectroscopy* **1**. In press. The spectra of polyatomic free radicals.

Rice, F. O. (1956). *J. Chem. Phys.* **24**, 1259L. Colors on Jupiter.

Rice, F. O., and Ditter, J. (1953). *J. Am. Chem. Soc.* **75**, 6066. Green sulfur, a new allotropic form.

Rice, F. O., and Freamo, M. (1951). *J. Am. Chem. Soc.* **73**, 5529. The imine radical.

Rice, F. O., and Freamo, M. (1953). *J. Am. Chem. Soc.* **75**, 548. The formation of the imine radical in the electrical discharge.

Rice, F. O., and Grelecki, C. (1957a). *J. Am. Chem. Soc.* **79**, 1880. The imine radical.

Rice, F. O., and Grelecki, C. (1957b). *J. Am. Chem. Soc.* **79**, 2679. The dimethylamino radical.

Rice, F. O., and Grelecki, C. (1957c). *J. Phys. Chem.* **61**, 824L. An active species formed in the electrical decomposition of dimethylamine.

Rice, F. O., and Grelecki, C. (1957d). *J. Phys. Chem.* **61**, 830L. The methyl imino radical.

Rice, F. O., and Scherber, F. I. (1955). *J. Am. Chem. Soc.* **77**, 291. The hydrazino radical and tetrazane.

Rice, F. O., and Sparrow, C. (1953). *J. Am. Chem. Soc.* **75**, 848. Purple sulfur, a new allotropic form.

Robinson, G. W., and McCarty, M., Jr. (1958a). *J. Chem. Phys.* **28**, 349L. Electronic spectra of free radicals at 4°K—NH_2.

Robinson, G. W., and McCarty, M., Jr. (1958b). *J. Chem. Phys.* **28**, 350L. Electronic spectra of free radicals at 4°K—HNO, NH and OH.

Robinson, G. W., and McCarty, M., Jr. (1958c). *Can. J. Phys.* **36**, 1590. Radical spectra at liquid helium temperatures.

Robinson, G. W., and McCarty, M., Jr. (1958d). "Informal Discussion on Free Radical Stabilisation." Faraday Society, Sheffield, England. Some diatomic and triatomic free radicals at 4.2°K.

Robinson, G. W., and McCarty, M., Jr. (1959). *J. Chem. Phys.* **30**, 999. Trapped NH_2 radicals at 4.2°K.

Schnepp, O., and Dressler, K. (1959). *J. Chem. Phys.* (Submitted for publication). The photolysis of ammonia in a solid matrix at low temperatures.

Schoen, L. J., and Broida, H. P. (1959). *J. Chem. |Phys.* (in press). Spectra emitted from rare gas-oxygen solids during electron bombardment.

Schoen, L. J., and Rebbert, R. (1959). *J. Mol. Spectroscopy.* **3,** 417. Electrical discharge induced luminescence of solids at low temperatures.

Schoen, L. J., Kuentzel, L. E., and Broida, H. P. (1958). *Rev. Sci. Instr.* **29,** 633. Glass dewars for optical studies at low temperatures.

Sowden, R. G., and Davidson, N. (1956). *J. Am. Chem. Soc.* **78,** 1291. Photochemical studies with rigid hydrocarbon solvents at low temperatures.

Symons, M. C. R., and Townsend, M. (1956). *J. Chem. Phys.* **25,** 1299L. Electronic spectrum of trapped ethanol radicals.

Terenin, A. (1943). *Acta Physicochim. U.R.S.S.* **18,** 210. Photochemical processes in aromatic molecules.

Thrush, B. A. (1956). *Proc. Roy. Soc.* **A235,** 143. The detection of free radicals in the high intensity photolysis of hydrogen azide.

Vegard, L. (1924a). *Nature* **113,** 716. The auroral spectrum and the upper atmosphere.

Vegard, L. (1924b). *Nature* **114,** 357. The light emitted from solidified gases and its relation to cosmic phenomena.

Vegard, L. (1924c). *Proc. Roy. Akad. Amsterdam* **27,** 113; *Communs. Phys. Lab. Univ. Leiden, No.* **168d.** Light emitted from solid nitrogen when bombarded with cathode rays and its bearing on the auroral spectrum.

Vegard, L. (1925). *Communs. Phys. Lab. Univ. Leiden, No.* **175;** *Det. Norske Vid. Akad. Skr. I, No.* **9.** The luminescence from solidified gases down to the temperature of liquid hydrogen and its application to cosmic phenomena.

Vegard, L. (1926a). *Communs. Phys. Lab. Univ. Leiden, Suppl. No.* **59 to 181.** The luminescence from solidified gases at the temperature of liquid hydrogen. Supplementary communication.

Vegard, L. (1926b). *Ann. phys.* **79,** 377. Das Leuchten verfestigter Gase und seine Beziehungen zu kosmischen Vorgängen.

Vegard, L. (1927). *Communs. Phys. Lab. Univ. Leiden, Suppl. No.* **62 to 181.** Further observations on the luminescence from solid nitrogen at the temperature of liquid hydrogen.

Vegard, L. (1929). *Communs. Phys. Lab. Univ. Leiden, No.* **200;** *Det. Norske Vid. Akad. Skr. I, No.* **7.** Continued investigations on the luminescence from solidified gases at the temperature of liquid hydrogen. Part 1. Luminescence produced by cathode rays of high velocity.

Vegard, L. (1930a). *Communs. Phys. Lab. Univ. Leiden, No.* **205a;** *Det. Norske Vid. Akad. Skr. I, No.* **13,** 1929. Part 2. Luminescence produced by canal rays.

Vegard, L. (1930b). *Det. Norske Vid. Akad. Skr. I, No.* **2.** Spectra from solidified gases and their interpretation.

Vegard, L. (1930c). *Ann. phys.* **6,** 487. Die Spektren verfestigter Gase und ihre atomtheoretische Deutung.

Vegard, L. (1930d). *Nature* **125,** 14L. New types of emission spectra.

Vegard, L. (1930e). *Det. Norske Vid. Akad. Skr. I, No.* **8.** The luminescence from solidified gases and its variation with the velocity of the exciting cathode rays.

Vegard, L., and Keesom, W. H. (1927). *Communs. Phys. Lab. Univ. Leiden, No.* **186.** On the luminescence produced by bombarding solidified gases with electric rays at the temperature of liquid helium.

Vegard, L., and Keesom, W. H. (1930a). *Proc. Roy. Akad. Amsterdam* **33**, 10. Continued investigations on the luminescence from solidified gases at the temperature of liquid helium.

Vegard, L., and Keesom, W. H. (1930b). *Communs. Phys. Lab. Univ. Leiden, No.* **205b**. Luminescence from solidified gases at the temperature of liquid helium.

Vegard, L., and Stensholt, S. (1935). *Det. Norske Vid. Akad. Skr. I, No.* **9**. The properties of the ε-system (Vegard bands) derived from new and previous measurement.

Vegard, L., Onnes, H. K., and Keesom, W. H. (1925). *Compt. rend.* **180**, 1084; *Proc. Roy. Akad. Amsterdam* **28**, 467, *Communs. Phys. Lab. Univ. Leiden, No.* **173d**. Émission de lumière par des gaz solidifiés à la température de l'hélium liquide et origine du spectre auroral.

Vegard, L., Onnes, H. K., and Keesom, W. H. (1926). *Communs. Phys. Lab. Univ. Leiden, No.* **183**. The luminescence of solidified gases at liquid helium temperature.

Wall, L. A., Brown, D. W., and Florin, R. E. (1959). *J. Chem. Phys.* **30**, 602L. Electron spin resonance spectra from gamma irradiated solid nitrogen.

Whittle, E., Dows, D. A., and Pimentel, G. C. (1954). *J. Chem. Phys.* **22**, 1943L. Matrix isolation method for the experimental study of unstable species.

Wiedemann, E. (1888). *Ann. Physik* **34**, 446. Über Fluorescenz und Phosphorescenz I. Abhandlung.

7. Electron Spin Resonance Studies of Trapped Radicals[1]

C. K. Jen

*Applied Physics Laboratory, The Johns Hopkins University,
Silver Spring, Maryland*

[1] Work supported by the Bureau of Ordnance, Department of the Navy, under NOrd 7386.

I. Introduction

The method of electron spin resonance (ESR) has been applied to the studies of trapped free radicals by combining the essential features of two previously developed techniques—that of trapping free radicals in rigid media (Lewis and Lipkin, 1942) and that of detecting electron spin resonance (Zavoisky, 1945). Since a free radical may be defined as a molecular fragment possessing an unpaired electron spin, the ESR method has the merit of detecting and studying a free radical by its most basic property. But the free spin property of a radical is also the principal reason for its extremely high chemical reactivity. Thus, radical stabilization by trapping techniques has become a powerful tool for the ESR method as well as for many other studies.

II. Elementary Principles of Electron Spin Resonance

A. SIMPLE ELECTRON SPIN RESONANCE

Consider an electron with the spin angular momentum $\mathbf{S}$. There is a magnetic moment $\mathbf{\mu}_s$ associated with the spin according to the following relation:

$$\mathbf{\mu}_s = -g_s\beta\mathbf{S} \tag{1}$$

where $\mathbf{\mu}_s$ is in electromagnetic units (emu) units, g_s a dimensionless factor, β the Bohr magneton, and $\mathbf{S}$ is in units of $\hbar$. The quantity g_s is variously called the g-factor or the spectroscopic splitting factor, which is experimentally determined to be 2.0023.

If the electron is placed in a magnetic field H_0 (along the z-axis), then the Hamiltonian for the magnetic energy of the electron is

$$\mathcal{3C} = g_s\beta\mathbf{S}\cdot\mathbf{H}_0 = g_s\beta S_z H_0, \tag{2}$$

which has the eigenvalue

$$W = g_s\beta M_s H_0 \tag{2a}$$

where M_S is the magnetic quantum number which can take only the value $\frac{1}{2}$ or $-\frac{1}{2}$ for an electron spin. Thus we have two energy levels, $W_{\frac{1}{2}}$ and $W_{-\frac{1}{2}}$, having a separation between them equal to $g_s\beta H_0$. Since there is a finite probability for a transition between these two energy levels, a change in the energy state can be stimulated by an external radio-frequency (rf) magnetic field (along the x-axis, say) under the resonance condition

$$h\nu_0 = g_s\beta H_0 \tag{3}$$

where ν_0 is the frequency of the wave at resonance, known as the Larmor frequency.

The aforementioned transition probability per unit time is given by the radiation theory as (Pake, 1956, p. 13)

$$p\left(-\frac{1}{2},\frac{1}{2}\right) = \frac{(\gamma H_1)^2}{4} f(\nu - \nu_0) \tag{4}$$

where $\gamma = g_s\beta/\hbar$, $H_1 =$ amplitude of rf magnetic field, and $f(\nu - \nu_0) =$ normalized shape function having a maximum value at $\nu = \nu_0$, such that $\int_0^\infty f(\nu - \nu_0)\,d\nu = 1$. If the total number of electrons be represented by N_0 and the distribution of population between the two states obeys the Boltzmann distribution, then

$$N\left(\frac{1}{2}\right) = N\left(-\frac{1}{2}\right) e^{-h\nu_0/kT} \tag{5a}$$

and

$$N\left(-\frac{1}{2}\right) - N\left(\frac{1}{2}\right) \doteq \frac{N_0}{2}\frac{h\nu_0}{kT} \tag{5b}$$

if $h\nu_0/kT \ll 1$ and T is the temperature of the lattice, which is here assumed to be in thermal equilibrium with the spins. The product of the quantities in Eqs. (4) and (5b), multiplied also by $h\nu$, gives the total power P absorbed by the spins from the rf field as follows:

$$P = \frac{\omega}{2}\left\{\frac{N_0\gamma^2\hbar^2 S(S+1)}{3kT}\,\omega_0 f(\nu - \nu_0)\right\} H_1^2 \tag{6}$$

where $S = \frac{1}{2}$ and $\omega = 2\pi\nu$. It should be remarked that Eq. (6) is actually correct for any value of S (integral or half-integral), even though it is here specifically derived for $S = \frac{1}{2}$. Equation (6) is to be identified with the power absorbed by a system having a complex susceptibility $\chi = \chi' - i\chi''$, i.e., $P = \omega/2\,\chi'' H_1^2$. The bracketed quantity in Eq. (6) can be expressed as $\chi'' = \chi_0\omega_0 f(\nu - \nu_0)$, where χ_0 is the usual static susceptibility.

The resonance condition in Eq. (3) and the spectral distribution of power absorbed expressed in Eq. (6) are illustrated in Fig. 1. The curve in

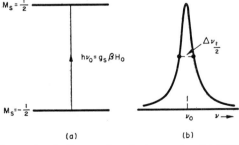

FIG. 1. (a) Energy level diagram for electron spin. (b) ESR spectral line.

Fig. 1(b) is known as a spin resonance spectral line. Since $f(\nu - \nu_0)$ is a normalized function, the peak value $f(\nu_0)$ should be inversely proportional to some measure of line width $\Delta\nu$, as shown in Fig. 1(b). By Eq. (3), passage through resonance can be equally well carried out by varying H instead of ν so that we would have a spectral line represented by

$$P \sim f(H - H_0).$$

Similarly, we can define a line-width parameter in terms of magnetic field as $\Delta H = 2\pi\Delta\nu/\gamma$.

Thus far, we have assumed that the lattice can take away energy from the excited spin system fast enough so that the Boltzmann distribution at the lattice temperature expressed in Eq. (5a), is maintained. However, this is not always the case. Let $n = n(-\frac{1}{2}) - n(\frac{1}{2})$ denote the instantaneous difference in population between the $M_s = -\frac{1}{2}$ state and the $M_s = \frac{1}{2}$ state. If the spin system is in thermal equilibrium with the lattice, then $n = n_0 = N_0 h\nu_0/2kT$ as given in Eq. (5b). The rate of decrease in n should be $2np(-\frac{1}{2}, \frac{1}{2})$ according to Eq. (4). On the other hand, the lattice will tend to restore the equilibrium at a rate proportional to $n_0 - n$. We shall designate this rate as $(n_0 - n)/T_1$ where T_1 is known as the spin-lattice relaxation time. The total rate of change in n is given by Eq. (7),

$$\frac{dn}{dt} = -\frac{(\gamma H_1)^2}{2}f(\nu - \nu_0)n + \frac{n_0 - n}{T_1}. \tag{7}$$

Putting $dn/dt = 0$ in the steady state, we have for n

$$n = \frac{n_0}{1 + \frac{(\gamma H_1)^2 f(\nu - \nu_0)T_1}{2}}. \tag{8}$$

Equation (8) says that the spin system will depart more and more from the thermal equilibrium with the lattice as the second denominator term assumes higher importance. As a result, the total power absorbed by the spins,

$$P = nh\nu P(-\frac{1}{2}, \frac{1}{2}),$$

reaches a constant as the rf (usually microwave) power is increased. This is known as the phenomenon of power "saturation." Also, it is noticed in Eq. (8) that the degree of saturation varies with $f(\nu - \nu_0)$, being highest at the spectral center $\nu = \nu_0$. By defining $\frac{1}{2}f(\nu = \nu_0)$ as T_2, which is interpreted as the spin-spin relaxation time, we have for n at $\nu = \nu_0$

$$n(\nu = \nu_0) = \frac{n_0}{1 + (\gamma H_1)^2 T_1 T_2} \tag{9}$$

which is a familiar expression used in Bloch's treatment of nuclear spin resonance (Bloch, 1946).

B. MAGNETIC HYPERFINE INTERACTIONS
OF AN UNPAIRED ELECTRON

When an electron with an unpaired spin is a part of an atom or a molecule, it has in general many electric and magnetic interactions with the nucleus or nuclei and other electrons. For simplicity, we shall consider a one-electron problem and study the magnetic interactions between the unpaired electron and only one "magnetic" nucleus. The electron may have, in addition to the spin and magnetic moment[2] as defined in Eq. (1), an orbital angular momentum $\mathbf{L}$ and its associated magnetic moment $\mathbf{\mu}_L$ under the following relation:

$$\mathbf{\mu}_L = -\beta\mathbf{L}. \tag{10}$$

Similarly, the magnetic moment of the nucleus is defined as

$$\mathbf{\mu}_I = -g_I\beta\mathbf{I} \tag{11}$$

where $\mathbf{I}$ is the angular momentum of the nucleus and g_I the nuclear g-factor referred to the Bohr magneton.

The Hamiltonian for the hyperfine (hf) interaction energy can be written as (Abragam and Pryce, 1951)

$$\mathcal{H}_{hf} = -g_Ig_S\beta^2\frac{8\pi}{3}\delta(\mathbf{r})\mathbf{I}\cdot\mathbf{S} + g_Ig_S\beta^2\left\{\frac{\mathbf{I}\cdot\mathbf{S}}{r^3} - 3\frac{(\mathbf{I}\cdot\mathbf{r})(\mathbf{S}\cdot\mathbf{r})}{r^5}\right\} - \frac{2g_I\beta^2}{r^3}\mathbf{I}\cdot\mathbf{L} \tag{12}$$

where $\delta(\mathbf{r})$ is the Dirac delta function and $\mathbf{r}$ is radius vector between the nucleus and the electron. The first term on the right-hand side of Eq. (12) describes the "contact" or isotropic hf interaction. The second and third terms are both anisotropic in nature and describe respectively the dipolar and orbital hf interactions. We shall discuss the first and second terms separately and the third term as a part of the more general picture.

C. ISOTROPIC HYPERFINE INTERACTION

The first term in Eq. (12) was first introduced by Fermi (1930) in a relativistic treatment of the atomic hyperfine interaction. The spin Hamiltonian for this term can be written as

$$\mathcal{H}_F = -\frac{8\pi}{3}g_Ig_S\beta^2\psi^2(0)\mathbf{I}\cdot\mathbf{S} \tag{13}$$

where $\psi^2(0)$ is the electronic density at the nucleus and F is the value of $\mathbf{F} = \mathbf{I} + \mathbf{S}$. It has a value only when the electron has a nonzero density directly at the nucleus itself, thus accounting for the name "contact." A

[2] For simplicity, we shall use the same notation g_S for a bound electron as for a free electron, thus overlooking a small correction due to relativistic mass change.

representative case is that of a hydrogen atom in the $^2S_{\frac{1}{2}}$ state, where the electronic s-function has a spherical symmetry and a finite density at the nucleus. By contrast, a non-s electron (p, d, f, etc.) would contribute nothing to this interaction because of zero density at the same nucleus. A generalization of this concept has been extended to the molecular radicals where a similar electron charge distribution around the magnetic nucleus can be treated in the same way.

The interaction in Eq. (13) causes a zero-field splitting of the electronic energy level into two hyperfine energy levels characterized by

$$F = I + \tfrac{1}{2} \quad \text{and} \quad F = I - \tfrac{1}{2},$$

respectively, having an energy separation between them,

$$\Delta W = -[(2I + 1)/2]8\pi/3 \; g_S g_I \beta^2 \psi^2(0).$$

In an external magnetic field $\mathbf{H}_0$, the total spin Hamiltonian is

$$\mathcal{3C} = g_I \beta \mathbf{I} \cdot \mathbf{H}_0 + g_S \beta \mathbf{S} \cdot \mathbf{H}_0 + \frac{2\Delta W}{2I + 1} \mathbf{I} \cdot \mathbf{S} \tag{14}$$

where the first and second terms on the right-hand side are the magnetic energies in the external field and the third term is the hyperfine energy equivalent to that in Eq. (13). The solution of Eq. (14) is given by the Breit-Rabi formula (Breit and Rabi, 1931; Nafe and Nelson, 1948)

$$W_{I\pm\frac{1}{2}} = -\frac{\Delta W}{2(2I + 1)} + g_I \beta M H_0 \pm \frac{\Delta W}{2}\left(1 + \frac{4Mx}{2I + 1} + x^2\right)^{\frac{1}{2}} \tag{15}$$

where M = magnetic quantum number of $\mathbf{F}$ and $x = (g_S - g_I)\beta H_0/\Delta W$. There are altogether $2(2I + 1)$ states with the transition between any two of them governed by the usual selection rules: $\Delta F = 0, \pm 1$ and $\Delta M = 0, \pm 1$. In a very high field, or more exactly when $x \gg 1$, Eq. (15) is reduced approximately to the following:

$$W_{M_I, M_S} \doteq M_I g_I \beta H_0 + M_S g_S \beta H_0 + M_I M_S A \tag{16}$$

where M_I and M_S are the magnetic quantum numbers of $\mathbf{I}$ and $\mathbf{S}$, respectively, and $A = 2\Delta W/(2I + 1)$. The selection rules for the limiting case are: $\Delta M_I = 0$ and $\Delta M_S = 0, \pm 1$. There are therefore $2I + 1$ uniformly spaced lines symmetrically placed around the center position defined by $h\nu_0 = g_S \beta H_0$. The separation between any two adjacent lines is equal to A/h in frequency units. The lines will have the same relative intensity because of equal matrix elements of transition. Figure 2 illustrates a simple hf spectrum of this type for $S = \frac{1}{2}$ and $I = 1$ and the passage through resonances by the variation of the magnetic field at a fixed frequency. In the general case (e.g., at low fields), however, the lines will not be uni-

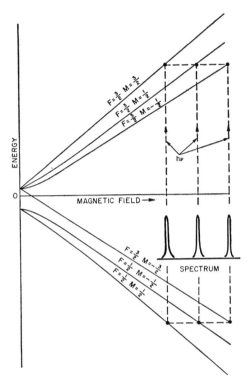

FIG. 2. Energy level diagram for $S = \frac{1}{2}$ and $I = 1$ and the hf spectrum in strong field.

formly spaced, nor will they have exactly the same relative intensity. Under these conditions, the quantities A (or ΔW) and g_s will have to be calculated by using Eq. (15) together with the experimental data.

The spectrum described above is said to have a hyperfine structure since the multiplicity of lines is a result of the hyperfine interactions between the nucleus and the electron. Specifically, an interaction of the contact type as in Eq. (13) is described by the term isotropic because of its independence on the orientation between the magnetic field and the molecular axes (a free atom, of course, does not have a preferred set of axes). Because of its simplicity and common occurrence in many free radicals, the isotropic hf interaction and its associated structure will claim most of our attention in the present treatment.

D. DIPOLAR HYPERFINE INTERACTION

The dipolar hyperfine interaction, as defined by the second term in Eq. (12), can be calculated classically as the potential energy between two magnetic dipoles having moments $\mathbf{u}_I$ and $\mathbf{u}_S$. In contrast to the isotropic nature of the contact interaction, this interaction is basically anisotropic as we shall soon see.

It is assumed here that the unpaired electron has an orbital angular momentum $\mathbf{L}$ as well as a spin. But the average value of $\mathbf{L}$ is zero in a non-linear polyatomic molecule, because no component of the orbital angular momentum can remain constant with this kind of symmetry (Van Vleck, 1932, p. 273). Thus we can leave out the effect of $\mathbf{L}$ for the present considerations. For a nonrotating molecule, $\mathbf{F} = \mathbf{I} + \mathbf{S}$ is a constant of motion which would assume the values of $I + \frac{1}{2}$ and $I - \frac{1}{2}$ in this case. Since, however, the hyperfine energy is usually rather small, the coupling between $\mathbf{I}$ and $\mathbf{S}$ is easily broken down in a strong magnetic field. We shall, for simplicity, assume that $\mathbf{I}$ and $\mathbf{S}$ are quantized independently along $\mathbf{H}_0 \to \infty$.

The hyperfine energy for the dipolar interaction can be shown (using the addition theorem for Legendre polynomials) to be (Frosch and Foley, 1952)

$$W_{\text{dip}} = -\frac{M_I M_S g_I g_S \beta^2}{2} \left\langle \frac{3 \cos^2 \alpha - 1}{r^3} \right\rangle_{\text{av}} (3 \cos^2 \theta - 1) \qquad (17)$$

where θ is the angle between $\mathbf{H}_0$ and a molecular axis of symmetry (say, along a unit vector $\mathbf{K}$) and α is the angle between $\mathbf{r}$ and $\mathbf{K}$. The term

$$\left\langle \frac{3 \cos^2 \alpha - 1}{r^3} \right\rangle_{\text{av}}$$

is obtained by averaging over the spatial part of the unpaired electronic wave function. It is noted that this average quantity would be zero if the electron is in an s-orbital but in general will not vanish for non-s orbitals.

The quantity W_{dip} in Eq. (17) is an anisotropic contribution to the hyperfine interaction energy because of the orientation-dependent term $(3 \cos^2 \theta - 1)$. For a non-s orbital electron, the total energy (magnetic plus hyperfine) is

$$W_{M_I, M_S} = M_I g_I \beta H_0 + M_S g_S \beta H_0 + M_I M_S A' \qquad (18)$$

where $A' = W_{\text{dip}}/M_I M_S$. It is seen that Eq. (18) depicts exactly the same type of spectrum as given by Eq. (16) except that the hf separation in this case (controlled by A') is a function of the orientation angle. This means that if all radical molecules are oriented in the same way in a single crystalline inert matrix, then A' can be determined from the observed hf structure which contains the information on the electron distribution. If, on the other hand, in a polycrystalline material the radical orientation is completely randomized, the whole hf pattern would be smeared into one broad line which may or may not be readily observable. However, if the radical should have a "tumble" motion, then W_{dip} can be either partially or completely averaged out so that the broadening due to the hf structure is either reduced or canceled.

The general case of an unpaired electron having an admixture of s and non-s character will have an hf coupling constant equal to

$$A(s) + A'(\text{non-}s)$$

which will bring about several different situations: (1) in a single crystal, a partially anisotropic hf spectrum; (2) in polycrystals, a spectrum determined by A but broadened by $A'(A \gg A')$, or an entirely smeared spectrum $(A \gtrsim A')$.

E. HYPERFINE INTERACTIONS FOR ELECTRONIC SPIN AND ORBITAL MOMENTS

If the electronic orbital angular $\mathbf{L}$ is not quenched as it is in a polyatomic nonlinear radical, then it becomes a factor in the hf structure of the ESR spectrum. For the purpose of the present discussion, we shall assume that the radical species is in the free state or trapped in a perfectly symmetrical crystalline medium (cf. the following section).

Let us start with an atom in a non-S state. The usually strong coupling between $\mathbf{L}$ and $\mathbf{S}$ makes each L-level split into a "spin-doublet" corresponding to $J = L \pm \frac{1}{2}$ which is the magnitude of $\mathbf{J} = \mathbf{L} + \mathbf{S}$. For ESR work, we are interested almost always in the lower of the two states. It has been stated before that the first term in Eq. (12) would vanish for a $L \neq 0$ state. It can be shown that the Hamiltonian including the magnetic energies can be written as (Ramsey, 1953, p. 25)

$$\mathcal{H} = g_J\beta\mathbf{J}\cdot\mathbf{H} + g_I\beta\mathbf{I}\cdot\mathbf{H} + \alpha\mathbf{I}\cdot\mathbf{J} \tag{19}$$

where

$$g_J = 1 + \frac{J(J + 1) + S(S + 1) - L(L + 1)}{2J(J + 1)},$$

the Landé g-factor,

$$\alpha = -g_I\beta^2 \left\langle\frac{1}{r^3}\right\rangle_{\text{av}} \frac{2L(L + 1)}{J(J + 1)}$$

and r = distance between the electron and the nucleus. The solution of Eq. (19) is given, again, by the Breit-Rabi formula in Eq. (15) with g_J and α substituted respectively for g_s and $A = 2\Delta W/(2I + 1)$. The hf spectrum can be described in exactly the same way as before. The addition of a nuclear quadrupolar term to Eq. (19), which has been omitted in this treatment, would make the spectrum somewhat more complicated but well analyzable.

The situation is considerably more involved for linear radicals, which include as special cases the diatomic radicals (e.g., OH, CH, etc.). Frosch and Foley (1952) have reduced Eq. (12) for this case to the spin Hamiltonian written as

$$\mathcal{3C}_{hf} = a\mathbf{I} \cdot \mathbf{S} + b(\mathbf{I} \cdot \mathbf{K})(\mathbf{S} \cdot \mathbf{K}) + c\Lambda\mathbf{I} \cdot \mathbf{K} \tag{20}$$

where a, b, and c are constants representing the averages of the various coefficients in Eq. (12) over the electronic wave function, $\mathbf{K}$ is a unit vector along the axis of symmetry, and Λ is the quantized component of $\mathbf{L}$ along the symmetry axis. Detailed evaluation of the term values for the hf interaction in Eq. (20) depends upon which particular coupling scheme is chosen for a given situation. We shall see, however, as far as the ESR hyperfine spectrum of a linear radical in strong magnetic field is concerned, the phenomenological treatment of a radical in an axial symmetrical crystalline field to be given in the following section will prove to be an equivalent approach to the subject.

F. Effect of Crystal Field on ESR Hyperfine Spectra

A trapped radical must of necessity have some electrostatic interaction with its trapping medium. The strength of interaction may be very strong as in an ionic crystal or rather weak as in a crystal of an inert element. The effect of the crystal field on the properties of an embedded radical can be studied by examining the nature of the field immediately surrounding the radical in a manner analogous to the treatment of paramagnetic ions. The potential energy function expressed in coordinates centering around the radical is (Van Vleck, 1932, p. 288)

$$V = ax^2 + by^2 + cz^2 \tag{21}$$

where a, b, and c are constants. Let us assume the radical electron has an orbital angular momentum $\mathbf{L}$ having a value $L = 1$ and for the moment overlook the considerations of the spin. If we take a simple p-orbital as the unperturbed ground state function and calculate the diagonal matrix element of V, we find that the original triply degenerate $(2L + 1)$ orbital states are now split into three distinct energy levels, if the constants a, b, and c are all unequal. At the same time, the diagonal matrix element of $\mathbf{L}$ is zero for all three components in space. This means that there is no net angular momentum along any direction. This phenomenon is known as the "quenching" of orbital angular momentum by assymetrical crystalline field. If, however, we have $a = b$ in Eq. (21), then there are only two distinct energy levels, one for $M_L = 0$ and the other for $M_L = \pm 1$. The double degeneracy of the levels for $M_L = \pm 1$ suggests that the angular momentum $\mathbf{L}$ has a constant component $(\Lambda = 1)$ along the z-direction, and indeed it does. Under almost all ordinary conditions, the state with $M_L = 0$ is lower than the state with $M_L = \pm 1$; and, if the thermal energy kT is much less than the energy difference, the low-lying $M_L = 0$ is the only one of importance. The $M_L = 0$ level, known as the orbital singlet state, has of course no net angular momentum. Lastly, the perfectly symmetrical case of $a = b = c$ does not split the levels at all and the radical will appear as if it is free in this field.

While the orbital angular momentum is quenched by an asymmetrical crystalline field, the electron spin, not being acted on by the electrical forces, would remain free and unquenched. Of course, if the orbital angular momentum is only partially quenched, there will be some lack of complete freedom on the part of the spin.

For the purpose of discussing the hf structure in the ESR spectrum of a trapped radical, we will choose the special case of axial symmetry, i.e., $a = b$ in Eq. (21). Further, for simplicity, assume the ground state is an orbital singlet. The spin Hamiltonian can be written as (Bleaney and Stevens, 1953, p. 134)

$$\mathcal{H} = DS_z^2 + \beta[g_{||}H_zS_z + g_{\perp}(H_xS_x + H_yS_y)]$$
$$+ AI_zS_z + B(I_xS_x + I_yS_y) \tag{22}$$

where z is along the axis of symmetry, D is a coefficient of zero-field splitting, $g_{||}$ and $g_{\perp}$ are respectively the g-factors along the directions parallel and perpendicular to the symmetry axis, A and B are the hyperfine coupling constants. The fact that anisotropy is already built in the system is shown by the factors $g_{||}$, $g_{\perp}$, A, B. They are designed here not only to take care of the anisotropic effects of the crystalline field, but also those of the molecule as evidenced by the dipolar interactions previously discussed. Under certain conditions, it is conceivable that the two anisotropic effects can be separated.

The solution of Eq. (22) is in general very complicated, but in strong fields it can be simplified to the following (omitting the effect of zero-field splitting and the unimportant nuclear magnetic field term)

$$W = M_s g\beta H_0 + M_I M_s K \tag{23}$$

where

$$g^2 = g_{||}^2 \cos^2\theta + g_{\perp}^2 \sin^2\theta$$
$$K^2 g^2 = A^2 g_{||}^2 \cos^2\theta + B^2 g_{\perp}^2 \sin^2\theta$$

and θ = angle between the symmetry axis and the applied field $\mathbf{H}_0$. Eq. (23), similar to many other equations discussed before, depicts a spectrum of $2I + 1$ hf lines of uniform spacing and equal intensity symmetrically placed around the center position defined by $h\nu_0 = g\beta H_0$. The quantity K/h is directly given by the spacing between adjacent lines in units of frequency. If this equation describes correctly a given situation, then the constants $g_{\parallel}$, $g_{\perp}$, A, and B can be readily determined from the experimental data on the angular variation of the spectral positions.

G. Nuclear Statistics and Hyperfine Spectral Intensities

When there are several identical magnetic nuclei in a radical, all the magnetic and hf terms in the preceding equations for a single magnetic nucleus can be simply modified by summing each term over all the identical nuclei. However, the distribution of the hf spectral intensities will depend upon the statistical weights which can be assigned to the individual hf levels by the well-known statistics (Herzberg, 1945, p. 50). In free radical problems encountered in practice, we are usually interested in the cases of either identical hydrogen or deuterium nuclei. Hence the total wave function $\psi = \psi_e \psi_v \psi_r \psi_n$ (electronic, vibrational, rotational, and nuclear) should be antisymmetric with respect to a symmetry operation (resulting in the exchange of any two identical nuclei) for identical hydrogen nuclei ($I = \frac{1}{2}$) and symmetric for identical deuterium nuclei ($I = 1$) according to the Fermi and Bose statistics in the respective cases. In the two examples of planar radicals which we will discuss later in the chapter (i.e., NH_2 versus ND_2 and CH_3 versus CD_3), ψ_e for a p_z orbital would change sign with a rotation around the twofold axis whereas ψ_v would not in the zero vibration state. This means that ψ_n should be symmetric for identical hydrogen nuclei and antisymmetric for identical deuterium nuclei in the lowest rotational state ($J = 0$). The situation is exactly reversed for the next higher rotational state ($J = 1$). We can then obtain the statistical weight for any hf magnetic level by counting the number of symmetric or antisymmetric nuclear wave functions that are associated with the same state.

Under the conditions, where either the even and odd rotational states have the same weight factor or the magnetic nuclei are equivalent in terms of hf coupling but nonidentical in terms of symmetry, then the statistical weight for a given hf level can be simply computed from the coefficient of a polynomial expressed as

$$(\alpha + \beta + \gamma + \cdots)^n \tag{24}$$

where α, β, γ $\cdots$ are the nuclear wave functions each corresponding to a different eigenvalue of m_I in $-I \leqq M_I \leqq I$ and n is the number of equivalent nuclei. Thus, for hydrogen nuclei, we have $(\alpha + \beta)^n$ giving the binominal coefficients for the statistical weights of the hf levels $M_I = -n/2$, $\cdots$, $n/2$. For deuterium nuclei, we have $(\alpha + \beta + \gamma)^n$, giving the polynomial coefficients for the statistical weights of the hf levels $M_I = -n,$ $\cdots, n.$

III. Experimental Techniques

A. Trapped Radical Samples for ESR Observation

Generally speaking, all the methods currently employed for the production and entrapment of free radicals are usable and have been used for ESR observation. Irradiation of solid or frozen chemicals by ultraviolet, X-, or gamma rays has been the most common source of radicals studied. Cold deposition of gaseous discharge products has been used principally for atomic and "small" radicals. High energy electron bombardment of frozen samples has been used on rare occasions to study a few light radicals. Polymerization and pyrolysis have been applied to high molecular weight organic radicals. In all except the last method, the entrapment of a radical depended upon the constraining action of a rigid medium and, in most cases, also the decrease of thermal activity at a low temperature (liquid nitrogen or helium). In the last-mentioned case (polymerization and pyrolysis), in which the radicals are formed at an appropriately elevated temperature but usually observed at room temperature, "self-trapping" is believed to be the mechanism of stability.

As mentioned above, many trapped radical experiments have been performed at low temperatures. In most of them, a low temperature cell is used to cool the sample down to the desired temperature, an appropriate arrangement for irradiation of the sample *in situ* or deposition of dissociated products on a cold surface, some means of inserting the sample holder (a glass ampule or a nonmetallic rod) into the microwave cavity, and a heat leak to raise the temperature to desired levels. As an illustration, Fig. 3 shows an arrangement specifically used in one experiment (Jen *et al.*, 1958) for the deposition of discharge products at liquid helium temperature (heat leak in the form of a heater coil not shown). For irradiation work, the discharge tube is replaced by a window across the external slit and a radiation source next to the window.

B. Microwave and Measurement Techniques

Most ESR spectrometers used by various workers are in the X-band microwave frequencies (around 9000 Mc/sec). The advantage lies in the convenient size of the microwave components and the ready availability of the steady magnetic field required for resonance measurements (around 3000 oersted). Some laboratories are equipped with the K-band (around 23,000 Mc/sec) spectrometers either for the main operation or as an additional unit for checking g-shifts and hyperfine coupling constants. We will

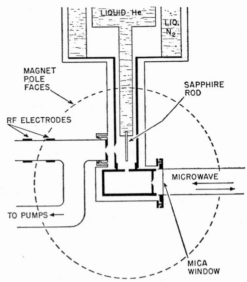

FIG. 3. A low temperature cell for deposition of discharge products (convertible for ultraviolet irradiation) and the microwave cavity.

discuss the general principles of technique and specific examples in the X-band unless otherwise specified.

1. Microwave Cavity

A microwave cavity, as a high-Q (quality factor) resonator, is used to respond to the minute changes in the microwave magnetic susceptibility of a small paramagnetic sample contained in the cavity. Referring to the discussion on the complex susceptibility $\chi = \chi' - i\chi''$ following Eq. (6), the effect of the real part χ' is to change the resonance frequency and that of the imaginary part χ'', the Q-factor of the cavity. A suitable arrangement can be made to measure either the dispersion (χ') or the absorption (χ'') and preferably not both at the same time. It is rather immaterial whether a reflection type (one coupling iris) or a transmission type (two coupling irises) of cavity is used. The sensitivity of these two types proves to be quite comparable. But it is essential, for high sensitivity, to obtain the highest possible Q (typically 5000 to 10,000 at room temperature) and to optimize the matching conditions for either type used.

2. Microwave Field and Frequency Stabilization

For an ESR experiment, the rf magnetic field {H_1 in Eq. (4)} at the sample should be perpendicular to the direction of the steady field. The rf magnetic field should be essentially homogeneous over the sample and is the strongest in intensity in the cavity stationary wave pattern. Aside from the homogeneity and the largest effect produced for a given power input, this arrangement serves to avoid any losses in the rf electric field, which is zero at maximum magnetic field. The quantity H_1 can be readily calculated from the measured power and the cavity Q-factor and is useful for calculations on line intensity and saturation effects. Under ordinary conditions, H_1 might have a value of the order of 0.1 oersted.

In practically all ESR experiments, passage through resonance is accomplished by varying the steady magnetic field at a fixed microwave frequency. With a sharp cavity width (say, 1 Mc/sec), the frequency of the microwave source, usually a klystron, should be stable within something like 10 to 20 kc so that it will be at the center of the cavity resonance. It is usually necessary to stabilize the source frequency with a self-correcting device. Where the absorption is the aim of the measurement, frequency stabilization with reference to the sample cavity itself has frequently been used. On the other hand, if the dispersion is to be observed, stabilization by an external cavity or some other means needs to be employed. In any case, the stabilized frequency can be measured very precisely by a frequency standard as a basic datum for the experiment.

3. Steady Magnetic Field and Field Modulation

The steady magnetic field should be sufficiently homogeneous over the sample that its effect on the line broadening is negligible. This condition can become stringent when a large gap distance between magnet pole pieces has to be used in order to accommodate the sample cavity and some associated apparatus (e.g., low temperature cell, etc.). Magnets with pole pieces up to 12 inches in diameter are not uncommon for this type of work. Magnetic fields are generally measured by proton resonance magnetometers, although some measurements of the field have been done relative to the resonance position of DPPH (diphenylpicrylhydrazyl).

Most workers in the field have resorted to the use of field modulation at a low frequency (several cycles to kilocycles) as a sensitive method of detection. The signal output at field modulation frequency is proportional to either dx'/dH or dx''/dH depending upon whether x' or x'' is to be measured without field modulation. The signal output is also proportional to the amplitude of the field modulation, which, however, should not be more than a small fraction of the line width in order to avoid distortion. The separation in field units between the maximum and minimum slopes is itself a convenient measure of the line width which can be correlated with the "half-width" measure for a known line shape.

4. Detection and Optimum Sensitivity

Microwave signals are generally detected first by a crystal rectifier. If a field modulation is used, the component at the modulation frequency is first amplified and finally detected by a phase-sensitive device. It is preferable not to use a very low modulation frequency, since crystal noise is almost inversely proportional to frequency. The crystal noise problem is solved in another way by some workers in the field by using a superheterodyne detection system where the first conversion of the signal is done at a high frequency (e.g., 30 Mc/sec).

The over-all sensitivity of an ESR spectrometer is determined by a multitude of factors such as high-Q cavity, low crystal noise, low noise amplification, narrow band-width, etc. Under favorable conditions, a practical optimum sensitivity for detecting 10^{12} spins with a line width of one oersted at room temperature has been reported.

For a given sample and apparatus, the detection sensitivity increases inversely as the temperature, since the static susceptibility has a temperature dependence of this form. Therefore a gain of 70 is generally expected by cooling the sample from room to liquid helium temperature.

5. Other Techniques

A method of "phase selection" related to field modulation can sometimes be used to great advantage in differentiating the spectral lines between radical species with rather different relaxation times. It has been found that there is often a phase shift between the applied field modulation and the output signal referring to the same frequency. This phase shift is some function of the relaxation times involved. If the phase shift between two spectral systems differs by $\pi/2$, for instance, then either signal can be selected with a complete elimination of the other merely by varying the phase of the reference signal in a phase detector. Less complete differentiation at other phase differences can also be very useful.

The method of "double spin resonance," attributable to Feher (1956), has the potentiality of being applied to the free radical work. It consists of observing the changes in an ESR signal when a nuclear resonance is simultaneously excited. This method can be profitably used to study the hf structure by nuclear excitation when such is unresolved in the regular ESR spectrum.

IV. Trapped Inorganic Radicals As Studied by ESR

We start the study of trapped radicals first with the inorganic ones because a simpler approach to the subject is afforded by dealing with a few atomic free radicals where the ESR spectra in the free state are well known and the theoretical interpretation is comparatively straightforward.

A. Trapped Hydrogen Atoms

Trapped hydrogen atoms were first observed by Livingston *et al.* (1955) in their studies on the ESR spectra of gamma irradiated frozen acids. An acid such as H_2SO_4, concentrated or dilute, was irradiated by Co^{60} gamma rays and examined at 77°K. The ESR spectrum showed a pair of dissimilar lines near the free electron position and two other lines symmetrically placed on either side of the central pair with a separation between the outer lines of about 505 oersted. The positions of the outer two lines, regarded as a doublet, suggested their origin as due to hydrogen atoms on the basis of known free-state data. This suggestion was confirmed by the existence of three lines in a frozen solution of the acid in D_2O, in agreement with the known triplet spectrum for deuterium atoms. A similar situation was found to be true for the acids $HClO_4$ and H_3PO_4, except that the central lines were considerably different from one acid to another.

Trapped hydrogen atoms were also observed by Delbecq et al. (1956) in their work on the ESR spectra of the so-called U_2-centers in an alkali halide crystal. First, U-centers were formed in KCl-KH when a KCl single crystal was heated in H_2 gas. These centers were supposed to be H^- ions in the negative ion vacancies of the crystal. After irradiation of the crystal by ultraviolet light, both the hydrogen doublet lines and the F-center resonance were observed. This phenomenon was interpreted by assuming first a conversion of the U-center into the U_1-center where the H^- ion was in an interstitial site, and then another conversion of the U_1-center into the U_2-center (H in the interstitial site) and the F-center (electron in the negative ion vacancy). The identification of the doublet, with a separation of about 500 oersted, as produced by hydrogen atoms was confirmed by the observation of the expected triplet spectrum in a KCl-KD crystal which was similarly processed and irradiated.

In another type of experiment, Jen et al. (1956) observed the ESR spectra of hydrogen atoms with a deposition of the discharge products in hydrogen on a surface at liquid helium temperature. The experiment was done at an X-band frequency. A doublet hf spectrum was observed with a separation of about 508.7 oersted. A typical low-field hydrogen line is shown in Fig. 4. A triplet hf spectrum was observed for the discharge products in deuterium with a separation of about 155.4 oersted between the end lines. Assuming that the doublet and triplet spectra were due to hydrogen and deuterium atoms, application of the Breit-Rabi formula gave consistent results for the g-factors and hf coupling constants: $g(H) \doteq g(D) \doteq g$ (free electron); $A(H)/A(D) \doteq g_I(H)/g_I(D)$, where the approximate equalities were true to a high order of precision. At the same time, the fractional deviation of the hf coupling constant of each atom trapped in its own molecular matrix from that of the free state was so small (lower by 0.23%) that the atom in this trapped state was interpreted to be very nearly free.

More recently Wall et al. (1959b) have observed the ESR spectrum of H atoms in gamma irradiated CH_4 at 4.2°K. By a similar process, they also

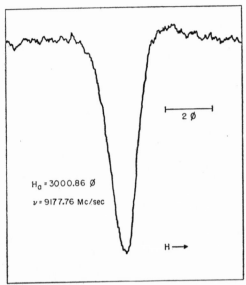

$H_a = 3000.86\ \phi$

$\nu = 9177.76\ Mc/sec$

$2\ \phi$

$H \longrightarrow$

FIG. 4. Low-field line of the doublet spectrum for H in H_2 matrix at 4.2°K. The symbol ϕ is used to denote oersted.

observed the spectra of H and D atoms in H_2 and D_2, respectively, except that the atomic yield was apparently lower here than in CH_4. In another case, Piette et al. (1959) observed the ESR spectrum of H atoms when solid hydrogen at 4.2°K was subjected to high energy electron irradiation. All these results on H atoms are in good agreement with those of the aforementioned experiments.

In summary, the principal results of the above-mentioned experiments[3] are tabulated in Table I. It is noteworthy that, while the g-factor of atomic hydrogen is essentially a constant under all conditions, the hf coupling constant shows some dependence on the nature of the trapping medium. We shall see later that this effect is in general to be expected when the matrix environment of a trapped radical is changed. It is also not difficult to understand the large breadth for the hydrogen lines in a KCl crystal because it is quite comparable to the usual hf broadening (due to K^+ and Cl^- nuclei) for color centers in the same crystal. There is no doubt, therefore, that we have been dealing in all these cases with hydrogen atoms in the $^2S_{\frac{1}{2}}$ state, which is somewhat perturbed by the trapping medium in each case.

[3] There is still another experiment on the ESR evidence of trapped hydrogen atoms in gamma-irradiated ice. But the situation is so controversial that we prefer to discuss it in another section of this chapter, where some general matrix effects will be taken up.

TABLE I

TABULATION OF TEMPERATURE (T), MICROWAVE FREQUENCY (ν), g-FACTOR (g_S),
HYPERFINE COUPLING CONSTANT (A), AND HALF-WIDTH $(\Delta H_{1/2})$ FOR SEVERAL
CASES OF TRAPPED HYDROGEN ATOMS, COMPARED WITH THE CASE OF FREE ATOM

Source of H	$T(°K)$	ν(Mc/sec)	g_S	A(Mc/sec)	$\Delta H_{1/2}$(oe)	Reference
Free atom	Room	9185	2.00226	1420.406	0.06	Beringer and Heald (1954)
Condensed discharge in H_2	4.2	9178	2.0023	1417.1	1.4	Jen et al. (1956)
γ-Irradiated H_2SO_4 in H_2O	77	23044	2.0024	1415	4	Livingston et al. (1955)
U_2-Center in KCl	80	9100	Free electron	1390	68	Delbecq et al. (1956)

B. TRAPPED NITROGEN ATOMS

Trapping of nitrogen atoms in the $^4S_{\frac{3}{2}}$ state was reported by Cole et al. (1957) and independently by Foner et al. (1958b). These two groups of workers used essentially the same method of depositing the discharge products of nitrogen on a liquid-helium-cooled surface and examined the ESR spectra of the frozen sample. They found a triplet spectrum which was expected of nitrogen atoms with $I(N^{14}) = 1$. A typical spectrum for this case is shown in Fig. 5, which is an example of the use of a different material for the matrix (in this case the ratio of H_2 to N_2 was about 100:1).

The hf splitting of the nitrogen atom is small enough that Eq. (16) gives an adequate description of the situation, except that S has now a value of $\frac{3}{2}$. The formulas for the calculation of g_S and A are as follows:

$$g_S = \frac{\nu}{(\beta/h)H_0} \tag{25}$$

$$A = (\Delta H/H_0)\nu \tag{26}$$

where ΔH = average separation of adjacent lines in field units.

The results calculated from the experimental data for the nitrogen atom are tabulated in Table II. Aside from some unexplained small discrepancies between the experimental results of the two cases of trapped atomic nitrogen, it is seen from Table II that the hf coupling constant for the trapped atom is appreciably larger than that of the free atom. By a study on the variation of the coupling constant in various matrices, this effect has been attributed to the influence of the matrix field (see Section VI).

Several groups of workers (Cole et al., 1957; Jen et al., 1958) have observed that, in addition to the triplet spectrum discussed above, there are

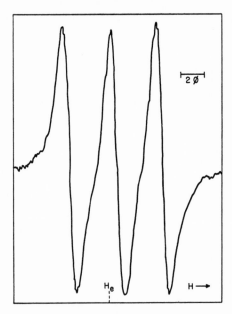

FIG. 5. Triplet spectrum of N^{14} in H_2 matrix at 4.2°K

TABLE II

TABULATION OF TEMPERATURE (T), MICROWAVE FREQUENCY (ν), g-FACTOR (g_S),
HYPERFINE COUPLING CONSTANT (A), AND HALF-WIDTH $(\Delta H_{1/2})$ FOR TWO CASES
OF TRAPPED NITROGEN ATOMS, COMPARED WITH THE CASE OF FREE ATOM

N	$T(°K)$	ν(Mc/sec)	g_S	A(Mc/sec)	$\Delta H_{1/2}$(oe)	Reference
Free atom	Room	9185	2.00215	10.45	0.09	Heald and Beringer (1954)
N in N_2 matrix	4.2	9397	2.0005	12.6		Cole et al. (1957)
N in N_2 matrix	4.2	9177.8	2.0020	12.08	2.3	Foner et al. (1958b)

also four resolvable satellite lines as shown in Fig. 6. Cole and McConnell (1958) explained the presence of the satellite lines in terms of zero-field splitting of the $^4S_{\frac{3}{2}}$ ground state of atomic nitrogen. They assumed the perturbation on the atomic energy state by the crystal field (axially symmetric) is of the form [cf. Eq. (22)]

$$D\{S_\lambda{}^2 - \tfrac{1}{3}S(S + 1)\} \qquad (27)$$

where S_λ is the component of the spin angular momentum along the crystal axis of symmetry λ. Let θ be the angle between the symmetry axis and the external magnetic field (along z). The transition energy corresponding to Eq. (27) is effectively zero for $S_z = \frac{1}{2} \leftrightarrow -\frac{1}{2}$ and a value centering around

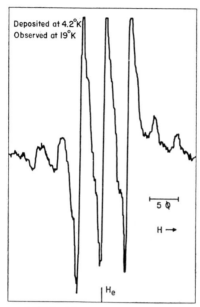

FIG. 6. ESR spectrum of N^{14} in N_2^{14} matrix at 4.2°K, showing satellite lines with the main triplet. ϕ denotes oersted.

$\pm D$ for $S_z = \frac{3}{2} \leftrightarrow \frac{1}{2}$ and $S_z = -\frac{1}{2} \leftrightarrow -\frac{3}{2}$ respectively in a polycrystalline material. As a result, there are three sets of triplet lines, with the principal triplet in the center and satellite triplets offset symmetrically from the center by the value of D. It so happens that D is approximately equal to the line spacing, thus making one line in each satellite triplet coinciding with the end lines of the principal triplet which, in Fig. 6, seem to be enhanced relative to the center line.

The aforementioned theory of the nitrogen satellite spectrum received some support from the ESR observations of Wall *et al.* (1959a) on gamma-irradiated solid N_2^{15}. They observed two strong lines symmetrically placed with respect to five weak lines, as shown in Fig. 7. The two strong lines may be called the principal doublet for N^{15} with $I(N^{15}) = \frac{1}{2}$ which is the exact counterpart of the principal triplet for N^{14}. If the central satellite line is ascribed to a different origin, then the remaining four lines can be regarded as two doublets displaced symmetrically from the center by a D-value for this case. This picture seems to be consistent with the above theory of zero-field splitting.

In another series of experiments, Wall *et al.* (1959b) observed the ESR spectra of N^{14} after irradiating the mixture of N_2 with H_2, Ne, A, or Xe. The N^{14} spectrum (triplet and satellite structure) observed in H_2, Ne, or Xe at 4.2°K was essentially the same as that in N_2 (Fig. 6). However, only

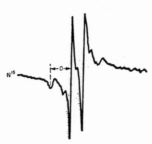

FIG. 7. ESR spectrum of N^{15} in gamma-irradiated N_2^{15}. Frequency = 9112 Mc/sec at 4.2°K. After Wall *et al.*, (1959a).

a strong triplet with broad lines but no satellite structure was obtained in argon at 4.2°K. When the same sample was irradiated at 20°K and observed at 4.2°K, two satellite triplets were seen to be only slightly displaced in opposite directions from the principal triplet. The results for N^{14} in A was reasonably explained by assuming a value for the splitting constant D approximately one-eighth that for N^{14} in N_2. But to explain the fact that the N^{14} spectrum appeared to be the same in H_2, Ne, or Xe as in N_2, the authors assumed that the N^{14} atoms were trapped in N_2 microcrystals which were not in solid solution with the matrix.

Although the theory of zero-field splitting seems to be very plausible, it does not necessarily mean that all nitrogen atoms in a polycrystalline sample are subject to this type of field. The fact that the principal triplet for N^{14} often can be observed without a visible trace of the satellites means that the nitrogen atoms under these conditions are not acted on by this rather specialized field. Also, it appears to be fairly difficult to account for the intensity distribution between the principal and satellite lines if all the nitrogen atoms are assumed to be subject to the same field.

C. Trapped Halogen Atoms

Trapped halogen atoms would make very interesting studies because of the new information we might have with these atoms in the ground $^2P_{\frac{3}{2}}$ state, for which the free-state data are available in the literature. The energy state for a halogen atom is bound to be spin degenerate on account of Kramers' degeneracy. Unfortunately, all experiments to date, done with polycrystalline materials, have failed in detecting the ESR spectrum of any halogen atom (Jen *et al.*, 1958). We can rule out the argument that there may not be any halogen atoms present in the samples tested. In the work of Cochran *et al.* (1959), the molecule HI was frozen in the matrix of argon and irradiated by ultraviolet radiation. While the ESR spectrum of H was found to be quite strong, the ESR spectrum of I has never been detected. It was most unlikely that the iodine atoms had all combined with each other while the hydrogen atoms had not. To explain the failure of observing a resonance for I, these authors assumed that the interaction between the matrix field and the orbital angular momentum produced a large enough magnetic anisotropy to broaden the resonance lines in a polycrystal beyond detection.

D. TRAPPED OXYGEN ATOMS AND MOLECULES

Attempts to observe the ESR spectra of either atomic or molecular oxygen in polycrystalline inert media have not been successful (Jen *et al.*, 1958). Recently Foner and Meyer (1958) used a single crystal of β-quinol clathrate to contain O_2 molecules which were isolated from each other by about 8 A. It has been previously shown by Meyer *et al.* (1958) from their measurements on the magnetic susceptibility of O_2 down to 0.25°K that a large zero-field splitting corresponding to about 4.15°K should be present. It was concluded that both high magnetic fields and high frequencies were required for the ESR observation. By using pulsed magnetic fields and liquid helium temperature, Foner and Meyer succeeded in observing resonances at 37 kMc/sec and 71 kMc/sec. This appears to be a nice confirmation of the theory of zero-field splitting and also serves to explain the failure of the attempts with polycrystalline materials at low frequencies and low fields.

Although O_2 is a stable molecule (like NO, NO_2, etc.) which would normally be outside of our present treatment of unstable radicals, it is profitable to consider this interesting case because it illustrates the phenomenon of zero-field splitting into a singlet ground level and the method that has been used to observe the ESR spectra. Moreover, the use of a clathrate compound to trap a radical may be important for further work of trapping radicals in a single crystal.

E. TRAPPED NH_2 AND ND_2 RADICALS

Foner *et al.* (1958a) observed the ESR spectra of NH_2 and ND_2 when 0.5% of NH_3 or ND_3 in argon at 4.2°K was photolyzed by ultraviolet radiation. Their result for NH_2 is shown in Fig. 8 and for ND_2 in Fig. 9. The nine-line spectrum for NH_2 is consistent with $[(2I(N^{14}) + 1][2I(H_2) + 1)]$ where $I(N^{14}) = 1$ and $I(H_2) = 2I(H) = 1$. The results were interpretable by the following energy equation in strong field for three magnetic nuclei, two of which are equivalent (omitting the nuclear magnetic energy terms which do not affect the transition energy),

$$W = M_s g_s \beta H + A M_s \sum_i m_i + B M_s M_I(N) \qquad (28)$$

where m_i is the magnetic quantum number of an individual hydrogen or deuterium nucleus. Application of Eq. (28) to the observed spectra of NH_2 and ND_2 gave a value for g_s approximately equal to that of the free

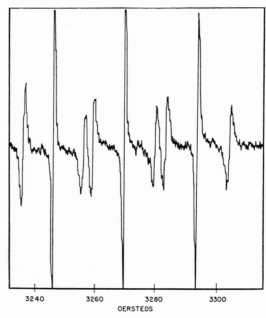

FIG. 8. ESR spectrum of NH_2 in argon at 4.2°K

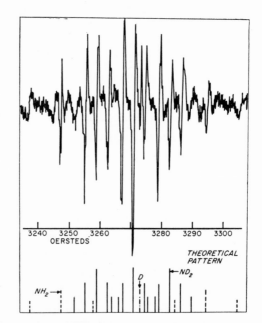

FIG. 9. ESR spectrum of ND_2 in argon at 4.2°K. Also present are spectra of D and NH_2 and weak traces of NHD.

electron in both cases and the following values for the hf coupling constants (in megacycles per second): $A(NH_2) = 67.0$, $A(ND_2) = 10.3$, $B(NH_2) = 28.9$, and $B(ND_2) = 33.8$. It is seen that the result $A(NH_2)/A(ND_2) = g_I(H)/g_I(D) = 6.5$ is as expected but the quantity B is not a constant contrary to expectation. The authors did not have a good explanation for this anomaly in the B constant.

McConnell (1958) pointed out, referring to the above observations on NH_2, that the near equality of the intensity for all the nine lines in the NH_2 spectrum (when the effect of line breadth is considered) means that (1) the NH_2 radical is freely rotating in the argon matrix at 4.2°K and (2) the statistical distribution of the radical population in various energy states is consistent with thermal equilibrium at 4.2°K. This interpretation seems to be inadequate on account of the following circumstances. We think condition (2) appears to be satisfied but condition (1) leads to contradictions which can be removed by assuming the radical is in a state of hindered rotation. The line of reasoning is briefly given in the following.

Figure 10 shows a diagram of the statistical weights of the hf lines due to hydrogenic nuclei (overlooking the nitrogen nuclear splitting which does not affect the present discussion) for NH_2 and ND_2 in the lowest (0_0) and next to the lowest (1_{-1}) rotational state. The states are labeled s or a according to the symmetry of the nuclear wave function being symmetrical or antisymmetrical in order that the over-all wave function can have the correct symmetry for the Fermi or Bose statistics (cf. Section I, G). The rotational energy difference ΔW is based on the data for free molecules. It is seen that if the conditions (1) and (2) are both satisfied, then only the intensity distribution for the ground 0_0 state should prevail at 4.2°K, since

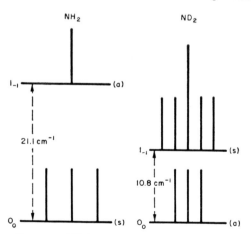

FIG. 10. Energy diagram of NH_2 and ND_2 lowest free rotational states and statistical weights for the hf spectra.

$e^{-\Delta W/kT}$ is about 7.2×10^{-4} for NH_2 and 2.5×10^{-2} for ND_2. It seems that the observed data on NH_2 are fairly consistent with this picture. On the other hand, each of the $M_I = \pm 2$ lines for ND_2 which can only come from the 1_{-1} state appears to be at least eight times stronger than the calculated intensity on the basis of free rotation. This phenomenon can be understood under the assumption that the effective rotational energy difference

$$[W(1_{-1}) - W(0_0)]$$

is narrowed to about 5 cm^{-1} as the molecular rotation is hindered by a potential barrier. If this assumption is valid, we would have to suppose that the rotation of NH_2 is also hindered. A rough calculation has shown[4] that the effective rotational energy difference for NH_2, resulting from the same potential barrier, is in the neighborhood of 17 cm^{-1}, which will make the center line in the hydrogen splitting of the NH_2 spectrum about 0.3 % stronger than the end lines. This difference in intensity is an order of magnitude smaller than the detectability of the experiment.

An alternative to the above explanation is to assume a nonequilibrium distribution of population among the rotational states. This is indeed the view taken by Robinson and McCarty (1958a, b), who noted that the population in state 1_{-1} was about 100-fold larger than the expected equilibrium value in their optical spectroscopic observations on NH_2 in argon at 4.2°K. They also observed the energy difference between the lowest rotational levels for trapped NH_2 to be substantially the same as for free-state molecules. As applied to the above-mentioned ESR work, the nonequilibrium interpretation can easily account for any contributions from the higher rotational state but seems to be in contradiction with the observed dependence of hf spectral intensity distribution upon the ratio of energy difference over temperature.

F. Trapped SiH$_3$ Radicals

The ESR spectrum of trapped SiH$_3$ radicals as shown in Fig. 11 (Cochran, 1959). This radical was obtained as a result of ultraviolet photolysis of SiH$_4$ in argon at 4.2°K. Since this quartet spectrum is similar to that of CH$_3$, we shall lump the discussion of this type of molecule and its ESR spectrum with that of CH$_3$, which will immediately follow. For the present, we only wish to record it under the subject of inorganic free radicals and also to explain how it was first obtained as a "spurious" spectrum. In the course of studying the deposition of the discharge products of hydrogen, Jen et al. (1958) observed a spectrum (the same as that shown in Fig. 11), and were not able to identify its origin. It was suspected that SiH$_3$ might be produced during the discharge of hydrogen in the silicon-rich discharge

---

[4] Private communication from S. M. Blinder, Applied Physics Laboratory, The Johns Hopkins University.

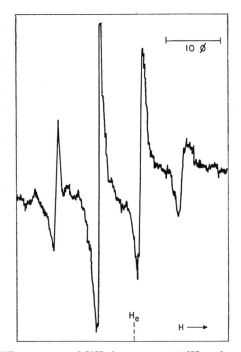

FIG. 11. ESR spectrum of SiH_3 in argon at 4.2°K. ϕ denotes oersted.

tubing. However, an experiment on the discharge in SiH_4 did not yield any positive result. Consequently, there had been some speculation that the spectrum might be due to H_3. The whole question remained uncertain until it was definitely settled by the photolysis experiment.

V. Trapped Organic Radicals As Studied by ESR

A large variety of organic radicals in the trapped state has been studied by the ESR method. The radicals have been produced in most cases by ultraviolet, X-, or gamma irradiation. With complex organic molecules, the molecular bonds can be broken up as a result of high energy radiation in so many different ways that it is often very difficult, if not impossible, to identify the radical species which is responsible for the observed spectra. We therefore prefer to lay emphasis on some of the simpler radicals produced mainly by ultraviolet irradiation and will attempt only a brief review of the more complicated cases.

A. TRAPPED METHYL RADICALS

Observation of methyl radicals has been reported by Smaller and Matheson (1958) on methane gamma irradiated at 4°K and measured at 20°K,[5]

---

[5] In a recent work, Wall et al. (1959b) observed the CH_3 spectrum in gamma-irradiated methane at about 4°K without having to raise the temperature to 20°K.

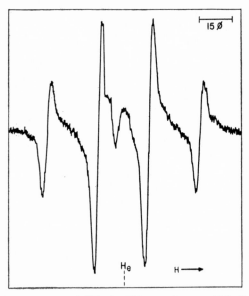

FIG. 12. Quartet spectrum of CH_3 in CH_4 matrix at 4.2°K. Small line near the center of the spectrum is due to a calibration sample of DPPH. ϕ denotes oersted.

by Rexroad and Gordy (1957) on EPA (ether, isopentane, and alcohol) solution ultraviolet-irradiated at liquid nitrogen temperature, and by Jen *et al.* (1958) on deposition of methane discharge products at 4.2°K. All agreed on the quartet structure of the spectrum, which is to be expected for the hf structure of three equivalent protons, but they differed somewhat on the line spacing which ranged from 22.9 to 26.7 oersted according to the different claims. The intensity distribution is generally 1:3:3:1 with some anomalies with respect to temperature to be discussed later.

A typical CH_3 quartet spectrum is shown in Fig. 12, which is the same whether observed from the discharge product of methane or from the ultraviolet-irradiated result of CH_3I, both at 4.2°K. Recently, Cochran *et al.* (Cochran, 1959) observed a seven-line spectrum for CD_3, produced from ultraviolet-irradiated CD_4, as shown in Fig. 13. The line separation in CD_3 is smaller than that in CH_3 by the ratio of the hydrogenic nuclear g-factors in accordance with expectation. The intensity distribution had a temperature dependence which will be discussed together with the similar situation for CH_3.

McConnell *et al.*[6] obtained the ESR spectrum of CH_3 with C^{13} isotopic substitution. The radical was produced by X-irradiation of polycrystalline methyl bromide and observed at 77°K. The C^{13} hf splitting amounted to 42 oersted.

[6] Private communication from Dr. H. M. McConnell.

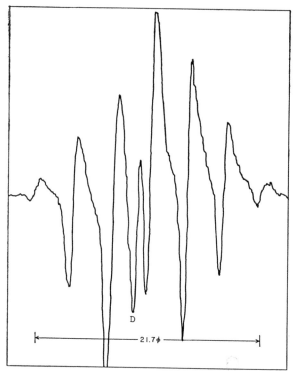

FIG. 13. ESR spectrum of CD_3 in CH_4 matrix at 4.2°K. D stands for deuterium atom center line; ϕ denotes oersted.

The combined weight of the above information (summarized in Table III) should leave scarcely any doubt about the reality of the CH_3 (or CD_3) ERS spectrum. What is really interesting and important is the answer to the questions: (1) What is the nature of the hf splitting? and (2) How much rotational freedom, if any, is there for the radical CH_3 in a rigid matrix? We shall first discuss question (1).

The CH_3 radical is a planar or very nearly planar molecule according to optical spectroscopic results (Herzberg and Shoosmith, 1956). The unpaired electron is in a π molecular orbital, which for simplicity can be taken as a p_z orbital having a direction perpendicular to the molecular plane and a node at the carbon nucleus. On this simple basis, it does not seem possible that the unpaired electron can have a finite density at the protons as required by the observation. A situation like this, which is rather common with many aromatic hydrocarbons, has led several workers (Weissman, 1956; Bersohn, 1956; McConnell, 1956; Jarrett, 1956) to invoke the role of configuration interaction, which in effect allows the unpaired electron to take on an admixture of the "s" orbitals around the protons attached to the

TABLE III
Experimental Conditions and Characteristics of
the ESR Quartet Spectrum of CH_3

Source of CH_3	$T(°K)$	ν(Mc/sec)	g_S	separation ΔH(oe)	$\Delta H_{1/2}$ (oe)	Reference
X-Irradiated $Zn(CH_3)_2$	77°	9000	Free electron	25		Gordy and McCormick (1951)
UV-Irradiated toluene in EPA	77°	9000	Free electron	27		Rexroad and Gordy (1957)
γ-Irradiated CH_4 at 4°K	20°	9350	2.0	27	3.5	Smaller and Matheson (1958)
Discharge of CH_4 in A	4.2°	9177.9	2.0024	22.9	3.7	Jen et al. (1958)
UV-Irradiated CH_3I in A	4.2°	9177.9	2.0024	22.9	3.7	Cochran and Bowers (1958)

carbon atom. McConnell and Chestnut (1958) have obtained a relationship which states that the hf coupling constant A for an attached proton is

$$A = Q\rho_c \qquad (29)$$

where ρ_c is defined as the "unpaired electron density" at the carbon atom and Q is a semiempirical constant having a value of -22.5 oersted for all CH bonds. This simple relationship has proved to be useful in a number of applications and is in good agreement with the observed results for CH_3 if we put $\rho_c = 1$. More recently, Karplus has dealt with the case of the methyl radical explicitly and obtained for $|Q|$ a value 27 oersted as an average result by a valence-bond calculation. The value of $|Q|$ was listed to be 23.4 oersted for one set of parameters of a planar molecule.

The question on the rotational freedom of a CH_3 in a rigid matrix is rather difficult to answer in quantitative fashion. However, an analysis of the intensity distribution of the CH_3 and CD_3 hf lines relative to the population distribution among the two lowest rotational levels may be helpful. Figure 14 shows an energy diagram for the CH_3 and CD_3 radicals with the statistical weights of the nuclear wave functions indicated for the hf transitions. It is seen that if the radical is freely rotating and the thermal equilibrium is established between the rotational levels at 4.2°K, then 99% of CH_3 and 83% of CD_3 would be in the ground rotational state and the intensity distribution should be approximately 1:1:1:1 for CH_3 and 0.2:2.2:4.4:5.4:-4.4:2.2:0.2 for CD_3 [distribution (1)]. If the equilibrium temperature is sufficiently high ($kT \gtrsim \Delta W$), then the intensity distribution will approach 1:3:3:1 for CH_3 and 1:3:6:7:6:3:1 for CD_3 [distribution (2)].

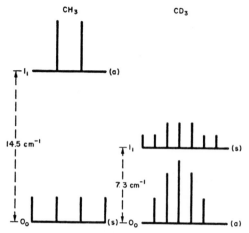

Fɪɢ. 14. Energy diagram of CH_3 and CD_3 lowest free rotational states (s for symmetric and a for antisymmetric nuclear wave functions) and statistical weights for the hf spectra.

The experimental results of Cochran *et al.* (1959) indicate that distribution (1) can prevail for a low binding matrix (e.g., H_2) at 4.2°K and distribution (2) can result if either radical is trapped in a matrix with larger binding energy (e.g., argon), particularly at a higher temperature. While definitive conclusions must await more detailed experimental data, it is tentatively believed that at 4.2°K, CH_3 shows almost free rotation in a less rigid matrix like H_2 and hindered rotation in a more rigid matrix like argon.

B. Trapped Nonplanar Aliphatic Radicals

The methyl radical discussed above has the unique position of being the first in the series of aliphatic radicals and also the rare distinction of being a planar molecule. Most other aliphatic radicals, unlike methyl and most aromatic radicals, are nonplanar in nature and their hf interactions are rather different in origin from those for planar molecules. The type of hf interaction that is often discussed in this connection is the so-called "hyperconjugation" which is a measure of the delocalization of the electron from one orbital to another due to the spatial overlap of wave functions (Coulson, 1953; Ingram, 1958, p. 115). The theory of hyperconjugation has not been sufficiently developed for us to make a quantitative application to various cases of interest. But some patterns of empirical evidence are gradually emerging to be effective testing grounds for the theory.

For the purpose of describing certain observed ESR spectra that are believed to be most probably associated with some radical species, we shall

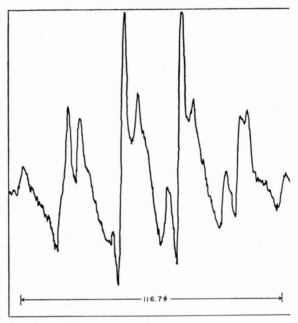

FIG. 15. ESR spectrum from photolysis of C_2H_5I in argon at 4.2°K. ϕ denotes oersted.

arbitrarily divide the nonpolar aliphatic radicals into the following groups:

Group A	Group B
(a) $CH_3\dot{C}H_2$	(c) $CH_3\dot{C}HCH_3$
(b) $CH_3CH_2\dot{C}H_2$	(d) $-CH_2\dot{C}HCH_3$
	(e) $-CH_2\dot{C}HCH_2-$

(a) Ethyl radical ($CH_3\dot{C}H_2$). Smaller and Matheson (1958) observed a quartet of triplets in gamma-irradiated ethane at 77°K. The average separation between lines was 26.3 oersted and that in each triplet 18.5 oersted. The same spectrum was observed by Cochran and Bowers (1958) with ultraviolet irradiation of 1 % ethyl iodide (CH_3CH_2I) in solid argon at 4.2°K. The ESR spectrum is shown in Fig. 15. Both groups of workers attributed the spectrum to $CH_3\dot{C}H_2$.

(b) n-Propyl radical ($CH_3CH_2\dot{C}H_2$). Cochran and Bowers observed a triplet spectrum after photolysis of 1 % n-propyl iodide in solid argon at 4.2°K. The ESR spectrum is shown in Fig. 16. The average separation between adjacent lines of this triplet was 34.8 oersted. In order to ascertain which hydrogen nuclei were responsible for the spectrum, α-deuterated n-propyl iodide ($CH_3CH_2CD_2I$) was photolyzed and observed in the same manner. The spectrum was the same as for the undeuterated compound.

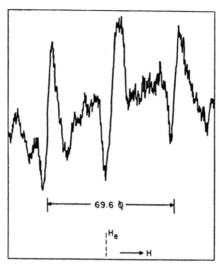

FIG. 16. ESR spectrum from photolysis of C_3H_7I in argon at 4.2°K. ϕ denotes oersted.

As a further check, β-deuterated n-propyl iodide ($CH_3CD_2CH_2I$) was again subjected to the same test. Here a five-line spectrum was observed with the line separation reduced by a factor of 6.5 from that of the original triplet spectrum. They concluded that the spectrum was most probably due to $CH_3CH_2\dot{C}H_2$, but the hf structure was almost entirely contributed by the β-hydrogens. If there were any contribution by the α-hydrogens, it must be less than the line width, which was about 4 oersted. The results in cases (a) and (b) are tabulated in Table IV.

The results by Smaller and Matheson (1958) for cases (c), (d), and (e) are listed in Table V.

In summary, it is seen that even though the evidence is still quite frag-

TABLE IV

RADICALS PROBABLY RESPONSIBLE FOR THE OBSERVED ESR FROM UV-IRRADIATED ALKLYL IODIDE AT 4.2°K

Case	Probable radical	$\Delta H_\alpha{}^a$(oe)	$\Delta H_\beta{}^a$(oe)	Compound irradiated	Reference
(a)	$CH_3\dot{C}H_2$	18.5	26.3	CH_3CH_3	Smaller and Matheson (1958)
				CH_3CH_2I	Cochran and Bowers (1958)
(b)	$CH_3CH_2\dot{C}H_2$	<4	34.8	CH_3CH_2I $CH_3CH_2CH_2I$	Cochran and Bowers (1958)

$^a \Delta H$ = separation between adjacent lines.

TABLE V

RADICALS PROBABLY RESPONSIBLE FOR THE OBSERVED ESR FROM GAMMA-IRRADIATED
ORGANIC COMPOUND S AT 77°K

Case	Probable radical	ESR lines	ΔH^a(oe)	Compound irradiated
(c)	$CH_3\dot{C}HCH_3$	8	25.0	Propane
(d)	$CH_3CH_2\dot{C}HCH_3$	7	29.3	n-Butane
	$-CH_2\dot{C}HCH_3$	7[b]	28.5	n-Hexane
(e)	$-CH_2\dot{C}HCH_2-$	6[b]	33.0	n-Octadecane
		6[b]	31.4	n-Octacosane

[a] ΔH = separation between adjacent lines.
[b] Some unresolved structure in lines.

mentary, there is a definite tendency for the group A radicals to have different hf coupling constants for the α- and β-hydrogens and for the group B radicals to have essentially the same hf coupling constant for both α- and β-hydrogens.[7]

C. TRAPPED ALCOHOL RADICALS

The ESR spectra of alcohol radicals in rigid media have been studied by many groups of workers. The radicals can be produced by X- or gamma-irradiation of frozen alcohol or by a method of secondary radical production devised by Ingram and Symons (1958). In the last-mentioned method, hydroxyl radicals were first produced by uv photolysis of hydrogen peroxide, and alcohol radicals were then formed as a secondary product when the hydrogen atom of the alcohol was abstracted by the hydroxyl radical. The observed ESR spectra and their postulated radical origin for several irradiated alcohols are given (Matheson, 1958) in Table VI.

It is seen from Table VI that it is difficult to form a clear-cut empirical rule governing the hf interactions between the unpaired electron and the protons attached to various carbon atoms. The case of the allyl alcohol radical (Ingram, 1958, p. 177) was believed to be successfully explained by McConnell's theory of configuration interaction applied to planar molecules including the concept of "negative spin density" (for the central carbon atom in the allyl alcohol radical). The other cases were believed to be qualitatively reasonable in terms of hyperconjugation but have not been quantitatively treated. There were, in addition, some complicated temperature effects on the number of lines as indicated in Table VI for the i-propanol radical and also effects on intensity distribution for the methanol and ethanol radicals as observed by Zeldes and Livingston (1959). But one general feature, which apparently held true for all ESR spectra of alcohol radicals,

[7] The statement for group B may have to be modified when more precise experimental data are available.

TABLE VI

POSTULATED RADICALS PROBABLY RESPONSIBLE FOR THE OBSERVED ESR FOR
SEVERAL IRRADIATED ALCOHOLS AT 77°K

Probable radical	ESR lines	ΔH^a(oe)	Compound irradiated
$\dot{C}H_2OH^b$	3	18	Methanol
$CH_3\dot{C}HOH^b$	5	22, 22	Ethanol
$CH_3\dot{C}HCH_2OH$	6^c	29, <14, 20	n-Propanol
$CH_3\dot{C}OHCH_3$	6 (77°K)		
	7 (110°K)	20, 0, 20	i-Propanol
$CH_2CH\dot{C}HOH$	4	12, 0, 12	Allyl alcohol
$\dot{C}H_2CHCHOH$			

^{a} ΔH = hf splitting for a proton attached to each carbon atom in the order of the formula from left to right. Each value indicates the hf splitting by a proton at each carbon atom in the order given by the formula.

^{b} Zeldes and Livingston (1959) stated that there is as much difficulty in deciding by ESR between CH_2OH and CH_2^+ as between CH_3CHOH and $C_2H_4^+$.

^{c} Structure present.

was that the hydroxyl proton did not contribute to the hf splitting. It is believed that this effect may be due to the rapid rotation of the hydroxyl radical or to the proton exchange through hydrogen bonding.

D. TRAPPED RADICALS FROM X- OR GAMMA-IRRADIATED ORGANIC COMPOUNDS

The ESR spectra of X- and gamma-irradiated organic compounds have been studied in a large number of cases. For instance, Gordy et al. (1955) studied the spectra of various X-irradiated amino acids (glycine, alanine, valine) and other organic acids like acetic and formic. They also studied the spectra of X-irradiated proteins, bone tissue, etc. They found that the ESR spectra of many complicated materials after irradiation presented rather simple patterns characteristic of simpler compounds. It has been the hope that such a study might supply clues for the breakdown processes occurring in biological systems. However, many of the conclusions reached have been highly speculative. Consequently, in this review we wish only to mention it in passing and recommend that the reader consult the original papers for further information.

E. SELF-TRAPPED RADICALS FROM POLYMERIZATION AND PYROLYSIS

It has been demonstrated that ESR lines can be observed when polymerization of certain monomers is in a gelled state or when the polymerized products precipitate out of the monomers. This phenomenon is believed to be due to the occlusion of the radical end in the polymer coil so that this active end is prevented from further interactions with other monomers. Such a theory has been strengthened by a recent systematic study on polymethylmethacrylate (Ingram *et al.*, 1958), where the observed overlapping four- and five-line spectra can be explained by a radical of the form

$$
\begin{array}{c}
\text{COOR} \\
| \\
-\text{CH}_2-\text{C}\cdot \\
| \\
\text{CH}_3
\end{array}
$$

which is the unit to be repeated in the polymer chain. The coexistence of the four- and five-line spectra has been found to be consistent with the qualitative notions of hyperconjugation and the rotational degree of freedom for the methyl group. This type of work indicates that the ESR method is in a position to add new information to polymer physics or chemistry.

It is an easily demonstrable fact that free radicals are produced and self-trapped if an organic material is pyrolyzed at a temperature of a few hundred degrees Centigrade (Ingram, 1958, p. 207). Charred sugar, for example, would give a strong ESR signal. The ESR spectrum of such a sample rarely shows any resolved hf structure and its line width can vary over a large range. The g-factor of a pyrolyzed sample is very nearly the same as that of a free electron. Although there is as yet no exact theory for the radical production and stabilization, it is believed that the phenomenon has much to do with the pyrolytic breakage of bonds at the edge of condensed carbon rings, whereupon the rings grow in size with new carbon-carbon bonds until, at a certain optimum size, the unpaired electron is stabilized in a nonlocalized orbital. Whatever the exact mechanism may turn out to be, this subject will no doubt be an interesting study of the self-trapping of radicals.

VI. Effects of Matrix Environment on ESR Spectra of Radicals

A. GENERAL DESCRIPTION

A trapped radical must necessarily have some sort of interaction with the trapping medium. The interaction may be so small under certain conditions that the radical appears to be very nearly free on the evidence of its ESR spectrum. On the other hand, the interaction may be so strong under certain other conditions that either the ESR characteristics are severely modified, or the spectrum cannot be observed in the usual instrumental range or not at all in a polycrystalline medium. Therefore, it is much more than a matter of passing interest that we should examine the effects of matrix environment on the ESR properties of trapped radicals.

The principal characteristics of an ESR spectrum are the following: g_s (electronic g-factor), A (hf coupling constant or constants), f (a line shape function, usually more simply described by the half-width $\Delta H_{\frac{1}{2}}$) and T_1 (spin-lattice relaxation time). The factor g_s is strictly a tensor quantity, but in isotropic medium it is describable simply as a scalar. It controls the "center of gravity" of the spectrum. In most cases of current interest in trapped radicals, g_s is very close to that of the free electron. However, the small deviation therefrom (Δg_s), which is in principle ascribable to a residual spin-orbital coupling, may be important for some refined consideration of interactions. The constant A is similar to g_s in being primarily a tensor quantity, but reducible to a scalar in isotropic medium. The deviation of A (ΔA) from the free-state value can be positive or negative just as Δg_s can be in a different fashion. The shape function f or the line-width parameter $\Delta H_{\frac{1}{2}}$ is a result of various agencies of line broadening. The "homogeneous" type of broadening is exemplified by the spin-spin interaction (T_2) between radicals in an isotropic medium. The "inhomogeneous" type is present in the hf broadening (when unresolved) or the broadening in a polycrystalline anisotropic medium (orbital and/or nuclear). The constant T_1 can come in as a factor in the shape function f, but it is an independent measure of the interaction between the electron spin and the lattice "phonons." In trapped radical systems at low temperatures, T_1 is usually long enough (of the order of milliseconds) that intensity saturation is a common occurrence.

In discussing the matrix effects, it would be desirable to specify the solid-state structure in the neighborhood of the radicals. Such a description, pertaining to the specific conditions of individual experiments, is not available. Based upon Peiser's experimental data on solid argon at liquid helium temperature (see Chapter 9), it is probable that the samples made under similar conditions consist of randomly oriented microcrystals. In many other cases, however, the matrices used by various workers are usually hydrocarbon solvents at liquid nitrogen temperature. Such matrices have often been described as "glasses" in amorphous states (Ingram, 1958, p. 172).

B. Nonpolar Matrices

Some studies (Jen *et al.*, 1958) have been made on the specific effects of a few nonpolar matrices like H_2, A, N_2, and CH_4 on the spin resonance properties of a few radicals like H, N, and CH_3. The results are shown in Table VII. It is seen that the effect of various matrices on g_s of these radicals is rather small. There is a tendency for g_s to be slightly depressed to a lower value at a higher matrix binding energy (i.e., in the direction from H_2 to CH_4). The magnitude of the deviation for the hf coupling constant is also a function of the character of the matrix, but what is more remarkable is that the sign of the deviation is negative for H and CH_3 and positive

TABLE VII

g-Factors, Hyperfine Coupling Constants (A), and Half-Width $(\Delta H_{1/2})$ of H, N, and CH$_3$ in Various Matrices[a]

Free radical species	Matrix	g_J	A (Mc/sec)	Deviation of A from free value	$\Delta H_{1/2}$(oe)
H	Free	2.002256(24)	1420.40573(5)		
	H$_2$	2.00230(8)	1417.11(20)	−0.23%	1.4
	A	2.00220(8)	1413.82(40)	−0.46%	2.5
	CH$_4$	2.00207(8)	1411.09(32)	−0.66%	4.7
N	Free	2.00215(3)	10.45(2)		
	H$_2$	2.00202(8)	11.45(8)	9.6%	1.6
	N$_2$	2.00200(8)	12.08(12)	15.6%	2.3
	CH$_4$	2.00203(8)	13.54(22)	29.5%	5.6
CH$_3$	Free	—	—		
	H$_2$	2.00266(8)	65.07(13)	—	1.4
	A	2.00203(8)	64.64(20)	—	3.7
	CH$_4$	2.00242(8)	64.39(18)	—	4.5

[a] The numbers in parentheses indicate experimental and/or conversion errors in the last figures of the associated values.

for N. We shall see later that the sign of ΔA is not an intrinsic property of a radical with respect to its matrix but is also dependent on the nature of the site. Nevertheless, the effect is consistent for the presumed same kind of site as in Table VII. The matrix effect on the line width appears to be one of progressive broadening with a more rigid matrix. A part of this line width may be due to spin-spin dipolar broadening of the radicals. Using the approximate formula $\Delta H_{1/2}$(dipolar) $\simeq 10^{-19} n$ oersted, where n denotes the number of spins per cubic centimeter we can calculate the density of radicals in a given sample. In the case of H in H$_2$ as listed in Table VII, the measured $\Delta H_{1/2}$ of 1.4 oersted puts an upper limit to the concentration of atomic hydrogen present in the sample, which gives for $\langle H \rangle / \langle H_2 \rangle$ about 0.02%. Actually, the dipolar contribution to the line width is approximately half of this amount. Lastly, some information on T_1 can be obtained by the power saturation effect as outlined in Eq. (9). For the case of H in H$_2$ at 4.2°K, the authors considered $T_1 \sim 10^{-3}$ sec to be a reasonable value. For a detailed discussion on this subject, the reader is referred to the original work (Jen et al., 1958).

It should be remarked that, although the line widths in the cases discussed above are relatively narrow, they are not typical of many radicals trapped in hydrocarbon glasses as reported by other workers (Ingram, 1958, pp. 123 to 126). In the latter experiments, done mostly around liquid nitrogen temperatures, a typical line width was reported to be about 12 oersted, which is almost an order of magnitude larger than the lowest values listed in Table VII. It was believed that this broadening was due to anisotropic hf interactions from protons in the solid matrix. The 12-oersted line width was considered to be already narrower than it might have been (about 25 oersted) if the same effect due to the protons in the radical molecule itself had not been averaged out. The fact that the hydrocarbon matrix (CH_4) listed in Table VII gives a line width less than one-half of the typical value indicates that line-broadening mechanism cannot be interpreted solely in terms of the above theory.

C. Polar Matrices

The case of hydrogen atoms in ice represents the only example of a strongly polar matrix that has been much studied. It is also a highly controversial case. Smaller et al. (1954) claimed the spectrum they observed in gamma-irradiated ice at 77°K and 350 Mc/sec was due to hydrogen atoms. The measured hf splitting was, however, only one-sixteenth of the free-state value for H. They explained this case and the parallel case of deuterium in terms of the dielectric constant of ice. On the other hand, Livingston et al. (1955) tried to detect the ESR spectrum of hydrogen atoms in gamma-irradiated ice at 77°K and microwave frequencies and found only a resonance around $g_s = 2$, which was not identifiable with the hydrogen atom. The use of a single crystal of ice or the irradiation of D_2O did not improve the situation. More recently, Piette et al. (1959)[8] conducted an ESR experiment at 9700 Mc/sec on ice irradiated at 4.2° and 77°K by the 40 Mev electrons from a linear accelerator. They found two lines appearing at ±250 oersted on either side of $g = 2$ at 4.2°K but not at 77°K. There were also some resonances closer to $g = 2$ which were present with some variation at both 4.2°K and 77°K. The doublet spectrum with a separation of 500 oersted between the lines was attributed to hydrogen atoms trapped in ice. This spectrum was observed to decay rapidly at about 20°K.

It seems scarcely possible to resolve all the conflicting pieces of evidence presented above. However, there is a satisfactory agreement between the experiments of Livingston et al. and Piette et al. because both groups maintained that the atomic hydrogen lines were not observable in ice at 77°K. The latter group's observation at 4.2°K of a hydrogen doublet having a separation of 500 oersted seems to be qualitatively consistent with the general pattern of hf interaction variation with strong binding matrices. The evidence presented for the H spectrum does not seem in any way contradictory to the published information from other workers. On the other hand, the result of Smaller et al. is in direct conflict with those of other groups and the basis of their interpretation is rather questionable.

[8] A private communication from these authors prior to the publication of their paper is gratefully acknowledged.

D. Multiple Trapping Sites for Radicals

Cochran *et al.* (1959) reported evidence of multiple trapping sites for hydrogen atoms in solid argon. They observed the ESR spectrum of atomic hydrogen after ultraviolet irradiation of 1 % hydrogen iodide in argon at 4.2°K. Instead of finding one doublet as in their earlier deposition experiments, they observed in this case three doublets, one of which coincided exactly with the deposition result. These three doublets were found to decay at widely different temperatures when the sample was subject to a slow warm-up. They interpreted this effect as evidence for three different sites for trapped hydrogen atoms in solid argon. One of the sites, in particular, showed in the ESR spectrum a positive shift for the hf coupling constant which was an indication, on a crude theoretical model, of a "cramped" site wherein the radical was acted upon by forces within a shorter range than at other sites. Occupation of such a site may depend upon the energy content of the free radical to get access to the site and finally be trapped.

Multiplicity of trapping sites will probably prove to be a more general phenomenon than has been established so far (Foner, 1959). The case of trapped nitrogen atoms in N_2 matrix, referred to earlier in this chapter, may be explained by assuming two different sites. Further knowledge along these lines, however, will have to come from more exact information on the crystalline structure as well as some detailed theoretical calculations.

From the restricted point of view of identifying radicals from their ESR spectra, the possibility of multiple sites could be a source of confusion. It would become a burden on the observer to have to distinguish between the spectra due to the same radical at different sites and those due to different radicals.

VII. Conclusions

In this chapter a review has been made of the rudimentary portion of the electron spin resonance studies of trapped radicals. For a more comprehensive coverage, the reader is referred to other review articles, books (Ingram, 1958; Bleaney and Stevens, 1953; Wertz, 1955), and original literature listed, although incompletely, at the end of this chapter.

Looking into the future, a wider use of single crystals for the trapping media, already begun by some, will broaden the scope of ESR studies. Many cases of magnetic anisotropy, biradicals, and excited-state molecules can then successfully be dealt with. Kinetic studies of trapped radical chemistry, radical concentration and diffusion, spin relaxation mechanisms, etc. for which the ESR method is well suited, tend to become increasingly important. Refined theories of hyperfine interactions will be in great demand for the interpretation of the large mass of experimental data.

ACKNOWLEDGMENT

The author wishes to express his indebtedness to his colleagues, Dr. S. N. Foner and Dr. E. L. Cochran, for many stimulating discussions on the subject, and to them and to Mr. V. A. Bowers for permission to include in this chapter unpublished materials, not all specifically mentioned, which have been the results of a joint effort.

REFERENCES

Abragam, A., and Pryce, M. H. L. (1951). *Proc. Roy. Soc.* **A205,** 135. Theory of the nuclear hyperfine structure of paramagnetic resonance spectra in crystals.
Beringer, R., and Heald, M. A. (1954). *Phys. Rev.* **95,** 1474. Electron spin magnetic moment in atomic hydrogen.
Bersohn, R. (1956). *J. Chem. Phys.* **24,** 1066. Proton hyperfine interaction in semiquinone ions.
Bleaney, B., and Stevens, K. W. H. (1953). *Repts. Prog. in Phys.* **16,** 108. Paramagnetic resonance.
Bloch, F. (1946). *Phys. Rev.* **70,** 460. Nuclear induction.
Breit, G., and Rabi, I. I. (1931). *Phys. Rev.* **38,** 2082. Measurement of nuclear spin.
Cochran, E. L. (1959). Fourth International Symposium on Free Radical Stabilization, *Natl. Bur. Standards (U. S.)* ESR spectra of trapped radicals produced by photolysis of Group IV and V hydrides.
Cochran, E. L., and Bowers, V. A. (1958). Paper presented at 134th Meeting of *Am. Chem. Soc.* Electron spin resonance of aliphatic free radicals produced by photolysis of the alkyl iodides at 4.2°K.
Cochran, E. L., Bowers, V. A., Foner, S. N., and Jen, C. K. (1959). *Phys. Rev. Letters* **2,** 43. Multiple trapping sites for hydrogen atoms in solid argon.
Cole, T., and McConnell, H. M. (1958). *J. Chem. Phys.* **29,** 451 L. Zero field splittings in atomic nitrogen at 4.2°K.
Cole, T., Harding, J. T., Pellam, J. R., and Yost, D. M. (1957). *J. Chem. Phys.* **27,** 593. Electron paramagnetic resonance spectrum of solid nitrogen afterglow at 4.2°K.
Coulson, C. A. (1953). "Valence." Oxford Univ. Press, London and New York.
Delbecq, C. J., Smaller, B., and Yuster, P. H. (1956). *Phys. Rev.* **104,** 599. Paramagnetic resonance investigation of irradiated KCl crystals containing U-centers.
Feher, G. (1956). *Phys. Rev.* **103,** 834. Observation of nuclear magnetic resonance via the electron spin resonance line.
Fermi, E. (1930). *Z. Physik* **60,** 320. Magnetic moments of atomic nuclei.
Fermi, E., and Segrè, E. (1933). *Z. Physik* **82,** 729. Theory of hyperfine structure.
Foner, S. N. (1959). Fourth International Symposium on Free Radical Stabilization, *Natl. Bur. Standards (U. S.)* Multiple trapping sites for hydrogen atoms in various matrices.
Foner, S. N., Cochran, E. L., Bowers, V. A., and Jen, C. K. (1958a). *Phys. Rev. Letters* **1,** 91. Electron spin resonance spectra of the NH_2 and ND_2 free radicals at 4.2°K.
Foner, S. N., Jen, C. K., Cochran, E. L., and Bowers, V. A. (1958b). *J. Chem. Phys.* **28,** 351. Electron spin resonance of nitrogen atoms trapped at liquid helium temperature.
Foner, Simon, and Meyer, H. (1958). See paper by Simon Foner presented at *Intern. Conf. on Magnetism, Grenoble, France*, 1958. High field antiferro- ferri- and paramagnetic resonance at millimeter wavelengths.

Frosch, R. A., and Foley, H. M. (1952). *Phys. Rev.* **88,** 1337. Magnetic hyperfine structure in diatomic molecules.

Gordy, W., and McCormick, C. G. (1951). *J. Am. Chem. Soc.* **78,** 3243. Microwave investigations of radiation effects in solids: methyl and ethyl compounds of Sn, Zn and Hg.

Gordy, W., Ard, W. B., and Shields, H. (1955). *Proc. Natl. Acad. Sci. U. S.* **41,** 983. Microwave spectroscopy of biological substances. I. Paramagnetic resonance in x-irradiated amino acids and proteins. II. Paramagnetic resonance in x-irradiated carboxylic and hydroxyl acids.

Heald, M. A., and Beringer, R. (1954). *Phys. Rev.* **96,** 645. Hyperfine structure of nitrogen.

Herzberg, G. (1945). "Infrared and Raman Spectra." Van Nostrand, New York.

Herzberg, G., and Shoosmith, J. (1956). *Can. J. Phys.* **34,** 523. Absorption spectrum of CH_3 and CD_3 radicals.

Ingram, D. J. E. (1958). "Free Radicals as Studied by Electron Spin Resonance." Butterworths, London; Academic Press, New York .

Ingram, D. J. E., and Symons, M. C. R. (1958). "Informal Discussion on Free Radical Stabilization," p. 1. Faraday Soc., Discussion, Sheffield, England. Electron resonance studies on radicals trapped in low temperature glasses and similar systems.

Ingram, D. J. E., Symons, M. C. R., and Townsend, M. G. (1958). *Trans. Faraday Soc.* **54,** 409. Electron resonance studies of occluded polymer radicals.

Jarrett, H. S. (1956). *J. Chem. Phys.* **25,** 1289. Hyperfine structure in paramagnetic resonance absorption spectra.

Jarrett, H. S., and Sloan, G. K. (1954). *J. Chem. Phys.* **22,** 1783. Paramagnetic resonance absorption in triphenylmethyl and dimesitylmethyl.

Jen, C. K., Foner, S. N., Cochran, E. L., and Bowers, V. A. (1956). *Phys. Rev.* **104,** 846. Paramagnetic resonance of hydrogen atoms trapped at liquid helium temperature.

Jen, C. K., Foner, S. N., Cochran, E. L., and Bowers, V. A. (1958). *Phys. Rev.* **112,** 1169. Electron spin resonance of atomic and molecular free radicals trapped at liquid helium temperature.

Karplus, M. (1959). *J. Chem. Phys.* **30,** 15. Interpretation of electron spin spectrum of methyl radical.

Lewis, G. N., and Lipkin, D. (1942). *J. Am. Chem. Soc.* **64,** 2801. Reversible photochemical processes in rigid media: the dissociation of organic molecules into radicals and ions.

Livingston, R., Zeldes, H., and Taylor, E. H. (1955). *Discussions Faraday Soc. No.* **19,** 166. Paramagnetic resonance studies of atomic hydrogen produced by ionizing radiation.

McConnell, H. M. (1956). *J. Chem. Phys.* **24,** 764. Indirect hyperfine interactions in paramagnetic resonance spectra of aromatic radicals.

McConnell, H. M. (1958). *J. Chem. Phys.* **29,** 1422. Free rotation of solids at 4.2°K.

McConnell, H. M., and Chesnut, D. B. (1958). *J. Chem. Phys.* **28,** 107. Theory of isotropic hyperfine interactions in π-electron radicals.

Matheson, M. S. (1958). "Informal Discussion on Free Radical Stabilization," p. 45. Faraday Soc., Discussion, Sheffield, England. Primary radiation chemical proceses in rigid media.

Matheson, M. S., and Smaller, B. (1955). *J. Chem. Phys.* **23,** 521. Paramagnetic species in γ-irradiated ice.

Meyer, H., O'Brien, M. C. M., and Van Vleck, J. H. (1958). *Proc. Roy. Soc.* **A243**, 414. The magnetic susceptibility of oxygen in a clathrate compound. II.

Nafe, J. E., and Nelson, E. B. (1948). *Phys. Rev.* **73**, 718. The hyperfine structure of hydrogen and deuterium.

Pake, G. E. (1956). *In* "Solid State Physics" (F. Seitz and D. Turnbull, eds.), Vol. 2, p. 1. Academic Press, New York.

Piette, L. H., Rempel, R. C., Weaver, H. E., and Flournoy, J. M. (1959). *J. Chem. Phys.* **30**, 1623. EPR studies of electron irradiated ice and solid hydrogen.

Ramsay, D. A. (1953). *J. Phys. Chem.* **57**, 415. The absorption spectra of free NH and NH_2 radicals produced by the flash photolysis of hydrazine.

Ramsey, N. F. (1953). "Nuclear Moments." Wiley, New York.

Rexroad, H. N., and Gordy, W. (1957). *Bull. Am. Phys. Soc.* [II] **2**, 227. Electron spin resonance of radicals produced by ultraviolet irradiation.

Robinson, G. W., and McCarty, M., Jr. (1958a). *J. Chem. Phys.* **28**, 349. Electronic spectra of free radicals at $4°K$—NH_2.

Robinson, G. W., and McCarty, M., Jr. (1958b). *Can. J. Phys.* **36**, 1590. Radical spectra at liquid helium temperatures.

Smaller, B., and Matheson, M. S. (1958). *J. Chem. Phys.* **28**, 1169. Paramagnetic species produced by γ-irradiation of organic compounds.

Smaller, B., Matheson, M. S., and Yasaitis, E. L. (1954). *Phys. Rev.* **94**, 202. Paramagnetic resonance in irradiated ice.

Van Vleck, J. H. (1932). "The Theory of Electric and Magnetic Susceptibilities." Oxford Univ. Press, London and New York.

Venkataraman, B., and Fraenkel, G. K. (1956). *J. Chem. Phys.* **24**, 737. Proton-deuteron hyperfine structure in paramagnetic resonance.

Wall, L. A., Brown, D. W., and Florin, R. E. (1959a). *J. Chem. Phys.* **30**, 602. Electron resonance spectra from gamma-irradiated solid nitrogen.

Wall, L. A., Brown, D. W., and Florin, R. E. (1959b). *J. Phys. Chem.* **63**, 1762. Atoms and free radicals by γ-irradiation at $4.2°K$.

Weaver, H. E., Rempel, R. C., and Piette, L. H. (1959). "$4.2°K$ Irradiation Studies for the Aerojet General Corporation." Contribution from Varian Associates, Palo Alto, California.

Weissman, S. I. (1954). *J. Chem. Phys.* **22**, 1378. Hyperfine splittings in polyatomic free radicals.

Weissman, S. I. (1956). *J. Chem. Phys.* **25**, 890. Isotropic hyperfine interactions in aromatic free radicals.

Weissman, S. I. (1958). *J. Chem. Phys.* **29**, 1189. On detection of triplet molecules in solution by electron spin resonance.

Wertz, J. E. (1955). *Chem. Revs.* **55**, 829. Nuclear and electronic spin magnetic resonance.

Winslow, F. H., Baker, W. O., and Yager, W. A. (1955). *J. Am. Chem. Soc.* **77**, 4751. Odd electrons in polymer molecules.

Zavoisky, E. (1945). *J. Phys. (U.S.S.R.)* **9**, 245. Spin-magnetic resonance in paramagnetic substances.

Zeldes, H., and Livingston, R. (1959). *J. Chem. Phys.* **30**, 40. Paramagnetic resonance study of irradiated glasses of methanol and ethanol.

8. Chemical and Physical Studies of Trapped Radicals

JAMES W. EDWARDS

Research and Engineering Division, Monsanto Chemical Company, Dayton, Ohio

I. Introduction

This chapter reviews recent physical and chemical studies involved in trapping of radicals from hydrogen, nitrogen, and oxygen. It has been the aim to include results obtained by physical methods other than electron spin resonance and ultraviolet, visible, and infrared spectroscopy—methods which are covered in detail in other parts of this volume—although results from these fields have been included in many instances in order to give a more complete description of the chemical systems. Mass spectrometry, being one of the most direct methods of observing free radicals, is reviewed herein, and numerous results from mass spectrometry are included in the discussion of the several radical systems. Although methods of calorimetry and measurement of magnetic susceptibility have not been fully developed, these techniques have added much to the understanding of trapping of radicals, particularly the work of Fontana (1958, 1959) on trapped nitrogen atoms. Results from several chemical studies have been included, inferences from these giving a measure of possible radical concentrations under various conditions.

II. Detection and Measurement of Free Radicals in the Gas Phase by Mass Spectrometry

The value of mass spectrometry in study of gas phase free radicals was demonstrated by Eltenton (1942, 1947) and Hipple and Stevenson (1943). Since then, workers have made notable advances in apparatus and technique (Lossing and Tickner, 1952; Foner and Hudson, 1953a). Recent review articles include those by Dunning (1955), Lossing (1957), and Beckley (1958); the latter includes a thorough review of the various combination reactor-ion sources, starting with the work of Eltenton. In this chapter we intend to cover only briefly the principles involved for the purpose of subsequent correlation with radical trapping studies and reactions of trapped radicals, thus omitting a vast amount of research that is contributing greatly to the field of chemical kinetics.

Free radical mass spectrometers are designed to bring the reactor or other source of radicals as close as possible to the ion source as, for example, having the components housed in adjacent chambers separated by a thin diaphragm and with an interconnecting orifice for effusion of gas. The path of the inlet gas is a straight line to the ionizing chamber, i.e., there are no turns in path as in the usual gas analysis spectrometer. By rapid pumping on the ion source, multiple passage of the gas and reaction products through the electron beam is reduced. Further, the filament is somewhat isolated in a separate chamber and also an easy-flow route from the filament to the pumping system is provided in order to reduce ion formation from pyrolysis products formed on the hot filament. Foner and Hudson (1953a, 1956a) have elaborated on this scheme by substituting a molecular beam system for the simple orifice and thus can operate the ion source at 10^{-7} to 10^{-6} mm Hg. Background ion currents are thereby reduced; moreover, by mechanically modulating the molecular beam and applying phase detection to the ion signal, background is virtually eliminated.

In the mass spectrometric method, gas containing radicals is introduced into the ion source on a path which is as nearly collision free as possible, and therein is subjected to electron bombardment. Below a characteristic minimum of electron energy no ionization takes place. As the energy of the electron beam is increased the parent molecular ion may be produced. With still more energy polyatomic species will be dissociated into numerous positive ions which will give an extended mass spectrum when accelerated through the analyzer and detecting sections of the spectrometer. The problem in detecting free radicals in gases is to distinguish between dissociative electron impact with a parent molecule capable of yielding the ion by the process

$$R_1R_2 + e \rightarrow R_1{}^+ + R_2 + 2e \tag{1}$$

and ionization of the free radical itself.

$$R_1 + e \rightarrow R_1{}^+ + 2e \tag{2}$$

The appearance potential $A(R_1{}^+)$ for formation of $R_1{}^+$ by Eq. (1) is

$$A(R_1{}^+) = D(R_1R_2) + I(R_1) + q$$

i.e., it is the sum of the dissociation energy, $D(R_1R_2)$, the ionization potential, $I(R_1)$ of R_1, and q, the excess kinetic or excitation energy of the particles. Stevenson (1951) has shown q to be negligible for $I(R_1) < I(R_2)$.

The most widely used method of mass spectrometric detection of radicals, introduced by Eltenton, depends then on the fact that the energy requirement of process (1) exceeds process (2) by the dissociation energy of R_1R_2, usually several electron volts. Thus the threshold electron accelerating voltage for appearance of $R_1{}^+$ will be less for a dissociated gas or reaction mixture than for the undissociated substances. Uncertainty in values of $A(R^+)$ arises in this method because of the distribution of electron energies resulting from emission from the relatively hot filament. Owing to this distribution the ion current does not rise sharply from zero at the appearance potential, but rises gradually according to an exponential function starting at a somewhat lower voltage. Field and Franklin (1957) have discussed various ways of treating the data to eliminate subjective errors, but the problem does not seem to be completely resolved.

For gas mixtures containing large concentrations of radicals, Lossing and Tickner (1952) employed a method of material balance. Electrons of 50 to 75 electron-volts' energy are used. Following usual methods of gas analysis, proportional contributions of all stable products of the ion current for $R_1{}^+$ are determined and subtracted from the observed current. The remainder is the ion current due to $R_1{}^+$ derived from radicals R_1.

III. "Active" Nitrogen

A. NATURE OF ACTIVE NITROGEN

During the past ten years there have been a number of investigations of active nitrogen by both physical and chemical approaches that have lead to considerable understanding of the phenomena associated with this substance. During this period further studies of active nitrogen have been made by spectroscopy, electron spin resonance, calorimetry, mass spectrometry, and chemical reactions of active nitrogen gas. This period has also witnessed development of trapping techniques which have been used extensively in the study of active nitrogen, especially by Broida and associates at the National Bureau of Standards.

The intense yellow afterglow of electrically dissociated nitrogen was first reported by Lewis (1900) and was later studied extensively by Rayleigh (Strutt, 1911a, b, 1912, 1915; Rayleigh, 1935, 1940a, b, 1942, 1947). It has come to be known as the Lewis-Rayleigh afterglow. Rayleigh named the glowing gas "active nitrogen" because of its chemical reactivity. Following Mitra (1945), the earlier work may be summarized as follows:

(1) The duration of the afterglow may be as long as 5 hours.

(2) The spectrum is discrete and involves only the first positive bands of nitrogen.

(3) Kinetics of decay of the afterglow appears to be second order with respect to "active centers" and also probably proportional to pressure.

(4) The temperature dependence appears to involve a negative temperature coefficient for the glow reaction.

(5) Among the active particles the postulated species include molecular ions and metastable atoms.

(6) Chemical reactivity is high, giving rise to nitrides, cyanogen (with hydrocarbons), and exciting the spectra of many atoms and molecules when mixed with it.

(7) The glow is enhanced when oxygen and certain other impurities are added in concentrations less than 1 %.

(8) The glowing gas from the discharge tube has a high electrical conductivity.

(9) Values ranging from near zero to 12.9 ev per mole have been reported for the energy content, but with the higher value now considered invalid.

The energy carrier in active nitrogen has in the past been attributed to various species. There now seems to be conclusive evidence that nitrogen atoms are mainly responsible, although there is also substantial evidence for excited molecular nitrogen. The most direct evidence for atomic nitrogen stems from mass spectrometric studies of appearance potentials by Jackson and Schiff (1955) and by Berkowitz et al. (1956). Jackson and Schiff obtained atom concentrations of 0.1 % to 1 % for nitrogen dissociated in a Wood's tube. The ionization efficiency curve obtained by these workers for mass 14 in the discharged gas shows two breaks indicative of three origins for mass 14. Of these, one would be by dissociative ionization of ground state molecular nitrogen as observed in the absence of the electrical discharge, leaving two sources of mass 14 existing as a consequence of the electrical discharge. The appearance potentials observed for the latter two processes were $14.7_2 \pm 0.2_6$ ev and 16.1 ± 0.3 ev. The lower value agrees with work of Berkowitz et al. (1956), who obtained 14.8 ev under a similar excitation process but did not observe the higher value. Both the 14.7

and 14.8 ev values are within the experimental error of the ionization potential of ground state (^{4}S) atomic nitrogen, 14.54 ev (Moore, 1949). This appearance of mass 14 at ionizing potentials which cannot be accounted for by other atomic transitions is taken as proof of existence of atomic nitrogen in the active gas.

The mass 14 appearance potential of 16.1 volts is of interest because of its relation to other evidence for active species other than atomic nitrogen in active nitrogen. According to Jackson and Schiff, several plausible mechanisms may account for this result; these are indicated by the following equations:

$$N(^4S) + e^- \rightarrow N^+(^1D) + 2e^- \tag{3}$$

$$N(^2D) + e^- \rightarrow N^+(^1S) + 2e^- \tag{4}$$

$$N_3 + e^- \rightarrow N^+ + N_2 + 2e^- \tag{5}$$

$$N_2{}^* + e^- \rightarrow N^+ + N + 2e^- \tag{6}$$

in which $N_2{}^*$ represents an excited metastable molecule.

There seems to be no clear choice among these alternatives. However, there are several other evidences of energetic species in active nitrogen in addition to the ground state atoms identified by Jackson and Schiff (1955). Before looking at these it should be noted that Berkowitz et al. (1956) did not obtain this same evidence for a second species in active nitrogen. A possible cause for this could be an unfavorable sampling arrangement into the mass spectrometer associated with possible slow removal of products from the ion source—a combination which could lead to removal of excess energy from the $N(^2D)$ and $N_2{}^*$ of processes (4) and (6) and to decomposition of N_3 if it is present in the first place. On the other hand, this would not affect process (3) and since the second species was not observed by Berkowitz et al., this process must not be involved. Thrush (1956) has given evidence for N_3 in products from flash photolysis of hydrazoic acid, and Milligan et al. (1956) have observed infrared absorption bands in products condensed at 4.2°K from glow discharges in nitrogen which they attribute to N_3. However, Harvey and Brown (1959) were unable to repeat these latter results. The question of the presence of N_3 in the discharged gases thus remains unresolved and needs further study, such as comparison of the visible absorption spectrum of discharged nitrogen to that of hydrazoic acid photolyzed according to the method of Thrush (1956). These data might also aid in understanding Rice's blue solid (Rice and Freamo, 1951).

The presence of excited nitrogen atoms in the afterglow has been reported by Tanaka et al. (1957), based on observation of ultraviolet absorption by ^2D and ^2P atoms. Concentrations were about 1/500 that of ^4S atoms so that these excited species would not account for the second species found by Jackson and Schiff, assuming that this species requires the same order of concentration as the ⁴S atoms in order to be observed. Broida and co-workers (Bass and Broida, 1956; Peyron and Broida, 1959) have observed ²D-⁴S emission from solid products condensed at 4.2°K from the nitrogen afterglow. This transition was observed in the solid during condensation, persisted after condensation for a few minutes, and then reappeared while warming up at about 10°K. This has been interpreted in terms of trapping of excited atoms by Herzfeld and Broida (1956) and by Herzfeld (1957). Quantitative estimates of the number of ²D atoms involved were not made; however, these results would seem to indicate that appreciable quantities were present in the gas phase.

An alternative explanation for the presence of excited atoms in the low-temperature solids condensed from active nitrogen may be described as the inverse of a photosensitized dissociation; i.e., an electronically excited molecule of nitrogen formed from two ground state atoms undergoes an inelastic collision with, and thereby excites, another ground state atom, e.g., by the processes

$$N(^4S) + N(^4S) \rightarrow N_2(A^3\Sigma_u^+) \tag{7}$$

$$N_2(^3\Sigma_u^+) + N(^4S) \rightarrow N_2(X^1\Sigma_g^+) + N(^2D) \tag{8}$$

Absence of the Vegard-Kaplan bands in Bass' and Broida's (1956) experiments would thus be a result of the facility with which this energy transfer can occur when the solid is warm enough to permit relatively rapid diffusion of atomic nitrogen, whereas the relatively slower diffusion expected in argon matrices (Becker et al., 1957) would be consistent with the fact that these bands are observed when the nitrogen is diluted with argon (Peyron and Broida, 1959).

A highly energetic species has been suspected in active nitrogen on the basis of the large heating effect observed by Rayleigh when metal foils were put into streams of active nitrogen (Rayleigh, 1940b, 1947). Temperatures attained by the foil indicated available energy of up to 12.9 ev per mole of the total nitrogen passed through the discharge. Gaydon (1947) suggested that the heating was due to the combined effect of cathode rays as well as active nitrogen, a suggestion which was later confirmed by Benson (1952), who repeated the metal foil experiments with discharged nitrogen. Benson placed a grounded aluminum elbow in the gas flow circuit to remove electrons from the discharge and was unable to observe the large heating effect; rather, he obtained an energy content of about 1/400 that observed by Rayleigh. He also showed that the intensity of the afterglow was not affected by removal of the charged particles, nor was the glowing gas affected by a magnetic field. The electron concentration in the glowing gas was found to be only about 10^9 per cubic centimeter, whereas the concentration of active particles was found to be about 10^{15} per cubic centimeter, assuming 9.7 ev as the excitation required to produce the afterglow. The concentration of particles excited to this extent, 1/350 of the nitrogen molecules, is in reasonable agreement with other measurements, e.g., with Rayleigh's estimate of 1/40 from chemical reactivity and 1/1300 from the intensity of the visible afterglow. As noted elsewhere in this chapter it seems of utmost importance to consider electron currents and general permeation of a system by a discharge when working with high voltages in low pressure systems. In this connection the reader is referred to the paper by Jackson and Schiff (1955) as an example of precautions needed to eliminate interfering side effects.

B. EXCITED MOLECULES IN ACTIVE NITROGEN

In the past there have been various suggestions regarding excited nitrogen molecules in the afterglow. Recent work bearing on this question includes the mass spectrometric work of Jackson and Schiff (1955) cited above, and studies by Evans and Winkler (1956), Freeman and Winkler (1955), Berkowitz et al. (1956), Kistiakowsky and Volpi (1958), Kistiakowsky and Warneck (1957), Kaufman and Kelso (1958), and Dressler (1959).

Winkler and co-workers have systematically studied the reactions of active nitrogen with a variety of compounds. In this work, atoms are produced by a high voltage condensed electrical discharge, giving concentrations of about 30 % as indicated by Wrede-gauge measurements. Experiments are carried out in a flow system at pressures of the order of 1 mm Hg with different relative flow rates in different experiments. With hydrocarbons the reaction zone is marked by a lilac flame; with halogen compounds, by an orange flame. Products are frozen out in a trap and analyzed. Hydrogen cyanide is the main product from carbon-containing compounds; yields rise linearly with flow rate of the substrate until a maximum is reached which depends on the rate at which nitrogen atoms reach the reaction zone. This maximum corresponds to complete removal of nitrogen atoms by the hydrocarbon. No ammonia or hydrazine is formed. This leads to the conclusion that attack is not by abstraction of hydrogen atoms as is usual in related radical reactions, but that attack is directly on the carbon atom.

In sharp contrast to the reactions of active nitrogen with hydrocarbons is the reaction with ammonia. Active nitrogen destroys ammonia, i.e., decomposes it to the elements, to an optimal extent, according to Freeman and Winkler (1955), which is independent of temperature over a wide range and which represents only about one-sixth of the activity of the nitrogen—

activity being determined from reactivity with ethylene to form hydrogen cyanide. After optimal reaction with ammonia, the gas is still reactive toward hydrocarbons, thus indicating that the ammonia decomposition is not directly dependent on nitrogen atoms and suggesting the presence of a second active species in quantities of approximately the same order of magnitude as the atom concentration.

Freeman and Winkler (1955) considered the N_3 radical as the possible second active component. Later, Evans and Winkler (1956), in a comprehensive discussion of active components in active nitrogen gave arguments which point to vibrationally excited ground state nitrogen molecules as the component reactive toward ammonia. They rule out electronically excited molecules in the $A^3\Sigma_u^+$ state mainly because of absence of Vegard-Kaplan bands in the afterglow (see below), although this is the lower state now generally accepted as involved in part of the afterglow emission (cf. Jennings and Linnett, 1958; Kistiakowsky and Warneck, 1957). The lifetime for radiation of the $A^3\Sigma_u^+$ state of nitrogen was found to be 10^{-3} to 10^{-4} sec by Muschlitz and Goodman (1953) and is considered reason that molecules in this state would not be sufficiently abundant for detection in chemical reactions. Recently, however, Lichten (1957) has carefully redetermined the lifetime of the $A^3\Sigma_u^+$ and $a^1\Pi_g$ states and concludes that the former workers had obtained the lifetime of the $a^1\Pi_g$ state instead of the A state and that the lifetime of the latter must be greater than 10^{-2} sec, a value in better accord with the nonappearance of the Vegard-Kaplan bands in the afterglow.

From the above it would appear that the A-state molecules could reasonably be the second active species. However, this conclusion is counter to conclusions of Kistiakowsky and Volpi (1958) on the reaction of nitrogen atoms with various gases. Reactions with H_2, CO, and NH_3 were found to be too slow for detection either at room temperature or near 250°C in a low pressure "stirred" reactor flow system with a mass spectrometer as the analytical device, the rate constants being less than 10^8 cc/mole sec. Addition of the gases causes no qualitative changes in the afterglow spectrum. In the presence of ammonia the intensity of the afterglow is reduced with the inverse of the intensity being linear in pressure of ammonia, and this occurs while the atomic nitrogen concentration remains unchanged. Recent studies indicate that the afterglow is caused indirectly through free nitrogen atoms via a mechanism involving combination of electronic ground state (^4S) atoms to give $N_2(^5\Sigma_g^+)$ which undergoes a radiationless, collision-induced transition to the $B^3\Pi_g$ state (Berkowitz et al., 1956; Kistiakowsky and Warneck, 1957). The radiative transition from this state to $A^3\Sigma_u^+$ then gives part of the afterglow bands; the rest of the afterglow bands are postulated to arise from $^5\Sigma_g^+$ via an unidentified Y-state. Since the nitrogen atom concentration is not affected by addition of ammonia while the afterglow is greatly reduced, the suggestion was made that ammonia decomposition is the result of an inelastic collision in the course of which the nitrogen molecule undergoes a triplet-singlet (e.g., $B^3\Pi_g$—$X^1\Sigma_g^+$) and the ammonia molecule a singlet-triplet transition (Kistiakowsky and Volpi, 1958). This would help explain quenching of the afterglow; however, additional effects on the nitrogen atom concentration may be expected to arise from the possible primary decomposition products of ammonia. For example, the excited state of ammonia may be one of the states involved in the photolytic decomposition below 2400 A, so that hydrogen atoms and amine radicals will be expected (Dixon, 1933; Duncan, 1935) as the direct decomposition products, according to the following equations:

$$N_2(B^3\Pi_g) + NH_3 \to N_2(X^1\Sigma_g^+) + NH_3^* \tag{9}$$

$$NH_3^* \to NH_2 + H \tag{10}$$

in which NH_3^* is an electronically excited state of ammonia. These decomposition products can then cause further decomposition of ammonia through radical attack on hydrogen atoms

$$H + NH_3 \to NH_2 + H_2 \tag{11}$$

$$H + NH_2 \to NH + H_2 \tag{12}$$

and also can react with nitrogen atoms and thus indirectly promote recombination via reactions such as

$$N + NH \to N_2 + H \tag{13}$$

$$N + NH_2 \to N_2 + 2H \text{ or } H_2 \tag{14}$$

It should be noted that the afterglow spectrum remained of normal appearance but with reduced intensity when ammonia was added (Kistiakowsky and Volpi, 1958), from which it should follow that the unidentified Y-state (Kistiakowsky and Warneck, 1957) nitrogen molecules should interact with ammonia to the same extent as the $B^3\Pi_g$ state molecules. A possible alternative mechanism for the decomposition of ammonia by active nitrogen would be by reaction with $N_2(^5\Sigma_g^+)$, the common precursor postulated for both the B- and Y-states (Kistiakowsky and Warneck, 1957), which must have a long lifetime in comparison to a period of molecular vibration (Berkowitz et al., 1956). This would account for reduction in intensity of the afterglow and would eliminate the requirement for equal rates of reaction of ammonia with both B- and Y-state nitrogen molecules.

There does not seem to be any evidence for existence of appreciable concentrations of $A^3\Sigma_u^+$ nitrogen molecules in active nitrogen. If significant quantities were present it seems that reaction with ammonia would occur as postulated for the other electronically excited nitrogen molecules. Worley (1948) failed to find an absorption spectrum due to transitions from $A^3\Sigma_u^+$ nitrogen molecules, suggesting a low concentration of molecules in this state. However, he suggested that interaction with mercury, which was present, may have reduced the lifetime (approximately 0.1 sec, Lichten, 1957) of this metastable state. The work of Broida and co-workers (cf. Peyron and Broida, 1959, for summary) has given an insight into the pronounced effect of several impurities on the afterglow of the solid condensed from dissociated nitrogen. Analogous effects should occur in the gas, and so it seems that further study of the absorption spectrum of the afterglow might well help explain several puzzling features of this substance.

Evidence for vibrationally excited ground state nitrogen in active nitrogen has been given by Kaufman and Kelso (1958), who studied the effect of added gases on the afterglow intensity. A fast flow system was used in which products from a microwave discharge were pumped through a long 2-cm i.d. Vycor tube at pressures of 1 to 3 mm Hg and at linear flow rates of about 10^3 cm/sec. Afterglow intensity could be measured at various points along the tube with a photomultiplier. Gases could be added at three points along the tube downstream from the discharge. Moderate amounts of helium, argon, nitrogen, and oxygen added past the discharge did not affect the normal, approximately exponential decay of the afterglow intensity of pure nitrogen. Comparable amounts of N_2O (approximately 2 %) produced a strong decrease in the afterglow intensity immediately past the mixing point, but resulted ultimately in a small increase in intensity far downstream. A considerable temperature rise was observed immediately past the mixing region—observed with glass-enclosed thermocouples introduced from the downstream end of the flow tube—whereas far downstream the temperature was actually somewhat lower than in the discharged nitrogen stream alone. A mass spectrometer analysis of the gas sampled far downstream did not detect any loss of N_2O nor formation of O_2 or NO. The heat evolution was estimated to be 2 kcal per mole of nitrogen and was little affected by increasing the amount of N_2O added. Without added gas, active nitrogen was always found to be above room temperature, corresponding to an evolution of heat in accord with slow degradation of excess vibrational energy and the heat of recombination of atomic nitrogen. The large evolution of heat immediately after addition of N_2O is then attributed to the efficacy of this gas for deactivating vibrationally excited nitrogen molecules, whereas the loss of glow intensity in this region is a result of the known negative temperature coefficient of the afterglow (Rayleigh, 1940a). Since N_2O causes rapid relaxation of the vibrationally excited nitrogen, the temperature some distance downstream falls below that observed for the discharged nitrogen without added N_2O and therefore explains the increased afterglow past the point where the temperature curves cross.

The normal decay of the species responsible for the heat release was investigated by adding N_2O at the three different gas inlets along the flow tube. The heat release was reduced 30 % and 50 % at points 20 and 50 cm downstream, respectively, by introducing N_2O at these points. The reduction corresponds to a half-life of about 0.050 seconds. This is time for about 3×10^5 collisions at a pressure of 1 mm Hg and is of the same order of magnitude as the number of collisions to deactivate vibrationally excited oxygen molecules (McCoubrey and McGrath, 1957; Lipscomb et al., 1956). This seems to be a reasonable value for collisional deactivation of vibrationally excited nitrogen by comparison to values obtained for oxygen in which the number of collisions for loss of a vibrational quantum has been found to be 5×10^5 for self-deactivation and 10^5 for deactivation by N_2 (McCoubrey and McGrath, 1957). Deactivation of O_2 by H_2O, NH_3, and NO_2 falls to the order of 10^2 collisions, showing an extremely high efficiency. Such an efficiency is also expected for N_2O, and is therefore consistent with the explanation of the heat release as an acceleration of vibrational relaxation of nitrogen molecules (Kaufman and Kelso, 1958).

Further support for presence of vibrationally excited nitrogen was obtained by Kaufman and Kelso (1958) by "titrating" the active nitrogen with nitric oxide (Kistiakowsky and Volpi, 1957) to the point of extinction of the afterglow but insufficient to produce the greenish-yellow air afterglow, i.e., sufficient NO was added to carry the reactions

$$NO + N \rightarrow N_2 + O \tag{15}$$

$$N + O + M \rightarrow NO^* \rightarrow NO + h\nu \tag{16}$$

to completion, but not enough to give the NO + O continuum (Kaufman and Kelso, 1957). Under these conditions downstream addition of N_2O caused a greater heat release than without NO, indicating that (a) the heat-releasing species is not affected by NO and (b) that the reaction N + NO produces more of the heat-releasing species, presumably vibrationally excited nitrogen. The latter would be analogous to reactions postulated by Lipscomb et al. (1956) to account for the spectroscopically observed vibrationally excited ground state oxygen produced in photochemical decomposition of NO_2 and ClO_2—cases in which there is clear evidence for long-lived excited states.[1]

Further study is needed to fully explain the above phenomena, which

[1] *Editor's Note:* Recently Dressler (1959) has shown conclusively, by vacuum ultraviolet absorption spectroscopy, that vibrationally excited N_2 is present in "active" nitrogen.

point to a second active species in active nitrogen; however, it seems possible to postulate a reasonable mechanism consistent with the observations discussed above as follows: Initially the glowing gas contains free nitrogen atoms and molecules in various states of vibrational and electronic excitation. Reaction of nitrogen atoms in ternary collisions yields $^5\Sigma_g^+$, $A^3\Sigma_u^+$, or $X^1\Sigma_g^+$ (ground electronic state) molecules. The loosely bound quintet state is the precursor of the upper states for the gaseous afterglow emission and exists in equilibrium with atomic nitrogen

$$N(^4S) + N(^4S) + M \leftrightarrows N_2(^5\Sigma_g^+) + M \qquad (17)$$

Because of the loose binding, the equilibrium rapidly shifts to favor free atoms as temperature is increased, thereby lowering concentration of quintet state molecules and accounting for the negative temperature coefficient of the afterglow. The lower state for the afterglow is $A^3\Sigma_u^+$ and in addition these A-state molecules are also formed directly by the reaction

$$N(^4S) + N(^4S) + M \rightarrow N_2(A^3\Sigma_u^+) + M \qquad (18)$$

giving A-state molecules in upper vibrational levels which may go to lower levels by collisional loss of vibrational energy or may undergo a collision-induced radiationless transition to upper vibrational states of the $X^1\Sigma_g^+$ state. Vibrationally excited ground state molecules are also formed directly from atomic nitrogen in ternary collisions

$$N(^4S) + N(^4S) + M \rightarrow N_2X^1\Sigma_g^+ + M \qquad (19)$$

There is thus seen to be the possibility of temporary storage of a large amount of excess vibrational energy. Assuming nitrogen molecules to behave much as oxygen, a relatively long relaxation time is expected for self-deactivation (McCoubrey and McGrath, 1957), at least for the ground state molecule, and probably also for the electronically excited A-state molecule. Molecules such as H_2O and NH_3 have a high efficiency for collisional deactivation of O_2—about 400 collisions to deactivate first vibrational level against 500,000 for self-deactivation of O_2 (McCoubrey and McGrath, 1957); these molecules, along with N_2O, which is a relatively efficient self-deactivator for vibrational energy (McCoubrey and McGrath, 1957), can therefore be expected to cause rapid vibrational relaxation of nitrogen molecules, thus accounting for the heat effect noted by Kaufman and Kelso (1958) in the case of N_2O and, if this latter effect is properly identified with vibrational relaxation, a similar effect for ammonia. The sensible temperature rise of the gas will result in a decrease of intensity of the afterglow with both gases as additives. This physical interaction does not account for the limited but significant decomposition of ammonia observed by Evans and Winkler (1956). However, it seems that this decom-

position might well occur by a process similar to the suggestion of Kistiakowsky and Volpi (1958) but involving interaction with $N_2A^3\Sigma_u^+$ instead
of with $N_2B^3\Pi_g$. Thus the sequence of reactions

$$N_2^* + NH_3 \rightarrow N_2 + NH_3^* \tag{20}$$

$$NH_3^* \rightarrow NH_2 + H \tag{21}$$

in which N_2^* is an A-state molecule in a low vibrational state and NH_3^*
is electronically excited ammonia. Sufficient energy would also be available
to decompose N_2O in an analogous fashion. This decomposition was not
detected (Kaufman and Kelso, 1958) by a mass spectrometric analysis of
the gas far downstream, but neither was ammonia decomposition found
mass spectrometrically (Kistiakowsky and Volpi, 1958), although Freeman
and Winkler (1955) had found reaction of active nitrogen with ammonia
to an extent corresponding to one-sixth of the activity of the nitrogen as
determined from its ability to form hydrogen cyanide on reacting with
ethylene.

In view of the reported paucity of ^2D nitrogen atoms in active nitrogen
(Tanaka *et al.*, 1957) it seems that the most plausible postulate regarding
the second source of mass 14 ions observed mass spectrometrically by
Jackson and Schiff (1955) is that the appearance potential at 16.1 ev is due
to nitrogen molecules possessing very high vibrational energies. Vibrationally excited A-state molecules as well as ground state molecules could well
contribute to this, with a criterion for observation in either case being high
purity of nitrogen used in the experiments. Traces of water, oxygen, and
other impurities may be detrimental because of the several possible mechanisms for deactivation of the excited species.

C. Thermal and Magnetic Studies at Low Temperatures

It was shown by Broida and Pellam (1954) that active nitrogen condensed
at 4.2°K glows with a brilliant green light and gives off occasional bright
flashes as long as the electrical discharge is operating. The glow persists
after the discharge is stopped, but with gradually diminishing intensity,
and then reappears upon warming to about 10°K (Peyron and Broida,
1959; Fontana, 1958). On further warming of the solid, the glow decreases
and at about 25°K a less intense blue-green glow is observed which persists to about 35°K. Similar glows have been produced by Hörl (1959)
using electron bombardment of solid nitrogen, and by Schoen and Rebbert
(1959) using a Tesla discharge to the solid, and by Wall *et al.* (1959) using
gamma irradiation from a Co[60] source. It should be mentioned that the
first studies of glowing nitrogen in the solid state were apparently those
of Vegard (1924a, b, 1926, 1930, 1935) in connection with elucidation of
the auroral glow of the upper atmosphere. Solid nitrogen at 4.2°K and at

20°K was irradiated with cathode rays and canal rays in these studies. The spectra of the resulting afterglows were analyzed by Vegard, many of the bands being correctly assigned to molecular transitions of N_2 and NO. McLennan and co-workers (1924, 1928, 1929) have made similar studies. Peyron and Broida (1959) have listed eleven identifiable spectral features in solid glowing nitrogen, showing correlations with Vegard's work as well as several recently discovered features.

Atomic nitrogen has been shown to be a principal energetic species in active nitrogen gas (see above), strongly suggesting the same for the glowing solid. The spectral data by Broida and co-workers (Chapter 6) at the National Bureau of Standards (Peyron and Broida, 1959, for example), have given evidence for atoms stabilized in the solid, and spin resonance studies have further corroborated this conclusion (Chapter 7). Estimates of the number of trapped atoms have been made by Broida and Lutes (1956) from measurements of the rate of temperature rise of the collection chamber, with the result that the N-atom content of the solid is approximately 0.2%. Minkoff et al. (1959) have estimated nitrogen atom content of the solid to be 4 to 6% based on heat release during warm-up and assuming all of the heat to arise from atomic recombination. Similar amounts of heat release were observed for electrically dissociated rare gases in the same calorimeter (Minkoff and Scherber, 1958), so that doubt exists as to the origin of heat in the experiments with dissociated nitrogen. Aside from the difficulties inherent in calorimetry of free radicals at low temperatures, there is the further complication that arises from heat evolution during phase changes of the substances condensed in metastable configurations on the cold surfaces (Peiser et al., 1958) (see also Chapter 9).

Very significant studies of trapped nitrogen atoms have recently been made by Fontana (1958, 1959), using thermal and magnetic measurements and correlation of the gross light effects with spectral investigations of similarly prepared deposits (Peyron and Broida, 1959). In the thermal experiments nitrogen atoms were produced by gas phase dissociation at about 1 mm pressure in a microwave discharge. From the discharge region the dissociation products traversed a Pyrex tube, 51 cm long and 15 mm outside diameter, to reach the flat Pyrex collector plate which was cooled in a liquid helium bath. A schematic diagram of the collecting chamber is shown in Fig. 1. The thermocouple junctions were inserted through holes drilled in the collector plate and were distributed so as to determine temperature gradients in the deposit as well as the effect of different path lengths of the leads back to the liquid helium bath. Temperatures were continuously observed with a recording potentiometer to an estimated accuracy of 0.5 to 1°K. Warm-up of the solid deposits was controlled by adjusting the level of liquid helium below the collector.

A typical temperature versus time record in one of these experiments is

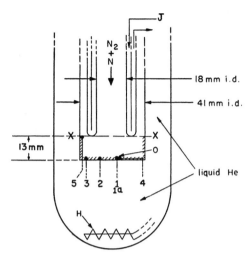

FIG. 1. Schematic diagram of the collector. O, active thermocouple junction; $X—X$, maximum level at which solid deposits; J, warm (room temperature) helium gas during deposition, evacuated during warm-up; H, 50-ohm heater; shaded area shows general distribution of the deposit; $1a$, single thermocouple used with 4.4-mm collector; 1 to 5, five thermocouples used with 2.0-mm collector. After B. J. Fontana, $J. Appl. Phys.$ **29**, 1667 (1958).

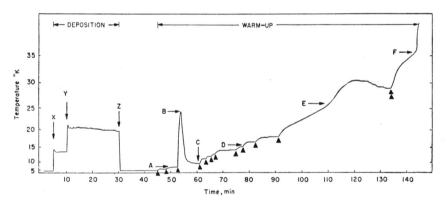

FIG. 2. Typical time versus temperature record with controlled warm-up. Measurements made with thermocouple $1a$ (Fig. 1). Nitrogen flow rate = 31.3 cc/min (STP). Total nitrogen deposited = 0.73 gm. X, gas flow started; no discharge. Y, microwave discharge started. Z, gas flow and discharge stopped. ▲ indicates lowering of the liquid helium level by roughly equal increments. After B. J. Fontana, $J. Appl. Phys.$ **29**, 1667 (1958).

shown in Fig. 2. The difference in deposition temperature with and without discharge (cf points Y and X, Fig. 2) was used to calculate the concentration of N atoms in the gas phase, results obtained being in reasonable agreement with the mass spectrometer results of Herron et al. (1958), who used closely similar dissociation techniques. At the end of deposition, the temperature quickly falls to 4.2°K and the green glow dies out. Then the system is slowly warmed up by lowering the helium level below the collector, whereupon a glow starts (point A, Fig. 2) and the temperature rises sharply even while the helium level is held constant. The temperature peak (ABC) coincides with a peak in the characteristic green glow, an emission which has been identified as the $^2D-^4S$ transition of atomic nitrogen. Assuming the heat to come from recombination of nitrogen atoms, it was estimated that the temperature rise from 9° to 22°K, taking 1.3 cal/gm as the specific heat of the solid, corresponds to 0.03 atom per cent nitrogen in the solid. Following the peak at B the glow dies out and can be reinitiated only by raising the temperature. However, no strong peaks normally occur beyond B although occasional temperature inflection points such as E appear and are accompanied by increased light emission, suggesting that exothermic processes continue as long as there is light emission. It is significant that continued warming is necessary to maintain or reinitiate the glow after passing point C. The peak (ABC) could be avoided by very slow warm-up, although the glow would start at about the same temperature. It could also be quenched after starting by lowering the temperature and then subsequently reinitiated several times by alternately raising and lowering the temperature, and thereby obtaining several lower peaks.

The features in the curve of Fig. 2 were found to be fairly reproducible. In a series of seven runs with constant microwave discharge conditions, but varying amounts of discharged nitrogen, the following temperatures were found for the characteristic features: initiation (A), 8.9° ± 0.4°K; first peak (B), 21.6° ± 1.6°K; reinitiation (D), 16.9° ± 1.9°K; final (F), 35.9° ± 0.6°K. The qualitative features of the warm-up were not appreciably changed over a range of deposition temperatures of about 6° to 31°K. It is somewhat surprising that the same behavior could occur after deposition at the higher temperatures, for it seems that the various processes should have run their course during deposition. Fontana (1958) suggests the possibility that the initiation may occur in warmer parts of the apparatus, but noted that a moving wave front of glow was never observed as might be expected if this were the case. Another possibility for initiation of the heat evolution and glow is suggested by an experiment (Broida and Peyron, 1958) which indicated that the active species could be distilled from one surface to another. Thus, in the present experiments, the glow may be initiated by distillation of nitrogen atoms from parts of the apparatus that warm up first.

Qualitative characteristics of the emission during deposition and warm-up were observed with narrow band-width filters, with results as given in Table I (see also Chapter 6).

TABLE I

EMISSION OF VISIBLE LIGHT FROM CONDENSED ACTIVE NITROGEN

Emitting species:	N	O	$N_2(?)$
Transition	$^2D - {}^4S$	$^1S - {}^1D$	(?)
Notation	α	β	B
Filter: Spectral region, A	5230	5578	4100
Band width, A	20	100	600

Temperature region[a]	Relative intensity[b]		
During discharge	S	S	S
$A \rightarrow B$, 9°–22°K	S	S	$W \rightarrow S \rightarrow W$
$D \rightarrow F$, 17–35°K	$S \rightarrow W$	$S \rightarrow W$	$W \rightarrow S \rightarrow W$
F, 35–36°K	n	n	W

[a] See Fig. 1.
[b] S = Strong; W = weak, n = nil.

The assignment of transitions given in Table I have been summarized by Peyron and Broida (1959), the α- and β-lines arising from electronically excited nitrogen and oxygen atoms, respectively, in the solid. Fontana (1958) suggested the B-bands arise from electronically excited N_2 formed from atomic nitrogen in the solid; however, Peyron and Broida (1959) attribute these bands to NO_2.

Magnetic susceptibility measurements were also made by Fontana (1959) in the above-described apparatus by adding a sensing coil around the deposition or collecting zone which thus permitted measurement of magnetic effects of the deposits using an ac mutual inductance method (deKlerk and Hudson, 1954). Although the arrangement was not ideal, it provided means for observing radical concentrations simultaneously with the thermal and light emission effects described above, thus adding considerably to the understanding of the complex phenomena involved. In this study an unexpected result was obtained on inserting a liquid nitrogen trap in the low pressure nitrogen feed line ahead of the discharge zone. When this was done the deposition temperature suffered a marked decrease—a decrease which did not occur when the nitrogen was passed to the collector without discharging—and also the gaseous nitrogen afterglow was found to decrease sharply on insertion of the trap. The effect of the trap was interpreted to mean that a large decrease in nitrogen atom concentration had occurred, presumably as a result of removal of traces of water. Use or nonuse of the trap thus made it possible to get widely different concentrations of atoms in the discharged gas stream.

Magnetic susceptibility measurements were made with "high" nitrogen atom concentrations, i.e., without passing the nitrogen through a liquid nitrogen trap before going through the discharge; however, no magnetic effects were observed; from this it was estimated that not more than 0.04 and possibly only 0.01 mole per cent atoms were present in the solid. In these experiments, over a range of flow rates of 2.7 to 90.0 cc/min (STP), the atom concentration was generally in the range 2 to 4 mole per cent, as estimated by Fontana's (1958) temperature-rise method. It was thus concluded that, with about 2 mole per cent or more of atoms, practically all recombine immediately upon condensation.

When lower concentrations of nitrogen atoms were used, i.e., by using a liquid nitrogen trap in the stream before the discharge, there was a marked change in magnetic behavior, the magnetic signal being observed to increase for short, random periods of time followed by abrupt decreases. These decreases occurred simultaneously with emission of brilliant yellow flashes of light and the appearance of sharp peaks on the record of temperature versus time. It appeared then that, under these conditions of low atom concentration in the gas stream, relatively large fractions of those atoms present could condense without recombination, but that at random times rapid or "catastrophic" recombination would be triggered. In a number of cases it was possible to follow the magnetic signal long enough to get an estimate of atom concentrations in the solid, with results that varied from 0.16 to 0.48 atom per cent. In these experiments, the difference between the deposition temperature with and without discharge was not great enough to permit an estimate in the atom concentration of the gas stream. However, the trend in the data indicated that the gas could contain only slightly more atoms than the resulting solid. These observations suggest that the maximum atom concentration that can be stabilized in the solid is near 0.5 mole per cent. Two experiments were described in which it was possible to turn off the discharge and gas flow after appreciable amounts of the paramagnetic material had condensed. In one case the concentration of nitrogen atoms was 0.17 mole per cent, based on susceptibility measurement during and immediately after deposition; however, this concentration rapidly decreased, falling by about 50 % in 2.5 minutes. In another experiment a concentration of about 0.21 mole per cent was indicated by the magnetic measurements. This concentration persisted about 30 sec, during which time the deposit cooled to 4.2°K, and then a brilliant flash of light was emitted after which the magnetic signal was no longer detectable, indicating that the concentration of atoms had fallen below about 0.01 mole per cent.

The strong variation of light emission which occurs coincidentally with the loss of paramagnetism and sharp rise of temperature was observed visually with the aid of narrow band-width filters (Table I). Effects observed were: (1) the disappearance of the green α-emission due to the (^{2}D–^{4}S) transition of atomic nitrogen; (2) a large increase in the β-emission, the ^{1}S–^{1}D) transition of atomic oxygen; and (3) an increase in the B-emission. The disappearance of the α-emission is consistent with the view that catastrophic recombination occurs, since excited and ground state nitrogen atoms should react equally readily during this rise in activity in the solid. The period of rapid reactivity is relatively short compared to the 20-sec half-life (Bass and Broida, 1956) of the ^{2}D–^{4}S transition in the solid. In view of the thermal and magnetic studies by Fontana (1958, 1959) it is probable that this reported 20-sec half-life corresponds to the half-life of a diffusion-controlled reaction (see below) rather than a pure radiation half-life.

In summary, the thermal and magnetic studies show that the stable solid condensed from active nitrogen contains about 0.01 to 0.04 mole per cent nitrogen atoms as an upper limit. This is the solid which exhibits the effects seen in the thermal studies. When the condensing gas stream has an excess of 1 to 2 mole per cent atoms, nearly all recombine during condensation. When the gas contains less than about 0.5 mole per cent atoms, from 0.2 to 0.5 mole per cent atoms can be stabilized in the solid for short periods. However, some of the experimental results indicate that recombination continues smoothly in this solid even after cessation of deposition and with the solid cooled to 4.2°K. This suggests that it is not the deposition process which limits the stable concentration of N atoms attainable (Fontana, 1959).

D. Mechanism of Solid Nitrogen Afterglows

The principal features of the solid nitrogen afterglows can be explained on the basis of trapped nitrogen and oxygen atoms and without the assumption that electronically excited atoms can exist for appreciable periods in the solid. The following theory of these features depends rather on the assumption that electronically excited nitrogen molecules have a radiative lifetime *in the solid* which is comparable to the half-life for the recombination reaction of nitrogen atoms, and that these molecules can readily transfer electronic energy to N and O atoms which subsequently radiate the α- and β-lines. For example, we assume that the dark, quiescent solid remaining at 4.2°K after cessation of deposition and decay of the afterglow contains about 0.01 mole per cent N atoms (Fontana 1959) and approximately an equal percentage of oxygen, in the form of atoms and oxides of nitrogen. These atoms are stabilized, i.e., unable to move. However, when the deposit is warmed to 9°K diffusion commences, with the nitrogen atoms migrating and reacting to give nitrogen molecules in ground ($X^1\Sigma$) and electronically excited states $A^3\Sigma$ and $^5\Sigma$,

$$N(^4S) + N(^4S) \rightarrow N_2 \tag{22}$$

or

$$N(^4S) + N(^4S) \rightarrow N_2^* \tag{23}$$

where N_2^* represents electronically excited molecules. Heat is also liberated causing further warming of the solid, thus further promoting the above and other reactions. Nitrogen atoms also react with nitrogen oxides, liberating oxygen atoms

$$N + NO \rightarrow N_2 + O \tag{24}$$

which may be in electronically excited states as formed and thus contribute directly to emission of the β-lines. Consequently the thermal peak observed by Fontana (1958) is the result of an exothermic reaction with a positive temperature coefficient—a commonly occurring type of reaction.

The enhancement of the yellow β-lines of oxygen during this thermal peak is explained, in addition to Eq. (24), as an interaction of oxygen atoms with excited nitrogen molecules N_2^*. This interaction involves transfer of energy followed by radiation by the excited atom O^* as indicated by the equations

$$O + N_2^* \rightarrow O^* + N_2^\ddagger \tag{25}$$

$$O^* \rightarrow O + h\nu \quad (\beta\text{-lines}) \tag{26}$$

or

$$O^*(N_2^\ddagger) \rightarrow O + N_2 + h(\nu_e \pm \nu_{\text{vib}}) \quad (\beta\text{-}, \beta'\text{-}, \beta''\text{-lines}) \tag{27}$$

in which $N_2^\ddagger$ is vibrationally excited N_2, and $O^*(N_2^\ddagger)$ signifies an intermediate radiative state possibly also giving rise to the primed β-lines. The strength of the β-emission during the thermal peak indicates that the oxygen atom has a particularly high efficiency for quenching electronic excitation of the nitrogen molecule. It is postulated that it is this high efficiency which, by direct competition with the radiative transition, prevents the Vegard-Kaplan (VK) bands from appearing in the solid glows of any but the purest nitrogen and in nitrogen diluted with argon (cf. Peyron and Broida, 1959, for example). This effect of oxygen on the VK-bands is a strong qualitative argument favoring the present theory. The α-lines arise in an analogous way to the β-lines, according to the present theory, i.e.,

$$N + N_2^* \rightarrow N^* + N_2^\ddagger \tag{28}$$

$$N^* \rightarrow N + h\nu \quad (\alpha\text{-lines}) \tag{29}$$

$$N^*(N_2^\ddagger) \rightarrow N + N_2 + h(\nu_e \pm \nu_{\text{vib}}) \quad (\alpha\text{-}, \alpha'\text{-}, \text{and } \alpha''\text{-lines}) \tag{30}$$

The α-lines are strong during the thermal peak—stronger than the β-lines. However, in the catastrophic recombination observed by Fontana (1958, 1959), β-lines seem to dominate.

Continuing discussion of the warm-up glows, as the reaction diminishes in speed, heat output falls, allowing the solid to cool to an environment temperature above that at which original glowing started (Fig. 2). At this point there is no light emission; this indicates, as Fontana suggested (1958), that diffusion has ceased and that there must be some sites in the solid at which atoms are more tightly bound. Thus an increase in temperature is needed to cause the glow to start again, and continual raising of the temperature is necessary to maintain the glow. During this temperature interval, 17 to 35°K (see Table I), both α- and β-lines continuously diminish in intensity and finally are extinguished at 35°K.

The B-bands, which are strong in the thermal peak, are weak immediately following the thermal peak, and increase on further warming. They then fall off again, persisting, however, longer than the α- and β-lines until the temperature reaches 36°K. Although Fontana indicates B-bands are from the N_2 molecule, they are attributed doubtfully to nitrogen dioxide, NO_2, by Peyron and Broida (1959). The bands are weak in N_2 containing <10 ppm O_2, get stronger for N_2 containing 0.1% O_2, but are not reported for 5% N_2-95% A mixtures—consistent, at least, with the NO_2 assignment which we assume is correct. During the warm-up, oxygen atoms are continuously regenerated from nitrogen oxides, as long as there are any nitrogen atoms, by reaction of Eq. (24), for example. Upon depletion of free nitrogen atoms the remaining oxygen atoms will be consumed by reaction with nitric oxide, the reaction

$$O + NO \rightarrow NO_2 \qquad (31)$$

being the final reaction emitting light and accounting for the change of the emission to blue (Table I). The reaction of Eq. (31) has been carried out in a reactor cooled with liquid helium by Ruehrwein et al. (1959), by mixing separate streams of gas near a surface cooled with liquid helium, with results that indicate that the reaction proceeds without an activation energy.

The thermal and principal light-emission effects of the warm-up of solid nitrogen discharge products are explained by the above mechanism. The combined thermal, magnetic, and light-emission effect attributed by Fontana to catastrophic recombination are likewise understandable on the above basis—the effects being more vigorous owing to greater concentration of atoms before onset of the exothermic reaction. A major problem remaining is to explain the subsidiary α- and β-emissions. The explanation implied in the paper by Peyron and Broida (1959), that these subsidiary, or primed, emissions are simply the α- and β-lines (see Table I), modified by either addition or subtraction of a vibrational quantum of the nitrogen molecule, does not seem satisfactory for several reasons. First we note that the *added* vibrational quantum should be longer than the subtracted quantum, but this is not the observed case. Also the subtracted quantum, corresponding to a vibrational quantum jump of $+1$ (2299 cm^{-1}, average) does not correspond to the vibrational frequency of the (2325 cm^{-1}) normal nitrogen molecule as determined by study of the Vegard-Kaplan bands emitted by the solid (Peyron and Broida, 1959).

Obviously much further work is needed to give a full explanation of these and other effects observed by the various investigators. It is hoped that studies will continue, for they offer the possibility of adding significantly to the basic understanding of the energetics of atomic and molecular processes.

IV. Hydrogen-Nitrogen Radicals

Rice and co-workers reported that when hydrazoic acid (HN$_3$) vapor is decomposed by any of several means, such as electrical discharge or thermal dissociation, and the products frozen out rapidly at around 77°K, a blue material is formed (Rice and Freamo, 1951, 1953; Rice and Grelecki, 1957). Mador and Williams (1954) also formed the blue solid by photolysis (2537 A) during and after deposition at 4° and 77°K. On warming to about 150°K, the blue material decomposes violently, the residue being ammonium azide. Rice and Freamo (1951) postulated that the blue color is caused by NH radicals stabilized in the solid and that the violent reaction on warm-up is due to rapid polymerization:

$$4NH \rightarrow NH_4N_3 \tag{32}$$

These reports have stimulated considerable subsequent investigation of the hydrogen-nitrogen system. Present evidence indicates that reaction (32) is not the cause of the violent reaction; but the cause remains unknown.

Foner and Hudson (1958a, b, and unpublished), using a free radical mass spectrometer (Foner and Hudson, 1953a), have investigated formation of radicals from HN$_3$ by electrical dissociation and from hydrazine, N$_2$H$_4$, by both electrical and thermal dissociation. A careful search was made for NH under conditions ascertained to be favorable for formation of Rice's blue solid (see below). However, in decomposing HN$_3$ the only species found in the mass range 14 to 100 were HN$_3$, N$_2$, NH$_3$, and a new compound which was established as diimide, N$_2$H$_2$. The lifetime of N$_2$H$_2$ was shown to be much longer than the substance responsible for the blue color. Thermal and electrical dissociation of hydrazine was reported to be very complex, giving several free radicals and higher order hydronitrogen compounds. In a subsequent report by Foner and Hudson (1958b) the following products were reported to be observed mass spectrometrically in electrically dissociated hydrazine: the radicals H, N, NH_2, and N_2H_3 and molecules H_2, NH_3, N_2H_2, N_2H_4, and N_3H_3. When these products are trapped in a liquid nitrogen trap and then allowed to evaporate by removing the coolant, N_2 comes off first and continues to be evolved during most of the warm-up period. Diimide and ammonia evaporate at the same time, and subsequently triazene [N_3H_3] and tetrazene [H_2N—N=N—NH_2] appear together, tetrazene going through its maximum before triazene. Hydrazine comes off last. Since tetrazene was not present in the gaseous products, it must have been formed in a reaction on the cold surface or in the condensed solid—a reaction involving, or initiated by, radicals originally present in the gas stream.

The evolution of nitrogen noted above may be explained in part by work of Rice and Scherber (1955). They observed that when pyrolysis products of hydrazine are condensed at 77°K a yellow material is formed which decomposes at 95°K, giving off pure nitrogen. They suggest the yellow compound is tetrazane (N_4H_6) formed during condensation at the low temperature by combination of two hydrazino radicals [$NH_2NH\cdot$] derived from the pyrolytic process. This decomposition may account for the nitrogen which Foner and Hudson observed to be given off. However, occlusion of gases with subsequent evolution on warming seems to be fairly common (Edwards and Hashman, 1957a; Ruehrwein et al., 1959; Hogg and Spice, 1957), so that chemical change may not be the sole cause in the present case.

In experiments in which ammonia was electrically dissociated in a microwave discharge (2450 Mc/sec), a yellow substance has been observed in the products condensed very near the discharge in a trap at 77°K (Edwards and Hashman, 1957b). Pressures in the system varied between 0.5 and 1.5 mm Hg. This material was a minor component along with a relatively large amount of ammonia. It could not be separated from the ammonia by vaporization and therefore was neglected at the time, although conditions of the experiments were varied in efforts to increase the yield. In view of the similarity of vapor pressure and color to N_2H_2 as reported by Foner and Hudson, N_2H_2 probably also was formed in this case. Electrically dissociated ammonia was condensed at 4°K without definite detection of the yellow substance, although the deposit was somewhat amber in color (Ruehrwein et al., 1959). When H_2 and N_2 were separately dissociated and then blended at a distance of only several centimeters from a liquid helium-cooled surface on which all the gases condensed, very low yields of ammonia were found. It is likely that small amounts of other compounds would have gone undetected. In these experiments at 77°K and at 4°K there was no sign of a blue deposit similar to that derived from hydrazoic acid.

Franklin *et al.* (1958b) have studied the decomposition of hydrazoic acid in an electrical discharge stimulated by a Tesla coil. Products, analyzed in a modified gas analysis mass spectrometer, were H_2, N_2, and NH_3, along with unchanged HN_3 and, probably, N_2H_2. Conditions were such that Rice's blue solid could be condensed on a cold finger situated downstream from the sampling leak; however, no radicals could be detected. It was concluded that the products observed were formed in the gas phase. This conclusion does not seem justified in view of the low pressures employed, the large volume through which the glow discharge may be presumed to have extended,[2] and the large number of wall collisions possible between the sample leak and the ion source—the latter estimated as 10^3 to 10^4 by the writer. An analysis of the blue solid was also made during warm-up by leading volatile products to the mass spectrometer. No volatile products were given off until the blue-to-white transition of the solid occurred. At this time a small amount of N_2 was observed, suggesting formation by a chemical reaction involving the blue material. Continuation of the warm-up gave HN_3 followed by NH_3, the latter presumably from decomposition of NH_4N_3. The ratio of the products, $N_2 : NH_3 : HN_3$ was found to be $0.0025 : 0.38 : 1.0$. Experiments were also performed to see if molecular nitrogen could be occluded during recondensation of the warm-up products, but results were negative, suggesting that the nitrogen was indeed formed by chemical reaction. In this connection it should be noted that occlusion of nitrogen from the gas phase would seem to be less likely than occlusion of nitrogen formed at the solid surface during or after condensation of the discharge products. Further, the possibility of formation of N_2 *in situ* by ion and electron bombardment of the Tesla discharge during and after condensation cannot be overlooked.

In this study Franklin *et al.* (1958b) observed evidence for formation of N_2H_2 in the electrical decomposition products of hydrazoic acid, hydrazine, and ammonia. With hydrazoic acid, the mass-30 peak was observed to increase by a factor of about 10 when the electrical discharge was turned on. They report also that the mass-30 peak has been observed in discharges in hydrazine and ammonia.

Matrix isolation studies in inert matrices at 20°K (Becker *et al.*, 1957) have given evidence for several intermediates in photolysis of hydrazoic acid. In matrices of nitrogen and argon, in which excess energy of the photon is believed sufficient to cause considerable diffusion and attendant reaction of radicals with other species, infrared spectra were observed for unstable intermediates which could be attributed to NH, NH_2, and N_3. An absorp-

---

[2] In interpreting data on reactions of radicals produced in electrical discharges, account must be taken of the tendency of high voltage discharges to permeate an entire gas-handling system when operating at low pressures—not merely where glow is bright. See Section III, A, calorimetry of active nitrogen, etc.

tion band was observed at 1093 cm^{-1} which could be associated with hydrazine; this was discounted since hydrazine was not detected in a mass spectrometric analysis of the product gas. Hydrazine is known to be strongly adsorbed on various surfaces, so it seems entirely possible that the band cited is from this compound, thus eliminating the discrepancy between this result and the X-ray detection of hydrazine azide in photolyzed hydrazoic acid as reported by Bolz et al. (1959). In the "harder", i.e., higher-melting matrix xenon at 20°K, the most unstable intermediates were not observed. This is understandable on the basis that diffusion is very limited in xenon and thus the radical NH formed in the primary dissociation act does not move far enough or with sufficient energy to cause reaction, but does move during warm-up to 70°K, forming ammonia, which with residual HN$_3$ gives the NH$_4$N$_3$ that was the observed product. Failure to observe infrared absorption of NH is attributed to the presumed inherent weakness of this single band in the high frequency regions (approximately 3110 cm^{-1}) as well as to the experimental difficulty of making measurements in this region where light scattering impairs the quality of the spectra. Becker et al. (1957) also suggest other possible species such as (NH)$_x$, N, and H, but discard these ideas in favor of simple trapping of NH and N$_2$ as formed. Becker et al. (1957) also photolyzed pure crystalline HN$_3$ at 66°K, detecting evolution of nitrogen in the process. Spectra obtained immediately after photolysis showed NH$_4$N$_3$. The absence of other products as well as lack of change on warm-up indicated formation of the one product directly without isolation and stabilization of appreciable intermediate species. This indicates a considerable mobility and reactivity of NH and possibly NH$_2$ radicals at 66°K in the HN$_3$ matrix.

Mador and Williams (1954) studied the stabilization of free radicals at 4°K and at 77°K from the electrical dissociation of hydrazoic acid. Infrared spectra showed strong N-H absorption, but none that could not be assigned to the residual HN$_3$ and the product NH$_4$N$_3$. The blue solids which were obtained showed structureless absorption bands at 3500 A and 6500 A, attributed to NH and NH$_2$, the latter presumably causing the blue color. The clear glassy solid formed when undissociated HN$_3$ was deposited at 77°K was shown to undergo a transition at 148°K, the same temperature at which the blue color disappeared from the condensed products of dissociation. Surprisingly the infrared spectrum of NH$_4$N$_3$ did not change appreciably at the transition, nor did other products appear—observations which indicate that concentrations of radicals stabilized in the solids at the lower temperatures must be very small. In other experiments by Mador and Williams (1954) hydrazoic acid was photolyzed at 2537 A during and after deposition at 4° and at 77°K, with resulting products the same as had been obtained from the vapor phase decomposition. A chain reaction for formation of NH$_4$N$_3$ was indicated by the quantum yields, and in different experiments the ratio of the relative intensities of the bands at 3500 A and at 6500 A were different, indicating that the two bands arise from different species. The wavelengths attributed to NH and NH_2 do not agree with those observed by Robinson and McCarty (1958a, b) for these radicals produced from hydrazine-argon mixtures by dissociation in a 15-Mc electrodeless discharge and condensation at 4°K. Although the wide discrepancy may be due to difference in matrices—argon versus HN_3 decomposition products—it seems reasonable to suppose that the predominant spectra observed by Mador and Williams may arise from a different species, e.g., the N_3 radical—a radical possible from HN_3 by an elementary reaction, but not from hydrazine.

Dows *et al.* (1955) also studied the infrared spectrum of products of electrical dissociation of hydrazoic acid condensed at low temperatures. Solids were condensed at 90°K and spectra were observed at several temperatures as the solids were allowed to warm up to 230°K. Absorptions due to HN_3, NH_4N_3, and NH_3 and to two reaction intermediates were observed. Intermediates were interpreted to be N_2H_2 and $(NH)_x$ with $x = 4$, and presence of NH in the solid at 90°K was inferred.

Thrush (1956) has observed the ultraviolet absorption spectrum of a polyatomic species in the flash photolysis of hydrazoic acid. He concluded that this is very likely due to the N_3 radical. The same spectrum was observed for deuterated hydrazoic acid, thus ruling out hydrogen-nitrogen compounds or radicals as the origin of this spectrum.

The available data indicate that the following radicals are formed by energetic stimuli applied to various hydrogen-nitrogen compounds in the gase phase: H, N, NH, NH_2, N_3, N_2H_3. The failure to find NH radicals by mass spectrometry of hydrazoic acid dissociation products suggests that it is highly reactive in the gas phase, rapidly abstracting hydrogen and going through the intermediate NH_2 to give ammonia as the main final product. Thrush (1956) has suggested a mechanism for formation of ammonia in photolysis which may also apply to pyrolysis and electrical dissociation:

$$HN_3 + h\nu \rightarrow HN + N_2 \tag{33}$$

$$HN + HN_3 \rightarrow NH_2 + N_3 \tag{34}$$

$$NH_2 + HN_3 \rightarrow NH_3 + N_3 \tag{35}$$

$$2N_3 \rightarrow 3N_2 \tag{36}$$

or

$$N_3 \rightarrow N_2 + N \tag{37}$$

$$N + N_3 \rightarrow 2N_2 \tag{38}$$

Formation of N_2H_2 could proceed by a radical abstraction reaction

$$NH + HN_3 \rightarrow N_2H_2 + N_2 \tag{39}$$

All the radicals in the above scheme have been identified spectroscopically in the gas phase (cf. Thrush 1956, for example), whereas Foner and Hudson (1958a) detected N_2H_2 mass spectrometrically.

Identification of the NH and NH_2 radicals in dissociation products of HN_3 condensed at low temperature is less certain. Study of HN_3 in a manner analogous to Robinson and McCarty's study of N_2H_4 would be significant in this respect, as would measurements of the electronic absorption spectra of photolyzed solids such as the solids studied by Becker *et al.* (1957). The possible presence of trapped N_3 radicals in matrix isolation studies of the latter workers and the discrepancy between the spectral studies of Robinson and McCarty and of Mador and Williams leaves room for the speculation that species other than NH and NH_2 are responsible for the gross spectra observed by the latter workers. It is difficult to understand, for example, absence of any reports of blue solids made from hydrogen-nitrogen compounds other than hydrazoic acid unless the blue is due to an entity such as N_3 which is specific to hydrazoic acid. Presence of N_3 could also account for N_2-evolution at the glass-solid transition at 148°K as noted by Franklin *et al.* (1958b). It should be noted, however, that Franklin *et al.* (1958a) have given evidence that N_3 must be very unstable, if it exists at all.

The large extent of the reaction of HN_3 dissociation products shown by mass spectrometry to take place in the gas phase leaves little to occur during and after condensation. That little reaction actually occurs after condensation was shown by several of the above studies, although some reaction might occur as indicated by the blue solid changing to white at 148°K and the small amount of nitrogen given off during this transition.

The work of Becker *et al.* (1957) attests to the highly reactive nature of NH, assuming of course, that the photolytic step is as given in Eq. (33) above for the gas phase photolysis. The implication of these results, then, is that the NH radical has sufficient reactivity to abstract hydrogen atoms from chemically stable molecules without the need for any activation energy at all. This is to suggest that this diradical can, by dynamic interaction on approach, undo the binding of certain bound hydrogen atoms.

Regarding the reactivity of NH, it is significant that this radical was not observed in the careful mass spectrometric investigations of Foner and Hudson (1958a, b) which were conducted under conditions that permitted detection and identification of other radicals such as H, N, NH_2, and N_2H_3 in dissociated hydrazine. It seems likely that NH must have at least brief existence in this mixture since all other possible dissociation fragments, both smaller and larger, are observed. On the other hand, presence of atomic nitrogen in this mixture suggests that nitrogen atoms are more or less passive toward bound hydrogen. This is consistent with conclusions reached by Evans *et al.* (1956) from studies of chemical reactions of active nitrogen.

They concluded that attack of nitrogen atoms on hydrocarbon compounds was directed toward the shielded carbon atoms, and not by the more usual abstraction of hydrogen atoms by radical attack. Atomic nitrogen does not appear to attack ammonia (Freeman and Winkler, 1955) (see discussion of active nitrogen above). Judging from its apparent long lifetime in decomposition products of hydrazine, it must not be very reactive toward the other N—H compounds and radicals.

V. "Active" Oxygen

A. INVESTIGATIONS OF ELECTRICALLY DISSOCIATED OXYGEN

"Active" oxygen can be produced by several types of electrical discharges (see Chapter 3). A number of earlier investigations have established oxygen atoms as the principal active species. For a summary of this work, see article by Linnett and Marsden (1956).

Recent mass spectrometric studies of active oxygen also indicate that large concentrations of electronically excited oxygen molecules are present. Foner and Hudson (1956b), using a free radical mass spectrometer previously described (1953a), observed a decrease of 0.93 volts in the appearance potential for mass 32 when the gas was electrically dissociated in a Wood's tube. Similar results were obtained in a microwave discharge. The decrease was attributed to the presence of oxygen molecules in the $^1\Delta_g$ state, which is separated from the ground state $^3\Sigma_g^-$ by 7882.39 cm^{-1} = 0.9772 ev (Herzberg and Herzberg, 1947). Concentration of excited molecules was estimated as 10 to 20 % based on comparison of initial slopes of the ion intensity curves with and without discharge. At high ionizing energy with the discharge on, the ion intensity of mass 32 was 30 % lower than with discharge off; it was indicated that this difference could be accounted for by the amount of atomic oxygen observed. This is about the highest reported degree of dissociation into atoms based on physical measurements, although similar results have been indicated by chemical reaction studies at low temperature (Ruehrwein and Hashman, 1959).

Active species in electrically dissociated oxygen have also been studied by Herron and Schiff (1958), using a free radical mass spectrometer. The change in the ionization curves between activated and nonactivated gas indicated the presence of ground state oxygen atoms (^{3}P) and excited O_2 molecules in the $^1\Delta_g$ state. Concentration of atomic oxygen was determined by an NO_2 titration method to be about 8 %, although considerably less than this amount was detected in the mass spectrometer. The concentration of excited O_2 molecules was estimated to be about 10 % on the basis of initial slopes of the ionization curves. In contrast to results of Foner and Hudson, Herron and Schiff observed an increase in mass 32 when the gas was discharged. This apparent anomaly was explained on the basis of the faster total rate of entry of oxygen into this mass spectrometer because of the more rapid effusion of atomic oxygen. Recombination of atoms inside the ion source resulted in a higher O_2 background during, rather than without, discharge. On the other hand, in the mass spectrometer used by Foner and Hudson, a molecular-beam sampling system was used in which the beam was mechanically modulated. Background signals were thus greatly reduced by applying phase detection to the ion signal, with the result that ambient gas had little effect on the output signal. Resolution of what at first seemed to be an inconsistency between the two investigations is thus seen to actually support the over-all interpretation. For complete consistency of the two investigations it is necessary to assume that $O_2(^1\Delta_g)$ has a lifetime considerably greater than the residence time in the mass spectrometer ion source.

B. Studies of Condensed Active Oxygen

Broida and Pellam (1954, 1955) condensed active oxygen on a Pyrex surface cooled with liquid helium. In contrast to the glows given off by condensed active nitrogen, no visible light was given off by the solid either during deposition or warm-up. Two variations of the deposition procedure gave different results. Using a clean, dry Pyrex surface as a collector, the active oxygen from a microwave discharge was deposited as thick layers of violet-colored solid matter. These layers peeled off, fell to the bottom of the apparatus, and melted at 54°K. When a small amount of water was first condensed in the apparatus and then followed by the active oxygen from a microwave discharge, a deposit having completely different characteristics was formed: a colorless, amorphous solid somewhat like thin, clear ice. When this solid was warmed above 4.2°K, but below 54°K, the transparent solid disappeared, leaving a white cloud in the chamber. Subsequently, the entire inner surface was seen to be coated with a purple solid, with the color density of the solid continuing to increase for about 30 seconds. From both procedures the material remaining above 54°K seemed to be an ozone-oxygen mixture.

In explanation of the clear solid (Broida and Pellam, 1954) obtained in the latter case, it was suggested that the solid contained free oxygen atoms which, on warming, reacted with O_2 to give ozone, O_3. This solid was studied further by Broida and Lutes (1956), who estimated, by interpretation of warming curves, that the O-atom content was about 1 %. Minkoff *et al.* (1959), in a calorimetric study, obtained heat releases from the condensed solid which correspond to 4 to 5 % free atoms if the heat release is due to the reaction

$$O + O = O_2 \quad \Delta H = -117 \text{ kcal/mole} \tag{40}$$

On the other hand, if heat release is due to the reaction

$$O + O_2 = O_3 \quad \Delta H = -25 \text{ kcal/mole} \tag{41}$$

then two to three times as many atoms would be indicated—say 11 to 12%. Doubt is cast on these results because similar heat effects were noted for argon, krypton, and nitrogen. In the case of nitrogen this corresponds to 4% atom content in the solid, a result that seems too high by two orders of magnitude, considering mainly the work of Fontana (1958, 1959, see above). Further doubt is cast because of the heat release that has been observed by Bolz et al. (1959) from undissociated gases of various kinds when condensed at low temperature—results that they trace to formation of metastable solids due to rapid crystallization.

The solids formed by condensation of active oxygen have been studied by Ruehrwein and Hashman (1959) and Ruehrwein et al. (1959), using Pyrex apparatus similar to that of Broida and Pellam (1954). It was shown that oxygen was about 40% dissociated into atoms in the microwave discharge used and that the amount of ozone formed was accounted for by the solid phase and surface reactions (40) and (41) above and with the ratio of the probability of reaction (40) to reaction (41) upon collision being equal to about 2.6. It was concluded that virtually all the ozone was formed during condensation of the gas. From the temperature (less than 20°K) at which the reactions occurred, it was estimated that the activation energy for ozone formation is no greater than about 1 kcal. Herron and Schiff (1958) have reported that ozone is not formed in the discharged gas at the pressures involved in these studies—less than 2 mm Hg.

It is interesting to note, however, that Ruehrwein and Hashman (1959) observed that certain of the deposits which yielded less than about 18% ozone were colorless and glassy, but became quite blue-colored on warming to 25°K. Color development was not reversible upon cooling and except in one case was gradual and uniform. In this one instance a sharp boundary between colored and colorless was observed to traverse the deposit from the top (warmer) to the lower end (cooler) at a rate of about 1 cm/sec. In spite of the deposit being spread over a considerable area, the heat released in this transformation was sufficient to raise the temperature about 2° momentarily from the starting temperature of 11°K. This temperature rise was detected with a thermocouple located outside the Pyrex containing vessel and thus may represent only a small fraction of the actual temperature rise of the solid oxygen. The behavior of this particular deposit, which yielded 12.9% ozone, is similar to the "catastrophic recombination" observed in nitrogen deposits by Fontana (see above). Except for the sudden development of color, this thermal behavior would not seem unreasonable for the 1% atoms stabilized in the solid as estimated by Broida and Lutes

(1956), and the higher estimates by Minkoff *et al.* (1959) would account qualitatively for both the total color change and the thermal effect. Harvey and Bass (1958), in infrared spectrometric studies of these solids, concluded that all ozone was formed on deposition, since no measurable increase in ozone concentration occurred on warm-up of the solids. From this they placed an upper limit of 3% on the concentration of oxygen atoms. In this latter study the conversion to ozone was not determined, so it is not known whether the observations apply to the low or the high side of the 18% limit noted by Ruehrwein and Hashman.

From the results available it does not seem possible to resolve the question of the upper concentration limit of oxygen atoms that can be stabilized in these solids. Most results can, however, be explained on the basis of a relatively low concentration—say 3% or less.

VI. Hydrogen-Oxygen System

A. HYDROGEN-OXYGEN RADICALS

The hydroxyl radical has been found spectroscopically in electrical discharges in water, hydrogen-oxygen mixtures, and in other gases (Lavin and Stewart, 1929; Bonhoeffer and Pearson, 1931; Oldenberg, 1935; Dousmanis *et al.*, 1955). This radical has been postulated as an intermediate in the oxygen-hydrogen and oxygen-hydrocarbon flame and other reactions (cf. Schumb *et al.*, 1955; Gaydon, 1957; Lewis and von Elbe, 1951). Hydroxyl radicals have been detected mass spectrometrically in hydrogen- and hydrocarbon-oxygen flames (Foner and Hudson, 1953a), in thermal decomposition of water vapor (Tsuchiya, 1954), and in the reaction of hydrogen atoms with molecular oxygen (Foner and Hudson, 1953b, 1955a; Ingold and Bryce, 1956).

Trapped hydroxyl and perhydroxyl radicals (HO_2) have been suggested to account for the electron spin resonance absorption of products condensed at 77°K from electrically dissociated water vapor (Livingston *et al.*, 1956). In similar studies, Gorbanev *et al.* (1957) attributed the paramagnetic resonance to the perhydroxyl radical. Spin resonance studies of gamma-irradiated ice at 77°K were interpreted in terms of both free hydrogen and hydroxyl radicals (Matheson and Smaller, 1955; Smaller *et al.*, 1954). On the other hand, Livingston and co-workers (1954, 1955) have obtained paramagnetic resonance spectra of gamma-irradiated aqueous solutions of mineral acids which have been interpreted to indicate the presence of free hydrogen atoms. Further study is needed to resolve the differences in these several results. The possible effect of trapped electrons and residual positively charged molecule ions in these systems does not seem to have been considered, although the primary effect of irradiation is expected to be ejection of electrons from neutral molecules. Trapping of the primary and secondary electrons at temperatures of the above experiments is expected in view of the appreciable electron affinity of water, reported by Gray and Waddington (1956) to be about 0.9 ev. Grossweiner and Matheson (1954) have postulated trapping of electrons to account for thermoluminescence of ice observed after X-irradiation at 77°K. They estimate a trap depth of 0.32 ev.

Hydroxyl radicals trapped in an argon matrix have been observed by means of ultraviolet (UV) spectroscopy by Robinson and McCarty (1958b). Argon and hydrazine containing water was passed through an electric discharge and the products condensed at 4.2°K. The UV spectrum contained a band at 32,090 cm^{-1} identified as the 0,0 band of OH.

The perhydroxyl radical (HO$_2$) has been shown to exist in the gas phase by Robertson (1954) and by Foner and Hudson (1953b). The radicals were formed by mixing electrically dissociated hydrogen with molecular oxygen; detection was by means of specially designed mass spectrometers (see above). Concentrations of the radicals were low—of the order of 1% or less—and thus insufficient to account directly for the relatively high yields of hydrogen peroxide obtainable from these mixtures by trapping at low temperatures, as discussed below. Foner and Hudson (1955b) later showed that relatively large concentrations of HO$_2$ could be made in the gas phase by reaction of hydrogen peroxide with electrically dissociated hydrogen peroxide or water. They proposed formation by the exothermic reaction

$$HO + H_2O_2 \rightarrow HO_2 + H_2O + 30 \text{ kcal} \tag{42}$$

the hydroxyl radicals being supplied in high concentration by the electrical discharge. In this study the H—O$_2$ bond dissociation energy was determined from measured appearance potentials and existing thermochemical data to be 47 ± 2 kcal per mole.

B. REACTIONS AT LOW TEMPERATURES

Jones and Winkler (1951) have pointed out the similarity of the products formed (and their behavior) from mixtures of atomic hydrogen and molecular oxygen, and from products of electrical dissociation of water vapor when either mixture is condensed in a cold trap at very low temperatures. A number of workers have investigated the H—O$_2$ reaction (Boehm and Bonhoeffer, 1926; Geib and Harteck, 1932; Rodebush et al., 1933, 1937; Badin, 1948; McKinley and Garvin, 1955; Hogg and Spice, 1957) and the reactions of dissociated water vapor (Rodebush et al., 1933, 1937, 1947; Jones and Winkler, 1951; Batzold et al., 1953; Frost and Oldenberg, 1936; Oldenberg and Rieke, 1939; Hogg and Spice, 1957) over ranges of temperature. In both systems the products condense as glassy solids when trapped at liquid nitrogen or hydrogen temperatures. When warmed to about $-115°C$ the "glass" appears to melt and begin to froth violently, evolving oxygen, and then to crystallize to an opaque solid, which, on further warming, melts to a concentrated aqueous solution of hydrogen peroxide. When the cold trap is maintained at higher temperatures, the yields of hydrogen peroxide and evolved oxygen decrease, until above about $-120°C$ only water is recovered, whereas oxygen evolution was not observed in products trapped above $-150°C$.

Chemical studies (Rodebush et al., 1947; Jones and Winkler, 1951; Badin, 1948; McKinley and Garvin, 1955) show conclusively that the hydrogen peroxide is not formed in the gas phase or by heterogeneous reactions in the pretrapping stage. In the atomic hydrogen-molecular oxygen reaction, studies with a free radical mass spectrometer also confirm that H_2O_2 is not formed in the gas phase at the usual low pressures (<1 mm) used in these studies (Foner and Hudson, 1955b). Although the perhydroxyl radical (HO_2), a probable intermediate, was detected at higher pressures (10 to 30 mm), its concentration ($\sim 1\%$) is not considered high enough to account for the high yields of peroxide usually obtained. The question then remains whether the reaction proceeds during condensation or if free atoms and radicals are trapped and subsequently react during warm-up when diffusion occurs in the solid. To account for "evolved oxygen" Geib and Harteck (1932) postulated that a metastable isomeric form of hydrogen peroxide (H_2O—O) existed which decomposed on warming. Jones and Winkler (1951) incorporated this hypothetical modification in their unified reaction mechanism for the atomic hydrogen-molecular oxygen reaction and the reactions of water dissociation products. Other workers have suggested that the evolution of oxygen is from decomposition of higher peroxides such as H_2O_3 (Ghormley, 1957) and H_2O_4 (Kruglyakova and Emanuel, 1952; Ghormley, 1957; Gorbanev et al., 1957), and other suggestions include stabilization and subsequent decomposition of radicals HO_2 and OH (McKinley and Garvin, 1955; Giguère, 1954; Ghormley, 1957; Allen and Stone, 1957). Several recent investigations to be discussed below indicate, however, that the radicals in the gas stream react during condensation at the surface to give the final stable products (H_2O_2, O_2, and H_2O) along with a relatively low concentration of trapped radicals.

Livingston et al. (1956) have observed the paramagnetic resonance absorption of solid deposits formed at $77°K$ from electrically dissociated water vapor. A similar absorption line was observed for the related substances (a) using D_2O instead of water, (b) using over 90% H_2O_2, (c) passing hydrogen through the electric discharge and blending in oxygen, and (d) passing ammonia through the discharge instead of hydrogen. All samples showed a pale yellow color which bleached at $-135°C$ concurrent with loss of paramagnetic resonance absorption. It was suggested tentatively that the spin resonance could be caused by hydroxyl and perhydroxyl radicals. The ratio of the number of molecules of hydrogen peroxide formed to unpaired electrons, i.e., the ratio of peroxide molecules to free radicals, was found to be 154 from water vapor, 91 from H_2O_2, and 125 from the blend of hydrogen atoms with oxygen molecules. These results suggest that most of the radicals react during condensation and further that the number of radicals present is not sufficient to account for subsequent evolution of oxygen. Thus reactions such as

$$OH + HO_2 \rightarrow H_2O + O_2 \qquad (43)$$

and

$$HO_2 + HO_2 \rightarrow H_2O_2 + O_2 \qquad (44)$$

cannot account for oxygen evolution, although these reactions may account in part for product formation during condensation. It will also be noted that the loss of paramagnetic resonance occurs at $-135°C$, appreciably below the temperature, $-115°C$, at which oxygen evolution occurs. In related studies of gamma-irradiated ice (Matheson and Smaller, 1955), paramagnetic resonance spectra were interpreted to indicate two radical species, H and OH, which, on annealing, disappeared at $-173°C$ and $-128°C$, respectively.

In a recent study of the reaction of atomic hydrogen with molecular oxygen at $4.2°K$, Edwards and Hashman (1957a, b) have obtained solids of empirical composition close to H_3O_4. On warming, these solids behaved similarly to those formed at higher temperatures, giving the evolved oxygen and leaving concentrated H_2O_2 on finally warming to room temperature. The apparatus and procedures used were similar to those given by Ruehrwein et al. (1959), important features being that all products were identified and mass balances obtained. The mole per cent of water in the water-peroxide mixtures obtained was between 3.5 and 30 in a series of nine runs, averaging 20 %, but in all cases too small to account for oxygen evolution via the reaction

$$H_2O\text{---}O \rightarrow H_2O + \tfrac{1}{2}O_2 \qquad (45)$$

as proposed earlier (Geib and Harteck, 1932), and in most experiments too small to be derived from the superoxides H_2O_3 and H_2O_4. The most convincing evidence for the nature of the solids came from "blank experiments" in which premixed hydrogen peroxide and oxygen were condensed in the reaction vessel at $4.2°K$. The solid deposits thus obtained were found to evolve oxygen at about $-115°C$, as had been found for the condensed products from dissociated water vapor and the reaction of hydrogen atoms with molecular oxygen. The mole ratio of evolved oxygen to hydrogen peroxide in these two experiments was 0.18 and 0.25 compared to values ranging from 0.29 to 0.37 (averaging 0.33) for nine runs involving reaction of electrically dissociated hydrogen with oxygen. Similar blank experiments were tried at higher temperatures (Edwards and Hashman, 1957b), specifically 77°K, under a variety of flow conditions. However, appreciable amounts of "evolvable" oxygen could not be formed. Foner and Hudson (1956a) have also attempted unsuccessfully to prepare the oxygen-evolving solid by condensing mixtures of hydrogen peroxide and oxygen at 77°K. In spite of these latter experiments, it seems that the best explanation of oxygen evolution from the various products is that it is evolved from a molecular complex with hydrogen peroxide—a complex not involving primary chemical bonds but being more of the nature of a clathrate-type compound or a mixed crystal (or glass). In reaching this conclusion, considerable weight is given to the fact that in the so-called blank experiments the oxygen evolution temperature is the same as in the experiments involving condensation of gas mixtures containing large concentrations of radicals.

Hogg and Spice (1957) have also studied the low temperature transitions of these solids, observing the stoichiometry, thermal effects, and glass formation, and have concluded that the glassy deposit is probably a mixed glass composed of water and hydrogen peroxide containing occluded oxygen. Devitrification of the glass at $-110°C$ then releases occluded oxygen. These workers make the plausible suggestion that the molecular oxygen which is occluded is that which is formed by reactions in the solid. This suggestion is apparently made on the basis that the amount of evolved oxygen in the products from water discharges increases with decreasing temperature, as indeed was indicated in the work of Jones and Winkler (1951).

Ruehrwein et al. (1959) found that at 4.2°K *less*, not more, evolved oxygen was obtained than at higher temperatures by other workers, i.e., when the inlet system was clean and coated with phosphoric acid. Conversions of 17 to 18 % of water to hydrogen peroxide were obtained under the conditions of the experiments, and only 2 to 4 % conversion to oxygen, all of which was of the "evolved" type. On the other hand, with the inlet system contaminated (presumably with Fe_2O_3), conversion to peroxide was down to 0.4 % of the water introduced and 31.4 % of this water was converted to molecular hydrogen and oxygen. Over half of this oxygen was later evolved on warm-up, thus demonstrating the occlusion of oxygen by water (20.89 mmoles H_2O + 0.11 mmole H_2O_2 + 2.77 mmoles O_2 in the solid condensate after vaporization of nonoccluded oxygen). The low yield of peroxide in this type of experiment and the improvements obtained by various successive treatments of the inlet tube suggest that the hydroxyl radicals are rapidly consumed by a heterogeneous reaction before impinging on the cold surface, e.g., by the reaction

$$2 \text{ OH} \xrightarrow{\text{wall}} \text{H}_2 + \text{O}_2 \qquad (46)$$

Rodebush *et al.* (1947) have previously proposed that this reaction is homogeneous. This reaction is thus shown to account for the presence of molecular oxygen in the water discharge products, and in view of the low mole ratio of evolved oxygen to hydrogen peroxide in the solids in the experiments at $4.2°K$, it seems that oxygen must not be formed during condensation of the solid. We are thus forced to look for an explanation of the occlusion other than by formation of oxygen by radical reactions in the solid. Sufficient data do not seem to be available to resolve this problem.

VII. Reaction of Hydrogen Atoms with Solid Olefins at Low Temperatures

An interesting new technique of low temperature chemistry has resulted from attempts to prepare free alkyl radicals in solids by exposure of olefins to H-atoms at low temperatures (Klein and Scheer 1958a, b, Scheer and Klein 1959). The experimental method used in studying these reactions consisted of exposing the olefin, uniformly deposited in a thin film (0.1 to 1 μ) at temperatures between 77 and $135°K$, to hydrogen atoms formed by dissociation on a hot ($1600°C$) tungsten ribbon filament. Reaction with the olefin was followed by observing pressure decrease in the 30 to 100 μ range with a thermocouple gauge.

Observations were made on the following olefins, listed in the order of decreasing rate of reaction with the H-atoms: propene, butene-1, isobutene, 3-methylbutene-1, 2-methylbutene-1, pentene-1, 3,3-dimethylbutene-1, and butadiene-1,3, with the last three having very slow rates. *trans*-Butene-2 and hexene-1 show no measurable rate at $77°K$, although the latter was found to react at higher temperatures (117 to $130°K$; Scheer and Klein, 1959).

The chemical processes which occur during the reaction can be inferred

TABLE II
ANALYSIS OF HYDROGENATION PRODUCTS

Olefin	Product	Percent
Propene	Propane	37
	Propene	58
	2,3-Dimethylbutane	5
Butene-1	*n*-Butane	56
	Butene (mostly butene-2)	40
	3,4-Dimethylhexane	4

from the mass spectrometric analyses of propene and butene-1 hydrogenations given in Table II. Hydrogen atom addition occurs at the terminal carbon of the olefin to form secondary alkyl radicals

$$H + -CH_2-CH=CH_2 \rightarrow -CH_2-\overset{\cdot}{C}H-CH_3 \tag{47}$$

The large spread in relative reactivities of the various olefins was attributed to minor variations in the activation energy for this step of the reaction— minor variations which are accentuated by the low temperature. Combination of two of the radicals from Eq. (47) leads to 2,3-dimethylbutane and 3,4-dimethylhexane in the propene and butene-1 hydrogenations, respectively, as in Eq. (48):

$$2(-CH_2-\overset{\cdot}{C}H-CH_3) \rightarrow \begin{array}{c} -CH_2-CH-CH_3 \\ | \\ -CH_2-CH-CH_3 \end{array} \tag{48}$$

Propane and n-butane in the products result from addition of a second hydrogen atom to the alkyl radicals, Eq. (49), and also from a radical-radical disproportionation reaction, Eq. (50). Olefin is reformed in the disproportionation giving rise to 2-butene in the case of 1-butene as the substrate.

$$-CH_2-\overset{\cdot}{C}H-CH_3 + H \rightarrow -CH_2-CH_2-CH_3 \tag{49}$$

$$2(-CH_2-CH-CH_3) \rightarrow -CH=CH-CH_3 + -CH_2-CH_2-CH_3 \tag{50}$$

In experiments with deuterium it was shown that isomerization of 1-butene to 2-butene occurs also by a mechanism involving hydrogen abstraction from a sec-butyl radical by a deuterium atom (Scheer and Klein, 1959), Eq. (51), to give the isomeric olefin and HD. With propene no HD was formed

$$CH_3-CH_2-\overset{\cdot}{C}H-CH_2D + D \rightarrow CH_3-CH=CH-CH_2D + HD \tag{51}$$

while in the case of 3-methyl-1-butene HD formed more readily than with 1-butene. This results in the following relative rates for abstraction of the hydrogen atom alpha to the free spin site: tertiary > secondary > primary.

Regarding stabilization of alkyl radicals formed at 77°K by the reactions discussed above, spin resonance measurements were made (Scheer and Klein, 1959), which, through lack of a detectable resonance signal, indicated that the steady state concentration of radicals must be less than 10^{12} spins, the detection limit of the instrument. This corresponded to a concentration of less than one free radical in 10^7 molecules.

REFERENCES

Allen, R. L., and Stone, F. S. (1957). *Nature* **180,** 752. Nature of the product condensed at low temperatures from dissociated peroxide vapor.

Badin, E. J. (1948). *J. Am. Chem. Soc.* **70**, 3651. (*Chem. Abstr.* **43**, 2079, 1949.) The reaction between atomic hydrogen and molecular oxygen at low pressures. Surface effects.

Bass, A. M., and Broida, H. P. (1956). *Phys. Rev.* **101**, 1740. Spectra emitted from solid nitrogen condensed at 4.2°K from a gas discharge.

Batzold, J. S., Luner, C., and Winkler, C. A. (1953). *Can. J. Chem.* **31**, 262. Reactions in dissociated peroxide vapor.

Becker, E. D., Pimentel, G. C., and Van Thiel, M. V. (1957). *J. Chem. Phys.* **26**, 145. Matrix isolation studies: infrared spectra of intermediate species in the photolysis of hydrazoic acid.

Beckley, H. D. (1958). *Angew. Chem.* **70**, 327. Mass spectrometric study of the reactions and properties of free radicals and atoms.

Benson, J. M. (1952). *J. Appl. Phys.* **23**, 757. Measurements of the physical properties of active nitrogen.

Berkowitz, J., Chupka, W. A., and Kistiakowsky, G. B. (1956). *J. Chem. Phys.* **25**, 457. Mass spectrometric study of the kinetics of nitrogen afterglow.

Boehm, E., and Bonhoeffer, K. F. (1926). *Z. physik. Chem. (Leipzig)* **119**, 385. The gaseous reactions of active hydrogen.

Bolz, L. H., Mauer, F. A., and Peiser, H. S. (1959). *J. Chem. Phys.* **30**, 349. Low temperature X-ray studies of Rice's blue material.

Bonhoeffer, K. F., and Pearson, T. G. (1931). *Z. physik. Chem. (Leipzig)* **B14**, 1. The capability of existence of the free hydroxyl radical.

Broida, H. P., and Lutes, O. S. (1956). *J. Chem. Phys.* **24**, 484. Abundance of free atoms in solid nitrogen condensed at 4.2°K from a gas discharge.

Broida, H. P., and Pellam, J. R. (1954). *Phys. Rev.* **95**, 845. (*Chem. Abstr.* **48**, 11932, 1954). Phosphorescence of atoms and molecules of solid nitrogen at 4.2°K.

Broida, H. P., and Pellam, J. R. (1955). *J. Chem. Phys.* **23**, 409L. A note on the preparation of solid ozone and atomic oxygen.

Broida, H. P., and Peyron, M. (1958). *J. Chem. Phys.* **28**, 725. Evaporation of active species trapped in a solid condensed from 'discharged' nitrogen.

deKlerk, D., and Hudson, R. P. (1954). *J. Research Natl. Bur. Standards* **53**, 173. Installation for adiabatic demagnetization experiments at the National Bureau of Standards.

Dixon, J. K. (1933). *Phys. Rev.* **43**, 711. The ultraviolet absorption bands of ammonia.

Dousmanis, G. C., Sanders, R. M., and Townes, C. H. (1955). *Phys. Rev.* **100**, 1735. Microwave spectra of the free radicals OH and OD.

Dows, D. A., Pimentel, G. C., and Whittle, E. (1955). *J. Chem. Phys.* **23**, 1606. (*Chem. Abstr.* **50**, 676, 1956.) Infrared spectra of intermediate species in the formation of ammonium azide from hydrazoic acid.

Dressler, K. (1959). *J. Chem. Phys.* **30**, 1621. Absorption spectrum of vibrationally excited N_2 in active nitrogen.

Duncan, A. B. F. (1935). *Phys. Rev.* **47**, 886. The ultraviolet absorption spectrum of ND_3.

Dunning, W. J. (1955). *Quart. Revs. (London)* **9**, 23. The application of mass spectrometry to chemical problems.

Edwards, J. W., and Hashman, J. S. (1957a). "Reaction of Oxygen with Atomic Hydrogen," *Abstr. of Papers of Am. Chem. Soc., 132nd Meeting, New York, 1957*, p. 42 S.

Edwards, J. W., and Hashman, J. S. (1957b). Unpublished results.

Eltenton, G. C. (1942). *J. Chem. Phys.* **10**, 403. The detection of free radicals by means of a mass spectrometer.

Eltenton, G. C. (1947). *J. Chem. Phys.* **15**, 455. The study of reaction intermediates by means of a mass spectrometer. I. Apparatus and method.

Evans, H. G. V., and Winkler, C. A. (1956). *Can. J. Chem.* **34**, 1217. The reactive components in active nitrogen and the role of spin conservation in active nitrogen reactions.

Evans, H. G. V., Freeman, G. R., and Winkler, C. A. (1956). *Can. J. Chem.* **34**, 1271. The reactions of active nitrogen with organic molecules.

Field, F. H., and Franklin, J. L. (1957). "Electron Impact Phenomena and the Properties of Gaseous Ions." Academic Press, New York.

Foner, S. N., and Hudson, R. L. (1953a). *J. Chem. Phys.* **21**, 1374. The detection of atoms and free radicals in flames by mass spectrometric techniques.

Foner, S. N., and Hudson, R. L. (1953b). *J. Chem. Phys.* **21**, 1608. Detection of the HO₂ radical by mass spectrometry.

Foner, S. N., and Hudson, R. L. (1955a). *J. Chem. Phys.* **23**, 1974. (*Chem. Abstr.* **50**, 2340, 1956.) OH, HO₂ and H₂O₂ production in the reaction of atomic hydrogen with molecular oxygen.

Foner, S. N., and Hudson, R. L. (1955b). *J. Chem. Phys.* **23**, 1364. Ionization potential of the free HO₂ radical and the H—O₂ bond dissociation energy.

Foner, S. N., and Hudson, R. L. (1956a). "Mass Spectrometric Studies of HO₂ and Some Related Free Radicals," Symposium on Free Radicals, Laval Univ., Quebec. Unpublished abstract.

Foner, S. N., and Hudson, R. L. (1956b). *J. Chem. Phys.* **25**, 601. Metastable oxygen molecules produced by electrical discharges.

Foner, S. N., and Hudson, R. L. (1958a). *J. Chem. Phys.* **28**, 719. Diimide—identification and study by mass spectrometry.

Foner, S. N., and Hudson, R. L. (1958b). *J. Chem. Phys.* **29**, 442. Mass spectrometric detection of triazene and tetrazene and studies of the free radicals NH₂ and N₂H₃.

Fontana, B. J. (1958). *J. Appl. Phys.* **29**, 1667. Thermometric study of the frozen products from the nitrogen microwave discharge.

Fontana, B. J. (1959). *J. Chem. Phys.* **31**, 148. Magnetic study of the frozen products from the nitrogen microwave discharge.

Franklin, J. L., Dibeler, V. H., Reese, R. M., and Krauss, M. (1958a). *J. Am. Chem. Soc.* **80**, 298. Ionization and dissociation of hydrazoic acid and methyl azide by electron impact.

Franklin, J. L., Herron, J. T., Bradt, P., and Dibeler, V. H. (1958b). *J. Am. Chem. Soc.* **80**, 6188. Mass spectrometric study of the decomposition of hydrazoic acid by the electric discharge.

Freeman, G. R., and Winkler, C. A. (1955). *J. Phys. Chem.* **59**, 371. The reaction of active nitrogen with ammonia.

Frost, A. A., and Oldenberg, O. (1936). *J. Chem. Phys.* **4**, 642. Kinetics of OH radicals as determined by their absorption spectrum. I. The electric discharge through water vapor.

Gaydon, A. G. (1947). "Dissociation Energies and Spectra of Diatomic Molecules." Wiley, New York.

Gaydon, A. G. (1957). "The Spectroscopy of Flames." Chapman & Hall, London.

Geib, K. H., and Harteck, P. (1932). *Ber.* **B65**, 1551. (*Chem. Abstr.* **26**, 5866, 1932.) A new form of H₂O₂ .

Ghormley, J. A. (1957). *J. Am. Chem. Soc.* **79**, 1862. Warming curves for the condensed product of dissociated water vapor and for hydrogen peroxide glass.

Giguère, P. A. (1954). *J. Chem. Phys.* **22**, 2085. (*Chem. Abstr.* **49**, 3712, 1956.) Spectroscopic evidence for stabilized HO_2 radicals.

Gorbanev, A. I., Kaitmazov, S. D., Prokhorov, A. M., and Tsentsiper, A. B. (1957). *Zhur. Fiz. Khim.* **31**, 515. (*Chem. Abstr.* **51**, 17430, 1957.) Paramagnetic resonance of the products formed at a low temperature for the dissociation of H_2O, H_2O_2, and D_2O vapors in a glow discharge.

Gray, P., and Waddington, T. C. (1956). *Proc. Roy. Soc.* **A235**, 481. Thermochemistry and reactivity of the azides II. Lattice energies of ionic azides, electron affinity and heat of formation of the azide radical and related properties.

Grossweiner, L. I., and Matheson, M. S. (1954). *J. Chem. Phys.* **22**, 1514. Fluorescence and thermoluminescence of ice.

Harvey, K. B., and Bass, A. M. (1958). *J. Mol. Spectroscopy* **2**, 405. Infrared absorption of oxygen discharge products and ozone at 4°K.

Harvey, K. B., and Brown, H. W. (1959). *J. Chim. phys.* **56**, 745. Étude aux infrarouges de certains solides condensés a porter de décharges en phase gazeuse.

Herron, J. T., and Schiff, H. I. (1958). *Can. J. Chem.* **36**, 1159. A mass spectrometric study of normal oxygen and oxygen subjected to electrical discharge.

Herron, J. T., Franklin, J. L., Bradt, P., and Dibeler, V. (1958). *J. Chem. Phys.* **29**, 230. Kinetics of nitrogen atom recombination.

Herzberg, L., and Herzberg, G. (1947). *Astrophys. J.* **105**, 353. Fine structure of the infrared atmospheric oxygen bands.

Herzfeld, C. M. (1957). *Phys. Rev.* **107**, 1239. Theory of the forbidden transitions of nitrogen atoms trapped in solids.

Herzfeld, C. M., and Broida, H. P. (1956). *Phys. Rev.* **101**, 606. Interpretation of spectra of atoms and molecules in solid nitrogen condensed at 4.2°K.

Hipple, J. A., and Stevenson, D. P. (1943). *Phys. Rev.* **63**, 121. Ionization and dissociation by electron impact: the methyl and ethyl radicals.

Hörl, E. (1959). *J. Mol. Spectroscopy* **3**, 425. Light emission from solid nitrogen during and after electron bombardment.

Hogg, M. A. P., and Spice, J. E. (1957). *J. Chem. Soc.* p. 3971. (*Chem. Abstr.* **52**, 846, 1958.) Nature of the low-temperature transition in hydrogen peroxide prepared by discharge-tube methods.

Ingold, K. U., and Bryce, W. A. (1956). *J. Chem. Phys.* **24**, 360. Mass-spectrometric investigations of the hydrogen-oxygen and methyl-oxygen reactions.

Jackson, D. S., and Schiff, H. I. (1955). *J. Chem. Phys.* **23**, 2333. Mass spectrometric investigation of active nitrogen.

Jennings, K. R., and Linnett, J. W. (1958). *Quart. Revs. (London)* **12**, 116. Active nitrogen.

Jones, R. A., and Winkler, C. A. (1951). *Can. J. Chem.* **29**, 1010. (*Chem. Abstr.* **46**, 2891, 1952.) Reactions in dissociated water vapor.

Kaufman, F., and Kelso, J. R. (1957). *J. Chem. Phys.* **27**, 1209. Excitation of nitric oxide by active nitrogen.

Kaufman, F., and Kelso, J. R. (1958). *J. Chem. Phys.* **28**, 510. Vibrationally excited ground-state nitrogen in active nitrogen.

Kistiakowsky, G. B., and Volpi, G. G. (1957). *J. Chem. Phys.* **27**, 1141. Reactions of nitrogen atoms. I. Oxygen and oxides of nitrogen.

Kistiakowsky, G. B., and Volpi, G. G. (1958). *J. Chem. Phys.* **28**, 665. Reactions of nitrogen atoms. II. H_2, CO, NH_3, NO, and NO_2.

Kistiakowsky, G. B., and Warneck, P. (1957). *J. Chem. Phys.* **27**, 1417. (*Chem. Abstr.* **52**, 5120, 1958.) Lewis-Rayleigh nitrogen afterglow.

Klein, R., and Scheer, M. D. (1958a). *J. Am. Chem. Soc.* **80,** 1007. The addition of hydrogen atoms to solid olefins at −195°.

Klein, R., and Scheer, M. D. (1958b). *J. Phys. Chem.* **62,** 1011. The reaction of H atoms with solid olefins at −195°.

Kruglyakova, K. E., and Emanuel, N. M. (1952). *Doklady Akad. Nauk S.S.S.R.* **83,** 593. The chain mechanism of the decomposition of hydrogen peroxide, and the existence of HO₂ radicals and of A. N. Bakh's higher hydrogen peroxide.

Lavin, G. I., and Stewart, F. B. (1929). *Proc. Natl. Acad. Sci. U. S.* **15,** 829. (*Chem. Abstr.* **24,** 1034, 1930.) Production of hydroxyl by the water-vapor discharge.

Lewis, B., and von Elbe, G. (1951). "Combustion, Flames, and Explosions of Gases." Academic Press, New York.

Lewis, P. (1900). *Ann. phys.* **2,** 459; *J. Chem. Soc.* (Abstr.) **78,** Part II, p. 702. Fluorescence and afterglow accompanying an electric discharge in nitrogen.

Lichten, W. (1957). *J. Chem. Phys.* **26,** 306. Lifetime measurements of metastable states in molecular nitrogen.

Linnett, J. W., and Marsden, D. G. H. (1956). *Proc. Roy. Soc.* **A234,** 489. The kinetics of the recombination of oxygen atoms at a glass surface.

Lipscomb, F. J., Norrish, R. G. W., and Thrush, B. A. (1956). *Proc. Roy. Soc.* **A233,** 455. The study of energy transfer by kinetic spectroscopy. I. The production of vibrationally excited oxygen.

Livingston, R., Zeldes, H., and Taylor, E. H. (1954). *Phys. Rev.* **94,** 725. (*Chem. Abstr.* **48,** 8041, 1954.) Atomic hydrogen hyperfine structure in irradiated acids.

Livingston, R., Zeldes, H., and Taylor, E. H. (1955). *Discussions Faraday Soc.* **19,** 166. Paramagnetic resonance studies of atomic hydrogen produced by ionizing radiation.

Livingston, R., Ghormley, J., and Zeldes, H. (1956). *J. Chem. Phys.* **24,** 483. Paramagnetic resonance observations on the condensed products of electric discharges through water vapor and related substances.

Lossing, F. P. (1957). *Ann. N. Y. Acad. Sci.* **67,** 499. Mass spectrometry of free radicals.

Lossing, F. P., and Tickner, A. W. (1952). *J. Chem. Phys.* **20,** 907. Free radicals by mass spectrometry. I. The measurement of methyl-radical concentrations.

McCoubrey, J. C., and McGrath, W. D. (1957). *Quart. Revs. (London)* **11,** 87. Energy transfer in gaseous collisions.

McKinley, J. D., Jr., and Garvin, D. (1955). *J. Am. Chem. Soc.* **77,** 5802. (*Chem. Abstr.* **50,** 3855, 1956.) The reactions of atomic hydrogen with ozone and with oxygen.

McLennan, J. C., and Shrum, G. M. (1924). *Proc. Roy. Soc.* **A106,** 138. On the luminescence of nitrogen, argon, and other condensed gases at very low temperatures.

McLennan, J. C., Ireton, H. J. C., and Samson, E. W. (1928). *Proc. Roy. Soc.* **A120,** 303. On the luminescence of solid N under cathode ray bombardment.

McLennan, J. C., Samson, E. W., and Ireton, H. J. C. (1929). *Trans. Roy. Soc. Can.* **23,** 25. (*Chem. Abstr.* **24,** 785, 1930.) Phosphorescence of solid A irradiated with cathode rays.

Mador, I. L., and Williams, M. C. (1954). *J. Chem. Phys.* **22,** 1627. (*Chem. Abstr.* **49,** 42, 1955.) Stabilization of free radicals from the decomposition of hydrazoic acid.

Matheson, M. S., and Smaller, B. (1955). *J. Chem. Phys.* **23,** 521. (*Chem. Abstr.* **49,** 10054, 1955.) Paramagnetic species in γ-irradiated ice.

Milligan, D. E., Brown, H. W., and Pimentel, G. (1956). *J. Chem. Phys.* **25,** 1080. Infrared absorption by the N_3 radical.

Minkoff, G. J., and Scherber, F. I. (1958). *J. Chem. Phys.* **28,** 992. Energy release from discharged monatomic gases trapped at 4°K.

Minkoff, G. J., Scherber, F. I., and Gallagher, J. S. (1959). *J. Chem. Phys.* **30,** 753. Abundance of active species trapped at 4.2°K from gaseous discharges.

Mitra, S. K. (1945). "Active Nitrogen—A New Theory." Indian Assoc. for Cultivation of Science, Calcutta, India.

Moore, C. E. (1949). *Natl. Bur. Standards (U. S.), Circ.* 467-I. Atomic energy levels as derived from the analyses of optical spectra.

Muschlitz, E. E., and Goodman, L. (1953). *J. Chem. Phys.* **21,** 2213. Lifetime of the $^3\Sigma_u^+$ state of nitrogen.

Oldenberg, O. (1935). *J. Chem. Phys.* **3,** 266. The lifetime of free hydroxyl.

Oldenberg, O., and Rieke, F. F. (1939). *J. Chem. Phys.* **7,** 485. Kinetics of OH radicals as determined by their absorption spectrum. V. A spectroscopic determination of a rate constant.

Peiser, H. S., Mauer, F. A., and Bolz, L. H. (1958). Private communication. X-ray diffraction studies of simple low temperature condensates.

Peyron, M., and Broida, H. P. (1959). *J. Chem. Phys.* **30,** 139. Spectra emitted from solid nitrogen condensed at very low temperatures from a gas discharge.

Rayleigh. (1935). *Proc. Roy. Soc.* **A151,** 567. Active nitrogen of long duration, law of decay, and of increased brightness on compression.

Rayleigh. (1940a). *Proc. Roy. Soc.* **A176,** 1. New studies on active nitrogen. I. Brightness of the afterglow under varied conditions of concentrations and temperature.

Rayleigh. (1940b). *Proc. Roy. Soc.* **A176,** 16. New studies on active nitrogen. II. Incandescence of metals in active nitrogen, and quantitative estimates of the energy liberated.

Rayleigh. (1942). *Proc. Roy. Soc.* **A180,** 123. Further studies on active nitrogen III and IV.

Rayleigh. (1947). *Proc. Roy. Soc.* **A189,** 296. The surprising amount of energy which can be collected from gases after the electric discharge has passed.

Rice, F. O., and Freamo, M. (1951). *J. Am. Chem. Soc.* **73,** 5529. (*Chem. Abstr.* **46,** 6026, 1952.) The imine radical.

Rice, F. O., and Freamo, M. (1953). *J. Am. Chem. Soc.* **75,** 548. (*Chem. Abstr.* **47,** 5291 1953.) The formation of the imine radical in the electrical discharge.

Rice, F. O., and Grelecki, C. (1957). *J. Am. Chem. Soc.* **79,** 1880. The imine radical.

Rice, F. O., and Scherber, F. (1955). *J. Am. Chem. Soc.* **77,** 291. The hydrazino radical and tetrazane.

Robertson, A. J. B. (1954). "Applied Mass Spectrometry." Institute of Petroleum, London.

Robinson, G. W., and McCarty, M., Jr. (1958a). *J. Chem. Phys.* **28,** 349. Electronic spectra of free radicals at 4°K—NH_2.

Robinson, G. W., and McCarty, M., Jr. (1958b). *J. Chem. Phys.* **28,** 350L. Electronic spectra of free radicals at 4°K—HNO, NH, and OH.

Rodebush, W. H. (1937). *J. Phys. Chem.* **41,** 283. (*Chem. Abstr.* **31,** 4571, 1937.) Reactions of oxygen and hydrogen at low pressures.

Rodebush, W. H., and Wahl, M. H. (1933). *J. Chem. Phys.* **1,** 696. The reactions of the hydroxyl radical in the electrodeless discharge in water vapor.

Rodebush, W. H., Wende, C. W. J., and Campbell, R. W. (1937). *J. Am. Chem. Soc.* **59,** 1924. (*Chem. Abstr.* **31,** 83894, 1937.) Formation of water and hydrogen peroxides at low pressures.

Rodebush, W. H., Keizer, C. R., McKee, F. S., and Quagliano, J. V. (1947). *J. Am. Chem. Soc.* **69**, 538. (*Chem. Abstr.* **41**, 3350, 1947.) The reactions of the hydroxyl radical.

Ruehrwein, R. A., and Hashman, J. S. (1959). *J. Chem. Phys.* **30**, 823. Formation of ozone from atomic oxygen at low temperatures.

Ruehrwein, R. A., Hashman, J. S., and Edwards, J. W. (1959). *J. Chem. Phys.* (to be published). Chemical reactions of free radicals at low temperature.

Scheer, M. D., and Klein, R. (1959). *J. Phys. Chem.* **63**, 1517. The double bond isomerization of olefins by hydrogen atoms at $-195°$.

Schoen, L., and Rebbert, R. E. (1959). *J. Mol. Spectroscopy* **3**, 417. Electrical discharge induced luminescence of solids at low temperatures.

Schumb, W. C., Satterfield, C. N., and Wentworth, R. L. (1955). "Hydrogen Peroxide." Reinhold, New York.

Smaller, B., Matheson, M. S., and Yasaitis, E. L. (1954). *Phys. Rev.* **94**, 202L. (*Chem. Abstr.* **48**, 7457, 1954.) Paramagnetic resonance in irradiated ice.

Stevenson, D. P. (1951). *Discussions Faraday Soc.* **10**, 35. Ionization and dissociation by electronic impact.

Strutt, R. J. (1911a). *Proc. Roy. Soc.* **A85**, 219. A chemically active modification of nitrogen, produced by the electric discharge. I.

Strutt, R. J. (1911b). *Proc. Roy. Soc.* **A86**, 56. A chemically active modification of nitrogen, produced by the electric discharge. II.

Strutt, R. J. (1912). *Proc. Roy. Soc.* **A87**, 179. A chemically active modification of nitrogen, produced by the electric discharge. IV.

Strutt, R. J. (1915). *Proc. Roy. Soc.* **A91**, 303. A chemically active modification of nitrogen, produced by the electric discharge. VI.

Tanaka, Y., Jursa, A., and LeBlanc, F. (1957). *In* "The Threshold of Space" (M. Zelikoff, ed.), pp. 89–93. Pergamon Press, New York. Vacuum ultraviolet spectra of the afterglows of pure N_2 and a mixture of N_2 and O_2.

Thrush, B. A. (1956). *Proc. Roy. Soc.* **A235**, 143. The detection of free radicals in the high intensity photolysis of hydrogen azide.

Tsuchiya, T. (1954). *J. Chem. Phys.* **22**, 1784. Mass-spectrometric detection of free OH radicals in the thermal decomposition products of H_2O vapor.

Vegard, L. (1924a). *Nature* **113**, 716. (*Chem. Abstr.* **18**, 2287, 1924.) The auroral spectrum and the upper atmosphere.

Vegard, L. (1924b). *Nature* **114**, 357. (*Chem. Abstr.* **18**, 3320, 1925.) The light emitted from solidified gases and its relation to cosmic phenomena.

Vegard, L. (1926). *Ann. phys.* **79**, 377. (*Chem. Abstr.* **20**, 2283, 1926.) The luminescence of solidified gases and its relation to cosmic processes.

Vegard, L. (1930). *Ann. phys.* **6**, 487. (*Chem. Abstr.* **25**, 29, 1931.) The spectra of solidified gases and their theoretical atomic meaning.

Vegard, L., and Stensholt, S. (1935). *Skrifter Norske Videnskaps-Akad. Oslo. I. Mat.-Naturv. Kl. No.* **9**, pp. 1–57. The properties of the ϵ-system (Vegard bands) derived from new and previous measurements.

Wall, L. A., Brown, D. W., and Florin, R. E. (1959). *J. Phys. Chem.* **63**, 1762, Atoms and free radicals by gamma irradiation at $4.2°K$.

Worley, R. E. (1948). *Phys. Rev.* **73**, 531. On the absorption spectrum of active nitrogen.

9. X-Ray Diffraction Studies and Their Crystal-Chemical Implications

H. S. PEISER

National Bureau of Standards, Washington, D. C.

I. Expectations from X-Ray Diffraction Studies

Reviewing scientific progress during the last few decades one must be impressed by the diverse and basic contributions made by X-ray crystallographic investigations. Solids have been studied by X-ray diffraction methods in chemistry, mineralogy, metallurgy, and engineering. These techniques have elucidated the phase structure and texture of technological materials from glasses to gems, from paints to polymers. They have thrown light on the chemical constitution of inorganic, organic, and biological compounds. In all these and other fields crystallographic evidence has been in the forefront. Was it not a foregone conclusion that X-ray crystallography would be equally useful in the study of low temperature solids containing trapped radicals? This was not quite so, unfortunately; the methods in this instance are decidedly limited.

Broida, Mauer, McMurdie, and other colleagues in the Free Radicals Research Program at the National Bureau of Standards, who planned in 1956 for X-ray work to be undertaken, had sober and accurate understanding of these limitations:

(1) If the concentration of the free radicals was likely to be less than 1%, it was virtually certain their direct influence on the X-ray diffraction effects could not be observed. X-ray diffraction data are insensitive to minor constituents. Unless one knows what to look for, less than 5% phase constituents or solid-solution replacements are commonly missed.

(2) The most precise and useful X-ray data are generally based on single-crystal diffraction studies. There seemed to be little hope, however, of growing single crystals containing trapped free radicals and no such work was planned. In the light of later work, however (Wall et al., 1958), single crystals might well be studied, during and after prolonged X- or gamma-ray exposure.

(3) The study of polycrystalline aggregates by X-ray "powder" methods alone seemed feasible. Best experimental results depend on the crystals being large compared with interatomic distances and the crystal strain being small. However, material with high strain and very small crystal size must be expected in vapor-deposited low temperature solids. Virtually amorphous solids, it was thought, might be regularly encountered.

(4) Hydrogen atoms, perhaps the most interesting of the free radicals, scatter X-rays but exceedingly weakly. This is because they have only a single, spread-out extranuclear electron. Only in the best X-ray work on well-crystallized solids could the effects of high concentrations of hydrogen atoms just be seen. Neutron diffraction is a more promising technique, but with presently available neutron fluxes an impossibly large sample would be needed.

(5) The determination of orientation textures (Section III, A) would be limited because of the need to choose X-ray diffractometry, rather than X-ray-camera photography, for the recording of the transient phenomena found in solids containing trapped radicals.

So much for the negative side of the picture. There was, however, the positive and substantial hope of elucidating the nature of the matrix structures. Under what conditions are condensed vapors polycrystalline, or are all these solids amorphous? Further questions immediately come to mind. Do amorphous solids crystallize on warm-up? Which are the crystalline phases formed, and what is their characteristic atomic or molecular pattern? Are there observable solid-solution effects? Conventional phase-diagram information even for binary mixtures is badly needed at low temperatures. What are the structural influences of deposition rate and temperature of the gas and of the condensation surface? What is the thickness of deposits? How efficient is the deposition process? If there is preferred orientation at surfaces, what kind of atomic planes are exposed to the impinging radicals? All these are relevant questions which can perhaps be answered by crystallographic study.

II. The X-Ray Equipment

The reader may have noticed that the low specimen temperature was not held a significant limitation to X-ray diffractometer studies. Modern Dewar designs, as described in Chapter 5, certainly justified optimism. The entry of the primary X-ray beam and the exit of diffracted beams over a wide arc, for instance, would have been a formidable obstacle a few years back. Nowadays, preformed 0.025-inch beryllium sheet windows are available which are capable of supporting a vacuum over large areas. The X-ray Dewar vessel that has served well is described fully elsewhere (Black et al., 1958). A summary of its chief characteristics will suffice here with reference to Figs. 1 and 2.

(1) The entire cold cell fits in place of the conventional specimen holder on an X-ray diffractometer with vertical goniometer axis (Fig. 2).

(2) Diffracted beams can be recorded by an electronic counter up to about 158° (2θ, being the angle of deviation). Alternative chart recording from a rate-meter circuit (Fig. 2, 5) and a scaling unit are provided for X-ray intensity determinations. Plots are made against time or diffraction angle, the precision being comparable with that of good room temperature equipment (Fig. 3).

(3) The flat specimen holder, made of gold-plated copper, is arranged for parafocusing (Peiser et al., 1955). The gold diffraction lines serve for reference and specimen-thickness determinations as mass per unit area. The gold deposit shows little preferred orientation.

(4) Wide holes within the copper specimen-holder block (Fig. 1, 17) are filled with coolant—usually liquid helium—from the inner Dewar. A fast vacuum pump can be used to pump on the coolants.

(5) The outer Dewar—usually liquid nitrogen or hydrogen—shields the entire inner Dewar except around the lower portion of the specimen holder, which is surrounded by a radiation shield (Fig. 1, 15). This is broken to allow visual observation through two ports (Fig. 2, 7). Where

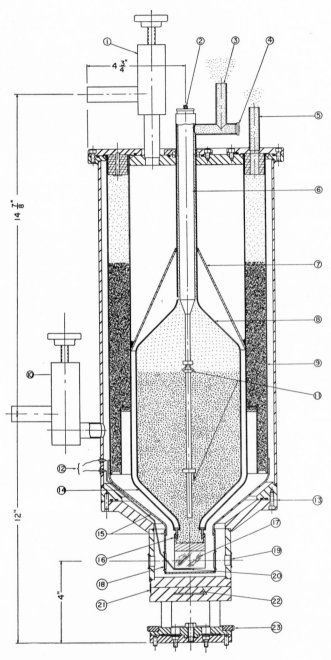

FIG. 1.

the X-ray beams pass to and from the specimen, the shield consists of only thin nickel foil (Fig. 1, *18*) which "filters" the Cu KαX-radiation used.

(6) Temperature measurement can be made by thermocouples (Chapter 5) embedded in the specimen and copper block. Arrangements exist for chart recording. The cold-junction is fixed to and rotates with the Dewar, but is not shown in either Figs. 1 or 2.

(7) Pressures can also be chart-recorded from a Philips gauge (not shown), connected to the space between the outer Dewar and the outer shell. This is connected to the volume surrounding the specimen. The inner Dewar, however, is surrounded mostly by a separate vacuum space. This ensures that vaporization of part of the specimen will not quickly lead to loss of coolant in the inner Dewar. The wall (Fig. 1, *14*) dividing the vacua constitutes the largest heat leak into the inner Dewar. When hydrogen is used as coolant in the outer Dewar the loss of helium from this source is, however, quite small.

(8) Vapors are introduced through a straight silica tube (Fig. 2, *6*) pointing directly at the cold specimen-holder surface. A waveguide (not shown) for microwave discharge can be fitted over the tube within 3 inches of that surface. Vacuum pumps and cold traps are conventional, as are the vapor flow gauges (Chapter 5) and ancillary X-ray equipment (Klug and Alexander, 1954; Peiser *et al.*, 1955).

"Powder" patterns can thus be observed at low temperatures with virtually no additional limitations. It is true that a Dewar for the crystallographic investigation of the very low θ region has not yet been built. Transmitted X-rays must be used for this purpose, so that the specimen holder itself would have to be transparent to X-rays, which presents an additional cooling problem. As this has actually been solved for the far more difficult electron-diffraction technique (Hörl and Marton, 1958), it is no exaggeration to say that "powder"-diffraction data at low temperatures can be as detailed and accurate as those from room temperature specimens.

FIG. 1. Section through the X-ray diffraction cryostat (after Black *et al.*, 1958). *1*, Valve for evacuating helium Dewar; *2*, helium fill-tube stopper; *3*, helium vent; *4*, electrical connector for helium-level indicator; *5*, nitrogen vent or fill tube (one not shown); *6*, helium fill tube; *7*, upper radiation shield; *8*, helium vessel; *9*, nitrogen vessel; *10*, valve for evacuating specimen chamber and nitrogen Dewar; *11*, resistors for helium-level indicator (two); *12*, thermocouple leads; *13*, demountable flange; *14*, inner vacuum wall; *15*, lower radiation shield; *16*, brass ring and soft-solder joint; *17*, copper block; *18*, nickel-foil X-ray windows 0.00035 inch thick; *19*, beryllium X-ray windows; *20*, thermocouple junction; *21*, translational-adjustment vernier; *22*, translational-adjustment slide; *23*, rotational-adjustment worm gear.

FIG. 2. View of the X-ray diffractometer equipped with a liquid-helium specimen cryostat (after Black *et al.*, 1958). *1*, X-Ray source slit; *2*, beryllium window; *3*, detector slit; *4*, graduated arc for measuring diffraction angles; *5*, diffractometer recorder; *6*, tube for introducing specimen; *7*, viewing port (two); *8*, supporting cables.

III. From Basic Theory

Familiarity with the concept of crystals having regularly repeating and three-dimensionally aligned atomic patterns is assumed. Crystallographic notation, symmetry groups, lattice theory, diffraction geometry, Fourier transforms, and a discussion of radial-distribution functions will be avoided in this chapter. For these the reader is referred to the texts by Klug and Alexander (1954), as well as by Peiser *et al.*, (1955).

Just a few other topics must be referred to here because of special relevance or inadequate treatment elsewhere.

A. POLYCRYSTALLINE AGGREGATES AND PREFERRED ORIENTATION

In elementary theory the perfect crystal is treated as having an infinitely repeating atomic pattern. It is self-evident that all real crystals have boundaries. Very perfect crystals do exist in which the lattice order—

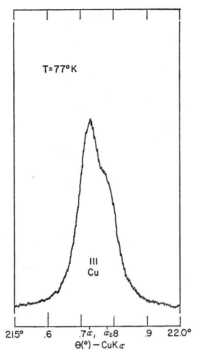

FIG. 3. Portion of low temperature diffractometer plot for copper to demonstrate resolving power under normal experimental conditions (note partial resolution even at low θ-values of α_1 and α_2 spectral lines differing in wavelength by only 0.004 A or 0.3%). After Black *et al.* (1958).

within 1 or 2 minutes of arc—persists three-dimensionally over macroscopic dimensions. More commonly, "single crystals" are composed of tiny "mosaics"—crystallites of linear dimension of around 10^{-4} to 10^{-6} cm within which the lattice order is very perfect. Adjacent crystallites are inclined to each other by angles up to about 1 degree. The boundaries of the mosaics tend to have concentrations of impurities and partially reflect thermal waves propagated by atoms. "Ideally imperfect" crystals of this type rarely have orientation spreads of more than 1 or 2 degrees covering all the mosaics.

In contrast with such "ideally imperfect" crystals, polycrystalline aggregates commonly have constituent crystals varying over all possible space orientations. A random orientation texture we can define as one in which the number of crystallites having a particular crystallographic direction within any element of solid angle is dependent only on, and proportional to, the number of such directions related by crystal symmetry. It is independent of the crystallographic direction chosen, of the orientation of the element of solid angle relative to the bulk shape of the aggregate, and of its directional history. This definition, ignoring preferred orientation probabilities between neighboring crystals, corresponds to common usage of the term. Even so, randomly oriented polycrystalline materials are rare. In the study of vapor-deposited low temperature solids preferred orientation appears to be the rule. That is to say that some crystallographic planes appear to align perpendicularly to the vapor-streaming direction or parallel to the surface on which deposition takes place. Thermal gradients on occasions may also exert an orienting influence.

B. Notes on the "Powder" Method

The number and position of the X-ray diffraction maxima depend on the over-all geometry of the atomic pattern unit (cell). The intensities of the diffraction lines, however, are determined by the kind, number, and positions of atoms in the cell. Only when the geometry of the cell is known can one assign the proper crystallographic symbols (plane "indices") to the X-ray diffractions. For this reason the determination of the cell geometry is often called "indexing."

For chemically simple crystals all this information can probably be deduced from really good powder data. When such is not available, skillful observations on changes in X-ray patterns with temperature and orientation can lead to successful indexing and often useful, partial structural information.

Whether a "powder" pattern has been successfully indexed or not, it is highly characteristic of the chemical phase of which it has been called the "fingerprint." It is here that perhaps a most significant contribution can be made by crystallography to trapped-radical research. An understanding or at least a recognition of the matrix phase must be basic to the description of solids containing free radicals. As will be shown in Section VI, this type of information is now available for a few trapped-radical-containing solids.

It has been shown that vapor-deposited, low temperature solids, though usually crystalline, often occur in modifications different from those obtained by cooling the same substances slowly from the melting point. In several instances phase mixtures have unexpectedly been found (e.g., see Section VI, F). At times their recognition from the "powder" patterns has been difficult and has depended on:

(1) gradual alterations in deposition conditions affecting the relative proportions of the constituents in the mixture (care is necessary to distinguish effects arising from preferred orientation);

(2) differences in transformation temperatures;

(3) discontinuities in dilatation curves of individual "powder" lines

[where transformation occurs to a modification which is crystallographically so closely similar that "powder" lines tend to occur in similar positions (e.g., see Section VI, F and H)].

Quantitative phase analysis depends either on a knowledge of the structures involved or on the examination of specimens containing known relative amounts of the phases. When the amount of a phase present is less than a few per cent it can often be missed altogether.

C. The Atomic Structure of Amorphous Materials

It is wrong to regard amorphous materials as being composed of atomic, ionic, or molecular species in complete disarray. After all, there is the same tendency for all solids to pack closely on the molecular scale, for chemical bonds to be satisfied, and for "coordination numbers" (numbers of nearest neighbors) and interatomic distances to be normal. Simple energy considerations, indeed, make it certain that the local order in the immediate vicinity of individual atoms is not greatly different from what it might be in crystals. What is different is the lack of lattice order in amorphous materials—there is no long-range order.

No sharp dividing line exists between materials one would call amorphous and those to which the term crystalline appears more appropriate. In a polycrystalline aggregate one might recognize a gradual decrease in crystal size. The X-ray "lines" would progressively broaden until eventually they would be called amorphous bands. In the absence of lattice order, X-ray diffraction can occur continuously with θ. This state of affairs is reached when the crystal dimensions are less than about 5×10^{-7} cm. For practical purposes this is a convenient order of size at which to place the division between amorphous and crystalline regions.

It must be remembered, however, that crystals are three-dimensional. We are sure to find instances where long-range lattice order is preserved in one or two dimensions, while in the other directions the order is lost after a few lattice repeats. It is convenient to call these one- and two-dimensional crystals. The effects seen on the X-ray "powder" pattern will be for some reflections to diffuse whereas others, according to directions of long-range order, remain sharp.

Our definition of an amorphous material can be simply: one that gives not even one "powder" line of intrinsic (as opposed to instrumental) half-height width less than about 1°. By analogy with ordinary glasses it seems appropriate to call low temperature solids giving such X-ray patterns "low temperature glasses." One modification of ice (Blackman and Lisgarten, 1957) is a well-established example of such a low temperature glass. It is useful to distinguish this type of amorphous solid from gel-like materials which have open aperiodic structures with submicroscopic voids. The distinction can be made from observations of very low-angle X-ray diffractions, which will be relatively strong for gels corresponding to their "structure" of dimensions larger than interatomic distances.

The X-ray diffraction intensities for materials integrated over all angles of observation are independent of the extent of long-range order. Since for crystalline materials the diffraction effects are concentrated at well-defined angles, the intensities to be measured are in general greater than for amorphous solids where they are spread over a range of deviation angles. The significant measurement of amorphous patterns is therefore experimentally a greater challenge. In theory, however, the diffractions are quantitatively related to the details of the short-range order. When there is preferred orientation in amorphous material, that is, when a structural feature in the short-range order (say a bond direction) is preferentially aligned with respect to the bulk shape of the specimen, additional diffraction information can be made available. Unrecognized orientation in amorphous solids, however, may lead to false interpretations of the X-ray pattern.

The degree of crystallinity is much less a characteristic of a phase than it is of the method of sample preparation and its subsequent history. The slower, in general, crystallization from melts and saturated solutions, the larger and more perfect the crystals formed. When depositing solids from low-pressure vapors, however, at surfaces far below the triple point, the opposite appears to hold. The slower the vapor stream, the more nearly the growth surface is held to the low substrate temperature and the more immediate is the loss of kinetic energy of the impinging molecules. Surface mobility is kept to a minimum, inhibiting formation of nuclei and the ordering of molecules into the equilibrium crystal arrays. For argon, nitrogen, oxygen, water, carbon dioxide, and other vapors deposited at 4°K the crystallinity is a function of deposition rate; the slowest rates producing the most nearly amorphous materials.

Discussion of amorphous materials should not be concluded without a warning against the visual classification of solids into crystalline and amorphous materials according to their opalescence or transparency. All one can say is that nonopaque solids in general appear optically translucent or clear if there are no refractive-index discontinuities such as grain boundaries of size larger than or comparable with the wavelength of light.

It is true that deposits which appear snowlike are generally crystalline. Shattered glasses, however, can give a remarkably similar appearance.

The converse, that clear solids are glasslike, is far less reliable still. Single crystals are generally transparent, as are compact polycrystalline materials whose grain size is appreciably less than the wavelength of light. This applies to very many of the materials containing trapped radicals.

IV. Disorder in Crystals

Cursory examination of low temperature, vapor-deposited solids has shown that amorphous solids are less commonly found than disordered crystalline materials. It is therefore important at the outset to discuss and classify the types of disorder in crystals.

A. DEFECT STRUCTURES

Departing only slightly from conventional treatment of this subject (Evans, 1939), there are fundamentally three types of defect structures, all of which might well be found in solids containing trapped radicals.

1. Substitution Type

In this, two or more chemically different atoms or atomic groupings occupy crystallographically equivalent sites. This is the most common of the effects that give rise to "solid solution." Usually there is, for any specific pair of molecular solvent and solute, a limit to the amount of substitution possible. This depends on temperature, the relative size and binding forces of the atoms involved, as well as the size of the cell. The X-ray pattern is affected owing to a change in average scattering power of atoms in the sites involved and to strains relative to the scattering planes.

Annealing of such solid solutions often leads to ordering of the substituted sites to relieve strains. The true pattern cell is now a multiple of the original one, and extra lines appear on the X-ray pattern, because of the "super-lattice." Strictly speaking, this amounts to a change of phase, but in practice the distinction is not always made, even when recognized. There is a continuous range between ordered and disordered solid solutions.

2. Vacancy Type

In this type of defect structure one or more crystallographically equivalent sites are only partially occupied. Changes in X-ray patterns are again observed: superlattices tend to be formed and similar order-disorder transformations occur. Substitutional solid solution often occurs in conjunction with the introduction of vacant sites. The concentration of vacancies is usually very limited.

3. Interstitial Type

In this a foreign chemical atom or group of atoms is occluded in holes of the structure at sites which are normally unoccupied. For example, in a structure of close-packed spheres of radius R additional atoms can be occluded without lattice expansion in the octahedral (6-coordinated) holes, if their radius is less than 0.41 R; or in the tetrahedral (4-coordinated) holes, if their radius is up to 0.23 R.

This type of defect structure also gives rise to solid-solution effects. It can cause changes in X-ray intensities, broadening of "lines," superlattice "lines," and additional "lines" due to change in pattern symmetry.

B. Defect Lattices

Defect structures are therefore crystal defects that consist of departures from the ideal atomic pattern at some lattice sites. These defects leave the lattice order essentially intact. Strictly interpreted the lattice postulate demands that every lattice point shall be precisely equivalent to every other. This is no longer quite obeyed in defect structures. There are local slight distortions from perfect over-all cell geometry. One must distinguish a different type of disorder in crystals in which locally the lattice order is abruptly broken. As will be seen these lattice defects are especially prone to introduce sites for the trapping of free radicals.

1. Plastic Deformation, Multiple Twinning, and Stacking Faults

Plastic deformation of crystals can be produced by influences other than externally applied macro stresses. In a mixture, for example, the transformation of one phase can induce plastic deformation in another. Temperature gradients in a specimen often cause plastic deformation on cooling or warming. Two mechanisms for deforming crystals plastically are well known: twinning and gliding.

Two portions of a crystalline solid are said to be twins if the structure patterns in the two portions are related by symmetry where no such symmetry exists in the structure itself. The "composition surface," along which the twins are joined, need not necessarily be geometrically related to the pseudosymmetry of the "twin law."

In plastic deformation of crystals by twinning, one portion of the crystal moves into a twin orientation relative to the original crystal. The change in shape is such as to decrease the stress. Twinning leads to an effective decrease of crystal size, observed by X-ray line broadening. The "reflections" from some lattice planes are exceptional in this respect. They include always those parallel to the twin plane (or perpendicular to the twin axis). For these, the twinned portions appear to have a continuous lattice. The concept of crystal size therefore becomes more complicated because it varies with the crystal plane studied.

The other mechanism for plastic deformation is by glide along "glide planes." This results in a parallel displacement by very many unit distances of one portion of the crystal, relative to another, along a microscopically observable boundary plane, which is always parallel to a lattice plane. Active parallel glide planes are usually separated by large distances compared with lattice spacings.

Repeated "polysynthetic" twinning is another common phenomenon of importance in low temperature studies. It may be explained by way of illustration in terms of close-packed structures. Cubic stacking of close-packed planes demands a regular cyclic sequence of three kinds of equivalent close-packed layers, such that a plane of symmetry is absent parallel to these planes. The sequence characteristic of hexagonal close-packing has a symmetry plane at every close-packed layer. Twinning on the close-packed cubic layer leads to a local hexagonal sequence. This type of twinning is therefore equivalent to a stacking fault. Its chief characteristic is that neighboring atomic layers are in normal relative positions. Only when the much weaker second-neighbor interactions are considered, is there any difference between the correct and faulty structures. The added energy associated with stacking faults is likely to be slight. At low temperatures the energy barrier to translation along the atomic planes, as would be necessary to correct a stacking fault, will be too great compared with available energy. Only on approaching the melting or triple point would one expect annealing to be effective in eliminating stacking faults.

There are, again, some X-ray diffractions whose sharpness can be shown to be unaffected by individual types of stacking faults. The quantitative broadening of the other "powder" diffractions is a difficult subject mathematically, but well treated by Wilson (1949). Further theoretical studies in this field are still needed.

Stacking faults are but one group of the structural faults that could occur. Some three-dimensionally faulted structures have already been observed, and limited experience has shown that each special case should be treated on its merits from a thorough knowledge of the idealized structure and the exact laws of faulting appertaining. In many instances it does not seem to matter whether a problem is approached from the standpoint of defect structures, superlattice coupled with some disorder, multiple twinning, or that of structural faults. Several of these in any specific instance are simply alternative ways of describing the identical crystal imperfections.

2. Dislocations

Within recent years yet another closely interrelated type of crystal imperfection, that of lines of lattice misfit or "dislocations," has been closely studied (Read, 1953), and this for excellent reasons. First, dislocation theory can account for many crystal properties which are hard to explain on the picture as so far given; and second, dislocations, as postulated by this theory, can actually be observed by a number of recent techniques.

Take first the edge dislocation—a plane of lattice points which terminates inside the crystal. Beyond the end of that lattice plane the adjacent planes bend to become nearest neighbors. If such an imperfection is trapped inside the crystal, a healing process is relatively difficult to visualize. The dislocation can, however, move by the terminated plane joining with part of a neighboring one, leaving the other portion of that neighboring plane in turn terminated.

The simple picture of glide deformation, as given above, can be shown to be in difficulties from energy considerations, unless portions of the crystals glide consecutively. The movement of dislocations provides the required picture to explain the observation of gliding in crystals. There are at any instant boundaries inside the crystal dividing portions showing a displacement (Burgers' vector), relative to one another, of either a unit lattice vector or a fraction of a lattice vector. One portion of the crystal in the latter event is in a twin configuration or can be described in terms of some other kind of fault position.

The Burgers' vector in edge dislocations is perpendicular to the defect line. Equally important are dislocation lines parallel to the Burgers' vector. They are called screw dislocations because the lattice forms a helix around the fault line. A crystal nucleus containing a screw dislocation will grow much faster (at least at high temperatures) than one made of perfect surfaces. This is because the accretion of new material to the growing surface can take place on faces with screw dislocations without ever having to start a new layer. This and several other classes of dislocations are self-perpetuating in growth.

The stress associated with a dislocation can be reduced by filling the gap with a foreign atom of suitable size. Dislocations therefore tend to collect impurities which in turn restrict the mobility of dislocations. The movement of dislocations, moreover, is geometrically strictly limited. Coplanar edge dislocations must remain coplanar. If they are of opposite sign, that is, having terminating lattice planes on opposite sides of the glide plane, they can attract and annihilate one another. Dislocations of like sign will have stress patterns that are mutually repelling. In the absence of external forces they will spread out evenly along the plane of dislocations. The lattice will show a bend at that plane. The boundaries of mosaics can be conveniently described as planes with a high dislocation density. Even grain boundaries in a polycrystalline aggregate can be so regarded.

V. Sites for Trapped Radicals

For a discussion of the basic crystal-chemical classification the reader is referred to Evans (1939). A brief discussion of packing density and clathrate structures only is needed here.

Crystal structures of all types surprise any student by the complexity of atomic patterns often found in equilibrium phases. They seem like ingenious solutions of the problem of fulfilling the geometric demands of all the atomic bonds as well as that of filling space efficiently. Any trapped radical will in such perfect structures be the source of much stress, lowering stability. To this the "clathrates" appear to provide an exception (Powell, 1956). In organic chemistry the C—H bond is purely covalent but OH and other semipolar groups produce intermolecular linkages stronger than "residual" (van der Waals') in character. The economic use of space in organic solids is often made difficult by awkward molecular shapes. When semipolar groupings add their demands for association of active spots on one molecule with similar groups on neighboring molecules at relatively short intermolecular distances and specific bond angles, open structures can arise with vacant spaces. Some of these structures can be stabilized by introducing into the voids molecular species of suitable size. Though not attached by primary bonds, they consolidate this "clathrate"-type structure through "residual" interatomic attractions. Radicals could clearly fulfill this stabilizing function, while remaining isolated from other reactive species as they would be trapped in cages of the matrix molecular framework.

The aim in the remainder of this section is, from well-established theory, to classify the possible sites for free radicals. Metallic structures are excluded because mobile electrons are available to pair at least partially with the free radicals. The storage of free radicals in such structures and their reclamation, fascinating though this subject may be, will not be discussed.

A. In Perfect Lattices

There must be at least one radical per lattice point in such perfect-lattice structures. One need distinguish merely between structures with or without diluent of stable molecules, atoms, or ions.

1. Pure Free Radicals

There is no energy barrier to be overcome for neighboring hydrogen atoms to react (except perhaps if their electrons are spin-aligned in a magnetic field, as described in Chapter 13). A crystal composed of hydrogen atoms alone cannot therefore exist under normal conditions. One suspects that the same might apply to the majority of simple radicals which form the principal subject of this book. Complex free radicals, especially those in which the unpaired electron is shielded by large stable groupings, will form normal crystals even at room temperatures. There is no doubt that many relatively simple ones could be crystallized and studied at low temperature, but this field remains almost completely unexplored.

2. Mixed Phases

If the free radical is associated regularly at every lattice point with one or more stable species one might call the phase, necessarily consisting of a mixture of molecular species, a "mixed phase." It must be distinguished from a "mixed crystal" in which solid solution occurs, that is, limited, random isomorphous replacement of one molecular species by another. Mixed phases can be classified further depending on the type of bond effecting crystal adhesion. In ionic structures the stringent demands of Pauling's rules (Evans, 1939) often lead to packing difficulties. Neutral molecules such as water and ammonia are known often to assist crystallization essentially as space fillers. Free radicals might well be expected to act similarly, but difficulties in the preparation of such mixed phases seem prohibitive. The same might apply to homopolar crystals in which the bond orientation leads to space not being economically filled. In van der Waals' crystals it is difficult to see how or why a mixed species should form unless an anisodesmic attachment of the radical to one stable molecule could be formed. By analogy with Evans' (1939) definition, "anisodesmic" here implies that the radical is more firmly bound to one molecule than to the rest of the structure. Herzfeld's (1957) N_3 species might, if it exists, form such an atomic pattern.

Clathrates with free radicals in all the cavities should be classified here in the subcategory of structures held by hydrogen bonds.

B. IN DEFECT STRUCTURES[1]

1. Substitutional

Isomorphous replacement of a matrix atom or molecule by a radical or by an anisodesmic radical-containing group will imply narrow limitations in the close correspondence of shape, size, and binding forces between solute and solvent groups. Radicals do not, however, resemble "stable" groups held in the structure by primary bonds. In this category, therefore, one needs to consider only crystals held by residual bonds. If for reasons discussed in Chapter 10, radical concentrations must be very low, the size tolerance may be relatively large, perhaps of the order of 15 to 20%, in all directions. The application of this type of rule is hampered by a lack of knowledge of the effective van der Waals' radii (Hartree, 1957). The effective size of truly free radicals will certainly be surprisingly large. None will be less than about 3 A in diameter and even hydrogen should be around 3.4 A. It is generally held that dissociated atoms such as H,

---

[1] For some interesting experimental results on hydrogen atoms in an argon matrix see Chapter 7.

N, or O could form limited substitutional solid solutions in the matrix of the corresponding diatomic molecules.

2. With Vacancies

Radical-containing structures with vacancies can be described as mixed phases with vacancy defects. Most clathrates, for example, do not ordinarily achieve 100 % occupancy of their cavities. Trapped-radical clathrates are likely to fall within this category, also. The preparative difficulties may be prohibitive for the chemically simplest radicals. Once made, there can be little doubt of their metastability.

3. Interstitial

The limitations on interstitial solute radii have been described in Section IV, A, 3 for four- and sixfold coordination. Other geometric restrictions are easily calculated from the relevant structures. The van der Waals' radii of atoms and radicals again apply (Hartree, 1957; and compare Section V, B, 1).

It is easily seen that in approximately close-packed structures, where sixfold coordinated holes are the largest, the radius limitation is severe. Dissociated atoms such as H, N, or O cannot fit into interstices of their own diatomic-molecular matrix structure and rather severe compression of the electron shell is needed for H, for example, to fit into the argon structure.

Interstitial sites are nevertheless of special interest because they occur in the simplest structures, and atoms once trapped in the interstices cannot touch one another.

C. IN DEFECT LATTICES

From simple crystal-chemical considerations it has thus been demonstrated that sites for radical species inside crystals are limited, and one might well gain the impression that the majority of materials containing trapped radicals at low temperature hold them in dislocation sites. From experimental evidence it is known that dislocations abound. There does not seem to exist any difficulty in so accounting for the concentrations so far achieved for most trapped radicals discussed in this book.

On crystal-chemical grounds one therefore feels disinclined to favor theories in which possible sites are homogeneously and regularly dispersed in the solids and in which random diffusion of radicals on warm-up is postulated. A radical trapped in a dislocation may well be shielded from contact with other radicals by the repulsion which like dislocations exert on one another. Radical concentrations may, moreover, be far higher in glide planes and at grain boundaries than corresponds to the bulk-average concentration.

D. In Glasslike Solids

From the discussion of amorphous materials (Section III, C) and that of dislocations (Section IV, B, *2*), a convenient definition of amorphous materials follows as having a dislocation density greater than about one in fifty lattice points.

If, furthermore, the opinion were held that radicals are most easily trapped at dislocations, then the search for high concentrations of simple radicals might well be focused on glasslike solids. Surprisingly, such appear to be rare for simple chemical compounds, even among vapor-deposited solids. Exceptions so far found include solids were oriented hydrogen bonding makes correct crystalline alignment of molecules difficult.

In practice, the admixture of free radicals alone exerts a strong influence toward crystallization. This is probably caused by the energy liberated from surface recombination of some radicals. It is immaterial whether one describes this as a thermal annealing process or whether one visualizes the newly formed vibrationally excited molecules as knocking neighboring molecules into their correct lattice sites.

VI. Some X-Ray Diffraction Observations

A. Argon

After brief surveys on vapor-deposited water, hydrogen peroxide, nitrogen, and hydrazoic acid (see below), it became clear that low temperature solids when deposited from vapor tend to form highly disordered structures, as had been anticipated. It was hoped that for argon good crystals would be found. The simple close-packed cubic structure (Dobbs and Jones, 1957) might be subject to stacking faults, but the simple atoms, which were pictured as having spherical symmetry, would readily fall into close-packed 12-coordinated positions.

It has been a matter for surprise and is of continued interest that the X-ray diffraction patterns at 4.2°K show considerable intrinsic line broadening (approximately $\frac{1}{4}°$ for 111, which reflection cannot be broadened by stacking faults; compare Section IV, B, *1*) corresponding to a very high dislocation density. This observation alone has a bearing on theories of vapor deposition. Many of these depend on surface migration of atoms (or molecules) after impingement. There is no doubt that such processes do occur at higher temperature, but if for argon at 4.2°K there were any atomic movement parallel to the surface, why would there not result crystals of high perfection? The observed disorder in argon is capable of explanation by theories postulating appreciable asymmetry of atomic forces or by theories which say that condensing atoms cause a displacement of neighboring surface atoms, a splash pattern which itself is frozen into the solid.

Vapor-deposited argon at 4.2°K tends to have a (111)-plane orientation parallel to the surface of deposition. The possibility of "epitaxy" on the specimen-holder surface must be discounted. Epitaxy is the growth of one crystal phase on the surface of another with an exact lattice fit of the two structures at the interface. Most experiments were performed on an essentially randomly oriented gold substrate. If epitaxy were responsible for the preferred orientation of argon, one would have to assume contact and parallel growth on only a proportion of the gold grains. The crystal size of the argon, furthermore, is so small that it is difficult to see how the orienting influence could be maintained throughout the argon film without an explanation for the orientation that is directly connected with the argon structure in geometric relation to the vapor-streaming direction or the temperature gradients.

On warm-up there is a gradual exothermic ordering process. When the bulk of the argon is still in contact with a surface at slightly above 30°K, the vapor pressure rises to around 10^{-3} mm of Hg, but subsequently well-crystallized unoriented argon appears, perhaps by partial redeposition. The experimental evidence does not yet prove whether the surface temperature, owing to the heat of ordering, rises well above that of the gold surface or whether the high dislocation density causes the argon to have an appreciably higher vapor pressure than that of the equilibrium phase. Annealed argon, recooled to 4.2°K, gives a cell edge $a = 5.308 \pm 0.01$ A; from this value the accurate van der Waals' radius for argon atoms at that temperature can be derived as 1.87(7) A (Bolz, Mauer, and Peiser, unpublished).

Molecular-nitrogen admixture causes increased line broadening at 4.2°K that cannot be simply explained by solid solution but must correspond to a greater dislocation density. The presence in the gas stream of free radical nitrogen, however, causes the vapor-deposited material to be almost fully ordered. This ordering influence by free radicals in the vapor stream has since been proved to be widespread (Section V, D). It was unexpected. If any changes in the X-ray pattern were anticipated at all they would have involved enhanced disorder due to frozen-in radicals that did not really fit isomorphously into the matrix structure. Actually it appears that this effect is in general greatly overshadowed by the annealing process due to the exothermic recombination of radicals at the surface. So effectively are atoms and molecules apparently knocked into equilibrium-phase positions by the recombining radicals that serious consideration should be given to this process for growing good crystals at very low temperatures.

B. NITROGEN

The structure of low temperature nitrogen, stable below 35°K, belongs to space group $Pa3$ with $a = 5.644 \pm 0.005$ A at 4.2°K (Bolz et al., 1959a). Unannealed vapor-deposited nitrogen gives very diffuse X-ray patterns with anomalous intensities that are certainly not explicable in terms of orientation. There is good evidence for 111 stacking faults, but these can also be shown to give an incomplete explanation of the observations. None of the anomalies have so far been explained in terms of misalignment of the N_2 molecular axes, which in the crystals should be inclined at the tetrahedral angle to the N_2 axes of all neighboring molecules. Near crystal boundaries this will not apply exactly and, in a sample containing very small crystals, considerable departures from the tetrahedral angle must be expected. Furthermore, if occasional N_2 molecules were to point in the wrong cube-diagonal direction, considerable distortions of the structure would arise that would affect the positions of neighboring molecules.

Attempts are being made to evaluate the exact electron-density distribution in the molecule from a careful structure determination of annealed nitrogen and hence to find the nature of the structural faults which appear to be special kinds of dislocations.

An exothermic ordering process with a steep increase of vapor pressure on warm-up begins from 12° to 20°K depending on deposition conditions. Relatively well-crystallized nitrogen redeposits on the surface unless the nitrogen vapor is removed by another cold surface. In the latter event the nitrogen transforms directly to the hexagonal ($P6_3/mmc$) high-temperature form.

Nitrogen deposited at 20°K shows more disorder than that deposited at 4.2°K and subsequently annealed at 20°K. This is evidence for believing that the annealing process in vapor-deposited nitrogen causes surface heating above the measured temperature. Yet the surface does not change to the high temperature form during annealing, so the temperature rise does not exceed the transformation temperature of 35°K. The vapor pressure observed in disordered nitrogen at 20°K is of the order of 2×10^{-5} mm of Hg. Although no measurements of the vapor pressure of stable low temperature nitrogen are available, it seems unlikely that it could reach that order of magnitude. There may thus be evidence here for the disordered vapor-deposited phase having a higher vapor pressure than the equilibrium phase.

Dissociated nitrogen atoms in the vapor cause ordering of the molecular nitrogen deposited, but this is not complete. The warm-up flashes occur at around the same temperature as ordering in undischarged nitrogen. Can one resist the speculation that the annealing process, which is known to eliminate most dislocations, is responsible for permitting trapped nitrogen atoms to meet and react?

The flashes observed during deposition are accompanied by large bursts of pressure during which the transient appearance of high temperature nitrogen stable above 35°K has not been detected. Yet the transformation is known to be a rapid one and the X-ray method is judged to be sensitive to the appearance of a 3×10^{-6}-cm layer of βN_2 for 0.5 second. Moreover, the trapped-radical content is in itself insufficient to account for the rise in pressure. Hence it must seem likely that the original nitrogen deposit is in a nonequilibrium state of appreciably higher vapor pressure than normal αN_2.

C. OXYGEN

Three equilibrium phases are known as follows (Wyckoff, 1951):

$$\alpha O_2 \xrightarrow{23.5°K} \beta O_2 \xrightarrow{43.4°K} \gamma O_2$$

γO_2 is isomorphous with αN_2; the other structures are not known. The most striking feature about the X-ray pattern of undischarged molecular oxygen is that it has much line broadening which places the deposit on the brink of being amorphous. On the definition here used (Sections III, C and V, D) this is the only glasslike low temperature solid found in condensed undischarged vapors containing no hydrogen.

When the oxygen vapor is passed through the microwave discharge, the resulting X-ray pattern shows an increase of disorder, presumably due to an appreciable admixture of ozone. On warm-up, crystallizations occur of βO_2 at 25°K and O_3 at 48°K. All O_2 can be pumped off at around 65°K and pure O_3 then remains on the specimen holder (Bolz, Mauer, and Peiser, unpublished).

Care is needed not to ascribe to free radical recombination the heat releases that are actually due to crystallization, which might well occur simultaneously. This is meant as a warning against simple calorimetry. It is true that normally a "blank" run on undissociated gas will produce a less-ordered solid. Therefore, errors due to neglecting heat of crystallization lead to underestimates of trapped-radical concentrations. For oxygen, however, the reverse is true.

More dangerous still are conclusions on trapped-radical recombinations in solids based on calorimetry without "blank" experiments. When hydrogen-bonded solids are involved, crystallization energies are relatively large. Structural information on such matrices is therefore especially relevant. The following four sections are devoted to hydrogen-bonded solids.

D. WATER

Apart from the crystal forms of ice that result when, by the application of high pressures, the water molecules are forced out of the low-coordination positions favored by the hydrogen bonds, there are two crystalline forms: one hexagonal, the other cubic. The latter is stable only around 150° to 173°K (Blackman and Lisgarten, 1957).

Vapor-deposited ice at 4.2°K is amorphous unless the deposition rate is high when there occurs incipient crystallization. This is especially true for films thicker than 0.1 cm, the surfaces of which are no longer effectively held to the substrate temperature during deposition. The modification here found is related to ordinary hexagonal ice by systems of faults, but these have not been elucidated. Ordering on warm-up occurs around 160°K, the hexagonal phase being formed. Crystallization of ice at 4.2°K becomes more marked when the water vapor is passed through a microwave discharge. This is the more remarkable because chemically only about a third of the water remains unreacted. Hydrogen peroxide also crystallizes, but only on warm-up (Section VI, E).

The evidence thus clearly shows that ordering of water molecules at low temperatures is difficult owing to the high activation energies involved. For low temperature solids one can probably generalize: when crystal-binding forces are stronger than "residual" and when structure changes are involved for which the atomic shifts are other than very small, the equilibrium phase in any given temperature range may be difficult to prepare. Slow cooling from the freezing point may be even less effective than vapor deposition.

E. HYDROGEN PEROXIDE

The structure of hydrogen peroxide is well known (Abrahams *et al.*, 1951). The geometry of the molecule is highly specific and a three-dimensional network of hydrogen bonds with parallel helices is formed.

Vapor-deposited hydrogen peroxide at 4.2°K and even 77°K is amorphous (Bolz, Mauer, Peiser, unpublished). On warm-up to 90°K, spontaneous strongly exothermic crystallization occurs, (Fig. 4), giving an oriented polycrystalline aggregate with preferred orientation. This can probably be explained in terms of hydrogen-bond helix orientation parallel to the specimen surface even in the amorphous material.

F. AMMONIA

Solid ammonia is cubic ($a = 5.082$ A at 77°K) (Bolz, Mauer, McMurdie and Peiser, unpublished) $P\ 2_13$. There can be no strong hydrogen bonding in this phase since every nitrogen atom has three symmetrically equivalent hydrogens and only one unshared pair of outer electrons.

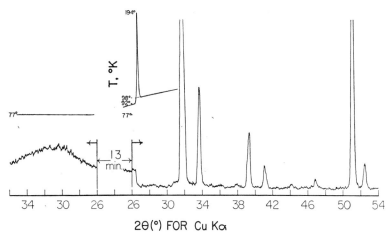

FIG. 4. The transformation of glasslike hydrogen peroxide to an oriented poly crystalline aggregate. Note the large temperature rise shown by a thermocouple embedded in the hydrogen peroxide. The temperature drop that follows is due to heat flow into the specimen holder. After equilibration the temperature of specimen and the massive specimen holder is significantly higher than it would have been had no transformation occurred.

At low temperatures one would expect a hydrogen-bonded structure to be more stable (Evans, 1939), that is, one in which each nitrogen has one long hydrogen neighbor, one short hydrogen neighbor, both participating in a hydrogen bond to other nitrogen atoms, and also two very close hydrogen neighbors that are not bound to other nitrogens.

Two new low temperature, low-symmetry phases have indeed been found, as well as an amorphous modification. The crystalline phases may have a temperature range of stability, although they cannot be made by slow cooling from the liquid. On warm-up they change irreversibly, and, it would appear, endothermically, into the cubic modification (Fig. 5).

Ammonia can thus have four distinct forms at 4°K, and extreme care is therefore needed in evaluating data on physical properties of solid ammonia and many other simple compounds at low temperatures.

Even X-ray examination does not appear to be a complete safeguard unless the interpretation is carefully performed. For example Vegard and Hillesund (1942) in studying differences between solid NH_3 and ND_3 obtained a value for the cube edge of the NH_3 structure which is in disagreement with our value and that of other investigators much beyond possible experimental error. The data as given agree excellently, however, on the assumption that their NH_3 was largely in one of the "new" crystal phases. Does this not throw doubts on the significance of conclusions reached by these authors on the similarity between NH_3 and ND_3?

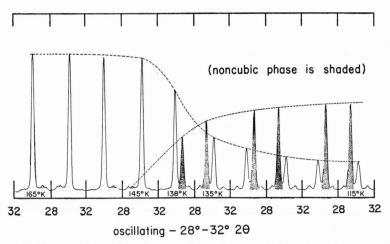

FIG. 5. The transformation of a new phase of ammonia into the cubic form. The same 2θ-region is scanned over and over again as warm-up proceeds from right to left. The pattern at first shows two strong lines, one due to the new phase, the other due to the cubic phase. Eventually only cubic ammonia remains.

G. HYDRAZOIC ACID

The crystallographic observations on Rice's Blue Material (Tesla-coil discharged hydrazoic acid deposited at 77°K) have been published (Bolz *et al.*, 1959b). The presence of electrons with unpaired spin has been confirmed by electron spin resonance (Gager and Rice, 1959).

The Blue Material differs fundamentally in three respects from other known solids with trapped simple radicals: (1) It is amorphous. (2) It is a solid held by essentially ionic forces. (3) It persists up to relatively high temperatures (about 150°K).

These may be closely related facts in conformity with a simple crystal-chemical picture. The amorphous character offers numerous sites for trapping, and the strong ionic forces provide a network that does not order till higher temperatures are reached.

X-ray evidence, as well as optical and mass spectroscopic observations (Mador and Williams, 1954; Franklin *et al.*, 1959) suggest that the name "imine radical" is not applicable to the principal chemical constituent of the solid.

H. DIBORANE

Vapor-deposited diborane is ordinarily crystalline (Bolz *et al.*, 1959c), exhibiting two distinct but related crystal forms. Microwave-discharged diborane contains hydrogen, and an attempt has been made either to observe an anisodesmic complex of hydrogen with diborane or to increase the hydrogen-atom concentration by dispersion in a low temperature glass of diborane (Bolz et al., 1959c). An amorphous material can indeed be formed by deposition at 4.2°K, but this is exceedingly unstable. As the amorphous character could be due to admixture of higher diboranes and even BH_3, the presence of which is suspected, there is no positive evidence for an anisodesmic complex. Nor is there proof of an enhanced hydrogen-atom concentration in the amorphous material compared with crystalline samples of discharged diborane, which have been found equally liable to a sudden and apparently spontaneous exothermic reaction process presumably due to hydrogen-atom recombination.

The crystallographic behavior of microwave-discharged diborane on warm-up is complex and will be described elsewhere, (Bolz et al., 1959c). The material is of some interest and would appear to deserve study by other techniques. It contains a boron-hydrogen phase that vaporizes below 120°K. This is the basis for believing that BH_3 may be present.

VII. Conclusion

The experimental results so far achieved through a crystallographic study of solids with trapped radicals are extremely modest. Often, however, this crystallographic evidence has seemed relevant to quite basic hypotheses although this evidence usually amounted to no more than a qualitative description of the degree of order in the matrix.

Complete disorder is rare in vapor-deposited solids. It seems to occur only when the deposition temperature is well below the triple point and when there is either a mixture of nonisomorphous molecules or when the energy barrier involved in the ordering process is large. This will occur chiefly for hydrogen-bonded solids and for molecules of large and unsymmetric shape. If they strike a cold surface they will stick with the chance orientation which they happen to have during impact. In the absence of the heat necessary to induce ordering, the material will remain amorphous.

Much more usual, however, are solids with some degree of long-range order, yet with a striking lack of complete three-dimensional lattice order. There is little doubt that it is the challenging and useful task for crystallographers to describe the type of faults—dislocations if that word is preferred—which characteristically describe the structural molecular geometry of the individual matrices involved.

In writing this chapter the author has selfishly discussed subjects that have been of interest to him for the past two years. He is deeply conscious that in doing so he has quite unreasonably ignored closely related studies, many of which, fortunately, are described in other pages of this book.

ACKNOWLEDGMENTS

The author is greatly indebted to his many colleagues who had a more than equal share in the experimental studies and the interpretative thoughts. The editors of this book, in leading the Free Radicals Research Program at the National Bureau of Standards, have often provided the vital stimulus or even the ideas themselves. The far-sighted financial support by the Department of the Army is also gratefully acknowledged.

REFERENCES

Abrahams, S. C., Collin, R. L., and Lipscomb, W. N. (1951). *Acta Cryst.* **4**, 15. The crystal structure of hydrogen peroxide.

Black, I. A., Bolz, L. H., Brooks, F. P., Mauer, F. A., and Peiser, H. S. (1958). *J. Research Natl. Bur. Standards* **61**, 367. A liquid-helium cold cell for use with an X-ray diffractometer.

Blackman, M., and Lisgarten, N. D. (1957). *Proc. Roy. Soc.* **A239**, 93. The cubic and other structural forms of ice at low temperature and pressure.

Bolz, L. H., Boyd, M. E., Mauer, F. A., and Peiser, H. S. (1959a). *Acta Cryst.* **12**, 247. A re-examination of the crystal structures of α and β nitrogen.

Bolz, L. H., Mauer, F. A., and Peiser, H. S. (1959b). *J. Chem. Phys.* **30**, 349. Low-temperature X-ray studies of Rice's blue material.

Bolz, L. H., Mauer, F. A., and Peiser, H. S. (1959c). *J. Chem. Phys.*, **31**, 1005. An exploratory study by low-temperature X-ray diffraction techniques of diborane and the products of a microwave-discharge in diborane.

Dobbs, E. R., and Jones, G. O. (1957). *Repts. Progr. in Phys.* **20**, 517. Theory and properties of solid argon.

Evans, R. C. (1939). "Crystal Chemistry." Cambridge University Press, London and New York. (A revised edition is to be published shortly.)

Franklin, J. L., Herron, J. T., Bradt, P., and Dibeler, V. H. (1959). *J. Am. Chem. Soc.* **80**, 6188. Mass spectrometric study of the decomposition of hydrazoic acid by electric discharge.

Gager, W. J., and Rice, F. O. (1959). *J. Chem. Phys.*, **31**, 564. Paramagnetic resonance spectra of active species. Blue material from hydrazoic acid.

Hartree, D. R. (1957). "The Calculation of Atomic Structures." Wiley, New York.

Herzfeld, C. M. (1957). *Phys. Rev.* **107**, 1239. Theory of the forbidden transitions of nitrogen atoms trapped in solids.

Hörl, E. M., and Marton, L. (1958). *Rev. Sci. Instr.* **29**, 859. Cryostat for electron bombardment and electron diffraction work.

Klug, H. P., and Alexander, L. E. (1954). "X-Ray Diffraction Procedures." Wiley, New York.

Mador, I. L., and Williams, M. C. (1954). *J. Chem. Phys.* **22**, 1627. Stabilization of free radicals from the decomposition of hydrazoic acid.

Peiser, H. S., Rooksby, H. P., and Wilson, A. J. C., eds. (1955). "X-Ray Diffraction by Polycrystalline Materials." Institute of Physics, London.

Powell, H. M. (1956). *Rec. trav. chim.* **75**, 885. Clathrate compounds.

Read, W. T., Jr. (1953). "Dislocations in Crystals." McGraw-Hill, New York.

Vegard, L., and Hillesund, S. (1942). *Avhandl. Norske Videnskaps-Akad. Oslo. I. Mat.-Naturv. Kl. No.* **8**, 1. Die Strukturen einiger Deuterium-verbindungen und ihr Vergleich mit denjenigen der entsprechenden Wasserstoffverbindungen.

Wall, L. A., Brown, D. W., and Florin, R. E. (1958). *J. Chem. Phys.* **30**, 602. Electron spin resonance spectra from γ-irradiated solid nitrogen.

Wilson, A. J. C. (1949). "X-Ray Optics." Methuen, London.

Wyckoff, R. W. G. (1951). "Crystal Structures." Interscience, New York.

10. Free Radical Trapping—Theoretical Aspects

J. L. JACKSON

National Bureau of Standards, Washington, D. C.

I. Introduction

As a result of the recent experiments on the trapping of reactive radicals at very low temperatures, several new and interesting theoretical problems have been suggested. In this chapter we will describe the work done on some of these problems and discuss their significance from the viewpoint of our understanding of the nature and properties of trapped radicals. The emphasis in the following discussions will be on the general physical ideas of the theories—details of the calculations will, whenever too involved, be omitted. As all of the work discussed here appears in journals, those interested in details may refer to the original papers.

There have generally been two completely different kinds of problems done under the general heading of theoretical trapped radicals research. One involves a detailed study of the microscopic properties of a particular radical trapped in a solid matrix; the other concerns itself with the statistical properties of mixtures of radicals and stable molecules at low temperatures. Each of these approaches yields its own distinctive kind of insight into the properties of mixtures that contain trapped radicals, and similarly each approach has its own difficulties and pitfalls. With a microscopic theory one attempts to describe in detail the energy levels and wave functions of trapped radicals in the solid. However, one is hampered by inadequate knowledge of the interactions between the molecules and radicals. The statistical theories are able in many cases to brush aside, rightly or

327

wrongly, one's ignorance of the detailed properties of the constituents of systems containing trapped radicals. These theories, however, are still faced with unique difficulties associated with the fact that a system with reactive radicals at low temperatures is not in equilibrium. One must therefore, in some way, according to some model, perform the relevant statistical averages over a reduced region of phase space (or with a nonequilibrium weighting).

In Section II we will describe a theory of the microscopic type that is concerned with the energy levels of nitrogen atoms in a matrix of nitrogen molecules. This work, which was performed by C. M. Herzfeld and his co-workers (Herzfeld and Broida, 1956; Herzfeld, 1957; Goldberg and Herzfeld, 1958) has the purpose of explaining particular spectral lines emitted in trapped radical experiments involving nitrogen and oxygen atoms in a nitrogen molecule lattice. The part of the theory dealing with nitrogen atoms appeared at the time of its publication to be in excellent agreement with experiment. Subsequent experimental results, however, disagreed with the theory. Nevertheless, even though the details proved to be wrong, the theory does show that the perturbations on the nitrogen atom are of the right order of magnitude to give a reasonable account of the experimental results. Furthermore, this theory is important in its influence on the development of our thinking about trapped radicals.

Section III is devoted to a description of the nonequilibrium statistical theories that have been proposed for obtaining average macroscopic properties of systems which contain radicals and stable molecules. It contains a discussion of the models developed to calculate the average number of radicals that may be trapped when mixtures of radicals and inert species are deposited from a beam. This approach is essentially based upon the idea that radicals are trapped when they are at lattice sites where they are isolated from other radicals. This leads to a statistical theory of the isolation process. Two different mathematical approaches to this theory have been published, one by Golden (1958) the other by Jackson and Montroll (1958). They both give essentially the same results—results which generally lead one to expect that high concentrations of radicals may be trapped.

In Section IV we will present a statistical theory of a dynamic nature, due to Jackson (1959a, b). It is a theory of chain reactions of nitrogen atoms in a nitrogen molecule lattice which is based upon a model that was suggested by experimental observations on such mixtures. The theory is relevant to the question of the percentage of nitrogen atoms that may be trapped in a nitrogen molecule lattice in that one obtains a critical concentration of nitrogen atoms such that higher concentrations will give rise to autocatalytic recombinative reactions. This theory predicts that the maximum percentage of nitrogen atoms that one may stabilize in a nitrogen molecule lattice is far less than that predicted by the isolation theories (about 0.1 % as compared with approximately 10%), which ignore mechanisms for driving radical mixtures toward equilibrium.

Finally in Section V, we will discuss generally what has been learned from these theoretical studies. We will also give some consideration to which directions seem most interesting and most profitable for future theoretical research along these lines.

It is no coincidence that essentially all of the problems discussed here which refer to a specific system involve nitrogen. Free radical experiments with nitrogen are uniquely spectacular in that there are brilliant glows and flashes of light which give dramatic visual evidence of the presence of active species. But these very same features have made nitrogen extremely interesting from a strictly scientific point of view. It has thus received a great deal of experimental attention and a correspondingly great amount of theoretical attention.

It should be noted that in this chapter we have excluded theoretical studies of the properties of isolated radicals (e.g., by molecular orbital or valence bond theory). We consider only theories where there is explicit reference to the fact that the radical is trapped in the solid and interacting with it in some way. Similarly, we are also excluding theoretical work on the properties of the matrices alone at low temperatures. Although the work done in both these areas is certainly relevant to the problems of trapped radicals, it is necessary to confine the subject matter in the stated way in order to set reasonable limits to the contents of this chapter.

II. A Microscopic Theory—Forbidden Transitions of Nitrogen Atoms

When nitrogen is passed through a microwave discharge and then deposited on a surface near 4.2° K, one observes flashes and glows at the surface. After the discharge is turned off, the glow at the surface persists for times of the order of a minute. Furthermore, if one waits until the deposited mixture is quiescent and then allows it to warm up, the deposited solid will emit a characteristic green glow. The details of the spectra were originally reported by Bass and Broida (1956).

One of the most notable spectroscopic features of this emission is a group of closely spaced green lines called the α-lines. As originally reported, this group consisted of five lines, which are tabulated according to their measured frequencies (in cm^{-1}) and approximate relative intensities in Table I. It is the α-group which gives rise to the observed afterglow, these lines having half-lives of over 10 seconds.

Herzfeld and Broida (1956) proposed that the α-group was a transition from the excited metastable states $2p^3$ ^2D of the nitrogen (N) atoms to the

TABLE I

THE ALPHA GROUP[a]

Lines			Intensities	
J	$|M_J|$	$\nu(\text{cm}^{-1})$	Observed	Calculated
3/2	1/2	19,173.1	2	1
3/2	3/2	19,151.4	8	3
5/2	1/2	19,120.6	10	10
5/2	3/2	19,096.9	6	8
5/2	5/2	19,077.2	4	5

[a] Herzfeld (1957).

ground states $2p^3$ ^4S. This transition in the free atom is highly forbidden as it requires both a transition from a doublet to a quartet and a change of the orbital angular momentum, $\Delta L = 2$. The half-life for the free atom is calculated to be about 20 hours (Ufford and Gilmore, 1950). The free atom has two levels in the excited states, $^2D_{5/2}$ and $^2D_{3/2}$; the ground ^4S states are degenerate. The corresponding transitions occur at 19,223 cm^{-1} and 19,231 cm^{-1}.

If the observed α-lines are to be identified with the $^2D - {}^4S$ transition, one must explain all of the ways in which the observed lines in the solid are different from the two transitions of the free atom. In particular, the theory must account for: (1) the change in the number and spacing of the lines; (2) the shift of the center of the groups by about 100 cm^{-1} to the red; and (3) the change in the lifetime.

Herzfeld and Broida (1956) indicate that a quadratic crystalline field can split the six degenerate $^2D_{5/2}$ states and four degenerate $^2D_{3/2}$ states into five doubly degenerate levels.[1] As the splitting produced in the ground ^4S state may be shown to be negligibly small compared to the splittings in the ^2D states, excited atoms perturbed by such a crystalline field would indeed give rise to five radiated lines. Estimates are made to indicate that one could obtain order of magnitude agreement in explaining the observed splittings, shift, and lifetime.

A detailed theory of this transition was presented subsequently in a paper by Herzfeld (1957). The effect of the lattice on the nitrogen atom is represented by a crystal field term in the Hamiltonian of the nitrogen atoms. The program which is carried out in this paper is that of finding the best crystal field potential from the viewpoint of giving rise to splittings and lifetimes for the ^2D states in closest agreement with the observed α-lines. The

---

[1] Systems with odd numbers of electrons will, by Kramers' theorem, retain at least a twofold spin degeneracy. This Kramers' degeneracy can be broken up only by placing the system in an external magnetic field.

crystal field determined in this way is then interpreted in terms of a geometrical model of an atom in the field of a nearest molecule. That is, a site where an atom would experience the required perturbing field is found. The Hamiltonian of a trapped atom is written,

$$H = H_f + V \tag{1}$$

where H_f is the free atom Hamiltonian and V represents the crystal field perturbation. H_f is written,

$$H_f = H_0 + E(s_i, l_j) \tag{2}$$

where H_0 contains the kinetic energies of the electrons and their Coulomb interactions with each other and with the nucleus. The second term $E(s_i, l_j)$ contains all of the spin-dependent interactions, namely the spin-orbit, spin-other orbit, and spin-spin interactions.

If one assumes that the crystal field perturbation on the nitrogen atom arises from the quadrupole field of a single closest molecule, then one may write the following expansion:

$$V = V_0 + V_1 + V_2 \tag{3}$$

with

$$V_1 = \alpha y' + \beta z'; \qquad V_2 = \gamma x'^2 + \Delta y'^2 - (\gamma + \Delta)z'^2 \tag{3a}$$

Here x', y', and z' are coordinates centered about the nitrogen nucleus which diagonalize the quadratic part of V. The x' direction is perpendicular to the plane of the molecular axis and atom. The linear term in x' does not appear in V, as the potential must be an even function of x'. The sums of the coefficients of x'^2, y'^2, and z'^2 must equal zero as the potential must be a solution of Laplace's equation.

The most important contribution to the splitting comes from the quadratic term, V_2; the linear term, V_1, is instrumental in decreasing the lifetime substantially. The constant term, of course, has no effect as it affects all levels equally. The matrix elements important in this calculation are indicated in Fig. 1.

It may be shown that the expectation of V_1 and V_2 vanishes for the ^2D states (as well as for the ^2P and ^4S states). The most important contribution to the splitting then arises from the V_2 coupling of the ^2D and ^2P states. This splitting may then be calculated by second-order perturbation theory.

This is done, taking into account not only V_2, but also the spin-other orbit and spin-orbit interactions. The final result for the five energy levels of the perturbed ^2D states depends on four parameters. Two, σ_2 and σ_0, depend upon V_2 and are determined by the γ and Δ which appear in V_2.

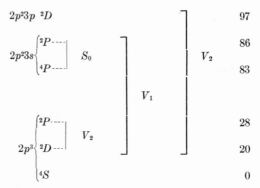

FIG. 1. Matrix elements important in the calculation: V_1 and V_2 are respectively the linear and quadratic parts of the crystal field. S_0 is the spin-orbit interaction. The numbers on the right are the energies of the levels in units of 1000 cm^{-1}.

The other two are free atom constants, ξ and η, respectively the spin-orbit and spin-other orbit coupling constants for the $2p^3$ configuration. For the free atom the best values of these constants are $\xi = 90$ cm^{-1} and $\eta = 0.715$ cm^{-1}. The results of this calculation for the levels (labeled according to J and M_J) are:

$$W\left({}^2D, \frac{5}{2}, \pm\frac{5}{2}\right) = W_0 - \frac{37}{5}\eta + \frac{8\sigma_2^2}{W}$$

$$W\left({}^2D, \frac{5}{2}, \pm\frac{3}{2}\right) = W_0 - \frac{37}{5}\eta + \frac{144\sigma_0^2}{5W} + \frac{24\sigma_2^2}{5W}$$

$$W\left({}^2D, \frac{5}{2}, \pm\frac{1}{2}\right) = W_0 - \frac{37}{5}\eta + \frac{72\sigma_0^2}{5W} + \frac{8\sigma_2^2}{5W} \tag{4}$$

$$W\left({}^2D, \frac{3}{2}, \pm\frac{3}{2}\right) = W_0 + \frac{110}{10}\eta + \frac{36\sigma_0^2}{5W} + \frac{5\xi^2}{4W} + \frac{6\sigma_0\xi}{W} + \frac{36\sigma_2^2}{5W}$$

$$W\left({}^2D, \frac{3}{2}, \pm\frac{1}{2}\right) = W_0 + \frac{110}{10}\eta + \frac{108\sigma_0^2}{5W} + \frac{12\sigma_2^2}{5W} + \frac{5\xi^2}{4W} - \frac{6\sigma_0\xi}{W}$$

Here W_0 is the unperturbed energy of the ^2D states in the absence of spin interactions and W is the difference between the ^2D and ^2P states of the $2p^3$ configuration, which is -7600 cm^{-1}.

It was not possible to fit the observed spacings by varying only σ_0 and σ_2. However, if one assumes that η may be different in the solid and if one then tries to fit the observed spacings by varying η, σ_0, and σ_2 one obtains excellent agreement between the splittings of the ^2D lines and the observed splittings of the α-lines for the choices $\eta = 3.46$ cm^{-1}, $\sigma_2 = 230$ cm^{-1}, and $\sigma_0 = 37$ cm^{-1}.

These parameters imply a shift toward the red of 37 cm^{-1} of the center of the lines, as compared to the two free atom lines. This is in the right direction but smaller than the observed shift of 102 cm^{-1}. It is, however, argued that configuration interaction could account for the rest of the shift, without causing as great a change in the splittings.

To calculate the theoretical intensities it is assumed specifically that all of the ^2D states are equally populated. Therefore, the intensity of the line will be directly proportional to the transition probability, or inversely proportional to the lifetime of the atoms in the particular perturbed ^2D levels. The lowering of the lifetime of the ^2D states is attributed principally to the second-order mixing in of the $2p^23s$ ^4P state into the $2p^3$ ^2D. This mixes via the V_1 coupling between the $2p^3$ ^2D and $2p^23s$ ^2P and the spin-orbit coupling between the ^2P and ^4P terms of the $2p^23s$ configuration. Although these couplings introduce only a very small amount of $2p^23s^1$ ^4P wave function into the ^2D state, this has a very great effect on cutting down the lifetime of the metastable ^2D states as the ^4P to ^4S transition is an allowed dipole transition. A detailed computation then leads to expressions for the transition probabilities in terms of the coefficients of V_1, α, and β. Assuming that the calculated transition probabilities are proportional to the observed intensities of the α-lines, one obtains reasonable agreement in the relative intensities when $|\beta/\alpha| = 2$. A comparison of the resulting calculated relative intensities with the observed ones is presented in Table I.

Now, with the desired crystal field parameters calculated, the next step is to find a site where the nitrogen atom will be in such a field. It is assumed that the nitrogen atom is closer to one molecule than to all others to such an extent that one may consider only the quadrupole field of that one molecule. Using the best value of the quadrupole moment, $Q = 1.3 \times 10^{-26}$ e.s.u. cm^2, the point where the most reasonable crystal field parameters were obtained was at a distance of 2.5 A from the center of the molecule and on a line from the center of the molecule which makes an angle of 50° with the molecular axis. Such a position was therefore suggested as the site of the nitrogen atom.

This theory at the time it was proposed gave a reasonable account of most of the experimental facts. Furthermore, it did so with a minimum of assumptions—one site with equal populations of excited ^2D states. Subsequent experimental results, however, have shown that this simple theory is inadequate. For one thing, more detailed searching has shown that the α-group has more than five component lines (Peyron et al., 1959). There are at least six lines within the group and possibly more. A single site for a nitrogen atom with a degenerate ^4S ground state could not account for more than five lines. One might still try to accept this explanation of five of the lines and attribute the other lines to atoms at other sites. However, a more fundamental discrepancy was obtained when the lifetimes of the separate components were individually measured. The assumption of the calculation was that the excited ^2D states were equally populated. The intensity difference was therefore attributed solely to differences in transition probabilities. The lifetimes of the lines should then have been inversely proportional to the intensities. This was not at all the case. Indeed, the most intense component has the longest lifetime. The explanation of the intensities of the components must therefore take into account the population differences of excited states.

Today, we therefore know that the situation is far more complicated than we had been led to believe by early experimental results. There is undoubtedly more than one site and there are population differences among the excited ^2D states. Recent identification of an intense group of lines in the infrared (Peyron et al., 1959) attributed to the $^2P - {}^2D$ transition in atomic nitrogen, indicates that the ^2D atoms which radiate probably arise as a result of this transition. The population difference of the ^2D states could then come from selection rules in the $^2P - {}^2D$ transitions.

Although the details of the theory discussed are incompatible with present knowledge of the α-lines, the theory does illustrate mechanisms which can give rise to the observed effects. It is important in that it shows how a reasonable crystal field can give rise to splittings of the observed order of magnitude and shorten the lifetime of the metastable ^2D states.

Similar techniques have been applied to explain a set of three lines observed when small amounts of oxygen are present in a nitrogen-deposition experiment (Goldberg and Herzfeld, 1958). These lines have been attributed to the transition $2p^4\ {}^1S - 2p^4\ {}^1D$ in atomic oxygen as perturbed by the crystal field. The observed lines are broad, of the order of 100 cm^{-1} wide. They occur at 18,016, 17,801, and 17,672 cm^{-1} as compared to a single line at 17,922 cm^{-1} for the free oxygen atom. The same procedure is applied as was used to explain the α-lines in nitrogen. A best location for an oxygen atom in the field of a nitrogen molecule is sought, with the criterion being that of getting the best agreement between computed and observed intensities, splittings, and lifetimes. The result obtained is that an oxygen atom on the perpendicular bisector of the N_2 axis and 2.3 A from the axis will account very well for the observed results.

It might be noted, however, that one question which is not answered by this theory, either as applied to oxygen or nitrogen, is why the atom should reside at the site determined in this *a posteriori* manner. Indeed, in the field of a single nitrogen molecule, an oxygen or nitrogen atom would not be expected to have a minimum of potential energy at the respective desired positions.

III. A Static Statistical Theory—The Statistics of Isolation

In this section the attempts to calculate averages relevant to mixtures containing trapped radicals will be considered. Any such calculation must in some way refrain from taking a full equilibrium average over all phase space, for at equilibrium, in low temperature ranges, there are no reactive free radicals. One must in some way confine the average over a restricted region of phase space. Such a restriction physically represents the fact that a gaseous radical mixture when deposited suddenly at low temperatures is quenched and its return to equilibrium is prevented (or at least greatly retarded).

A nonequilibrium statistical theory of this type has been applied to the problem of computing the average number of radicals that may be trapped in typical deposition experiments. Two independent mathematical approaches, based upon similar physical ideas, have been worked out (Jackson and Montroll, 1958; Golden, 1958). In both methods atoms and molecules are assumed to reside at sites of perfect lattices. The basic idea of these theories is that radicals are trapped in the lattice when they are at sites where they are surrounded by inert species, i.e., molecules (triatomic molecules are excluded). Thus no two free radicals may be nearest neighbors in the lattice.

Both methods consider all the different ways in which a gaseous mixture, initially with A_0 atoms and M_0 molecules, can be put down on a lattice under the condition that when two atomic free radicals are nearest neighbors they will recombine and form a molecule. As the deposition and recombination processes are assumed to be random, all of the different final configurations of atoms and molecules are taken to be equiprobable. This means that complete quenching is assumed—no greater weight being assigned to configurations with lower energy (fewer atoms).

Having posed this model, the problem is to average the number of trapped atomic radicals over all final configurations. Exact solution of this problem has to date not been possible except in the unrealistic case of a one-dimensional lattice (Jackson and Montroll, 1958). The two works cited essentially present two different approximate methods of solving this problem for a three-dimensional lattice. The method of Golden consists in writing an approximate expression for the partition sum of a mixture with at most A_0 atoms (the number of atoms in the beam), wherein there is no weighting of the terms in the sum as to energy. This involves obtaining an approximate expression for the number of ways in which M molecules and A atoms can be distributed in the lattice for values of $A \leq A_0$, with $2M + A = 2M_0 + A_0$. The average of A over these configurations is evaluated approximately by finding that value of A for which there are the greatest number of different ways of distributing the A atoms and M molecules, i.e., the average of a random variable is replaced by the value of the variable which maximizes the statistical density function of the variable. The result is a function of an effective coordination number, Z', which Golden introduces in obtaining his expression for the number of configurations with A atoms and M radicals. The result for the average number of trapped radicals for reasonable values of Z' is of the order of magnitude of 10% (provided the percentage of atoms in the beam was sufficiently greater than 10%).

It should be mentioned that this is a greatly simplified account of the work presented in Golden's paper. He also considers the effect of different assumptions with regard to the deposition process and calculates the effect of adding an inert diluent to the mixture.

Jackson and Montroll (1958) introduce a different approximation to evaluate the average number of trapped radicals implied by this model of trapping by isolation. They consider a typical lattice site at which an atomic radical has been deposited. It is assumed that it may recombine only with one of its Z nearest neighbors in the lattice. The probability that it remains a trapped radical is then the probability that it does not recombine with any of its nearest neighbors. The simplifying assumption which enables one to obtain an approximate expression for this probability is that of treating the binding processes of any atoms which are the nearest neighbors of a particular one as independent processes. To illustrate how this may be used to obtain an equation for the average number of trapped radicals, let us consider the simple, but extreme, case of a beam consisting entirely of atoms depositing on a lattice. We imagine the atoms coming down on the lattice sites and forming bonds with nearest neighbor atoms at random. The probability that a particular atom remains unbound is equal to the probability that all its Z nearest neighbors bind to other atoms. The probability that one nearest neighbor of the atoms we are considering will bind to one of its other $Z - 1$ neighbors is $(Z - 1)/Z$ times the probability that it becomes bound. If we call f the probability of an atom remaining unbound and use the assumption of the independence of binding processes of the Z nearest neighbor atoms we obtain the equation,

$$f = \left[\left(\frac{Z - 1}{Z} \right) (1 - f) \right]^Z \tag{5}$$

The solution, of course, depends upon the coordination number of the lattice, Z. For the values $Z = 6$, 8, and 12, the values of f which solve Eq. (5) are, respectively, 0.14, 0.12, and 0.10. These three choices of Z correspond respectively to the number of nearest neighbors in the simple cubic, body-centered cubic, and face-centered cubic lattices.

The theory was also carried through as a function of concentration of atoms in a beam which contained molecules and inert atoms as well. The over-all result of this theory is that if there is a high enough percentage of atoms in the beam, the expected trapped atom concentration is of the order of 10 %—the same order of magnitude as is obtained from the Golden theory. Both methods lead to the result that adding an inert diluent might increase the ratio of trapped atoms to molecules (of the atom) but would not increase the percentage of trapped radicals, calculated as a fraction of the total number of molecular species (radicals, molecules, and diluent atoms).

There have recently been two more elaborate calculations of the number of trapped radicals predicted by this model. These calculations accept the isolation model but attempt to use more accurate statistical methods than those of Jackson and Montroll (1958). The indications are that the approximation of Jackson and Montroll (1958) gives percentages that are somewhat too high. Thus Chessin (1959) using Ising model techniques gets 6.5 % for the trapped radical concentration in a face-centered cubic lattice, as compared to 10 % by Jackson and Montroll. R. Kikuchi, in an unpublished calculation, using a statistical method wherein the bonding of adjacent pairs of atoms, rather than adjacent atoms, are treated independently, obtained 7.1 % for the corresponding concentration.

The difficulty with the type of theory dealt with in this section is that it is not clear to what physical systems it applies. As we shall see in the following section, it seems likely that for concentrations considerably smaller than the 10 % of the isolation theories, one is not justified in disregarding, for nitrogen at least, mechanisms by which the mixture is driven toward thermodynamic equilibrium. It would still be interesting to see if these theories are applicable to systems of somewhat less reactive radicals.

IV. A Dynamic Statistical Theory—Chain Reactions of Trapped Nitrogen Atoms

In this section a theory of chain reactions of nitrogen atoms in a matrix of nitrogen molecules (Jackson, 1959a,b) will be described. The theory proposes a simple model for the propagation of chain reactions and leads to a critical concentration of nitrogen atoms in the solid, such that higher concentrations will be unstable. This critical concentration is estimated both on the basis of a rough *a priori* model, and an *a posteriori* method, wherein a theoretical formula is applied to experimental results.

The theory was motivated by the detailed observations of Fontana (1958, 1959) on explosive recombination reactions in mixtures of nitrogen atoms and molecules at liquid helium temperature. Fontana deposited nitrogen from a microwave discharge onto a 4.2° K surface and measured the number of trapped radicals present by measuring the magnetic suscepti-

bility of the deposited material. The essential result was that the susceptibility builds up approximately linearly with time during deposition until
an explosion occurs, at which time the susceptibility suddenly returns to its
initial value at the beginning of the deposition, indicating the recombination of almost all the atoms. In general, the higher the concentration of
nitrogen atoms in the beam, the shorter the time between explosions. At
concentrations of the order of 0.2 % N atoms in the beam, the time between
explosions was of the order of a minute, while with beams with 2 % N
atoms there was no noticeable over-all increase in the susceptibility during
deposition (which was interpreted as meaning that the atoms recombined
essentially instantaneously).

These experiments suggested that the trapped N atoms in the solid
could chain-react and that the surface of the solid inhibited the chain reaction. Thus with lower concentrations of atoms in the beam, it takes a
longer time between explosions because a larger critical size of the material
must be built up in order for the mixture to be unstable.

In setting up a model for the chain reactions, earlier thermal studies on
deposited nitrogen radical mixtures proved suggestive (Fontana, 1958). According to these studies there was no observable recombination of nitrogen
atoms in a nitrogen molecule lattice when the temperature was kept below
$10°$ K.[2] Between $10°$ K and $36°$ K atoms would recombine. When the temperature was kept constant at some intermediate value (between $10°$ K
and $36°$ K), the observed glow, which is used as an indicator of recombinations, would decay, and resume only when the mixture was warmed up
further. These experiments suggested the view that the atoms are in traps,
in the solid, which have a range of binding energies corresponding to temperatures between $10°$ K and $36°$ K.

With these experiments in mind, the following simplified model for chain
reactions of nitrogen atoms trapped in a nitrogen molecule lattice was proposed. The atoms are thought of as being randomly located in traps in the
lattice. The number of atoms per unit volume is called n; the number of
molecules per unit volume, M; the initial temperatures of the lattice, T_0.
For simplicity, all traps are assumed identical with a binding energy corresponding to an average temperature, T_f.

When two atoms recombine they liberate an amount of energy, W, which
is assumed to heat up the lattice in the neighborhood of the recombining
atoms. That volume about the recombining atoms, which is heated up above
the temperature T_f, is called V_f. Trapped atoms within this volume are
assumed to be liberated from their traps. The average number of trapped
radicals freed by a recombination is nV_f. These atoms may then move
through the solid. They may lose energy and return to an unoccupied trap,

---

[2] This refers to the warm-up part of a deposition experiment, under conditions of
low concentrations of N atoms so that chain reactions were not possible.

or they may encounter another atom and combine with it. It is assumed that the average probability that one radical so freed will give rise to another recombination is proportional to the concentration of nitrogen atoms in the lattice, and this probability is written $\xi n/M$. Here ξ is a numerical parameter which determines the likelihood of a single mobile atom recombining. The average number of secondary reactions which result from a single initiating reaction is $nV_f \cdot \xi n/M$. It is clear that when this number is greater than unity, the mixture is unstable and an initiating recombination may give rise to an explosion. When this number is less than unity, according to this model a fluctuation which gives rise to a recombination will die out and the mixture will be stable. The concentration at which this number becomes equal to unity is called the critical concentration, n_c, and is given by

$$n_c = \sqrt{\frac{M}{\xi V_f}} \tag{6}$$

As the effect of the surface in inhibiting these proposed chains has been ignored in this discussion, this critical concentration represents the maximum concentration of nitrogen atoms that one could stabilize in an infinite volume. For a finite volume, where the inhibiting effect of the surface must be considered, higher stable concentrations may be obtained.

On the basis of this model, two general directions of inquiry are followed. First, a rough *a priori* estimate of the order of magnitude of n_c is made in order to ascertain whether or not the expression for n_c is generally consistent with experimental results. Secondly, a formal theory of the chain reactions is presented to describe some of the details of the processes that occur. The formal theory, when used with experimental results is shown to yield *a posteriori* methods for evaluating the parameters.

To estimate n_c as given in Eq. (6), we must obtain estimates for V_f and ξ. An order of magnitude estimate of V_f is obtained by considering the energy of recombination, W as the strength of a classical point source of heat which propogates away from the recombining atoms. The volume V_f is that volume about the recombining center wherein the temperature at some time becomes greater than T_f. Assuming constant conductivity and specific heat the result is

$$V_f = \frac{4\pi}{3} \left(\frac{3}{2\pi e} \right)^{3/2} \frac{W}{\rho c(T_f - T_c)} \tag{7}$$

where ρ is the density of the mixture, c is an average value of the specific heat in this temperature region, and e is the base of natural logarithms.[3] If

[3] In Jackson (1959a) the factor $\pi^{3/2}$ is omitted in the denominator of this equation. The percentages estimated in that paper are therefore in error, being too small by a factor $\pi^{3/4} = 2.36$.

one writes the average molar specific heat, c_m, in the following way

$$c_m = \alpha L k \tag{8}$$

where L is Avogadro's number and k Boltzmann's constant, then α is a number of order of magnitude unity. The result for the fractional concentration takes the form

$$\frac{n_c}{M} \cong 1.8 \sqrt{\frac{\alpha k (T_f - T_0)}{\xi W}} \tag{9}$$

A rigorous attempt to estimate ξ would be extremely difficult. The probability of a mobile atom recombining with another atom depends on the rate of diffusion and the energy loss of the atom as it moves through the solid and upon the cross section for recombination. A simple way of thinking about the recombinations is introduced (Jackson, 1959a) which leads to the estimate that ξ is of the order of magnitude of the coordination number of the solid molecular lattice. As the equilibrium lattice of nitrogen below 36° is face-centered cubic, the coordination number is 12.

The specific heat of nitrogen at these temperatures indicates that a reasonable mean for α is 1.5. According to the interpretation of Fontana's experiments, T_f should lie between 10° K and 36° K. Thus, $T_f - T_0$ may be estimated to be of the order of magnitude of 20° K. Using the most "reasonable" estimates of the parameters that enter into Eq. (9), the result is

$$n_c/M \cong 0.7\% \tag{10}$$

This figure is generally of the right order of magnitude from the viewpoint of interpreting Fontana's experiments. A beam with 2% N atoms would, when deposited, be expected to give rise to almost instantaneous recombination, whereas deposited beams with lower concentrations would explode only after a large enough volume was collected. As explosions were observed with less than 0.2% N atoms the result of 0.7% is almost certainly too high.

This result is also suggestive with regard to the question of the maximum percentage that one can stabilize. Although there is much arbitrariness in selecting the values of the parameters that occur in Eq. (9), it is shown that even when extreme choices are made (Jackson, 1959a) the estimates of n_c/M do not go much higher than 2%. Thus, even though these estimates are approximate, they indicate strongly, that the values of the order of 10% obtained from the static isolation theories would not be applicable to nitrogen.

In a later paper (Jackson, 1959b), a formal description is given of the way in which this type of chain reaction progresses. The theory is worked out by two mathematical methods—one wherein the reaction is treated as a succession of discrete steps involving recombinations and freeing of trapped radicals, and the other involving partial differential equations for the average number of radicals.

The formal theory is applied in the following three ways:

1. It is used to give a detailed justification of the criterion used for stability.

2. It is used to discuss the stability of mixtures as a function of size.

3. It is applied to the situation where radicals are produced by irradiating nitrogen kept at liquid-helium temperatures. The theory predicts a class of "saturation curves" for the number of trapped radicals as a function of time as well as a value for the maximum, or saturation concentration, n_s, which is obtainable by such an experimental procedure.

As the mathematical method involving discrete sequences is used only in the first and last applications, whereas the method which works with the equations for the average concentrations yields all of the results, only the latter method will be described here.

Equations are derived for n, the average number of trapped atoms per unit volume, and m, the average number of free or mobile radicals per unit volume. The mean lifetime of a radical in its mobile state is denoted by θ. According to the assumptions of the model this means that $(m/\theta) \cdot (\xi n/M)$ mobile radicals enter into recombination reactions per unit time and $m/\theta[1 - (\xi n/M)]$ mobile radicals become trapped radicals per unit time. As each reaction converts nV_f trapped radicals into mobile ones, there are, on the average, $(m/\theta) \cdot (\xi n/M) - nV_f = m/\theta[(n/n_c)^2]$ trapped radicals destroyed and mobile radicals created per unit time. The general equations for rates of change of n and m also must allow for the possibility that mobile radicals may be produced, perhaps by radiation. A source term, κ, stands for the number of mobile radicals produced per unit volume. Taking into account the diffusion of mobile radicals puts a term, $D\nabla^2 m$ (where D is a diffusion constant), into the equation for the rate of change of m.

The equations which are obtained when all these mechanisms for change are combined are

$$\frac{dm}{dt} = \kappa - \frac{m}{\theta}\left[1 - \left(\frac{n}{n_c}\right)^2\right] + D\nabla^2 m \tag{11a}$$

$$\frac{dn}{dt} = \frac{m}{\theta}[1 - \beta n/n_c - (n/n_c)^2] \tag{11b}$$

where $\beta = \xi n_c/M$.

To apply these equations to obtain criteria for stability, κ is assumed zero and the initial behavior of solutions is considered as a function of the initial n, when the mobile radical concentration fluctuates from $m = 0$. The existence of a solution for m, which is initially exponentially increasing is taken as a criterion for instability. The change of n with time is neglected in this investigation of the stability of the solution of Eq. (11a). The boundary condition used is that m is zero at the surface. This is not intended to be taken literally, rather it is the simplest way of representing the effect of the surface in inhibiting the chain reactions. If one then attempts to write an eigenfunction of Eq. (11a) as $m = F(x, y, z)\, T(t)$, one obtains

$$\frac{dT(t)}{dt} = -\frac{T(t)}{\theta}\left[1 - \left(\frac{n}{n_c}\right)^2 + \theta D\lambda^2\right] \tag{12}$$

where

$$\nabla^2 F(x, y, z) = -\lambda^2 F(x, y, z) \tag{13}$$

Here λ^2 is an eigenvalue of Eq. (13). For the mixture to be stable we must have

$$1 - \left(\frac{n}{n_c}\right)^2 + \theta D\lambda^2 > 0 \tag{14}$$

for all eigenvalues λ. To apply this result to a deposition experiment, the solution for an infinite plane slab of thickness l is written. The result for the eigenfunctions and eigenvalues is

$$\begin{gathered} F_j(x) = \sin(j\pi x/l) \\ \lambda_j = j\pi/l \end{gathered} \qquad (j = 1, 2, 3, \cdots) \tag{15}$$

Defining the critical thickness, l_c, such that greater lengths are unstable, Eq. (15) leads to

$$l_c = \frac{\pi\sqrt{D\theta}}{\sqrt{(n/n_c)^2 - 1}} \tag{16}$$

It may be directly seen that the larger n is, the smaller the thickness at which a mixture will become unstable. For values of n only slightly greater than n_c the critical length becomes quite large and approaches infinity as $n \to n_c$. For $n < n_c$, Eq. (16) does not apply as the mixture is always stable [the inequality Eq. (14) is always fulfilled].

This equation (Jackson, 1959b) is applied to deposition experiments, wherein some indication of l_c as a function of n is obtained. Using these results Eq. (16) is used as an equation for n_c, the critical concentration. On this basis a rough estimate of 0.1% is obtained for n_c/M.

In applying Eqs. (11a) and (11b) to an irradiation experiment, the term $D\nabla^2 m$ is neglected and solutions are obtained for n and m initially equal to zero. An approximate solution is obtained for the experimental situation $\kappa\theta/n_c \ll 1$, i.e., when the number of radicals produced in the mean lifetime of a mobile radical is small compared to n_c. Saturation curves for the concentration as a function of time are then obtained essentially as a function of one parameter, $\beta = \xi n_c/M$. If the following definitions are introduced:

$$\tau = \frac{\kappa t}{n_c}; \qquad y = n/n_s \tag{17}$$

$$n_s = n_c(\sqrt{1 + \beta^2} - \beta)$$

then the solution which represents the build-up of trapped radical concentration (proportional to y) as a function of time (proportional to τ) is

$$\tau = y + \beta \ln (1 - y) \left(1 + \frac{y}{f(\beta)}\right)^{f(\beta)} \tag{18}$$

where

$$f(\beta) = \frac{\sqrt{1 + \beta^2} + \beta}{\sqrt{1 + \beta^2} - \beta}. \tag{18.1}$$

All curves saturate at $y = 1$, which means that the saturation concentration is n_s as given in Eq. (17). It may be seen that n_s is always smaller than n_c. In Fig. 2, some representative saturation curves are shown. In general for small β values the curve stays linear most of the way up and then sud-

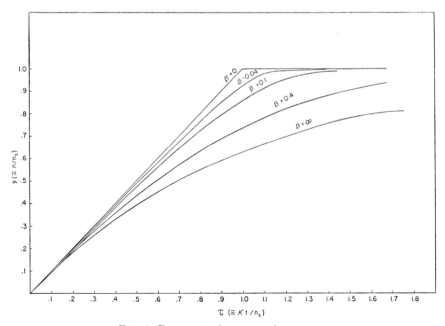

FIG. 2. Representative saturation curves.

denly saturates. For large values of β, the curves become nonlinear much earlier and the saturation is a much more gradual process. An experimental saturation curve could be used to measure β.

It is noted that the results of the two experiments discussed, one involving deposition, the other irradiation, yield experimental values of n_c and β, respectively. These two measurements in principle determine the two parameters introduced in setting up the model ξ and V_f according to:

$$\xi = \beta(n_c/M)^{-1}; \qquad MV_f = (\beta n_c/M)^{-1}$$

This theory of chain reactions, although it does give a qualitative explanation of some experimental results, is obviously too crude to be expected to be quantitatively correct. For one thing, the fact that the temperature is not treated in detail is an omission of this theory which a more exact theory would have to take into account. A more detailed theory would, however, be extremely difficult both because of the additional mathematical complications, and also because one is not at present on a secure footing with respect to the details of the physical processes that occur in frozen radical mixtures. From this viewpoint there is perhaps a virtue to such a very simple picture as the one described in this section.

V. Conclusion

As detailed theoretical investigations of systems with trapped radicals have been undertaken only within the last few years, it really should not be too surprising that the theoretical results which have been discussed here are so meager and so tentative. Furthermore, in view of the lack of familiarity with trapped radical systems and the resulting lack of intuition as to the important physical processes in such systems, it is understandable that there should have been a certain amount of groping for proper models to describe their behavior. Nevertheless, on the positive side, the theoretical work herein described not only has done much to point to profitable directions for future research but has already provided many insights into the nature of trapped radical mixtures.

The detailed theory originally presented to explain the α-lines in nitrogen-deposition experiments proved inadequate to explain all of the complexities of the actual experimental situation. It did, nevertheless, show qualitatively how the observed effects probably come about owing to perturbations of the nitrogen atom by the lattice. The theory of the α-lines, it should be recalled, was a justification of the identification of these lines with the nitrogen atom transition—an identification which is no longer questionable. Another positive achievement of this type of theory is the apparently satisfactory account it gives of the β-lines due to the oxygen atom.

The theory of chain reactions of trapped nitrogen atoms in a nitrogen molecule lattice is undoubtedly highly oversimplified. However, it provides a generally satisfactory way of thinking about the observed autocatalytic recombination reactions. At least for the present, the theory is in reasonable agreement with experiment. An important consequence of the chain reaction theory is its implication with respect to the possibility of obtaining high concentrations of trapped radicals—high enough to be interesting from the viewpoint of storing energy. If the interpretation of Fontana's deposition experiment given by the theory is correct, then not only are the maximum concentrations pitifully small (with regard to energy storage) but what is worse, going to temperatures lower than 4° K would profit very little. Of course, these considerations are limited in that they apply only to mixtures of nitrogen atoms and nitrogen molecules and to the kind of solid formed when the mixture is deposited from the gas. One still cannot definitely say whether, using different matrices or different radicals, it still might be possible to obtain mixtures which store appreciable amounts of energy. In view of the extremely small maximum percentage obtained on the basis of the chain reaction analysis, one must at least say that such a prospect now seems rather unlikely.

The statistical theories of isolation of trapped radicals in matrices are apparently not applicable to highly reactive systems such as nitrogen, where the radicals are more firmly trapped. The isolation theory is a perfectly good theory—waiting for a suitable physical system.

At present there are great quantities of experimental results which have not been given detailed theoretical explanation. It is to be hoped that these data will be used to obtain a more precise picture of the nature and properties of trapped radical mixtures. This approach, of working from experimental results rather than via *ab initio* theories seems at present to be more profitable from the viewpoint of making progress in understanding these systems. For example, it would appear worth while to re-examine the theory of the spectrum of the nitrogen atoms in light of the new, detailed results of Peyron et al. (1959). Hopefully, a theoretical analysis of these results would yield information on the nature of the radiating atoms. Of particular interest theoretically are the lines which are apparently simultaneous transitions involving an electronic jump in an N atom and a single vibrational quantum in a nitrogen molecule. Theoretical analysis of these lines could certainly tell us much about the relationship between the excited atom and the molecules in the lattice.

Recently, Cole and McConnell (1958) have measured the zero-field splittings in the ESR spectrum of atomic nitrogen in a nitrogen molecule lattice at 4.2° K. This, too, might prove to be a useful tool in locating the N atom in the lattice. If one were able to calibrate this splitting as a function of external electric field (perhaps by a molecular beam experiment) one could then look for a site in the nitrogen lattice where the field would be of the right magnitude.

The present state of spectroscopic information is obviously far in advance of the theoretical analyses of these data. However, with regard to results in the kinetics of processes involving trapped radicals, the experimental situation is far more difficult. So much so that refinements, even on a theory as approximate as the one of chain reactions, would seem to be of marginal value until a greater degree of control and reproducibility could be obtained in the relevant experiments. It appears that the control of the deposition process would, from the viewpoint of obtaining reproducibility and under-standing, be a subject worthy of both experimental and theoretical study. With improved knowledge of the nature of the deposited solid, as a function of the temperature, flow rate, and composition of the gas, one could go more deeply into the properties of the mixtures as determined by their structure.

REFERENCES

Bass, A. M., and Broida, H. P. (1956). *Phys. Rev.* **101,** 1740. Spectra emitted from solid nitrogen condensed at 4.2°K from a gas discharge.

Chessin, P. L. (1959). *J. Chem. Phys.* **31,** 159. Free radical statistics.

Cole, T., and McConnell, H. M. (1958). *J. Chem. Phys.* **29,** 451. Zero field splittings in atomic nitrogen at 4.2°K.

Fontana, B. J. (1958). *J. Appl. Phys.* **29,** 1668. Thermometric study of the frozen products from the nitrogen microwave discharge.

Fontana, B. J. (1959). *J. Chem. Phys.* **31,** 148. Magnetic study of the frozen products from the nitrogen microwave discharge.

Goldberg, H., and Herzfeld, C. M. (1958). *Bull. Am. Phys. Soc.* [II] **3** (3), 180. Crystal field theory of the spectrum of O atoms trapped at 4°K.

Golden, S. (1958). *J. Chem. Phys.* **29,** 61. Free radical stabilization in condensed phases.

Herzfeld, C. M. (1957). *Phys. Rev.* **107,** 1239. Theory of the forbidden transitions of nitrogen atoms trapped in solids.

Herzfeld, C. M., and Broida, H. P. (1956). *Phys. Rev.* **101,** 606. Interpretation of spectra of atoms and molecules in solid nitrogen at 4.2°K.

Jackson, J. L. (1959a). *J. Chem. Phys.* **31,** 154. Dynamic stability of frozen radicals, Pt. I—Description and application of the model.

Jackson, J. L. (1959b). *J. Chem. Phys.* **31,** 722. Dynamic stability of frozen radicals, Pt. II—The formal theory of the model.

Jackson, J. L., and Montroll, E. W. (1958). *J. Chem. Phys.* **28,** 1101. Free radical statistics.

Peyron, M., Hörl, E. M., Brown, H. W., and Broida, H. P. (1959). *J. Chem. Phys.* **30,** 1304. Spectroscopic evidence for triatomic nitrogen in solids at very low temperatures.

Ufford, C. W., and Gilmore, R. M. (1950). *Astrophys. J.* **111,** 580. Multiplet intensities for the lines $^2D - {}^4S$ of NI.

11. Trapped Radicals in Astrophysics[1]

BERTRAM DONN[2]

Department of Physics, Wayne State University, Detroit, Michigan

I. Introduction

A considerable part of the mass of the galaxy, especially in the spiral arms, consists of diffuse gas and dust not associated with stars. The sun and many other stars have various forms of matter associated with them, much of the material being at considerable distances from the stars. In the atmosphere of cool stars, temperatures are low enough for molecules to form in abundance, and the formation of solid particles under certain conditions may be possible.

For the present discussion the significant feature of these regions of space is the occurrence of low temperatures which allow chemical processes, including radical reactions in solids, to become significant. As is characteristic of astronomy, the temperature range greatly exceeds that normally encountered in the laboratory. Recent laboratory techniques, however, have extended the low temperature limit to about 1°K, as has been discussed in detail, in previous chapters, and shock tube studies have extended

[1] Most of the research on which this chapter is based has been supported by the National Science Foundation. Some of the work was carried out as consultant to the Free Radical Research Section of the National Bureau of Standards and supported by the Department of Defense.

[2] *Present address:* Goddard Space Flight Center, National Aeronautics and Space Administration, Washington, D. C.

the upper limit to several thousand degrees. Pressures in the regions where astronomical chemistry is important are always extremely low.

The best-known areas of cosmic chemistry and the ones principally studied have been concerned with objects available to the laboratory. These are samples of the earth's crust and meteorites. Deductions about the chemistry of the earth's interior have also been made. In recent years the earth's atmosphere has come in for extensive analysis, stimulated by the availability of rocket data of the upper atmosphere and by the International Geophysical Year (I.G.Y.) program in general. Spectroscopic and theoretical investigations of the atmospheres of the other planets have also been carried out.

This brief sketch indicates the scope of cosmic chemistry. For the earth and meteorites the chemistry is essentially complete, and for the most part geochemistry analyzes the end result of a long and complex series of processes which have taken place from 0°C to a few thousand degrees.

The interstellar medium is by far the largest system where chemical processes still are significant. Another extensive area for investigation lies much closer and includes comets and associated phenomena in the solar system. These two areas have important characteristics in common. In both, solid matter is important and temperatures are below about 150°K. Atoms and radicals are dominant in the interstellar and interplanetary gas. They also most probably played a major role in the primeval solar nebula, directly or indirectly through their recombination products. Consequently, in the largest cosmic systems where chemical processes play a significant role, the conditions are such that we are concerned with the behavior of free radicals at low temperatures.

II. Chemistry of Interstellar Matter

A. Composition and Density

Interstellar space presents a fascinating and complex series of physico-chemical problems. Some of the most fundamental of these are closely related to the subject of this volume—namely, the formation and trapping of free radicals.

It is now well recognized that there is a considerable quantity of matter dispersed throughout interstellar space (Greenstein, 1951; Aller, 1954; Dufay, 1957a). Most of this consists of an extremely low density gas with a composition that appears to follow the usual cosmic abundance ratios (Stromgren, 1948). Table I lists the relative abundances of the more common elements according to Suess and Urey (1956).

Some of these elements have been undetected because of the inaccessibility of their resonance spectrum lines, which tend to lie in the ultraviolet

TABLE I

ABUNDANCES OF THE MORE COMMON ELEMENTS[a]

Element	Abundance $(Si = 10^6)$	Density in space $(atom/cm^3)$
H	4×10^{10}	1
He	3×10^9	1×10^{-1}
O	2×10^7	5×10^{-4}
Ne	9×10^6	2×10^{-4}
N	7×10^6	2×10^{-4}
C	3.5×10^6	1×10^{-4}
Si	1×10^6	2.5×10^{-5}
Mg	9×10^5	2×10^{-5}
Fe	6×10^5	2×10^{-5}
S	4×10^5	1×10^{-5}
A	1×10^5	2.5×10^{-5}
Al	9×10^4	2.5×10^{-5}
Ca	5×10^4	1×10^{-6}
Na	4×10^4	1×10^{-6}
Ni	3×10^4	1×10^{-6}

[a] According to Suess and Urey (1956).

(UV). Although the noble gases will not take part in reactions, their relatively large concentration can have significant effects. Finally we note that carbon, silicon, and the metals will be mostly ionized because their ionization potentials are lower than for hydrogen. The corresponding ionizing radiation, therefore, will not be depleted by the photoionization of hydrogen.

In addition to the relative cosmic abundance, Table I shows the average concentration of the elements in space. This column clearly displays the extremely low mean densities. The concentrations may get larger by factors of 10 or 100 in the regions of higher-than-average density.

In addition to the atoms so far observed, two diatomic radicals are known, CH and CN. They have concentrations of about $10^{-8}/cm^3$ (Bates and Spitzer, 1951). The diatomic molecules of the most abundant atoms, H_2, O_2, and N_2 are to be expected but unfortunately under interstellar condition they produce no absorption lines in the currently observable region of the spectrum (McKellar, 1941; Herzberg, 1955). Molecules of some less abundant atoms and a number of diatomic radicals of common elements in space have lines in the visible region, and an extensive search for them would be extremely valuable.

In addition to the gaseous stratum in the galaxy, a dust component consisting of small particles also occurs (Aller, 1954; Greenstein, 1951; Dufay, 1957a), the two being rather thoroughly mixed. The only available means of analyzing the dust depends upon deriving its optical properties from the manner in which the small particles interact with starlight. Only very limited information can be obtained in this way.

If one assumes that the solid particles in space interact with light through classical electromagnetic theory expressed by Maxwell's equations, the following results are obtained. The effective diameter of the particles is a few tenths of a micron. Further, the particles are most likely dielectric or nonconducting (van de Hulst, 1946, 1948) although the case is not definitely settled (Guttler, 1952). Because they polarize light, they must be elongated rather than spherical. Finally, their alignment in space can most plausibly be explained by attributing magnetic properties to the grains (Davis, 1955).

It has, however, been pointed out (Platt, 1956) that, considering quantum effects in much smaller-sized particles, the general optical characteristics of the obscuring medium can also be explained. With such grains, diameters of 10 to 100 A are possible. More detailed studies of this hypothesis need to be carried out. It is in general agreement with certain theoretical considerations of particle formation (Platt and Donn, 1956).

For the micron-sized grain, the density in clear regions of space is 1 particle in 10^{12} cm³ and may rise to 1 particle in 10^9 cm³ in more concentrated regions producing high obscuration of starlight. About one hundred times as many of the smaller particles are required for the same optical effect although they possess only one-tenth as much mass.

B. Temperatures in Space[3]

Among the characteristic features of interstellar space are the large deviations from thermodynamic equilibrium which occur. As a consequence, results may not be simply carried over from terrestrial laboratory data. Nor is it clear how to modify the usual results because we have insufficient theoretical knowledge of the effect of the nonequilibrium conditions. Fortunately, temperatures may be ascribed to the different degrees of freedom with a fair degree of approximation, especially for translational energy. These results are given in Table II. The degree of ionization, however, must be determined separately for each atom (Stromgrem, 1948), and an ionization temperature cannot be assigned.

The black body temperature is based on the energy density of integrated starlight. Much higher color temperatures for the radiation occur because of the great excess of shortwave radiation, as we are dealing with highly diluted starlight. HI regions are regions where hydrogen is neutral, and HII regions are those in which hydrogen is ionized. The high propor-

[3] See Dufay (1957b).

TABLE II

TEMPERATURES IN SPACE (°K)

Black body	3°
Color temperature of radiation	5000°IR, 15,000°UV
Vibration temperature of molecules	3°
Excitation temperature of atoms	3°
Kinetic temperature of gas, HI region	100°
Kinetic temperature, HII region	1000 to 10,000°
Temperature of grain	20°

tion of ultraviolet galactic light will be effective in causing ionization and molecular dissociation.

C. FORMATION AND COMPOSITION OF GRAINS

A study of the solid particles in space can be divided into a number of closely related parts which at this stage must be considered separately (Donn, 1953). As the two preceding sections show, the low temperature behavior of atoms and free radicals must play a significant role. That chemical reactions may enter the problem was first suggested by ter Haar (1944), although he did not consider them in his analysis. Shortly after, van de Hulst (1948, 1949) explicitly introduced chemical reactions to build up molecules on the grain. Because of the low activation energy of many free radical reactions and the high gas temperature, 10,000°K, used in his calculations, he concluded that the grain would consist primarily of hydrogen-saturated compounds, water, ammonia, and methane.

The subject was subsequently restudied with particular regard to chemical problems (Donn 1953, 1954), and it was shown that these were more complex than had been assumed. A significant feature for the subject of this volume are the very low temperatures which occur in the problem. In HI regions where chemical processes are most likely, the gas kinetic temperature is only 100°K. The grain temperature, about 20°K, also has a dominant role. For ordinary surface-catalyzed reactions chemisorption is an important step (Taylor, 1951). With the very low collision frequency in space, adsorption is necessary to bring the particles together, although here physical adsorption is to be expected and the catalytic effect may be small.

Table I shows the overwhelming effect of atomic hydrogen for chemistry in space. One sees also that the next important elements are oxygen, nitrogen, and carbon. Hydrogen atoms strike 1 cm^2 of surface at a rate of 5×10^4 per second, whereas for the next three elements the collision frequency is about one per second. For other elements the frequency may be obtained by multiplying the hydrogen frequency by the abundance ratio over the square root of the mass. A micron-sized grain will be struck by about three hydrogen atoms per day.

By the type of apparently fortunate coincidences which occur frequently in science, at the time when the significance of low temperature free radical chemistry for interstellar matter was recognized, new techniques were being applied to study such phenomena in the laboratory. Such investigations are still in their early stages, and theoretical and experimental results are still too incomplete to warrant definite conclusions.

The recent work on stabilizing radicals at low temperatures is extremely relevant to interstellar matter and furnishes strong evidence that some small fraction of the grain will consist of uncombined radicals. The laboratory data discussed in the first part of this book indicate that the radical concentration in bulk solids is below 1 %. Because of the extremely slow deposition rate, the high radical concentration in the interstellar gas, and the small size of the grain, the radical concentration in the grain need not necessarily be the same as for the terrestrial experiments.

At present the only theoretical discussion available for carrying out an extrapolation from gross laboratory dimensions to the very small grains is the analysis by Jackson in Chapter 10. Surface effects are neglected in that treatment, and Jackson remarks that higher concentrations would be expected before instability develops in small volumes. For nitrogen atoms in a molecular nitrogen matrix, Jackson's analysis leads to a radius of about 10^{-7} cm around a pair of recombining atoms as the volume in which the temperature becomes high enough for the radical traps to be destroyed. As the maximum radius of the condensed grain is about 10^{-5} cm, the particle, particularly during the early stages of growth, would be only slightly larger than the effective unstable zone around a recombined molecule. In such situations the correction suggested by Jackson may become appreciable.

The experimental data on reactions of atoms and radicals of astronomical interest will now be reviewed. These data establish a reasonably firm basis for estimating the chemical properties of cosmic matter at low temperatures, although much remains to be done.

Experiments on condensing radicals from the gas or producing them by irradiation of solids lead to essentially the same results. Hydrogen atoms have been produced and retained in solids at 77°K (Livingston et al., 1954). Nitrogen atoms have been maintained until 35°K (Broida and Pellam, 1954). Oxygen has not been extensively studied, but there are indications (Broida and Pellam, 1955) that oxygen atoms can exist at temperatures of astrophysical interest. Robinson and McCarty (1958a, b) have identified NH_2, NH, OH, and HNO by their optical spectra at 4.2°K. The hydroxyl radical has also been reported at 77°K, but there is some uncertainty and conflict associated with the interpretation of the electron spin resonance (ESR) spectra, as Franklin and Broida (1959) point out.

Unstable species from photodissociation of diazomethane have been observed by Milligan and Pimental (1958), one of which they propose as the long-sought-for CH_2 radical. In addition a trapped unstable tautomeric form of diazomethane was also suggested. A red glow appeared between 20 and 25° and lasted until 30°K, indicating the limits of stability.

The original work by Rice and associates (Rice and Freamo, 1951, 1953) on condensing reactive material showed that some unstable species producing a blue color was condensed at 77°K and remained stable up to 148°K. The composition of the active substance still remains obscure (Franklin and Broida, 1959).

A number of other organic radicals have been identified or their existence suggested in low temperature inert matrices including CH_3 at 77°K. These radicals have been discussed in the summary by Franklin and Broida (1959).

Only small concentrations of radicals can be stabilized, as the determinations of concentration clearly indicate. Paramagnetic resonance spectroscopy has always given concentrations of the order of 0.1 %. Some thermal measurements have led to concentrations of several per cent (Minkoff et al., 1959), but these results appear to include sources of heat other than atom recombination.

In discussing the related problem of the composition of comets, it was pointed out (Donn and Urey, 1956, 1957) that in addition to trapped radicals, reactive molecules would be present in low temperature cosmic matter. There has since been increasing evidence that a complex array of unsaturated molecules is to be expected.

The most extensively studied problem has been that on the blue material from dissociated hydrazoic acid. A clear indication of the experimental difficulties involved is the present uncertainty over the nature of the blue deposit referred to earlier. For our purpose, however, there are definite conclusions regarding the products derived by careful warming-up of the deposit. In that case one finds some HN_3, NH_4N_3, and hydrogen, but little nitrogen.

It has been known for some time that when oxygen is subjected to a discharge and then condensed in a liquid nitrogen trap, a considerable yield of ozone is obtained. Similarly with water vapor, a high proportion of hydrogen peroxide is found (Urey and Lavin, 1929). Recent work (Scheer and Klein, 1959) indicates that when an atomic hydrogen stream strikes an oxygen film at 20°K, the warm-up yields appreciable amounts of hydrogen peroxide. The experiments by Milligan and Pimentel (1958) indicate similar results for simple organic radicals.

There seems ample evidence to warrant the conclusion that an appreciable concentration of unsaturated, reactive molecules will be found in solid particles in space that have grown by condensation of atoms and radicals. Only very small proportions of trapped radicals are found in macroscopic solids, but larger concentrations may occur in small grains in space. The consequences of these chemical species for astronomical problems involving low temperature material need to be considered. Some of the consequences are treated elsewhere in this chapter.

Further work on reactions of free radicals in or on solids at low temperatures is extremely important for astronomical problems of the type considered in this discussion. Fortunately, a number of such studies are under way, some with the astronomical application directly in mind.

A completely different model for the interstellar grain has been proposed by Platt (1956) in which the optical properties depend upon quantum processes rather than classic electromagnetic theory. In this hypothesis the grain itself is a complex radical in which the individual atoms, ions, and electrons combine to form a particle in which the atoms are all chemically bonded. Because of the random growth at low temperatures, unsaturated bonds would exist. These, together with the generally unbalanced charges that will develop, mean the particle would not have filled energy bands as are required in a dielectric. By applying the free electron theory of conjugated molecules (Ruedenberg and Scherr, 1953) to this case, Platt was able to deduce optical properties in general agreement with observations. Because of the small size, 20 A, and relatively few atoms, there would be statistical fluctuations in the dimensions of the three axis sufficient to account for the polarization, if alignment takes place by the Davis-Greenstein mechanism. Because of their structural properties, these particles will tend to be paramagnetic also and the alignment process should apply.

The type of structure proposed by Platt does not seem unreasonable, as a preliminary analysis of the problem has indicated (Platt and Donn, 1956). A comprehensive theory of chemical processes in space does not yet exist. Indeed a complete formulation of the problem has not appeared. A brief discussion has been given by ter Haar (1944) and the writer has also discussed it on several occasions (Donn, 1953, 1954; Platt and Donn, 1956), but no detailed investigation has yet been published.

III. Chemistry of the Solar System

A. Interplanetary Dust

To a large extent the situation with regard to dust in the solar system at distances several times the radius of the earth's orbit is similar to that described in the preceding section. Table III shows the temperature a black body would attain at various distances from the sun.

TABLE III

BLACK BODY TEMPERATURES IN THE SOLAR SYSTEM

Distance (astronomical units)	Planet positions	Temperature (°K)
1	Earth, 1	300
	Mars, 1.5	
2		215
4		150
	Jupiter, 5.2	
8		110
	Saturn, 9.5	
10		90
	Uranus, 19	
20		65
	Neptune, 30	
40	Pluto, 40	45
100		30

The astronomical unit (A.U.) is defined as the mean distance of the earth from the sun. To give a better reference for the temperature variation in space, the positions of the planets are included with their mean distances shown. Intermediate temperatures may be calculated from the expression (Russell *et al.*, 1938)

$$T(°K) = \frac{300}{\sqrt{r}}$$

where r is in astronomical units.

In the initial stages of the solar system before the planets formed, densities were much higher than now. At 1 A.U., current density estimates are about 10^3 atoms per cubic centimeter (Whipple and Gossner, 1949), nearly all hydrogen. Before the sun and the planets formed, the mean density was about 10^{10} times greater or about 10^{-11} gm/cm^3, but the temperature would have been near 3°K.

It would take us too far afield to go further into the development of such a system. A discussion of this problem and additional references will be found in Urey (1952). Again it should be emphasized that the discussion as to the chemical composition of the material applies here also because of the extremely low initial temperatures. The initial particles would be a mixture of saturated and unsaturated compounds, perhaps in equal abundance with a small concentration of radicals.

As the sun evolved and became hot, the matter within about 2 A.U. would react to form more stable compounds. Beyond the orbit of Pluto little change would occur, as Table III shows. In addition the reduction in density as the planets developed would increase the stability of the nonequilibrium species.

When the sun attained its present characteristics, additional means of increasing the reactive constituents would arise. For the ultraviolet radiation and the energetic corpuscular emission from the sun would ionize and dissociate matter. The formation of radicals and active species in general has been fairly extensively investigated. Other chapters of this book describe the use of such means of activation to radical trapping in solids. Because grains in space remain undisturbed for extremely long periods, they can be expected to build up a concentration of reactive material independently of the original composition.

Most of the quantitative studies of chemical effects of corpuscular radiation have been in gases or liquids. Very few data on rates are available for solids at low temperatures. At present, therefore, only rough estimates are available. G values, or the number of products produced per 100 ev of absorbed radiation, are roughly the same for a wide number of chemical systems and for protons, alpha rays, X-rays, and gamma rays over a considerable range of energies in aqueous solution (Hart, 1954). Wall et al. (1959) have found G equal to 0.4 for Co^{60} gamma rays for the production of stabilized nitrogen atoms at 4.2°K. Previous studies have shown that considerable recombination occurs, and, therefore, the total production of radicals would be higher.

In aqueous solutions the yield of H_2O_2 runs about one-third that of the decomposition of H_2O at about 20 Mev and increases to about two-thirds at 0.29 Mev for light particles (Hart et al., 1956). These results are in general agreement with hydrogen peroxide concentrations around 50% found in the warm-up of condensed discharge products of water (Ghormley, 1957). Systematic studies of radiation chemistry of pertinent solids are needed for cosmic chemistry.

A rough indication of the results to be expected for the interplanetary dust can be obtained from the data recorded in Table IV.

TABLE IV

RADIATION CHEMISTRY OF INTERPLANETARY DUST

Flux of solar corpuscular emission[a]	10^6 protons/cm² sec
Particle energy[a]	10^5 ev
Energy flux	10^{11} ev/cm² sec
G (radical production)	1 to 10
Radical production	10^9/cm² sec
Radical production (grain radius 10^{-4} cm)	10^{-7}/sec
Radical production per grain per 10^6 years	10^6
Radical concentration per 10^6 years	10^{-6} gm/gm

[a] Kiepenheuer, K. O. (1953) in "The Sun" (G. P. Kuiper, ed.), p. 437. Univ. Chicago Press, Chicago, Illinois.

The figure for the total radical production in a million years should be increased somewhat to allow for UV radiation, low energy cosmic radiation, and the possibility of high energy particles trapped in interplanetary space by magnetic fields. It would appear from these rough calculations that radiation chemical effects could become significant for about 0.1-μ dust particles in astronomical times. These effects would be divided between trapped radicals and recombination products.

B. COMETS[4]

At the present time it appears that among the most significant objects for the study of low temperature cosmic chemistry are those remarkable and mysterious celestial objects, the comets. A growing interest has produced many observable data on comets and also led to the development of new ideas and theories about their nature. When these are taken together with recent work on low temperature chemistry, one finds strong indications that these objects are indeed a cosmic low temperature laboratory (Donn, 1959). There is good hope that intensive study may not only reveal current behavior, but serve as a means to analyze chemical and physical processes that occurred in the early stages of the solar system.

When comets are seen near the sun, they consist of three parts; the tail, the coma or head, and the nucleus. Because of the extremely low density, no reactions can take place in the head or tail except possibly very close to the nucleus (Delsemme and Swings, 1952; Haser, 1955; Donn and Urey, 1957). It is the unobservable nucleus which forms the essential permanent part of the comet as it revolves about the sun. The diffuse gas in the prominent head or tail comes from the nucleus by processes which increase in intensity as the sun is approached.

By examining the spectra of comets and the ejection of matter from the nucleus in terms of recent work in low temperature free radical chemistry, a reasonable model for comet structure and activity can be obtained (Whipple, 1950; Haser, 1955; Donn and Urey, 1956, 1957). Comet orbits are such that nearly all the time the comet is so far from the sun that the black body temperature would be below 150°. As comets are small objects, the nuclei having been estimated at a few kilometers, no appreciable radioactive or gravitational heating would have occurred. When comets do get close to the sun, the loss of volatile gases will still keep the temperature below about 150°K.

The nucleus of a comet is too small to be observed, but when comets first become visible at 4 to 5 A.U. they already have coma about 10^4 km in diameter. Table IV shows that they have been ejecting matter at tem-

[4] For a general description see N. Bobrovnikov in "Astrophysics" (J. A. Hynek, ed.). McGraw-Hill, New York, 1951, or Kometen by K. Wurm, V54, Encylopedia of Phys., (S. Flugge, ed.) Springer, Berlin, 1959.

TABLE V

COMET HEAD SPECTRA

Component	Distance at which first observed (A.U.)
Continuum-dust	When spectrum is first observed
CN	3
C_3 , NH_2	2
C_2	1.8
OH, NH, CH	1.5
Na	1
Fe, Ni	$\sim$0.01

peratures near 100°K. The known data on comet material obtained from spectroscopic observations are summarized in Table V (Swings and Haser, 1956).

One of the characteristics of comets is their tendency to flaring or jet action. In the former a nearly spherical halo suddenly forms and gradually disappears, whereas the latter phenomena refer to emission of matter in narrow streams, generally toward the sun. In all cases velocities are low, usually about 0.5 to 1 km per second.

Whipple (1950) proposed a comet model in which the nucleus consisted of a loosely compacted solid composed primarily of solid water, ammonia, and methane, and about 25 % meteoric matter. In order to overcome some of the difficulties with this model, Haser (1955) and Donn and Urey (1956, 1957) modified the structure to include a significant fraction of free radicals. The suggestions by Donn and Urey (1957) also included an appreciable amount of thermally stable but reactive molecules such as H_2O_2 , C_2H_2 , C_2H_4 . Because of the low experimental concentration of radicals and the low production rate in interplanetary matter estimated in the previous section, we must now assign the chemical effects to the reactive recombination products rather than trapped radicals.

Chemical energy of reactive species preserved at low temperatures seems to be the most reasonable way of explaining comet activity at low temperatures. For this to happen, some nonthermal means of causing chemical reactions must be found. For comet Schwassmann-Wachmann I, which has had numerous flares while always beyond the orbit of Jupiter, some correlation of flares and terrestrial magnetic storms produced by streams of solar corpuscular radiation has been found (Richter, 1954). However, the general association between cometary and solar activity is still unsettled (Beyer, 1952; Houziaux and Battiau, 1957).

The mechanism proposed for the activation of unstable cometary matter depends upon the heating effect of a stream of high energy particles. Each particle produces a small high temperature volume in which the material is no longer frozen (Yankwich, 1956; Seitz and Koehler, 1956). The mobility enables the molecules to react, and the energy thus released causes further reaction. A detailed comparison of this process with observations based on results of hot-atom chemistry has been published (Donn and Urey, 1957). Because of the extrapolation to the uncertain structure of of comet nuclei, this hypothesis for the instigation of comet activity is very tentative. Experimental work to study chemical effects in reactive material by energetic atoms such as that described here needs to be performed. The over-all model and mechanism seem to be very useful for accounting for comet activity and suggesting additional observations and laboratory experiments.

C. PLANETS, SATELLITES, AND METEORITES

The disk of Jupiter shows a number of bands parallel to its equator and also various colored areas. Most prominent is the great red spot. Yellows and browns predominate, but other colors have also been reported (Peek, 1958). Saturn is generally similar although, being twice as distant, it is more difficult to observe. These markings must be atmospheric features such as clouds. In such a case it is difficult to explain the colors since simple compounds of the abundant elements generally form white solids.

As a result of the colors obtained by condensing various radicals, Rice (1956) has proposed a trapped radical mechanism for the color on Jupiter. Ultraviolet radiation from the sun or electrical storms was suggested as the source of the radicals. More recently Papazian (1959) has proposed that high energy protons trapped in van Allen-type radiation belts around Jupiter may be the means of radical production. In this case the analysis of Section III, A would apply. Since a very small concentration of radicals in some cases can produce an intense color (Mador, 1954), a mechanism with only low efficiency can work here.

As the lunar surface is also exposed to intense ultraviolet and energetic corpuscular radiation from the sun, Platt (1958) has suggested that the properties of the lunar surface may be modified from their normal condition. The absence of an atmosphere should increase the tendency of the lunar surface to resemble radiation-damaged material with distorted structures which are highly absorbing in the visible. The writer is aware of no information on radiation effects on rocks or lunar-type dust. Small, highly reactive radicals are not likely for such material, and the high daytime temperatures, up to 100°C, would certainly prevent any build-up of reactive material.

Meteorites have undergone a complex history, as Urey has indicated (Urey and Donn, 1956), and some rapid melting process appears to be necessary. In order to account for the rather short time scale of that heating, a chemical heating mechanism was proposed, based on the reactive comet model mechanism. As the material condensed to form the larger asteroidal objects and possibly planets also, the initial condensate would be a mixture of stable and reactive molecules plus a small concentration of radicals. If a reaction were set off after they had grown to many kilometers, considerable heating would be produced, sufficient for the reduction of iron and the melting of iron and silicates. After cooling, more material of the same composition as the original would accumulate and the process would be repeated. This process may have taken place many times until asteroidal- or planetary-sized objects were formed.

This mechanism is rather speculative, as is the prediction of the behavior of large accumulations of the reactive material.

On the other hand, it seems almost certain that the solar system started from material of the reactive composition considered and that the structure of the meteorites calls for rapid heating.

The continued observations of celestial objects for which chemistry plays a significant role and the concomitant laboratory investigations of simulated cosmic matter should give new insight in many astronomical problems. Simultaneously, there will develop an understanding of new areas of chemistry with the heretofore terrestrial restrictions removed. With the prospect of space research and observation beyond the earth's atmosphere now opening, the prospects for cosmic chemistry are indeed bright.

REFERENCES

Aller, L. H. (1954). *In* "Astrophysics—Nuclear Transformations, Stellar Interiors and Nebulae," Chapter 6. Ronald Press, New York.

Bates, D. R., and Spitzer, L. (1951). *Astrophys. J.* **113**, 441. The density of molecules in interstellar space.

Beyer, M. (1952). *Mém. soc. roy. sci. Liège, Collection in 4° 13*, 236. Brightness of comets and solar activity.

Broida, H. P., and Pellam, J. R. (1954). *Phys. Rev.* **95**, 845. Phosphorescence of atoms and molecules of solid nitrogen at 4.2°K.

Broida, H. P., and Pellam, J. R. (1955). *J. Chem. Phys.* **23**, 409. Preparation of solid ozone and atomic oxygen.

Davis, L., Jr. (1955). *In* "Vistas in Astronomy" (A. Beer, ed.), Vol. I, p. 336. Pergammon Press, London. Theories of interstellar polarization.

Delsemme, A. H., and Swings, P. (1952). *Ann. astrophys.* **15**, 1. Hydrates of gas in cometary nuclei and in interstellar grains.

Donn, B. (1953). *Astron. J.* **58**, 38 (Abstr.). The problem of the evolution of interstellar grains.

Donn, B. (1954). *Mém. soc. roy. sci. Liège, Collection in 4° 15*, 571. Some chemical problems of interstellar grains.

Donn, B. (1959). *Astron. J.* **64**, 126(Abstr.). Comets and the chemistry of matter in space.

Donn, B., and Urey, H. C. (1956). *Astrophys. J.* **123**, 339. On the mechanism of comet outbursts and the chemical composition of comets.

Donn, B., and Urey, H. C. (1957). *Mém. soc. roy. sci. Liège, Collection in 4°* **18**, 124. Chemical heating processes in astronomical objects.

Dufay, J. (1957a). "Galactic Nebulae and Interstellar Matter," p. 129. Philosophical Library, New York.

Dufay, J. (1957b). "Galactic Nebulae and Interstellar Matter," p. 237. Philosophical Library, New York.

Franklin, J. L., and Broida, H. P. (1959). *Ann. Rev. Phys. Chem.* **10**, p. 145. Trapped energetic radicals.

Ghormley, J. A. (1957). *J. Am. Chem. Soc.* **79**, 1862. Warming curves for the condensed product of dissociated water vapor, and for hydrogen peroxide glass.

Greenstein, J. L. (1951). *In* "Astrophysics" (J. A. Hynek, ed.), p. 256. McGraw-Hill, New York. Interstellar matter.

Guttler, A. (1952). *Z. Astrophys.* **31**, 1. Über die Materie der interstellaren Korner.

Hart, E. J. (1954). *Ann. Rev. Phys. Chem.* **5**, 139. Radiation chemistry.

Hart, E. J., Ramler, W. J., and Rocklin, S. R. (1956). *Radiation Research* **4**, 378. Chemical yields of ionizing particles in aqueous solutions: effects of energy of protons and deuterons.

Haser, L. (1955). *Compt. rend. acad. sci.* **241**, 742. The conservation of free radicals at low temperature and structure of the nuclei of comets.

Herzberg, G. (1955). *Mém. soc. roy. sci. Liège* **15**, 291. Laboratory investigations of the spectra of interstellar and cometary molecules.

Houziaux, L., and Battiau, L. (1957). *Bull. acad. roy. méd. Belg.* [5] **43**, 171. Note sur les relations entre les activités solaire et cométaire.

Livingston, R., Zeldes, H., and Taylor, E. H. (1954). *Phys. Rev.* **94**, 725. Atomic hydrogen hyperfine structure in irradiated acids.

McKellar, A. (1941). *Publs. Dominion Astrophys. Observatory, Victoria, B. C.* **7**, 251. Molecular lines from the lowest states of diatomic molecules composed of atoms probably present in interstellar space.

Mador, I. L. (1954). *J. Chem. Phys.* **22**, 1617. The stabilization of methyl radicals.

Milligan, D. E., and Pimentel, G. C. (1958). *J. Chem. Phys.* **29**, 1405. Matrix isolation studies: possible infrared spectra of isomeric forms of diazomethane and of methylene, CH_2.

Minkoff, G. J., Scherber, F. I., and Gallagher, J. S. (1959). *J. Chem. Phys.* **30**, 753. Energetic species trapped at 4.2°K from gaseous discharges.

Papazian, H. (1959). *Publs. Astron. Soc. Pacific* **71**, 237. The colors on Jupiter.

Peek, B. M. (1958). "The Planet Jupiter." Faber and Faber, London.

Platt, J. R. (1956). *Astrophys. J.* **123**, 486. On the optical properties of interstellar dust.

Platt, J. R. (1958). *Science* **127**, 3313. On the nature and color of the moon's surface.

Platt, J. R., and Donn, B. (1956). *Astron. J.* **61**, 11 (Abstr.). On the formation and optical properties of sub-micron sized interstellar grains.

Rice, F. O. (1956). *J. Chem. Phys.* **24**, 1259. Colors on Jupiter.

Rice, F. O., and Freamo, M. (1951). *J. Am. Chem. Soc.* **73**, 5529. The imine radical.

Rice, F. O., and Freamo, M. (1953). *J. Am. Chem. Soc.* **75**, 448. Formation of the imine radical in electrical discharge.

Richter, N. (1954). *Astron. Nachrichten* **281**, 241. Die Helligkeitsausbrüche des Kometen 1925II und ihre Zusammenhänge mit der Sonnentätigkeit.

Robinson, G. W., and McCarty, M., Jr. (1958a). *J. Chem. Phys.* **28**, 349. Electronic spectra of free radicals at 4°K—NH_2.

Robinson, G. W., and McCarty, M., Jr. (1958b). *J. Chem. Phys.* **28**, 350. Electronic spectra of free radicals at 4°K—HNO, NH, and OH.

Ruedenberg, K., and Scherr, C. W. (1953). *J. Chem. Phys.* **21**, 1565. Free-electron network model for conjugated systems I. Theory.

Russell, H. N., Dugan, R. S., and Stewart, J. Q. (1938). "Astronomy," Vol. 2, p. 540. Ginn, New York.

Scheer, M., and Klein, R. (1959). *J. Chem. Phys.* **31**, 278. Reaction of atomic hydrogen with solid oxygen at 20°K.

Seitz, F., and Koehler, J. S. (1956). *In* "Solid State Physics" (F. Seitz and D. Turnbull, eds.), Vol. 2, p. 307. Academic Press, New York. Displacement of atoms during irradiation.

Stromgren, B. (1948). *Astrophys. J.* **108**, 242. On the density distribution and chemical composition of the interstellar gas.

Suess, H. E., and Urey, H. C. (1956). *Revs. Modern Phys.* **28**, 53. Abundances of the elements.

Swings, P., and Haser, L. (1956). "An Atlas of Representative Cometary Spectra." Institute d'Astrophysique, Liège, Belgium.

Taylor, H. S. (1951). *Discussions Faraday Soc.* **8**, 1. Catalysis: retrospect and prospect.

ter Haar, D. (1944). *Astrophys. J.* **100**, 288. On the origin of smoke particles in the interstellar gas.

Urey, H. C. (1952). "The Planets," p. 114. Yale Univ. Press, New Haven, Connecticut.

Urey, H. C., and Donn, B. (1956). *Astrophys. J.* **124**, 307. Chemical heating for meteorites.

Urey, H. C., and Lavin, G. I. (1929). *J. Am. Chem. Soc.* **51**, 3286. Some reactions of atomic hydrogen.

van de Hulst, H. C. (1946). *Recherche Astron. Observatoire Utrechte* **11**, Part 1. The optics of spherical particles.

van de Hulst, H. C. (1948). *In* "Centennial Symposia," p. 73. Harvard Coll. Observatory, Cambridge, Massachusetts. Evolution and physics of solid particles.

van de Hulst, H. C. (1949). *Recherche Astron. Observatoire Utrechte* **11**, Part 2. The solid particles of interstellar space.

Wall, L. A., Brown, D. W., and Florin, R. E. (1959). *J. Phys. Chem.*, **63**, 1762. Atoms and free radicals by γ-irradiation at 4.2°K.

Whipple, F. L. (1950). *Astrophys. J.* **111**, 375. A comet model I: the acceleration of comet Encke.

Whipple, F. L., and Gossner, J. L. (1949). *Astrophys. J.* **109**, 380. An upper limit to the electron density near the earth's orbit.

Yankwich, P. E. (1956). *Can. J. Chem.* **34**, 301. Chemical effects of the nuclear transformation N^{14} (n, p)C^{14}.

12. Trapped Radicals in High Polymer Systems

H. MORAWETZ

Polymer Research Institute, Polytechnic Institute of Brooklyn, Brooklyn, New York

I. Introduction

The concept of the "trapping" of a free radical is subject to some variation in its definition. Presumably, such trapping eliminates the normal characteristic high reactivity of these species. Nevertheless, some residual reactivity will remain under any given set of conditions and the factor by which the normal reactivity has to be reduced to make the free radical qualify as a "trapped radical" will be rather arbitrary. In addition, it will not suffice to speak of reduced radical reactivity, but we have to specify the type of reaction we are concerned with. Such reactions fall into two broad classes: reactions of two radicals with each other, resulting in a decrease of the total radical concentration, and reactions of radicals with molecules containing no unpaired electrons, in which the population of radicals is preserved. In the following discussion we shall take the broadest view of radical trapping. Included in this classification are all effects due to media rich in high polymeric substances which reduce the disappearance rate of a given type of free radical.

II. The Determination of the Rate Constants for Radical Reactions in Polymerizing Systems

Although trapped radicals may be produced in a wide variety of polymeric materials, their occurrence has been studied most extensively in the products of vinyl polymerization. In this case the trapped radicals need not be introduced into the polymer by such treatment as photolysis or exposure to ionizing radiation, but they may owe their origin to the reaction by which the polymer was formed. Most of the early evidence of radical trapping during vinyl polymerization was based on characteristic deviations from the normal kinetic pattern. Free radical polymerization kinetics have been discussed in detail by Mark and Tobolsky (1950), by Flory (1953a), by Burnett (1954), and by Bamford *et al.* (1958). We shall review here only some features pertinent to our discussion.

Free radicals are formed in the polymerizing system by initiation reactions which may be thermal, photochemical, or the result of ionizing radiation. They may be derived from the monomer itself or a labile initiator such as a peroxide or diazo compound. Initiation is followed by a chain propagation reaction in which one monomer at a time adds to the growing free radical:

$$
\text{H} \left(-\text{CH}_2 - \underset{\underset{\text{Y}}{|}}{\overset{\overset{\text{X}}{|}}{\text{C}}} - \right)_n -\text{CH}_2 - \underset{\underset{\text{Y}}{|}}{\overset{\overset{\text{X}}{|}}{\text{C}}}{}^{\boldsymbol\cdot} + \text{CH}_2 {=} \underset{\underset{\text{Y}}{|}}{\overset{\overset{\text{X}}{|}}{\text{C}}} \rightarrow
$$

$$
\text{H} \left(-\text{CH}_2 - \underset{\underset{\text{Y}}{|}}{\overset{\overset{\text{X}}{|}}{\text{C}}} - \right)_{n+1} -\text{CH}_2 - \underset{\underset{\text{Y}}{|}}{\overset{\overset{\text{X}}{|}}{\text{C}}}{}^{\boldsymbol\cdot}
$$

$$(1)$$

In the simplest case the kinetic chains are terminated by a reaction of two chain radicals, resulting either in their combination or in hydrogen transfer. Schematically the process is represented by

		Rate	
Initiation:	$C \rightarrow R$	I	(2a)
Propagation:	$R + M \rightarrow R$	$k_p(M)(R)$	(2b)
Termination:	$2R \begin{array}{l} \longrightarrow \text{p} \\ \longrightarrow 2\text{p} \end{array}$	$k_t(R)^2$	(2c)
			(2d)

where C, R, M, and p represent initiator, radical, monomer, and "dead" polymer. Under usual conditions a steady state radical concentration is attained rapidly, given by

$$d(R)/dt = I - k_t(R)^2 = 0$$
$$(R) = (I/k_t)^{1/2}$$

(3)

The polymerization rate is then

$$-d(M)/dt = (k_p/k_t^{1/2})(M)I^{1/2}$$

(4)

while the average number of monomer units per chain $\bar{P}$ is given by

$$\bar{P} = \frac{-d(M)/dt}{d(p)/dt} = s[k_p/k_t^{1/2}][(M)/I^{1/2}]$$

(5)

where s is a factor varying from 0.5 when all termination involves proton transfer to 1.0 when termination occurs exclusively by radical combination. If s is known, e.g., from work with C^{14}-labeled initiators, (Arnett and Peterson, 1952; Bevington et al., 1954a,b) the rate of chain initiation I may be eliminated between Eqs. (4) and (5), giving the ratio $k_p/k_t^{1/2}$.

In practice the situation is complicated by the general occurrence of chain transfer

$$R\cdot + SY \rightarrow RY + S\cdot$$

(6)

which affects $\bar{P}$ but not the steady state radical concentration or the polymerization rate, provided $S\cdot$ is sufficiently reactive. Solvents, initiators, and the monomer may all act as chain transfer agents. Taking account of all these possibilities, Johnson and Tobolsky (1952) have shown that a plot of $1/\bar{P}$ against $[-d(M)/dt]^2$ has a slope of $k_t/sk_p^2(M)^2$, so that the ratio k_t/k_p^2 may still be obtained if the factor s is known.

The resolution of the ratio k_t/k_p^2 into the individual rate constants was first achieved by Burnett and Melville (1945) and by Bartlett and Swain (1945), who used a rotating sector to expose the polymerizing system to intermittent illumination. If the sector blocks the light source during half the time, the polymerization rate will be only half that observed under steady illumination, provided the light and dark periods t' are long compared to the mean lifetime of the free radicals τ. However, if $t'/\tau \ll 1$, the system behaves as if the light intensity had been reduced to one-half and the polymerization rate becomes, according to Eq. (4) $\sqrt{1/2}$ times the steady illumination rate. If t' and τ are of comparable magnitudes, the ratio of polymerization rates under intermittent and steady illumination is a function of t'/τ and τ can therefore be obtained from the experimental

TABLE I

RATE CONSTANTS (LITER-MOLE^{-1} SEC^{-1}) AT 60°C AND ACTIVATION ENERGIES FOR
CHAIN PROPAGATION AND TERMINATION[a]

Monomer	k_p	ΔE_p (kcal/mole)	k_t	ΔE_t (kcal/mole)
Styrene	176	7.8	7.2×10^7	2.4
Methyl methacrylate[b]	573	6.3	2.4×10^7	2.8
Methyl acrylate	2090	7.1	0.95×10^7	5
Vinyl acetate	3700	7.3	15×10^7	5.2

[a] Data from Matheson et al. (1951b).

[b] The original data of Matheson et al. (1949) were recalculated to take account of later information about the nature of the termination reaction (Bamford et al., 1958).

data. Since the steady state radical concentration must be $(k_t \tau)^{-1}$, we obtain by substitution into (2b)

$$-d(M)/dt = k_p(M)/k_t\tau \tag{7}$$

With $k_t/k_p{}^2$ already known, k_p and k_t may be calculated.

Typical results obtained in the careful measurements of Matheson et al. (1951b) are given in Table I.

For typical polymerization experiments with monomer conversion rates of 1 to 10% per hour, the propagation rate constants listed above lead to steady state radical concentrations of 10^{-9} to 10^{-7} moles per liter. The corresponding lifetimes of kinetic chains $\tau = [k_t(R)]^{-1}$ range from 0.1 to 10 seconds.

In recent years a number of additional methods have been developed for the estimation of τ. These depend on following the polymerization reaction either during the build-up of the radical concentration at the outset of irradiation or during the decay of the radicals when irradiation is discontinued (Bamford and Dewar, 1948; Majury and Melville, 1951; Grassie and Melville, 1951; Bengough and Melville, 1954).

If the termination is second order in free radicals, the monomer conversion at the outset of the illumination is

$$-\Delta(M)/(M_0) = (k_p/k_t) \ln \cosh (t/\tau) \tag{8a}$$

or, with $t \gg \tau$,

$$-\Delta(M)/(M_0) \approx (k_p/k_t)[t/\tau - \ln 2] \tag{8b}$$

During the period in which the radical concentration decays after illumination is discontinued, the monomer consumption is given by

$$-\Delta(M)/(M_0) = (k_p/k_t) \ln [1 + t/\tau] \tag{9}$$

However, unlike the rotating-sector method, measurements of the polymerization during the nonsteady state phase are in principle applicable also when the kinetic chain termination is first order with respect to radicals, as may be the case not only in the presence of inhibitors, but also when a radical-trapping mechanism is operative.

III. The Gel Effect

A. THE TERMINATION RATE CONSTANT AT HIGH POLYMER CONCENTRATIONS

It has long been observed that a number of free radical polymerizations tend to accelerate at conversions of 10 to 30 %. Since the reaction is strongly exothermic (15 to 20 kcal per mole), it was at first suspected that the effect is due to a failure to maintain a constant temperature as the increasing viscosity of the medium reduces the rate of heat transfer. However, Schulz and Blaschke (1941) demonstrated that the acceleration is not eliminated by rigorous thermostating. Norrish and Smith (1942) reinvestigated the phenomenon in the polymerization of methyl methacrylate and observed acceleration up to five to ten times of the initial polymerization rate. At the same time it was shown that the molecular weight of the polymer increases with the increasing reaction rate. This eliminated the possibility that the acceleration might be due to an increase in the chain initiation rate and suggested that the reduced probability of interactions involving two chain radicals in the highly viscous medium leads to a growth of the kinetic chain length. This phenomenon is commonly described in the literature as the "gel effect" or the "Trommsdorff effect," after an investigator who carried out a parallel study in Germany.

Norrish and Smith (1942) believed that the macroscopic viscosity of the system is the controlling factor in the gel effect, and they found that a fivefold increase in the initiation rate (leading, presumably, to a reduction in the chain length by a factor of $\sqrt{5}$) delayed the onset of the accelerating phase from 14 to 26 % conversion. Higher polymerization temperatures had qualitatively a similar effect. Very significant results were obtained from experiments in which the methyl methacrylate was diluted with 40 volume per cent of various solvents. Thermodynamically good solvents eliminated the increase of both the polymerization rate and the chain length of the polymer formed, while these effects persisted when the diluent was a bad solvent. This, too, was thought to be a consequence of the higher viscosity of the systems containing bad solvents.

The ideas advanced by Norrish and Smith were borne out by later investigations. Trommsdorff *et al.* (1948) showed that when either poly-(methyl methacrylate) or another polymer were added to methyl methacrylate monomer, the gel effect made itself felt from the outset of the polymerization. They found also that when the monomer was allowed to polymerize until the accelerated reaction phase was reached, then shock-cooled from 60° to −20°C and warmed again after 14 hours to the original temperature, the same reaction rate was resumed. These experiments suggest that even during the accelerated reaction phase the radical concentration attains fairly rapidly its steady state value governed by the instantaneous values of the propagation and termination rate coefficients. Schulz and Harborth (1947) carried out a careful study in which they compared the factor by which the polymerization rate was accelerated at any given time with the relative increase in the molecular weight of polymer formed at that instant. Their results (and the more precise reinterpretation of these by Schulz) confirms the assumption that the acceleration of the polymerization is fully accounted for by the increase of the kinetic chain length. In the presence of an inhibitor, the kinetic chain length is determined by the relative reactivity of a radical with the monomer and the inhibitor, and the gel effect is entirely eliminated.

The determination of absolute reaction rate constants by the rotating sector method has been carried to high polymer conversions by Matheson et al. (1949, 1951a). Their results show that in the case of the bulk polymerization of methyl methacrylate at 30°C the termination rate constant at 33 % conversion is reduced by a factor of 200 from its initial value, whereas the propagation rate constant remains almost unchanged (Matheson et al., 1949). For styrene, Matheson et al. (1951a) report only a 40 % acceleration of the polymerization rate at 44 % conversion and 50°C, but Fujii (1954a) found an acceleration by factors of 5.2 and 16.9 at conversions of 38 % and 60 % at 25°C. A particularly pronounced gel effect was found in the polymerization of methyl acrylate at 30°C, with the rate accelerating sharply almost from the beginning of the reaction (Matheson et al., 1951b; Bengough and Smith, 1958).

Recently it has become possible to study the gel effect by following directly the increase in the radical concentration by electron spin resonance (ESR) spectroscopy. Bresler et al. (1959) studied in this way the photoinitiated polymerization of methyl methacrylate at 25°C, and they found that the radical concentration increased sharply at around 40 % monomer conversion, reaching the remarkably high value of 1.5×10^{-3} moles per liter.

In interpreting the decrease in the value of k_t at high polymer conversions, it must be remembered that the rate constants for reactions involving high polymers may remain unaltered even when the polymer-containing system attains extremely high viscosities (Flory, 1953b). This is so since a highly viscous medium will not only slow down the diffusion of reagents toward each other, but also their diffusion apart. The reaction will then not be diffusion controlled if the activation energy is such that the number of collisions of the reactive centers, while they are close together, is small compared to the average number which is required for reaction to occur. It is thus the low activation energy for radical-radical recombination reactions which produces the early drop in k_t with increasing polymer content of the system. (According to Melville, 1956, even the ΔE_t values given in Table I are too high and the activation energy for termination may not exceed 1 kcal per mole.)

Although the decrease in the termination rate constant at high polymer concentrations is qualitatively understood, attempts to relate the effect to "viscosities" and "diffusion coefficients" (Vaughan, 1952; Schulz, 1956) are rather misleading. Benson and North (1959) have recently observed that the termination rate coefficient at the outset of the polymerization of methyl methacrylate is inversely proportional to the viscosity of the solvent—yet the termination coefficient changes little during the first 10 % of the polymerization, in spite of a relatively large change in the macroscopic viscosity of the system. At higher concentrations the chains become so heavily intertwined that the diffusion of polymer molecules as a whole no longer contributes significantly to the collision frequency of radicals with each other. Diffusion of the radicals depends then on the change of configuration of relatively short portions of the chain molecules. This view is strikingly confirmed by Schulz's data (1956) which show that the relative increase in the polymerization rate over its initial value depends less and less on initiator concentration (and thus on the polymer chain length) as the polymerization proceeds to very high conversion. The Brownian motion of chain segments is known to be independent of the total chain length, and it does not cease when the polymer is cross-linked and its macroscopic viscosity becomes infinite. The decay of radicals attached to a cross-linked matrix was observed by Melville and was followed recently by Atherton et al. (1958) with the use of electron spin resonance (ESR) spectroscopy. Similar experiments by Bresler et al. (1959) showed that when a sample of methyl methacrylate photopolymerized at 25°C to 60 % conversion is removed from the light source, about 80 % of the radicals disappear within a few minutes whereas the remaining ones are apparently stable indefinitely. These observations are consistent with the model suggested by Fox and Loshaek (1953) according to which the radicals in such systems may be assigned to finite volume within which they may move and encounter other similarly restrained radicals.

Yet another complicating factor was uncovered in Fujii's studies of the polymerization of styrene (1954a) and methyl methacrylate (1954b). At high conversions the polymerization rates were found to be proportional to a higher power of the initiation rate than the $I^{1/2}$ dependence required by Eq. (4). This means that the order of the kinetic chain termination reaction with respect to the chain carrying radicals must be lower than 2. Fujii estimated it as 1.7 to 1.8 and 1.4 to 1.6 for styrene at 60% and 68% conversion and as 1.6 to 1.8 for methyl methacrylate at 60% conversion. These observations indicate that a fraction of the radicals become inactive by a monomolecular mechanism involving apparently an occlusion of the macromolecular chain end in regions where it is sterically inaccessible. Some doubt has been cast on Fujii's conclusions by recent studies of Bengough and Melville (1959b), who found that at 25% conversion the rate of methyl methacrylate polymerization depends on a lower than 0.5 power of the irradiation intensity. The significance of this finding is not understood and no such anomalies were found in butyl acrylate polymerization (1959a).

B. THE PROPAGATION RATE CONSTANT AT HIGH POLYMER CONCENTRATIONS

Schulz and Harborth (1947) first called attention to the fact that the rate of methyl methacrylate polymerization may decay extremely fast at very high conversions. They postulated that eventually not only the reaction of radicals with each other but also their reaction with the monomer molecules is sharply slowed down. As a result, the polymerization of methyl methacrylate proceeds only to a limiting conversion of 86% and 91.5% at 50°C and 70°C. This phenomenon has been studied in detail by Bengough and Melville (1955) and their data for vinyl acetate polymerization are given in Table II. Similar data were obtained by Melville for methyl methacrylate and butyl acrylate.

The decrease in the rate coefficient for chain propagation occurs generally much later than the corresponding decrease in the rate coefficient for chain termination. The activation energy for propagation is much larger than that for the termination reaction (see Table I). This, as we have seen, means that the diffusion of the reactive centers has to be slowed down much further before the reaction rate becomes diffusion controlled. Also, the

TABLE II
RATE CONSTANTS FOR THE POLYMERIZATION OF VINYL ACETATE AT 25°C[a]

% Conversion	k_p	k_t
4	895	240
23	1290	126
46	1980	90
57	238	6.7
65	87	1.15

[a] Data from Bengough and Melville (1955).

propagation reaction involves only one center attached to a chain molecule, when the second reagent is small. In concentrated polymer solutions the system has, on the molecular level, a network structure which offers little hindrance to the motion of small molecules, although diffusion is slowed down very drastically with increasing particle size (Grün, 1947). Schulz (1956) has presented impressive evidence to show that the decay in the propagation rate coefficient is related to the glass transition point of the polymer.

C. THE FORMATION OF RADICALS IN CONCENTRATED POLYMER SOLUTIONS

The interpretation of the experimental data in studies of the gel effect were all based on the assumption that high concentrations of polymers do not interfere with the rate of radical formations by thermal or photochemical decomposition of the initiator. This assumption may be questioned since Breitenbach and Frittum (1958) have found peroxides embedded in polystyrene to be much more stable than in solution. The effect was interpreted by the high probability that the primary radicals will recombine before they diffuse apart. Norman and Porter (1955) observed that benzyl radicals form easily on photolysis of toluene in organic glasses, but no benzyl radicals were detected on irradiating bibenzyl. These results show that the probability of recombination depends on the mass of the primary radicals. Such effects of the viscosity of the system on the rate of radical formation are probably not restricted to polymers in bulk but make themselves felt, to a lesser extent, in concentrated polymer solutions, since Aboul-Saad (1958) found that a tenfold increase in solution viscosity is sufficient to decrease to one-half the rate of initiation in a free radical polymerization.

The same effect has been demonstrated most convincingly by Dolgopolsk et al. (1958) who studied the decomposition of an unsymmetrical diazo compound. Additions of polymer to the solution of this reagent increased substantially the yield of the product of the recombination of the primary radicals. Dolgopolsk et al. concluded that the cage effect operating in these experiments must also favor the recombination of radicals formed by chain scission in bulk polymers, contributing substantially to polymer stability against thermal degradation.

IV. Heterogeneous Polymerization

A. EMULSION POLYMERIZATION

When a water-insoluble monomer is emulsified in an aqueous medium and free radicals are generated in the aqueous phase, much more rapid rates of formation of high molecular weight polymer can be obtained than is

possible in homogeneous systems. Smith and Ewart (1948) showed that
the characteristics of emulsion polymerization result from the growth of
chains in discrete emulsified particles of polymer swollen with monomers
which are so small that the existence of more than one radical in a particle
would result in almost immediate chain termination. Under these conditions, each radical entering a monomer-polymer particle will either start or
stop a kinetic chain. Thus half the particles will carry growing polymer
chains at any one time, whatever the concentration of the particles and the
rate of radical production. The effect may be illustrated taking as an example polystyrene particles containing 40% monomer at 60°C. Neglecting
the gel effect, the data in Table I would predict for a growing chain to add
700 monomer units per second. If the particle diameter is 0.2μ, a content
of two free radicals would be equivalent to a radical concentration of
8×10^{-7} moles per liter and a mean time for radical-radical reaction of
0.02 seconds. If radicals are generated at such a rate that they enter the
particles at mean intervals of the order of seconds, this rate will control the
lifetime of the growing radicals which can be made much longer than in
homogeneous systems. We are using here the oil-water interface which
separates the radicals from each other, as a means for "trapping" them
with respect to the termination reaction. Bianchi et al. (1957) have obtained spectacular results by irradiating a monomer-polymer emulsion
with flashes of light at equal time spacings. Under these conditions the
chains initiated by radicals formed in one flash will be terminated only by
those generated in the next one. Thus all chains grow for the same length
of time, and if this time is long compared to the mean time required for the
addition of one monomer unit, the polymer will have an unusually narrow
distribution of molecular weights.

B. Polymerization in Precipitating Media

A systematic study of the effect of a gradually decreasing solvent power
of the medium on the rate of a radical polymerization has been carried out
by Chapiro (1950), who studied the behavior of styrene mixtures with
various alcohols. When the polymer precipitates in a highly swollen state,
the reaction rate is much less than in homogeneous solution, since the
radicals are being concentrated by the separation of a polymer-rich phase
resulting in a corresponding shortening of the kinetic chains. However, when
the polymer is very insoluble in the medium, it will separate in the form
of compact small particles and the polymerization rate tends to increase.
This behavior may be partly due to the state of particle subdivision as discussed in the previous section, but the decrease in the effective termination
rate constant characteristic of systems with high polymer concentration

would tend to extend the size of the particles which contain a single growing chain.

Similar conditions are encountered generally wherever the polymer is insoluble in its monomer and the polymerization is carried out in the absence of a solvent. Bengough and Norrish (1950) noted that the rate of vinyl chloride polymerization increases steadily during the formation of the insoluble polymer and that high initial rates may be obtained by adding polymer to freshly distilled monomer. These observations were interpreted by assuming chain transfer to the surface of the polymer, resulting in radicals which can propagate and which transfer eventually to a monomer. Chapiro (1956a) studied the gamma ray-initiated polymerization of vinyl chloride and found that the radicals in the precipitated polymer remained active for several days. In a later study Bengough and Norrish (1953) found that the polymerization of vinylidene chloride at 47 to 75°C is subject to a similar autoacceleration as observed with vinyl chloride. However, Burnett and Melville (1950) found no such effect in the temperature range of 15 to 35°C. This discrepancy is probably due to the fact that at the lower temperature the swelling of the polymer is so low that trapped radicals in the polymer phase are unable to propagate.

Radical trapping in precipitating polymer has been studied most extensively with acrylonitrile. If the acceleration in the polymerization rate is really due to the accumulation of long-lived propagating radicals in the polymer produced, the extent of the polymerization observed, after monomer irradiation is discontinued, should increase with the polymerization time. Bamford and Jenkins (1953), as well as Bensasson and Bernas (1957, 1958) found, in fact, that the aftereffect at 25°C increased with the amount of previously formed polymer. Bamford and Jenkins (1953) provided also a striking demonstration of the presence of trapped radicals in the precipitated polymer: If the precipitated polymer is heated with acrylonitrile monomer or with styrene, some of the monomer is polymerized very rapidly as the temperature approaches 60°C. This "activity" of the polyacrylonitrile is destroyed by exposure to atmospheric oxygen. If the active polymer is heated in contact with acrylonitrile only to 40°C, some polymerization is induced without, however, impairing the extent of a subsequent reaction at 60° (Bamford and Jenkins, 1955).

These results are undoubtedly a consequence of the gradual increase in the swelling of the polymer as the temperature is raised. This would first increase the propagation and later the termination rate to significant values and also explain why the molecular weight of the polymer formed first increases and then drops off with rising temperature (Bamford and Jenkins, 1953; Thomas and Pellon, 1954).

The polymerization rate was found by Thomas and Pellon to be proportional to the 0.82 power of the initiation rate, as against the square root dependence predicted by Eq. (4), and this was interpreted by a superposition of a bimolecular radical-radical termination and a unimolecular radical "burial" process. Bamford and Jenkins (1955) estimated the concentration of the radicals occluded in the "active polymer" by reaction with diphenyl-picrylhydrazyl (DPPH), but later studies based on paramagnetic resonance (Bamford *et al.*, 1955; Ingram *et al.*, 1958; Bamford *et al.*, 1959) led to higher values of the trapped radical concentration. Ingram *et al.* found also that inferences of radical trapping from polymerization kinetics were not always reliable: it was found, for instance, that poly(vinylidene chloride) gave no ESR spectrum, although Bengough and Norrish (1953) had interpreted the autoacceleration of vinylidene chloride polymerization in terms of radical trapping. The difference in the behavior of polyacrylonitrile and poly(vinylidene chloride) reflect obviously differences in the rate at which occluded radicals may decay over longer periods of time.

The mechanism of the radical trapping in the polymerization of acrylonitrile has been studied in detail by Bamford and his collaborators. They found (Bamford and Jenkins 1955; Bamford *et al.*, 1959) that the probability that a given radical is trapped in unit time rises with the polymerization rate, suggesting that the trapping mechanism may involve coalescence of the growing chains. A surprisingly large fraction of the radicals formed during the polymerization are trapped: with a mean polymerization rate of 15 % per hour and a final 17 % conversion at 25°C, 13 % of the radicals formed were found trapped in the precipitated polymer. Contrary to expectation, the addition of up to 8 % of dimethylformamide (which swells the polymer) increased the radical concentration. Bamford *et al.* (1959) reported also that a copolymer of acrylonitrile with methyl methacrylate prepared in a heterogeneous system had an ESR spectrum indicating only acrylonitrile radicals. Ivin (1959) has pointed out that the known reaction rate constants of the two radicals with the two monomers would have led to the expectation that methyl methacrylate radicals would be present in higher concentration, and it was concluded that the probability of occlusion in a given medium must vary with the nature of the radical.

V. Radicals Trapped in Bulk Polymers

A. Low Molecular Weight Radicals in a Polymer Matrix

The restricted mobility of molecules incorporated in polymers in bulk makes it possible to employ polymers as a medium for the trapping of free radicals, avoiding the inconvenience attached to the use of low temperatures required in most trapped radical investigations. As mentioned earlier, it has to be kept in mind that small molecules may be little impeded in their motion through high polymers, though the bulk viscosity may be extremely high (Grün, 1947). However, at ordinary temperatures many polymers are below their glass transition temperature and have, therefore, very high internal viscosities which make them suitable media for radical trapping. Bijl and Rose-Innes (1955) irradiated with ultraviolet light at room temperature p-phenylenediamine and N,N-tetramethyl-p-phenylenediamine incorporated into poly(methyl methacrylate). They demonstrated by paramagnetic resonance measurements that the concentration of the radicals formed remained unchanged for a month. Another demonstration of radical trapping was provided by Breitenbach and Frittum (1958), who decomposed photochemically benzoyl peroxide incorporated into polystyrene and showed that the trapped radicals could be used subsequently to initiate acrylonitrile polymerization.

Schneider (1955) has pointed out that the ESR spectrum of low molecular weight free radicals incorporated into a glassy polymer matrix shows a hyperfine structure up to much higher radical concentration than would be the case in a low viscosity medium. This behavior reflects the inability of the polymer structure to support long-range exchange effects.

B. TRAPPED POLYMERIC RADICALS

The ESR spectrum of a polymer subjected to ionizing radiation was first reported by Schneider et al. (1951). They interpreted their results obtained with poly(methyl methacrylate) in terms of trapped electrons. However, Wall and Brown (1956) reported chemical evidence for high concentrations of trapped radicals in irradiated poly(methyl methacrylate) and Fraenkel et al. (1954) found that the radicals trapped in the gel produced by the polymerization of glycol dimethacrylate have the same ESR spectrum. This suggests that the spectrum must be due to radicals of the type

$$
\begin{array}{c}
CH_3 \\
| \\
R\!-\!CH_2\!-\!C\cdot \\
| \\
COOCH_3
\end{array}
\qquad (10)
$$

The spectrum has been described by Combrisson and Uebersfeld (1954), by Schneider (1955), and by Abraham et al. (1958), who showed that poly(methyl methacrylate), poly(ethyl methacrylate) and poly(methacrylic acid) give, on irradiation, very similar spectra. The spectrum appears to be due to the superposition of two sets of lines, the "A set" at 0, ± 26 and ± 51 gauss relative to the center with relative intensities of $1:4:6:4:1$ and a weaker "B set" at ± 39 and ± 13 gauss with relative intensities $1:3:3:1$. Ingram et al. (1958) and Abraham et al. (1958) interpret the spectrum as due to two conformations of the radical, in which the unpaired electron may couple equally to four and three protons, respectively. However, Bresler *et al.* (1959) heated a sample of poly(methyl methacrylate) containing trapped radicals from 25°C to 60°C, and they observed that the "B set" of the ESR spectrum was intensified at the expense of the "A set." They concluded that the transformation was due to chain transfer to the terminal vinylidene group of a poly(methyl methacrylate) chain terminated by disproportionation

$$
\begin{array}{ccccc}
\mathrm{CH_3} & \mathrm{CH_3} & & \mathrm{CH_3} & \mathrm{CH_3} \\
| & | & & | & | \\
{-}\mathrm{CH_2}{-}\mathrm{C}^{\cdot} & + \ \mathrm{C}{=}\mathrm{CH}{-} & \rightarrow & {-}\mathrm{CH_2}{-}\mathrm{CH} & + \ \mathrm{C}{=}\mathrm{C}{-} \quad (10a) \\
| & | & & | & | \\
\mathrm{COOCH_3} & \mathrm{COOCH_3} & & \mathrm{COOCH_3} & \mathrm{COOCH_3}
\end{array}
$$

Ingram *et al.* pose the question, how is it possible for the irradiated polymer to contain radicals of one type only? They suggest that the primary process of chain rupture leads to two different radicals:

$$
\begin{array}{cccc}
\mathrm{CH_3} & \mathrm{CH_3} & \mathrm{CH_3} & \mathrm{CH_3} \\
| & | & | & | \\
\mathrm{RCH_2}{-}\mathrm{C}{-}\mathrm{CH_2}{-}\mathrm{C}{-}\mathrm{R'} & \rightarrow \ \mathrm{RCH_2}{-}\mathrm{C}^{\cdot} & + \ \mathrm{R'}{-}\mathrm{C}{-}\mathrm{CH_2} & (11) \\
| \qquad\quad | & | & | \\
\mathrm{COOCH_3} \ \ \mathrm{COOCH_3} & \mathrm{COOCH_3} & \mathrm{COOCH_3}
\end{array}
$$

However, these radicals will tend to split off monomer units and the second radical will be rapidly converted to a more stable structure by the addition of a monomer

$$
\begin{array}{cccc}
\mathrm{CH_3} & \mathrm{CH_3} & \mathrm{CH_3} & \mathrm{CH_3} \\
| & | & | & | \\
\mathrm{R'C}{-}\mathrm{CH_2} & + \ \mathrm{CH_2}{=}\mathrm{C} & \rightarrow \ \mathrm{R'}{-}\mathrm{C}{-}\mathrm{CH_2}{-}\mathrm{CH_2}{-}\mathrm{C}^{\cdot} & (12) \\
| & | & | & | \\
\mathrm{COOCH_3} & \mathrm{COOCH_3} & \mathrm{COOCH_3} & \mathrm{COOCH_3}
\end{array}
$$

Schneider (1955) found that irradiation of poly(methacrylate) at liquid nitrogen temperature led to a spectrum without fine structure which changed irreversibly to the complex spectrum on warming. The original low temperature spectrum was interpreted as being due to trapped electrons. This interpretation is consistent with the accepted mechanism of the formation of radicals by ionizing radiation involving first the loss of an electron followed by secondary processes leading to free radicals.

Another ESR spectrum of a polymer radical which seems to be well established (Schneider, 1955; Abraham and Whiffen, 1958) is that of irradiated polystyrene. The spectrum has a triplet structure with satellites at ±17 gauss and an intensity ratio of 1:2:1. Abraham and Whiffen believe that the trapped radicals have the structure

$$
\begin{array}{c}
\qquad\quad \cdot \\
{-}\mathrm{CH_2}{-}\mathrm{C}{-}\mathrm{CH_2}{-} \\
| \qquad\qquad\qquad\qquad (13) \\
\mathrm{C_6H_5}
\end{array}
$$

and that steric restrictions in the chain conformation prevent two of the four neighboring methylene hydrogens to couple with the odd electron. This interpretation is preferred to one in terms of

$$\begin{matrix} \overset{\displaystyle .}{-CH-CH_2} \\ | \\ C_6H_5 \end{matrix} \qquad (14)$$

since these energetic radicals would be expected to rearrange rapidly to more stable structures. It should be noted that Fraenkel et al. (1954) also observed three peaks in ESR spectra of radicals trapped in the gel obtained on polymerizing divinylbenzene. In such a system the original radical would have the structure

$$\begin{matrix} H \\ | \\ -CH_2-\overset{\displaystyle .}{C} \\ | \\ C_6H_4-CH- \\ | \end{matrix} \qquad (15)$$

but this might rearrange by subsequent chain transfer to a more stable form analogous to (13).

The ESR spectrum of irradiated polytetrafluoroethylene (Teflon) was first reported by Ard et al. (1955) who observed eight peaks. However, Rexroad and Gordy (1959) have reported recently that the result is extremely sensitive to traces of oxygen and when the Teflon sample is rigorously degassed prior to irradiation, the ESR spectrum consists of two partially superimposed quintets. The spectrum was interpreted as being due to four fluorine nuclei with a coupling of 33 gauss and one with a coupling of 92 gauss, and the radical was assigned the structure

$$\begin{matrix} F & F & F \\ | & | & | \\ -C-C-C- \\ | & \cdot & | \\ F & & F \end{matrix} \qquad (16)$$

On cooling the irradiated sample, Gordy (1955) found that the fine structure of the spectrum disappears, reflecting, apparently, a phase transition in the polymer.

A careful study of the effect of the physical state of a polymer on the stability of radicals formed during irradiation was carried out with polyethylene by Lawton et al. (1958a, b). The authors conclude that radical trapping may occur under three conditions:

(1) Radicals are trapped within the crystallites of the polymer. Here it appeared that not only the fraction of crystalline material as estimated from X-ray diffraction is important, but that the type of crystal may vary

in its effectiveness for radical trapping. In particular, crystallites in straight-chain, high density polyethylene were a better trapping medium than crystallites of branched-chain, low density polyethylene.

(2) In the amorphous phase, radical trapping occurs below the glass transition temperature.

(3) When the polymer is heavily cross-linked, radicals are trapped because they may be attached to the network structure in positions from which no other radicals are accessible.

The radicals were found to be stable for long periods of time at the temperature of liquid nitrogen. At 25°C, high density polyethylene lost 68% of the radicals in 2700 hours; the decay in low density, branched-chain polyethylene was considerably faster. The decay was found to be due to radical-radical recombination, resulting in cross-link formation. Dole and Keeling (1953) and Dole *et al.* (1954) have cited evidence indicating that the trapped radicals migrate along the polymer chain before they disappear by mutual interaction.

Ohnishi and Nitta (1959) found that the decay of radicals produced in poly(methyl methacrylate) by γ rays followed second order kinetics with a rate constant of 0.1 liter-mole^{-1}-sec^{-1} at room temperature. With an irradiation rate of 7.1×10^5 rep. per hour, the steady state radical concentration attained after long time irradiation was 5.8×10^{-4} M.

Abraham and Whiffen (1958) reported ESR spectra for a number of additional irradiated high polymers. The G-values[1] for radical formation ranged from 0.2 to 2.5. Only in the case of poly(vinyl alcohol) and starch was there strong evidence that more than one kind of radical is present. Chemical evidence was also presented for trapped radicals in irradiated poly(vinyl chloride) (Chapiro, 1956b) and nylon (Bevington and Eaves, 1956).

C. Reaction of Trapped Radicals with Oxygen

Free radicals are known to react generally with oxygen, and the rate of the reaction of radicals trapped in a polymer matrix would be expected to be essentially determined by the rate of oxygen diffusion. The peroxy radical may or may not react further in a more or less complex reaction sequence.

Schneider *et al.* (1951) and Combrisson and Uebersfeld (1954) noted that the decay of the ESR spectrum of irradiated poly(methyl methacrylate) was accelerated by exposure to the atmosphere, and it must be concluded that the peroxy radical breaks down rapidly in this case. At the other extreme is irradiated Teflon, in which the radicals add oxygen to form a

---

* The G-value is the yield per 100 electron-volts of absorbed energy.

very stable species, whose spectrum has been described by Ard *et al.* (1955) and by Abraham and Whiffen (1958) as consisting of a single peak, whereas Schneider reported it as a triplet. Rexroad and Gordy (1959) have recently found that admission of oxygen to a previously irradiated Teflon sample results in the production of a new peak in the spectrum, without a corresponding elimination of the peaks characterizing the unoxygenated radicals. The implication here is that the oxygen adds to a type of radical whose spectrum cannot be observed. When high density polyethylene is irradiated *in vacuo* at 25°C and exposed subsequently to the atmosphere, a single-line strongly asymmetric ESR spectrum of the peroxy radical is rapidly built up (Abraham and Whiffen, 1958). However, infrared spectroscopy reveals that the carbonyl content of such samples builds up slowly over many hundred hours (Lawton *et al.*, 1958a). It would then appear that the secondary processes (in which further oxygen was found to be consumed) are very slow under the restraint of the crystalline regions. In low density, branched-chain polyethylene whose crystallites are, apparently, much less perfect, no significant concentrations of peroxy radicals are detected after exposure of irradiated specimens to the atmosphere (Ballantine *et al.*, 1959).

VI. Initiation of Graft Polymerization by Trapped Radicals

In recent years great interest has been aroused by the possibility of producing branched polymers in which the "backbone" of the macromolecule consists of one type of monomer units and the branches are formed by sequences of another monomer. Such products are obtained very conveniently by irradiating a polymer with ionizing radiation and allowing the free radicals formed during irradiation to react with a suitable monomer. The literature in this field has been summarized by Chapiro (1957). Two procedures may be employed:

(1) The polymer may be in contact with the monomer (and more or less swollen by it) during the irradiation.

(2) The polymer may be irradiated alone and brought into contact with the monomer at a later time, before the radicals have disappeared. A variation of this technique involves the exposure of the irradiated polymer to oxygen and thermal decomposition of the peroxides formed to radicals acting as graft initiators (Chapiro, 1958).

When irradiating a swollen polymer in the presence of excess monomer, it is advantageous to use systems in which the radiation sensitivity of the polymer greatly exceeds that of the monomer, so that little of the monomer is lost by homopolymerization in the solution phase. An analysis of the process is complicated, since the radiation sensitivity of the polymer, the monomer diffusion rate, and the values of the propagation and termination rate coefficients are all subject to considerable variations during the course of the grafting.

The largest amount of experimental work has been carried out on grafting to polyethylene. Chen *et al.* (1957) found that the relative increase in weight of polyethylene sheets immersed in styrene and exposed to ionizing radiation became faster with increasing thickness of the sheets. This strange phenomenon was later interpreted by Chen (1958), who pointed out that if the reaction is diffusion controlled, the polymerization may be *faster* in the interior of the sheets where the monomer concentration is *lower* due to the "gel effect" (see Section III, A). This interpretation was supported by the finding that polystyrene occluded in the polyethylene had a very high molecular weight. The role of the "gel effect" in graft polymerization has recently also been discussed by Chapiro (1958) and by Hoffman, *et al.* (1959).

Entirely different conditions are encountered when Teflon is irradiated in the presence of various monomers. Teflon is very highly crystalline and it cannot be swollen by the monomers. It was, therefore, hardly surprising that Chen *et al.* (1957) could attain only surface grafting to Teflon, leaving the bulk of the material unchanged. However, Chapiro (1959) reinvestigated this problem and proved that a homogeneous graft of styrene to Teflon may be obtained, provided the irradiation rate is sufficiently low so that the monomer is not used up by polymerization while it diffuses through the grafted surface layer. If this condition is satisfied, the monomer will attack deeper and deeper layers of the specimen, which swells more and more as its crystallinity is disrupted. The grafting may be continued until the sample eventually disintegrates.

Much less information is available on grafting to preirradiated polymers. Magat and Tanner (1955) have demonstrated the grafting of various monomers to preirradiated cellulose and cellulose acetate. Work along that line has been expanded by Immergut *et al.* (1958). Ballantine *et al.* (1959) found, contrary to previous assertions (Chapiro, 1958), that styrene will graft to vacuum-irradiated polyethylene. The trapped radical content was higher in high density polyethylene (as found also by Lawton *et al.* (1958a)), but in spite of this, grafting at 22°C was faster to low density samples because of their more extensive swelling in the monomer.

Some typical examples of the technical utility of irradiation-grafted polymers are surveyed by Chen *et al.* (1957). Styrene grafted to polyethylene may be sulfonated, yielding permselective membranes of low electrical resistivity, retaining essentially the desirable mechanical properties of the parent polyethylene. The solvent resistance of silicone rubbers may be improved by grafting of acrylonitrile. The adhesion to the inert surface of Teflon is greatly increased by surface grafting of styrene. It is interesting in this context that with the use of ionizing radiation of low penetrating power the location of trapped radicals can always be restricted to the surface layer, so that grafting may be carried out in a manner which will modify only the surface properties of the polymer (Chapiro, 1957). Oster et al. (1959) have shown that the same result may be obtained by ultraviolet irradiation of polymers containing various sensitizing agents.

VII. Radical Polymerization in the Solid State

A number of crystalline monomers such as acrylamide, acrylic and methacrylic acid, and vinyl stearate polymerize when exposed to ionizing radiation (Restaino et al., 1956). The early interpretation of this reaction was invalidated by the observation of Adler and Ballantine (1958) that much of the reaction takes place over very long periods of time after the samples have been removed from the irradiation source. A detailed study of the course of the polymerization has been carried out by Fadner et al. (1959) and Morawetz and Fadner (1959) who exposed their samples at $-80°C$ to relatively light doses of 200,000 to 800,000 rep. Under these conditions, the polymerization during irradiation is negligible and the change of polymer yield and molecular weight can be followed conveniently after thermostating the exposed samples at various temperatures.

The most significant observations were: (1) Although the polymerization rate decays rapidly, the reaction does not stop even after several months following irradiation and the yield does not converge to a limit. (2) The ratio of polymer yield and weight-average molecular weight changes very little over long periods of time. (3) The molecular weight of the polymer seems to be a function, at any given temperature, of the polymerization time only with little, if any, dependence on irradiation dose.

The linear dependence of the polymer yield, for long polymerization times, on the logarithm of the time is consistent with chain termination by the interaction of two growing chain radicals, but this agreement may be accidental, since the ESR absorption decays much more slowly than the polymerization rate. With acrylamide, the ratio of yield and polymer molecular weight lead to an initial radical concentration of the order of $10^{-4}M$ at an irradiation dose of 200,000 rep, giving a G-value for radical formation of 0.44. The propagation rate coefficient calculated from the initial polymerization rate at 25° is 0.007 liter-mole^{-1} sec^{-1} as against a value of 18,000 found by Dainton and Tordoff (1957) for acrylamide polymerization in aqueous solution. This difference is a measure of the restraint imposed by the crystal lattice on the radical reactivity, more than 8 seconds being required for the addition of a monomer unit to a chain radical embedded in the crystal.

An interesting result was obtained when following the polymerization of

preirradiated solid solutions of acrylamide containing 5% of the isomorphous saturated analog, propionamide. The reaction rate was unaffected, but
the molecular weight of the product became independent of time on reaching a limiting value. This indicates extensive chain transfer by the propionamide, a compound which would normally not be expected to be an
efficient chain transfer agent. It must be concluded that whenever the
growing chain radical encounters in its path a molecule of propionamide,
it is held by the lattice forces in its vicinity for times long enough to render
chain transfer reasonably probable. The crucial importance of the nature
of the crystal lattice is also demonstrated by the fact that preirradiated
potassium acrylate polymerizes rapidly at 0°C, while the sodium and
lithium salts react at comparable rates only at 110°C and 150°C, respectively.

The irradiation of clathrate complexes of monomers with urea and
thiourea leads to a very rapid polymerization in the solid state. Apparently,
the radicals formed may move relatively unimpeded in the channels of the
host helices (Classen, 1956; Brown and White, 1958).

References

Aboul-Saad, I. A. (1958). Ph.D. Thesis. Polytechnic Institute of Brooklyn, Brooklyn,
New York. Effect of viscosity on radical formation and recombination.
Abraham, R. J., and Whiffen, D. H. (1958). *Trans. Faraday Soc.* **54**, 1291. Electron
spin resonance spectra of some γ-irradiated polymers.
Abraham, R. J., Melville, H. W., Ovenall, D. W., and Whiffen, D. H. (1958). *Trans.
Faraday Soc.* **54**, 1133. Electron spin resonance spectra of free radicals in irra-
diated polymethyl methacrylate and related compounds.
Adler, G., and Ballantine, D. S. (1958). Private communication.
Ard, W. B., Shields, H., and Gordy, W. (1955). *J. Chem. Phys.* **23**, 1727. Paramagnetic
resonance of X-irradiated Teflon. Effects of absorbed oxygen.
Arnett, L. M., and Peterson, J. H. (1952). *J. Am. Chem. Soc.* **74**, 2031. Vinyl polymeri-
zation with radioactive aliphatic isobisnitrile initiators.
Atherton, N. M., Melville, H. W., and Whiffen, D. H. (1958). *Trans. Faraday Soc.* **54**,
1300. The estimation of free radicals trapped in a polymer gel by electron spin
resonance spectroscopy.
Ballantine, D., Glines, A., Adler, G., and Metz, D. J. (1959). *J. Polymer Sci.* **34**, 419.
Graft copolymerization by pre-irradiation technique.
Bamford, C. H., and Dewar, M. J. S. (1948). *Proc. Roy. Soc.* **A192**, 309. A method for
determining velocity constants in polymerization reactions and its application to
styrene.
Bamford, C. H., and Jenkins, A. D. (1953). *Proc. Roy. Soc.* **A216**, 515. Studies in
polymerization. VI. Acrylonitrile: the behavior of free radicals in heterogeneous
systems.
Bamford, C. H., and Jenkins, A. D. (1955). *Proc. Roy. Soc.* **A228**, 220. Studies in polymerization. IX. The occlusion of free radicals by polymers. Physical factors determining the concentration and behavior of trapped radicals.
Bamford, C. H., Jenkins, A. D., Ingram, D. J. E., and Symons, M. C. R. (1955).

Nature **175,** 894. Detection of free radicals in polyacrylonitrile by paramagnetic resonance.

Bamford, C. H., Barb, W. G., Jenkins, A. D., and Onyon, P. F. (1958). "The Kinetics of Vinyl Polymerization by Radical Mechanisms." Academic Press, New York.

Bamford, C. H., Jenkins, A. D., Symons, M. C. R., and Townsend, M. G. (1959). *J. Polymer Sci.* **34,** 181. Trapped radicals in heterogeneous vinyl polymerization.

Bartlett, P. D., and Swain, C. G. (1945). *J. Am. Chem. Soc.* **67,** 2273. The absolute rate constants in the polymerization of liquid vinyl acetate.

Bengough, W. I., and Melville, H. W. (1954). *Proc. Roy. Soc.* **A225,** 330. A thermocouple method for following the non-stationary state of chemical reactions. I. The evaluation of velocity coefficients for vinyl acetate, methyl methacrylate and butyl acrylate polymerizations.

Bengough, W. I., and Melville, H. W. (1955). *Proc. Roy. Soc.* **A230,** 429. A thermocouple method for following the non-stationary state of chemical reactions. II. The evaluation of velocity coefficients and energies of activation for the propagation and termination reactions for the initial and later stages of the polymerization of vinyl acetate.

Bengough, W. I., and Melville, H. W. (1959a). *Proc. Roy. Soc.* **A249,** 445. Polymerization of butyl acrylate at high conversion.

Bengough, W. I., and Melville, H. W. (1959b). *Proc. Roy. Soc.* **A249,** 455. Polymerization of methyl methacrylate at high conversion.

Bengough, W. I., and Norrish, R. G. W. (1950). *Proc. Roy. Soc.* **A200,** 301. The mechanism and kinetics of the heterogeneous polymerization of vinyl monomers. I. The benzoyl peroxide catalyzed polymerization of vinyl chloride.

Bengough, W. I., and Norrish, R. G. W. (1953). *Proc. Roy. Soc.* **A218,** 149. The mechanism and kinetics of the heterogeneous polymerization of vinyl monomers. II. The benzoyl peroxide catalyzed polymerization of vinylidene chloride.

Bengough, W. I., and Smith, A. C. K. (1958). *Trans. Faraday Soc.* **54,** 1553. Evaluation of velocity coefficients for the early stages of the polymerization of methyl acrylate.

Bensasson, R., and Bernas, A. (1957). *J. chim. phys.* **54,** 479. Sur certains caractères de la polymérisation radiochimique de l'acrylonitrile II. Le post-éffet.

Bensasson, R., and Bernas, A. (1958). *J. Polymer Sci.* **30,** 163. Polymérisation radiochimique de l'acrylonitrile. Quelques observations sur le post-éffet.

Benson, S. W., and North, A. M. (1959). *J. Am. Chem. Soc.* **81,** 1339. The effect of viscosity on the rate constants of polymerization reactions.

Bevington, J. C., and Eaves, D. E. (1956). *Nature* **178,** 1112. Use of irradiated polymers as initiators of polymerization.

Bevington, J. C., Melville, H. W., and Taylor, R. P. (1954a). *J. Polymer Sci.* **12,** 449. The termination reaction in radical polymerizations. I. Polymerizations of methyl methacrylate and styrene at 25°.

Bevington, J. C., Melville, H. W., and Taylor, R. P. (1954b). *J. Polymer Sci.* **14,** 463. The reaction in radical polymerizations. II. Polymerizations of styrene at 60° and of methyl methacrylate at 0° and 60° and the copolymerization of these monomers at 60°.

Bianchi, J. P., Price, F. T., and Zimm, B. H. (1957). *J. Polymer Sci.* **25,** 27. Monodisperse polystyrene.

Bijl, D., and Rose-Innes, A. C. (1955). *Nature* **175,** 82. Preparation of solid solutions of free radicals at room temperature.

Breitenbach, J. W., and Frittum, H. (1958). *J. Polymer Sci.* **29,** 565. Freie Radikale in fester Phase.

Bresler, S. E., Kazbekov, E. N., and Saminskii, E. M. (1959). *Vysokomolekularnye Soedinenia*, **1**, 132. A study of macroradicals in polymerization and degradation processes. I. (In Russian; English abstract.)

Brown, J. F., Jr., and White, D. M. (1958). *Abstracts of papers presented at 133rd meeting, Am. Chem. Soc. San Francisco, 1958*, p. 14R. Stereospecific polymerization in canal complexes.

Burnett, G. M. (1954). "Mechanism of Polymer Reactions." Interscience, New York.

Burnett, G. M., and Melville, H. W. (1945). *Nature* **156**, 661. Propagation and termination rate coefficients for vinyl acetate polymerization.

Burnett, J. D., and Melville, H. W. (1950). *Trans. Faraday Soc.* **46**, 976. The photopolymerization of vinylidene chloride.

Chapiro, A. (1950). *J. chim. phys.* **47**, 747. Sur la polymérisation des composés vinyliques amorcés par les rayons γ.

Chapiro, A. (1956a). *J. chim. phys.* **53**, 512. Sur la polymérisation du chlorure de vinyle amorcé par les rayons γ.

Chapiro, A. (1956b). *J. chim. phys.* **53**, 895. Action des rayons γ sur les polymères à l'état solide. III. Irradiation du chlorure de polyvinyle.

Chapiro, A. (1957). *Ind. plastiques mod. (Paris)* **9** (2), 34. Irradiation des solutions de polymères. Synthèse des copolymères greffés.

Chapiro, A. (1958). *J. Polymer Sci.* **29**, 321. Synthèse des copolymères greffés à partir des polymères ayant subi l'action des radiations ionisantes. I. Caractères généraux de la reaction de greffage sur le polyéthylène préirradié.

Chapiro, A. (1959). *J. Polymer Sci.* **34**, 481. Préparation des copolymères greffés du polytetrafluoroéthylène (Teflon) par voie radiochimique.

Chen, W. K. W. (1958). Graft Polymers Derived by High Energy Radiation. Ph.D. Thesis. Polytechnic Institute of Brooklyn, Brooklyn, New York.

Chen, W. K. W., Mesrobian, R. B., Ballantine, D. S., Metz, D. J., and Glines, A. (1957). *J. Polymer Sci.* **23**, 903. Studies on graft copolymers derived by ionizing radiation.

Classen, H. (1956). *Z. Elektrochem.* **60**, 982. Polymerisation in einer Kanaleinschlussverbindung.

Combrisson, J., and Uebersfeld, J. (1954). *Compt. rend.* **238**, 1397. Détection de la résonance paramagnétique dans certaines substances organiques irradiées.

Dainton, F. S., and Tordoff, M. (1957). *Trans. Faraday Soc.* **53**, 489. The polymerization of acrylamide in aqueous solution. II. The effect of ferric perchlorate on the X- and γ-ray initiated reaction.

Dole, M., and Keeling, C. D. (1953). *J. Am. Chem. Soc.* **75**, 6082. Long-range migration of chemical activity in the solid state.

Dole, M., Keeling, C. D., and Rose, D. G. (1954). *J. Am. Chem. Soc.* **76**, 4304. The pile irradiation of polyethylene.

Dolgopolsk, B. A., Erusalemskii, B. L., Milovskaia, E. B., and Belonovskaia, G. P. (1958). *Doklady Akad. Nauk SSSR., Phys. Chem. Section*, **120**, 783. The cage effect and the thermal stability of polymers. (In Russian).

Fadner, T. A., Rubin, I., and Morawetz, H. (1959). *J. Polymer Sci.*, **37**, 549. The mechanism of free radical polymerizations in the crystalline state.

Flory, P. J. (1953a). "Principles of Polymer Chemistry," Chapter IV. Cornell Univ. Press, Ithaca, New York.

Flory, P. J. (1953b). "Principles of Polymer Chemistry," pp. 76–78. Cornell Univ. Press, Ithaca, New York.

Fox, T. G., and Loshaek, S. (1953). *J. Am. Chem. Soc.* **75**, 3544. Cross-linked polymers. I. Factors influencing the efficiency of cross-linking in copolymers of methyl methacrylate and glycol dimethacrylates.

Fraenkel, G. K., Hirshon, J. M., and Walling, C. (1954). *J. Am. Chem. Soc.* **76**, 3606. Detection of polymerization radicals by paramagnetic resonance.

Fujii, S. (1954a). *Bull. Chem. Soc. Japan* **27**, 216. Effect of conversion on the mechanism of vinyl polymerization. I. Styrene.

Fujii, S. (1954b). *Bull. Chem. Soc. Japan* **27**, 238. Effect of conversion on the mechanism of vinyl polymerization. II. Methyl methacrylate.

Gordy, W. (1955). *Discussions Faraday Soc.* **19**, 182. In discussion.

Grassie, N., and Melville, H. W. (1951). *Proc. Roy. Soc.* **A207**, 285. A refractometric method for following the non-stationary state of chemical reactions.

Grün, F. (1947). *Experientia* **3**, 490. Diffusionsmessungen an Kautschuk.

Hoffman, A. S., Gilliland, E. R., Merrill, E. W., and Stockmayer, W. H. (1959). *J. Polymer Sci.* **34**, 461. Irradiation grafting of styrene to high pressure and low pressure polyethylene films.

Immergut, E. H., Hughes, C., and Sarvas, A. (1958). *Abstracts of papers presented at 133rd meeting, Am. Chem. Soc. San Francisco, 1958*, p. 5R. Physical and molecular structure of graft polymers.

Ingram, D. J. E., Symons, M. C. R., and Townsend, M. G. (1958). *Trans. Faraday Soc.* **54**, 409. Electron resonance studies of occluded polymer radicals.

Ivin, K. J. (1959). *J. Polymer Sci.* **34**, 194. In discussion.

Johnson, D. H., and Tobolsky, A. V. (1952). *J. Am. Chem. Soc.* **74**, 938. Monoradical and diradical polymerization of styrene.

Lawton, E. J., Balwit, J. S., and Powell, R. S. (1958a). *J. Polymer Sci.* **32**, 257. Effect of physical state during the electron irradiation of hydrocarbon polymers. I. The influence of physical state on the reaction occurring in polyethylene during and following irradiation.

Lawton, E. J., Powell, R. S. and Balwit, J. S., (1958b). *J. Polymer Sci.* **32**, 277. Effect of physical state during the electron irradiation of hydrocarbon polymers. II. Additional experiments and discussion pertaining to trapped radicals in hydrocarbon polymers.

Magat, E. E., and Tanner, D. (1955). "Procédés de modifications de fibres ou de pélicules de polymères naturels ou naturels régénerés." Belgian Patent 546,817.

Majury, T. G., and Melville, H. W. (1951). *Proc. Roy. Soc.* **A205**, 323, 496. A dielectric constant method for following the non-stationary state in polymerization.

Mark, H., and Tobolsky, A. V. (1950). "Physical Chemistry of High Polymeric Systems. Interscience, New York.

Matheson, J., Auer, E. E., Bevilacqua, E. B., and Hart, E. J. (1949). *J. Am. Chem. Soc.* **71**, 497. Rate constants in free radical polymerization. I. Methyl methacrylate.

Matheson, M. J., Auer, E. E., Bevilacqua, E. B., and Hart, E. J. (1951a). *J. Am. Chem. Soc.* **73**, 1700. Rate constants in free radical polymerization. III. Styrene.

Matheson, M. J., Auer, E. E., Bevilacqua, E. B., and Hart, E. J. (1951b). *J. Am. Chem. Soc.* **73**, 5395. Rate constants in free radical polymerization. IV. Methyl acrylate.

Melville, H. W. (1956). *Z. Elektrochem.* **60**, 276. Kinetics of polymerization reactions in viscous systems.

Morawetz, H., and Fadner, T. A. (1959). *Makromol. Chem.*, **34**, 162. The kinetics and mechanism of the solid state polymerization of acrylamide.

Norman, I., and Porter, G. (1955). *Proc. Roy. Soc.* **A230**, 399. Trapped atoms and radicals in rigid solvents.

Norrish, R. G. W., and Smith, R. R. (1942). *Nature* **150**, 336. Catalyzed polymerization of methyl methacrylate in the liquid phase.

Ohnishi, S.-I., and Nitta, I. (1959). *J. Polymer Sci.* **38**, 451. Rate of formation and decay of free radicals in γ-irradiated poly(methyl methacrylate) by means of electron spin resonance absorption measurements.

Oster, G., Oster, G. K., and Moroson, H. (1959). *J. Polymer Sci.* **34**, 671. Ultraviolet induced crosslinking and grafting of solid high polymers.

Restaino, A. J., Mesrobian, R. B., Morawetz, H., Ballantine, D. S., Dienes, G. J., and Metz, D. J. (1956). *J. Am. Chem. Soc.* **78**, 2939. γ-ray initiated polymerization of crystalline monomers.

Rexroad, H. N., and Gordy, W. (1959). *J. Chem. Phys.* **30**, 399. Electron spin resonance studies of irradiated Teflon: Effects of various gases.

Schneider, E. E. (1955). *Discussions Faraday Soc.* **19**, 158. Paramagnetic resonance in X-irradiated plastics and in plastic solutions of free radicals.

Schneider, E. E., Day, M. J., and Stein, G. (1951). *Nature* **168**, 645. Effect of X-rays upon plastics. Paramagnetic resonance.

Schulz, G. V. (1956). *Z. physik, Chem.* (*Frankfurt*) **8**, 290. Über die Polymerisationskinetik in hochkonzentrierten Systemen. Zur Kinetik des Trommsdorffeffektes an Methylmethacrylat.

Schulz, G. V., and Blaschke, R. R. (1941). *Z. physik, Chem.* (*Leipzig*) **B50**, 305. Orientierende Versuche zur Polymerisation des Methacrylsäuremethylesters.

Schulz, G. V., and Harborth, G. (1947). *Makromol. Chem.* **1**, 106. Über den Mechanismus des explosiven Polymerisationsverlaufes des Methacrylsäuremethylesters.

Smith, W. V., and Ewart, R. H. (1948). *J. Chem. Phys.* **16**, 592. Kinetics of emulsion polymerization.

Thomas, W. M., and Pellon, J. J. (1954). *J. Polymer Sci.* **13**, 329. Kinetics of acrylonitrile polymerization in bulk.

Trommsdorff, E., Köhle, H., and Legally, P. (1948). *Makromol. Chem.* **1**, 169. Zur Polymerisation des Methacrylsäuremethylesters.

Vaughan, M. F. (1952). *Trans. Faraday Soc.* **48**, 576. Diffusion controlled reactions during the bulk polymerization of styrene.

Wall, L. A., and Brown, D. W. (1956). *J. Research Natl. Bur. Standards* **57**, 131. Chemical activity of γ-irradiated polymethyl methacrylate.

13. Trapped Radicals in Propulsion

MAURICE W. WINDSOR

Space Technology Laboratories, Los Angeles, California
Guest Scientist, Free Radicals Research Program,
National Bureau of Standards, Washington, D. C.

I. Introduction and History

Energy manifests itself in many forms and there are also many ways in which energy may be used to propel a vehicle. In this chapter we shall discuss the possibilities of using trapped radicals as a source of energy and we shall specifically devote our considerations toward their potential use for propelling rockets. There is both logical and historical justification for this restriction. Historically, the avid desire of the rocket men for more and more powerful chemical fuels provided impetus for much of the recent activity in free radical research. Furthermore, the rocket engine is the simplest of all prime movers, and it is therefore a logical choice of propulsion system on which to focus our attention.

A. ENERGY OF CHEMICAL REACTIONS

Conventional chemical systems for propelling rockets are limited in energy to the energy released by the rearrangement of chemical bonds when two molecules react together. In general, the reactants are themselves perfectly stable molecules when isolated from each other, but their atoms can be rearranged to give new molecules containing stronger bonds than those in the original molecules. The energy left over, or heat of reaction, appears as kinetic energy of the products. It is apparent, however, that this net energy is usually only a small fraction of the total chemical binding energy of the atoms in the molecule. For example, take the reaction between hydrogen and fluorine, one of the most vigorous reactions known:

$$H_2 + F_2 = HF + HF + 128.4 \text{ kcal}$$

The heat of reaction is 128.4 kcal/mole, but the binding energy of the H_2 molecule is 103 kcal/mole, and that of F_2 is 35.6 kcal/mole. That of HF is 133.5 kcal/mole. We can think of the reaction in the following terms: It is necessary to expend 103 kcal to dissociate the H_2 into atoms and 35.6 kcal to dissociate the F_2. When the atoms are rearranged to make two HF molecules we get back 133.5 kcal for each H—F bond, making a net gain of 128.4 kcal. This is shown in Fig. 1. In actual fact the reaction almost certainly goes over a smaller potential hill than corresponds to complete dissociation, but this does not alter the net amount of energy that can be extracted from a reaction. In this case then, the energy of reaction is only 48% of the total chemical binding energy. Similar figures are given by other systems.

The chemist's role in making rocket propellants is, then, to select molecules with very strong bonds like HF, H_2O, N_2, CO, CO_2, take them apart,

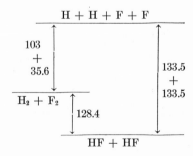

FIG. 1. The reaction between hydrogen and fluorine. Numbers denote kilocalories per mole.

and put them together into more weakly bound molecules like H_2, F_2, O_2, HNO_3, N_2H_4, C_2H_5OH. To do this, energy must be expended, and this energy is stored as potential energy in the reactive molecules so made. Such molecules come fairly high up on a diagram like Fig. 1, and when they react they tumble down into the more stable arrangements and give up this stored energy as heat.

Of course, the more energy one packs into a molecule, the harder it is to stop it either from blowing apart of its own accord or from reacting with virtually everything with which it comes into contact. Thus energetic propellants like red fuming nitric acid (RFNA) and liquid oxygen (LOX) have to be handled with extreme care and often at very low temperatures.

B. FREE RADICALS

One may ask what is the maximum amount of energy one can store in a molecule. The answer is surprisingly simple and at the same time paradoxical. The limit is reached when you have a system of free atoms, just as we have a system of free H atoms and free F atoms at the top of Fig. 1. Of course, with more complex molecules like CH_4 and NH_3 and H_2O, there are several halfway houses before reaching a completely atomic system. These correspond to $CH_3 + H$, $NH_2 + H$, or $H + OH$. In other words, when such molecules react, the system never goes through a completely atomic state because there are subassemblies which stick together and retain their identity during a reaction. These units are called free radicals. They are characterized by having an unpaired electron so that they cannot rearrange within themselves to form a stable molecule. Consequently they are extremely reactive. For the sake of generality we can include free atoms under the generic name "free radicals," atoms being the extreme case of complete dissociation.

It appears that the first suggestion that free radicals might be used as rocket fuels was made by Zwicky (1943) in his morphological analysis of chemical energy sources for propulsion. Since then a great deal of effort has been spent on studying free radicals, their formation and their properties, and this is discussed at length in other chapters of this book.

Of course, free radicals have been important in combustion for many years. Although usually present only in minor amounts, their exceptional reactivity confers on them a major role in determining the mechanism and rate of a chemical reaction.

However, from the aspect of energy storage this high reactivity is a major obstacle, and in order to stabilize free radicals we have to "trap" them at very low temperatures. A "trapped" free radical sounds something of a contradiction, and in fact it is true that when we catch a free radical in a trap it is no longer free. Nevertheless, by going to very low temperatures,

we can make everything so quiet thermally that even a quite weak trap
will suffice. The trap usually consists of a cage of relatively inert molecules
which effectively immobilize each radical and separate it from other radicals.
On warming up the system, the cages collapse, the radicals recombine,
and energy is released in the form of heat or light. It was probably the
observations of Broida and Pellam (1954) on nitrogen frozen at 4.2°K
after passage through an electrodeless discharge which provided inspiration
for the possibility of using trapped radicals as rocket fuels. The frozen
nitrogen emitted a greenish-blue glow with brilliant blue flashes. The
emission was attributed to the presence of free nitrogen atoms in the solid
lattice.

From the propulsion viewpoint however, it turns out that it is one thing
to sit down and cook up a superfuel with pencil and paper. It is another
thing to make it and store it and handle it and do so in ton quantities.
With free radicals, it is enough of a problem just to make them and store
them even in small quantities, let alone worry about how to handle ton
quantities. But the potential payoff is sufficiently high with the right
systems that, even though it is a long shot, it is worth the try.

II. Rocket Fundamentals[1]

Before discussing the relative merits, for propulsion, of trapped radicals
compared to other fuels, it will be helpful to look at some of the fundamental principles of rocket theory.

A. Escape Velocity

In any space-travel mission, "escape velocity" always plays an important
role. This is the velocity a vehicle must have at the earth's surface to
coast away from the earth and escape completely into outer space. It is
also equal to the velocity a body initially at rest at infinity would acquire
in falling toward the earth, if there were no atmosphere to retard it. For
the earth, escape velocity is 11.2 km/sec (7 miles/sec or about 25,000 mph).
This figure is derived from the fact that the energy required to leave the
earth is gR per unit mass, g being the acceleration due to gravity and R,
the radius of the earth. Equating this to the kinetic energy $\frac{1}{2}MV^2$ we get

$$V_{escape} = \sqrt{2gR}$$

The expression gR per unit mass helps give us a feeling for the task of
escaping from the earth. Since R is 4000 miles, escaping from the earth is
like climbing a hill 4000 miles high against a force of one gravity all the
way. Looked at another way, it takes as much effort to send *1 pound* to

[1] For further reading, see Clarke (1951a, b) and Sutton (1949).

infinity as it does to hoist a 10-ton truck to the top of the Empire State Building. It is obvious then that for every pound of payload put into space, many hundreds of pounds of fuel must be carried, let alone the weight of the structure needed to hold this fuel.

B. Mass Ratio and Exhaust Velocity

Because it takes so much energy to put a pound in space, a loaded rocket, about to take off, is very largely fuel. This is because of the relative puniness of present-day fuels. The final velocity V_f attained by a rocket is directly related to the exhaust velocity V_e of its engine and the amount of fuel it carries. We get the relationship by integrating the equation of motion

$$V_e dm = m \, dV \quad \text{or} \quad dm/m = dV/V_e$$
$$\ln M_0/M_T = V_f/V_e \quad \text{or} \quad M_0/M_T = \exp(V_f/V_e)$$

where M_0 is the initial weight, M_T the empty weight when all fuel has been used, and ln means logarithm to the natural base e.

M_0/M_T is called the mass ratio and is given the symbol R. It measures the relative weight of fuel compared to engines, structure, and payload. There are very definite limitations on the size of R, imposed by the structural strength of materials. One can get some feeling for the practical implications of high mass ratios by observing that an ordinary hen's egg has a mass ratio of about 10. This means that a "fully loaded" egg weighs ten times more than the empty shell. Looked at another way, 90 % of the total weight is useful contents. Rockets are made of metal, not eggshell. On the other hand, neither can they afford to be so fragile. It is therefore all the more remarkable that mass ratios appreciably greater that 10 are now in common use. The Atlas ICBM has a take-off weight in excess of 250,000 pounds. The dry weight (less payload) is less than 15,000 pounds, and the payload is something over 2500 pounds. From these recently released figures one can calculate that an Atlas at the moment of take-off is about 93 % fuel and liquid oxygen by weight. It is in effect a thin-skinned metal balloon. This is not obtained without some sacrifices, as those who have watched movies of abortive rocket launchings will readily realize. Although the structure is strong enough to sustain stresses in the vertical direction, if it happens to topple over, as in an unsuccessful launching, the whole rocket crumples and collapses like a house of cards. Weight saving has reached the point where a modern rocket, loaded with fuel, cannot support its own weight in a horizontal position.

The final velocity (all fuel burned) attained by a single-stage rocket is, then, given by, $V_f = V_e \ln R$. With R equal to 13, this means that V_f cannot be greater than 2.5 V_e. The highest exhaust velocities currently attainable are about 2.5 km/sec using kerosene and liquid oxygen (JP4 and LOX). This leads to a figure of 6.25 km/sec for the limit of a single-stage rocket, which is about three-quarters of the speed needed to set a satellite in orbit.

This simple limitation explains several things. It explains why satellite-launching vehicles consist of several stages in which a smaller rocket rides piggyback on a larger one before making its own leap into space. It also explains why it takes a 10-ton rocket to orbit a 20-pound satellite. The IGY rocket Vanguard is a three-stage vehicle with each stage making up the payload (about 10%) of the previous stage. The complete rocket thus weighs three factors of 10 more than the satellite it puts into orbit. The Atlas ICBM does somewhat better in putting about 4 tons into orbit for a starting weight of something over a hundred tons. The 4 tons, however, includes the weight of engines and structure. It does this using effectively two stages, the first stage comprising two booster engines which drop off early in the ascent.

We can rest assured that giant multistage vehicles are with us as long as we must rely upon conventional chemical fuels to propel our rockets.

C. Rocket Engines and Fuels

We have seen in Section II, A that velocities exceeding 11 km/sec are needed to escape from the earth and still higher velocities are necessary for interplanetary voyages. We have seen in Section II, B that the restriction of practically attainable mass ratios limits the final velocity of a rocket to something over twice the exhaust velocity of the motor, or about 6 km/sec. Let us now examine the characteristics of the rocket motor and the relation between exhaust velocity and the fuel it uses.

1. Specific Impulse

The merit of a propellant is usually described in terms of its specific impulse, I_{sp}, which is the thrust produced for each pound of propellant burned per second. It is measured in units of "lb sec/lb," usually contracted to just "sec." Since the thrust is equal to the exhaust velocity times the *mass* of propellant ejected per second, we have,

$$I_{sp} = \frac{\text{thrust}}{weight \text{ ejected per second}} = \frac{V_e \times \text{mass per second}}{\text{weight per second}} = V_e/g$$

Taking $g = 981$ cm/sec², typical figures for LOX and JP4 would be V_e, 2.5 km/sec, and I_{sp}, 255 sec.

2. Exhaust Velocity, Chamber Temperature, and Molecular Weight

The exhaust velocity depends in turn upon the heat released by the chemical reactions in the combusion chamber and the efficiency with which this heat is converted into directed kinetic energy by expansion through the nozzle of the rocket motor. A complete expression for V_e involves the expansion ratio (chamber pressure to ambient pressure), the ratio of specific heats, γ, of the gas mixture and the gas constant, R, as well as the chamber temperature, T, and the mean molecular weight, M, of the exhaust gases. However, the expression is not very sensitive to either expansion ratio or to γ and a satisfactory approximate relation, using an expansion ratio of 20:1, is

$$V_e \simeq \tfrac{1}{4}\sqrt{T/M}$$

With T in °K this gives V_e in km/sec. For example with $T = 3000°$K and M about 30, we get $V_e = 2.5$ km/sec.

The importance of this equation is that it shows that the two basic criteria for high performance are high chamber temperature and low molecular weight. With conventional fuels where the exhaust gas is a mixture of H_2, H_2O, CO, and CO_2 and sometimes N_2, we have very little control over M, which will usually be somewhere around 30. It is interesting to note that optimum performance is usually obtained with a fuel-rich combination because of the effect of hydrogen in the hydrocarbon fuel in reducing the average molecular weight, even though this also occasions a lower combustion temperature.

The effect of molecular weight on exhaust velocity is shown in Table I for a series of gases initially at 3000°K and 300 psia. The overwhelming superiority of hydrogen is immediately apparent. Given an independent source of energy, the best possible performance is obtained by using it to heat hydrogen to a high temperature. The importance of this conclusion in evaluating free radical systems as rocket propellants will be seen in the next section.

TABLE I

EFFECT OF MOLECULAR WEIGHT ON EXHAUST VELOCITY[a]

Gas	Molecular weight	Exhaust velocity[b] (km/sec)
H_2	2	7.60
He	4	4.30
HF	20	2.39
H_2O	18	2.99
CO	28	2.10
N_2	28	2.09
CO_2	44	2.12
BF_3	68	1.82

[a] After Table 2 of Baum et al. (1956).
[b] Conversion, 60%; initial temperature, 3000°K.

D. Summary and Conclusions

(1) Escape velocity is 11.2 km/sec. Orbital velocity is 8 km/sec. Many hundreds of pounds of chemical fuel are needed to accelerate 1 pound of useful payload to such high speeds.

(2) It is not feasible to design a rocket which will reach a speed greater than 2.5 times the exhaust velocity of its engine. Such a rocket has a mass ratio of 13 and when loaded would be 93 % propellants by weight.

(3) Since the best propellants in current use give exhaust velocities of about 2.5 km/sec, the best velocity attainable by a *single-stage* rocket is about 6 km/sec. Two- and three-stage vehicles are therefore necessary to achieve orbital velocity and escape velocity.

(4) The exhaust velocity of a rocket motor depends mainly on the chamber temperature and the molecular weight of the exhaust gases according to the relation $V_e \simeq \frac{1}{4}\sqrt{T/M}$. With conventional chemical propellants we have little control over M, which usually lies close to 30. However, given an independent source of energy, hydrogen, by virtue of its low molecular weight, would be by far the best choice of propellant.

III. Free Radicals as Energy Sources[2]

A. Limits of Orthodox Chemical Systems

The last section showed that the basic practical criterion by which the merit of a propellant combination is measured is its specific impulse. A high specific impulse requires a high chamber temperature and a low molecular weight. A useful single parameter which effectively combines both these criteria is the energy released, ϵ, per unit weight of propellant. If all this energy could be converted with 100 % efficiency into kinetic energy of motion of the exhaust products, a velocity of $\sqrt{2\epsilon}$ would be obtained. This immediately enables us to calculate an upper limit to the exhaust velocities of conventional chemical systems. The performance of a series of propellants is summarized in Table II. The orthodox chemical system with the highest energy per gram is H_2/O_3 with $\epsilon = 3.83$ kcal/gm. The theoretical maximum velocity is 5.66 km/sec or, in terms of specific impulse, 577 sec. In practice, since the rocket is a heat engine, it is impossible to convert all the heat into kinetic energy, and 50 to 60 % is the best figure that can be achieved. Although higher conversions could be achieved by increasing the expansion ratio, so lowering the exhaust temperature, the advantage gained would be more than offset by the extra weight of the extended nozzle that would be needed. Therefore we have to divide the above velocity by $\sqrt{2}$ or 1.4, giving a figure of about 400 sec as the practical upper limit for chemical propellants.

[2] For further reading see, Baum *et al.* (1956), Szego and Mickle (1958), Mickle *et al.* (1958).

TABLE II

PERFORMANCE OF ROCKET PROPELLANTS

Propellant	Mixture ratio[a]	Energy (kcal/gm)	Percentage conversion	Specific impulse (I_{sp})	Exhaust velocity (V_e) (km/sec)
Conventional					
LOX-JP4	2.6	2.05	50	306	3.00
LOX-alcohol	1.5		50	242	2.37
N_2O_4-N_2H_4[b]	1.4	1.59	61.5	293	2.87
RFNA-aniline	3.0		50	221	2.17
LOX-N_2H_4[b]	0.9	1.88	66.5	330	3.23
F_2-N_2H_4[b]	2.4	2.47	62.3	366	3.59
LOX-H_2[b]	7.0	3.15	51.5	376	3.68
F_2-H_2[b]	15.0	3.16	59.3	402	3.94
O_3-H_2	8.0	3.83	50	408	4.00
Free radical[c]					
H	Performance	52.1	50	1520	14.7
CH	calculated for	10.9	50	714	7.0
N	hypothetical	8.0	50	595	5.8
BH	propellant	6.2	50	520	5.1
NH	composed of	5.2	50	465	4.6
BH_2	100% free ra-	5.2	50	465	4.6
O	dicals	3.7	50	401	3.9
NH_2		2.9	50	365	3.6

[a] Oxidant: fuel, by weight.
[b] Carter (1958).
[c] Baum et al. (1956).

This level of performance has not yet been attained. The practical systems presently in use are kerosene and liquid oxygen, ethanol and liquid oxygen, and red fuming nitric acid (RFNA) and aniline. The specific impulses of these, shown in Table II, lie in the range 250 to 300 sec. Boron-based fuels give useful, but only marginal, improvement and, although better than kerosene and LOX, they fall far short of liquid H_2 and LOX. However, although from an energetic point of view, the best chemical systems are H_2-O_2, H_2-O_3 and H_2-F_2, there are also other factors which have to be taken into account. All the above combinations involve liquid hydrogen whose very low density (7% that of water) demands large fuel-tank volume. Furthermore, liquid ozone is extremely unstable and liable to explode spontaneously. With fluorine, the exhaust gases are toxic and corrosive HF, which would present a further hazard with ground-fired rockets. Nevertheless, rocket motors have been test-run on all the above combinations, and liquid fluorine is now available by the tank carload.

All this not withstanding—even given the ability to run large motors on H_2-F_2 and achieve I_{sp}'s of 400—it will still take two stages to orbit a satellite and to achieve escape velocity. Big improvement though this is, flying to the moon will never become like flying the Atlantic as long as we have to rely on conventional chemical fuels.

B. Free Radicals as Propellants

1. Energy Content

In principle at least, free radicals allow us to surmount this energy barrier or, more appropriately perhaps, creep around it. The dissociation energy of nitrogen N_2 is 225 kcal/mole. Recombination of pure monatomic nitrogen would give an ϵ of 8 kcal/gm, leading to a specific impulse of 595 sec. The dissociation energy of hydrogen at 103 kcal/mole is less than half that of nitrogen, but owing to its very low molecular weight, a propellant consisting of 100 % monatomic hydrogen would have an ϵ of 51.5 kcal/gm and a specific impulse of 1520 sec.

Other radicals such as NH, CH, OH, and BH also possess high energies of reaction and recombination. The energy content of these and other free radicals is given in Table II together with that of the H_2-O_3 system, which represents the best conventional chemical propellant. One sees immediately that only the low molecular weight species H, CH, N, BH, NH, and BH_2 are worthy of closer consideration. Other radicals cannot provide enough energy per unit weight to qualify. In particular, notice that oxygen atoms even at 100 % concentration give a figure only slightly better than our best chemical propellant H_2-O_3 . This is due mainly to the high molecular weight 32, of the molecular O_2 exhaust product.

2. Performance as Propellants

Since most work in recent years has been done on them, let us examine the characteristics of H, N, and NH as propellants. Because of their inherent instability, it is most unlikely that free radicals will ever be stabilized in the pure form. Furthermore, we have already seen that, given an independent source of energy, the most effective way to use it is to heat a working fluid of hydrogen gas. Therefore, let us consider the behavior of H, N, and NH mixed in various proportions with molecular hydrogen. This has been done for a pressure ratio of 20 by Scheller et al. (1957), and their calculations are shown graphically in Figs. 2 and 3.

In Fig. 2, the specific impulse and the combustion temperature are shown as a function of the molar percentage of atomic nitrogen for a mixture of atomic nitrogen and molecular hydrogen. Maximum performance is at-

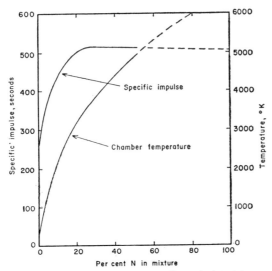

FIG. 2. Rocket performance of N-H₂ system. Gas admitted to combustion chamber at 300°K; chamber pressure, 300 psia.

tained at 30 % atomic N with an I_{sp} of 515 sec. Beyond this the I_{sp} falls off gradually while the chamber temperature continues to rise. The striking fact emerges that not only is no advantage to be gained by going beyond 30 % in atomic N concentration, but that 100 % atomic N actually yields a worse performance than 30 % used to heat hydrogen. This illustrates emphatically the importance of achieving a low average molecular weight.

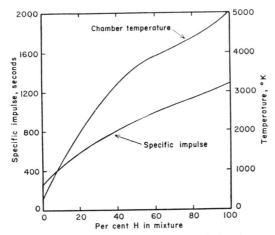

FIG. 3. Rocket performance of H-H₂ system. Gas admitted to combustion chamber at 300°K; chamber pressure, 300 psia.

So important is this molecular weight factor that should atomic nitrogen be used to heat molecular nitrogen instead of hydrogen, a 50 % mixture yields an I_{sp} of only 260 sec, half that of a corresponding N—H mixture and no better than a conventional mixture of kerosene and liquid oxygen. The behavior of the imine radical (NH) in H_2 is similar to that of atomic nitrogen. Optimum performance is reached at 40 % NH with an I_{sp} of 450 sec, falling to 400 sec at 100 % NH.

It is obvious from the foregoing that the best system will be H-H_2 since both the monatomic hydrogen fuel and the molecular hydrogen working fluid have the lowest possible molecular weights. This expectation is borne out by the data in Fig. 3. The performance rises smoothly with the percentage of H in the mixture from 230 sec for pure H_2 to 1280 sec for 100 % pure monatomic hydrogen.

Even at 10 % H an I_{sp} of 500 is obtained, and this at the very modest chamber temperature of about 1200°K.

Incidentally, even at 0 % H, an I_{sp} of 230 sec is given. This is because, for lack of specific heat and heat of vaporization data on condensed free radicals, Scheller et al. have taken 300°K as the temperature at which the propellant enters the combustion chamber. It is interesting to observe then that H_2 gas on its own at 20 atmospheres and 300°K is a quite respectable propellant, since by expanding and thus cooling it, some of its heat content may be converted into thrust. The snag of course is that tankage volume for compressed H_2 gas at 300°K would be prohibitively large and if one uses liquid hydrogen one immediately has to provide a large quantity of heat to vaporize it and warm it up to 300°K. However, it does show that worth-while, specific impulses may be obtained with hydrogen even at temperatures which, by rocket engine standards, are exceedingly low.

3. Further Combustion with Oxygen

There arises the possibility of further increasing the performance of a free radical mixture containing hydrogen by combustion with oxygen. No advantage can be gained by doing this, however, because of the increase in molecular weight caused by the addition of oxygen. This increase more than offsets the extra energy produced by the combustion. To illustrate this, Scheller et al. give calculations for a mixture containing 17 % by volume of atomic H in molecular H_2. This gives an I_{sp} of 545 sec and a temperature of 1740°K. Addition of one part of O_2 to six of H_2 results in a *drop* in performance to 425 sec while T rises to 3200°K.

We conclude then that the optimum way of using a free radical fuel is to use its energy of recombination to heat a working fluid of H_2. This is because of the importance of low molecular weight in achieving a high exhaust velocity. Furthermore, again because of this factor, the best system of all is H-H$_2$, with the highest performance being given by 100 % monatomic H.

4. Other Radicals and Atoms

A number of other radicals present themselves for evaluation. These include O, OH, CH, CH$_2$, CH$_3$. In all cases, however, the recombination energy per unit weight is less than that of nitrogen and the molecular weight of the products leads to inferior performance. We are therefore justified in concentrating our attentions on the radicals N and NH and particularly on monatomic hydrogen.

5. Ions and Metastable Species

In addition to atoms and free radicals there exist other species, some of which possess even higher energy referred to their normal combined state. These include ions and molecules or atoms in metastable states. One example of the latter is the triplet excited state of the helium atom in which the two electrons of the helium atom have been unpaired and one of them promoted to the ^2S level. Triplet helium can be formed in an electric discharge and it is well known spectroscopically. Although ordinary helium is completely inert, triplet helium should be very reactive chemically and at low temperature might combine with hydrogen to form HeH, helium hydride.[3] McGee and Wharton (1958) claim recently to have indirect evidence of such a compound at 4°K. A fuel consisting of 100 % triplet helium would give an I_{sp} of 2900 sec, and HeH would yield 2400 sec, according to their estimates.

C. EXPERIMENTAL ACCOMPLISHMENT IN TRAPPING RADICALS

For full details of experiments on trapped radicals and methods of measuring their concentration, reference should be made to the earlier chapters. Apart from a discussion of a novel proposal for stabilizing pure monatomic hydrogen and some other speculations on high energy species of low molecular weight, only a brief summary will be given here.

The main condition for stabilizing free radicals is the provision of a matrix which must be, (a) inert, so that it does not react with the radicals, and (b) rigid, so that diffusion and recombination are prevented. These conditions lead to two important conclusions: (1) Only dilute mixtures of radicals in an inert matrix will be stable. A maximum concentration would appear to be 10 to 14 %, depending on the type of packing that obtains, the criterion being simply that every radical shall be surrounded by a com-

---

[3] A method for the alleged production of helium hydride has even been patented (see Warnke, 1934).

plete shell of matrix molecules (Golden, 1958; Jackson and Montroll, 1958). There are other theoretical arguments (see Chapter 10) based on the possibility of a recombination chain reaction, which predict very much lower values, of the order of a few tenths of 1 %, for the maximum concentration. (2) The rigidity requirement for the matrix means low temperatures and in the case of H atoms and N atoms in matrices of N_2 or H_2 , temperatures of 4°K and below are necessary. This requires the use of liquid helium, and to get below 4°K the pressure over the liquid helium must be reduced.

Actual experimental accomplishments for the trapping of the radicals of most interest to us here, N, NH, and H, are as follows:

N atoms: About 0.3 % in solid N_2 (Fontana, 1959).

NH radicals: No measurements available.

H atoms: Wall *et al.* (1959) report a figure of 0.15 % for H atoms in gamma-irradiated methane and 0.0006 % for H atoms in solid hydrogen. In the latter case, H atoms disappeared on standing at 4.2°K with a half-life of about 19 hours. Very recently, using an independent method in which the amount of heat released by recombination is simulated electrically, Windsor (1959a) has obtained estimates of 0.1 to 0.2 % for H atoms in discharged hydrogen deposited on a surface below 4°K. Furthermore, periodic releases of heat were observed during deposition indicating that such concentrations were stable only for short periods of time (a few seconds to a few minutes depending on flow rate). This instability is attributed to the existence of large temperature gradients across the growing deposit, the very presence of trapped atoms causing a distortion of the crystal lattice thus lowering the thermal conductivity by several factors of ten.

We conclude then, that maximum stable concentrations of trapped N atoms and H atoms presently achievable are a few tenths of 1 %. Furthermore, the thermal phenomena just described may constitute a fundamental limitation to attempts to trap higher concentrations.

D. Spin-aligned H Atoms

Since monatomic hydrogen is the only system that promises really dramatic increases in specific impulse, and since experimental accomplishments to date are so meager, it is worth examining some novel proposals and new approaches to the problem of stabilizing hydrogen in the atomic form.

One idea which is currently being actively investigated by Fite (1958) and also in the author's laboratory at the National Bureau of Standards (Windsor, 1959b) is the possibility of preventing or retarding recombination by the use of spin-aligned atoms. The argument runs as follows. Every hydrogen atom possesses a single, unpaired electron. The spin of this electron, which also gives rise to a magnetic moment, can point either up

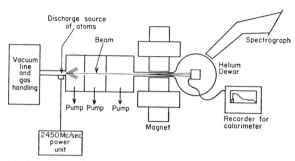

FIG. 4. Apparatus for condensing spin-aligned hydrogen atoms.

or down relative to an external magnetic field or to the field of another atom. When two atoms approach each other they must, in order to form a stable H_2 molecule, do so with their spins opposed so that an electron-pair bond may be formed in accordance with the Pauli principle. Should their spins be parallel, only an unbound, repulsive triplet state of H_2 is formed which immediately breaks up again into atoms. The proposal is, then, to prepare hydrogen atoms all with their spins parallel and attempt to condense them at temperatures below 4°K.

The experimental arrangement is shown schematically in Fig. 4. Atoms generated in a microwave discharge effuse through a narrow vertical slit into a chamber evacuated by a high speed diffusion pump. By the use of further slits a well-collimated atomic beam is obtained which passes through a very inhomogeneous magnetic field. This is the well-known Stern-Gerlach experiment (1922 and sequence), and the beam is spatially resolved into its two components in each of which all the atoms have their spins aligned parallel to each other (see also Ruark and Urey, 1930). One of these beams enters a helium Dewar and impinges on a surface kept at about 1°K. The presence of trapped atoms is sought for by observing the heat of recombination when the sample is allowed to warm up. Up to the time of writing no positive results have been achieved. The central question is whether spin alignment can be retained during collision with the surface and the process of deposition.

A theoretical treatment of a system of spin-aligned H atoms has been made by Jones et al. (1958). They conclude that, because of coupling between the electron spin and the nuclear spin which would cause spin flipping to occur, a magnetic field of 2×10^5 gauss would be needed at 1°K to prevent spin-aligned hydrogen atoms from recombining. By going to 0.1°K, the field could be reduced to 2×10^4 gauss. This gloomy prediction is feebly illuminated by the somewhat remote chance that condensed monatomic hydrogen might be ferromagnetic and thus provide its own magnetic field.

They also predict that, under the conditions stated, atomic hydrogen should form a liquid with a density of 0.04.

Despite the discouraging theoretical predictions, nothing is yet known experimentally about the behavior of spin-aligned atoms at very low temperature, and with respect to rocket fuels, this approach is probably the most promising line of attack.

E. Engineering Considerations

Discussion of the problems of handling and using free radical fuels is really premature but perhaps not without some value, since no fuel, however powerful, is of much use if it is either impossible to use or carries heavy penalties in tankage volume or in toxicity.

It may be taken for granted that free radical fuel would be solid and would be stable only at very low temperatures. This means that either a slurry must be made of the solid dispersed in some working fluid, preferably liquid hydrogen, or the fuel must be used as a solid propellant. The trouble with liquid hydrogen is its low density (0.07), requiring fuel tanks about fourteen times bigger than those for a conventional hydrocarbon fuel. Large penalties in structural weight are thereby incurred. Liquid hydrogen has other disadvantages. It boils at so low a temperature, $20°K$, that everything else except helium would be frozen solid in liquid hydrogen. To avoid the risk of solid particles plugging orifices and jamming valves, all air and moisture must be scavenged from the lines before introducing the liquid hydrogen. Mixtures of solid air or solid oxygen and liquid hydrogen may be explosive. The low temperature also raises other problems. Choice of tank and piping materials is difficult since most metals become brittle at these temperatures and their strength falls to less than a third of its room temperature value. Efficient insulation of the fuel tanks to reduce evaporation loss would be required, and if, as is likely, it were necessary to maintain the fuel at temperatures below the melting point of solid hydrogen in order to prevent radical recombination in the fuel tanks, this could easily prove to be the most formidable task of all. Recombination once begun would spread catastrophically. The problem is not unlike trying to store nitroglycerin inside a furnace, and the penalties for failure would no doubt be somewhat similar.

Another major problem is also a cooling problem, but several thousand degrees removed—the problem of cooling the combustion chamber. Modern rocket motors are able to tolerate the high combustion chamber temperatures, around $3000°K$, only by the techniques of regenerative cooling. The chamber walls are kept from melting by circulating one of the propellants through passages in them on its way to the injectors. This circulation also preheats the propellant. Any such preheating of a free radical fuel would probably lead to an explosion in the supply lines. This means the use of a separate coolant, perhaps also liquid hydrogen, but containing no trapped radicals. This problem of chamber cooling is considered in detail in the report by Scheller *et al.* (1957).

The other alternative to be faced is that it might be necessary to use free radical fuel as a solid propellant. Nobody has yet had the courage seriously to consider a large rocket of *solid* hydrogen. However, theoretical estimates have been made of the burning rate (Nichols, 1958) of solid free radical fuels. Numbers of the order of 2000 ft/sec are obtained. If these are valid, "combustion" is too weak a word to describe the burning of a free radical fuel: "detonation" would be more appropriate.

It is apparent that the technological problem of handling and usage might well outweigh the energetic advantages offered by free radical fuels. The ideal free radical fuel, which is still an optimist's dream and may never be otherwise, would be a stable liquid form of monatomic hydrogen, perhaps a ferromagnetic system of spin-aligned atoms or a high-density metallic form of hydrogen produced only at very high pressures but subsequently stable (Wigner and Huntington, 1935). If this liquid could be stored safely but would release its energy on vaporization, not only could the size of present-day rockets be scaled down a factor of 10, but talk of round-trip voyages with landings on the moon and planets would lose much of its present air of wishful thinking.

F. CONCLUSIONS

(1) The maximum specific impulse attainable *in principle* with conventional chemical propellants is about 400 sec. Our present capabilities are something less than 300 sec.

(2) There is no doubt that selected free radical systems, if available, would offer substantial improvements. Monatomic hydrogen could give specific impulses exceeding 1000 sec.

(3) However, the characteristics of such fuels are likely to make their handling and use at best formidable and at worst impossible.

(4) Research accomplishments to date hold out little hope of storing free radicals in the concentrations required.

(5) Monatomic hydrogen is the only system which offers really dramatic gains in performance. Further basic research on free radicals as fuels is worthwhile, but in view of this, should focus on attempts to stabilize monatomic hydrogen. Promising avenues of attack involve the study of spin-aligned atoms and attempts to prepare metallic hydrogen or such compounds as helium hydride.

IV. Other Exotic Methods of Propulsion

A. Types of Space Mission

There are two basic types of space mission, each with its own special requirements: (1) Ascent from the surface of a planet into an orbit or to achieve escape velocity. (2) Travel between orbits around different planets without ever descending to a planetary surface. We might call this "space cruising."

These two phases of space travel are analogous to city driving, on the one hand, and turnpike cruising on the other. There are also similarities in the power requirements. For climbing out of the earth's gravitational potential well, we need a motor capable of delivering a high thrust for a short time, much as we need a low gear to overcome friction and get started in city traffic. The high thrust is needed to combat a planet's gravity, which must be exceeded before the vehicle will lift from the surface. This task demands a propulsive system, such as a conventional chemical rocket motor, which ejects a large amount of mass with a relatively low exhaust velocity. If the velocity is too high we are throwing away kinetic energy in the exhaust gases when the rocket is moving slowly, with a consequent loss of efficiency. A desirable figure is about 10 km/sec, i.e., comparable to orbital and escape velocities. This is, nevertheless, about four times higher than our present capabilities.

For cruising between orbits there is no minimum thrust requirement. This is because a body in orbit has no weight. It travels in a perpetual state of "free fall," freely responding to the forces acting on it. Only when a body resists the pull of gravity does the phenomenon of weight appear. Consequently even an acceleration of $1/1000$ g will gradually build up a quite high velocity when applied over a period of many days or even months. We see then that the power requirements demanded of an engine for interorbital cruising are much less stringent than for climbing up the steep hill from earth to orbit. Once this is done we can, as it were, shift into a much higher gear and increase speed quite gradually.

For deep-space travel we need then an engine which ejects only small amounts of mass, but at a relatively high velocity. The high velocity is needed so that journeys to the outer planets can be made in a reasonable length of time, say one year (3×10^7 seconds). An acceleration of 2×10^{-3} g (2 cm/sec^2) applied over a period of six months would produce a terminal velocity of 300 km/sec (200 miles/sec), making it possible to travel 3000 million miles in a year, accelerating for half the time and decelerating for the final six months. An exhaust velocity of 300 km/sec and a mass ratio of 2.7 would meet this requirement. If a lower efficiency can be tolerated for the sake of saving on weight of propellant carried, this velocity could be

TABLE III
CHARACTERISTICS OF VARIOUS PROPULSION SYSTEMS

Propulsion system	Exhaust velocity (V_e) (km/sec)	Specific impulse (I_{sp}) (sec)	Thrust (g)	Remarks
Chemical	4.0	400	>1	
Nuclear (fission)	5.6	560	>1	At 2000°K
	7.6	775		At 3000°K
Thermonuclear (fusion)	20	2040	~2	
Ion propulsion	10^4 to 10^5	10^6 to 10^7	10^{-3}	
Plasmajet	10 to 100	10^3 to 10^4	>1	
Magnetohydrodynamic	200	20,000	0.5	
Solar sailboat	3×10^5	3×10^7	10^{-3}	
Trapped radicals				
10% H in H_2	4.4	450	>1	
20% H in H_2	5.8	600	>1	
100% H	12.4	1280	>1	

increased to perhaps 1000 km/sec. A rocket engine with these characteristics has already been designed. It is called an ion rocket.

B. TYPES OF PROPULSION SYSTEMS

The characteristics of the various types of propulsion systems are summarized in Table III. Chemical rockets produce accelerations greater than 1 g and can ultimately give a specific impulse in the region of 400 sec. Let us now look at the other systems.

1. The Nuclear Rocket

The nuclear rocket, using a fission-type reactor to heat hydrogen gas, gives an I_{sp} of 560 sec at 2000°K, probably the limit for a motor based on a conventional solid-fuel reactor since even the more refractory uranium compounds, such as the oxides, melt at temperatures around 2500°K. One way around this difficulty would be to seal the uranium in cans of a higher-melting material such as tungsten. If 3000°K could be reached, an I_{sp} around 800 sec would be obtained. Other schemes entail the use of a homogeneous gas-phase reactor in which the fissile material is injected into the working fluid, but unless an exceedingly cunning method can be devised for recovering the fuel from the exhaust gas, this method would be prohibitively wasteful of fissile material. Perhaps the biggest snag is that once an atomic motor has been fired it is irrevocably altered and becomes a highly radioactive and hazardous piece of machinery. Furthermore it cannot be shut down too rapidly for fear of melting the reactor. Because of these inherent obstacles it seems more likely that fission will first be used as a source of power for ionic propulsion or other electrical propulsion systems.

2. The Thermonuclear Rocket

At a temperature of a hundred million degrees hydrogen catches fire and burns to the nuclear ash helium. In so doing it releases a vast amount of energy. This is the fusion reaction which makes possible the H bomb and which helps power the sun. No material walls can contain such temperatures but, because the gas is completely ionized, the resulting plasma can be controlled with magnetic fields.

A thermonuclear rocket will probably be a pulsed device in which an intense electrical pinched discharge is used to ignite the gas. A small fraction of the fusion energy will be extracted electromagnetically to power the next cycle, and the major fraction, residing in the fast-moving charged particles, will be converted by a magnetic nozzle into kinetic energy. Exhaust velocities of 20 km/sec appear to be feasible, but first we must achieve a self-sustaining fusion reaction.

Clauser (1958) concludes that, by analogy with Whittle's turbojet engine, the first application of fusion power may be for propulsion rather than in power stations. Although both present formidable problems, it is easier to design a system in which most of the energy is used as kinetic energy as in the thermonuclear rocket, than one in which the major part of the energy must be extracted as electrical power.

3. The Ion Rocket

In the ion motor, a stream of metal ions is accelerated to very high velocities by an electrostatic field. Heavy ions give the best efficiency. Because it has a low ionization potential and can be ionized thermally, cesium is the usual choice. In principle, velocities of 10^4 or 10^5 km/sec are feasible, but the jet is extremely tenuous and the thrust would be only about 10^{-3} g. The ion rocket is suited, therefore, only for acceleration out of a satellite orbit and deep-space missions and would achieve a high propulsive efficiency only at vehicle speeds of 10,000 to 100,000 km/sec. Although many design studies have been made of ionic propulsion systems, better and lighter sources of electrical power, perhaps based on fission or fusion, are needed for such a system to become really feasible. And if the fusion reaction is successfully harnessed it is more likely to be used directly than for powering an ion rocket.

4. Plasmajets and Magnetohydrodynamics

By confining an electric arc with a vortex of inert gas or water, thus increasing its density, temperatures of 50,000°K have been reached in the laboratory. This is the principle of the plasmajet. If lighter means of generating electric power are found, such a device could find propulsive applications. It might, for example, be used for upgrading the thermal energy from a reactor to kinetic energy at a much higher temperature, with a consequent gain in specific impulse. Power requirements are in the 100-megawatt range, and with present technology the weight penalties entailed in putting a steam generating plant and a turbo alternator into a space rocket are absurd.

Perhaps the most readily envisaged use of a plasmajet is as a giant booster in which the power equivalent of Hoover Dam is used for a few seconds to give a space vehicle its initial acceleration to around 1000 mph. To store 7.5×10^8 joules a condenser bank of capacity 0.6 farad at 50 kv and copper bus bars 4 inches in diameter to carry the current are needed. The volume needed for this would be a cube of side 150 feet. If a launching site in a deep canyon were chosen, these might be laid with guide rails for perhaps a thousand feet up the canyon wall.

As well as controlling ionized gas streams, as in the thermonuclear rocket, magnetohydrodynamic (MHD) principles may also be used to provide acceleration. Gas velocities of 200 km/sec (500,000 mph) in magnetically driven shock waves have been produced in the laboratory (Kantrowitz, 1958). But as with ionic propulsion the application of MHD awaits more efficient means of generating electrical power.

5. The Solar Sailboat

By unfurling giant sails in space, one could use the wind of photons from the sun to tack about the solar system. Calculations indicate that a sail with an area of 20 acres and thinner than the thinnest plastic film kitchen wrap would be needed.

6. The Free Radical Rocket

Where do free radicals stand in relation to these other exotic means of propulsion. The closest comparison is perhaps with the nuclear rocket with H atoms replacing the reactor as a source of heat. Trapped H atoms in H_2 in concentrations of 10 % or more compare favorably in specific impulse with the nuclear rocket and do not suffer the disadvantages of radioactivity. Pure monatomic hydrogen yielding specific impulses in excess of 1000 provides better performance than the fission rocket and approaches the performance of a thermonuclear engine.

The nuclear rocket presents formidable engineering problems, but the power source is by now well known and harnessed in the nuclear reactor. In the case of the thermonuclear rocket the source of power, the H bomb, is again by now well known but not as yet tamed for peaceful use. The free radical rocket is still one step behind. Nobody has yet made an FR bomb and, to match the weight penalties of heavy electrical machinery suffered by the other systems, the free radical rocket carries the bulky cross of hydrogen's very low density.

For free radicals the crux of the problem is concentration. So far there is no evidence that amounts exceeding fractions of 1 % can be stored. Unless some radically new breakthrough occurs at the basic research level, the chance of ever running a rocket on monatomic hydrogen, or any other trapped radical, is about the same as the author's chance of finding lakes of liquid helium in the next place when once he shuffles off this mortal coil. But this won't stop him from trying.

REFERENCES

Baum, L., Graff, H., Hormats, E. I., and Moe, G. (1956). Research on ultra-energy fuels for rocket propulsion. Aerojet-General Corp., Azusa, California, *Rept. No.* **1149**.

Broida, H. P., and Pellam, J. R. (1954). *Phys. Rev.* **95**, 845. Phosphorescence of atoms and molecules of solid nitrogen at 4.2°K.

Carter, J. M. (1958). *Astronautics* **3**, 26 (Sept.). How high for chemical fuels.

Clarke, A. C. (1951a). "Interplanetary Flight." Temple Press, London; Harper, New York.

Clarke, A. C. (1951b). "The Exploration of Space." Temple Press, London; Harper, New York.

Clauser, M. (1958). *In* "Conference on Extremely High Temperatures," p. 209. (H. Fischer and L. C. Mansur, eds.), Wiley, New York.

Fite, W. (1958). General Atomics, San Diego, California. Private communication.

Fontana, B. J. (1959). *J. Chem. Phys.*, **31**, 148. Magnetic study of the frozen products from the nitrogen microwave discharge.

Gerlach, W., and Stern, O. (1922). *Z. Physik* **9**, 349. Der experimentelle Nachweis der Richtungs Quantelung im Magnetfeld.

Golden, S. (1958). *J. Chem. Phys.* **29,** 61. Free radical stabilization in condensed phases.

Jackson, J. L., and Montroll, E. W. (1958). *J. Chem. Phys.* **28,** 1101. Free radical statistics.

Jones, J. T., Jr., Johnson, M. H., Mayer, H. L., Katz, S., and Wright, R. S. (1958). Aeronutronics Systems Inc., Glendale, California, *Publ. No.* **U-216**. Characterizations of hydrogen atom systems.

Kantrowitz, A. (1958). *Astronautics* **3**, 18 (Oct.). Introducing magnetohydrodynamics.

McGee, H. A., and Wharton, W. W. (1958). *Chem. Eng. News* **36**, 32.

Mickle, E. A., Clauson, W. W., and Szego, G. C. (1958). Free radicals as high energy propellants (unpublished). General Electric Co., Cincinnati, Ohio, Flight Propulsion Lab. Dept.

Nichols, P. L., Jr., and Hanzel, P. C. (1958). Stanford Research Institute, Menlo Park, California. Private communication.

Ruark, A. E., and Urey, H. C. (1930). "Atoms, Molecules and Quanta," p. 89. McGraw-Hill, New York.

Scheller, K., Thacher, H. C., Jr., and Bierlein, J. A. (1957). (Wright Air Development Center) *ASTIA Document No.* **118101.** Free radicals as fuels.

Sutton, G. P. (1949). "Rocket Propulsion Elements." Wiley, New York.

Szego, G. C., and Mickle, E. A. (1958). *Astronautics* **3,** 36 (Jan.). A look at free radicals.

Wall, L. A., Brown, D. W., and Florin, R. E. (1959). *J. Phys. Chem.* **63,** 1762. Atoms and free radicals by γ-irradiation at 4.2°K.

Warnke, C. J. (1934). U. S. Patent No. **1,967,952.** Method and apparatus for treating gases.

Wigner, E., and Huntington, H. B. (1935). *J. Chem. Phys.* **3,** 764. On the possibility of a metallic modification of hydrogen.

Windsor, M. W. (1959a). Unpublished results.

Windsor, M. W. (1959b). *Chem. Eng. News* **37,** 38. Magnetic field separates H atoms.

Zwicky, F. (1943 ff.). See *Astronautics* **2,** 45 (Aug. 1957). Propellants for tomorrow's rockets. (This paper of 1957 refers to earlier works by Zwicky.)

14. Trapped Radicals in Radiation Damage

R. S. ALGER

United States Naval Radiological Defense Laboratory, San Francisco, California

I. Introduction

In this chapter the role played by free radicals in the mechanism of radiation damage will be examined. Generally, radiation effects are discussed according to the type of material involved, e.g., metals, semiconductors, inorganic dielectrics, or organic compounds. Free radicals are usually found in this last group, the organic compounds, which offer the greatest latitude for free radical activity. Although most of our interest will be with this group, there are several reasons for including some examples from the inorganic category. First, there is the matter of nomenclature. Reactions involving high energy ionization or excitation in the three types of inorganic materials mentioned have been described with fair success in terms of the band picture of solids and the language of imperfections in crystals. Portions of this concept and language have been carried into the realm of polymers, and will be encountered in the comparison of theories at the end of the chapter. Secondly, the extension of the discussion of trapped radicals to trapped atoms suggests analogous defects in such well-ordered structures as the alkali halide crystals. In the following sections dealing with the initial radiation-induced reactions and their effect on optical and chemical properties, the behavior of trapped radicals and atoms in both simple inorganic and organic systems are discussed in order to establish points of similarity and to emphasize fundamental differences. After a comparison of the nomenclature appropriate to the two systems, the experimental evidence regarding radicals and radiation effects is described, first for simple compounds and crystals, then for polymers. Finally, various mechanisms for the induced reactions are considered along with methods for reducing radiation effects.

A. NOMENCLATURE

The terminology of imperfections generally employed in discussing radiation effects in inorganic crystals is listed in Table I. The crystal is described in terms of a very nearly perfect lattice containing an assortment of the resident defects. Following the synthesis of Seitz (1952), the high energy radiations are treated as transient defects which can interact with either the normal lattice or the resident imperfections. Figure 1 shows an energy-level diagram applicable to an alkali halide crystal containing the resident imperfections of interest in this chapter. Ionization and excitation are indicated by electron transitions to the appropriate energy levels.

When an organic crystal is considered, defects comparable to the resident imperfections can be described as indicated in the right-hand column of Table I. Trapped radicals or atoms are electrically neutral, chemically active species which may be colored, paramagnetic, and photosensitive. Strictly speaking, the hole is a vacant state near the top of the filled electron-energy band, but in an alkali halide crystal the hole results from the presence of a neutral halogen atom in the lattice. It is this atom which resembles the radical. Our primary concern will be with these radicals and atoms and their production by the transient imperfections. Defects such as vacancies in the crystal lattice, interstitial ions or molecules, and lattice dislocations have meaning only in crystal structures, and are prevalent in both organic and inorganic crystals. Phonons are the quantized bits of acoustical energy which can be generated mechanically or thermally and for all practical purposes are always present. Excitons and excited molecules represent degrees of excitation short of ionization. In keeping with the band picture of alkali halide crystals, the holes and excitons are considered to be quite mobile, whereas a comparable mobility in their molecular

TABLE I

CLASSIFICATION OF IMPERFECTIONS IN SOLIDS[a]

Resident defects

Inorganic	Organic
Holes and electrons	Molecular ions, radicals and electrons
Excitons	Excited molecules

Phonons
Vacant lattice sites and interstitial atoms
Foreign atoms
Dislocations

Transient defects

Photons: X-rays, gamma rays, etc.
Charged particles: electrons, beta and alpha particles, protons
Neutral particles: neutrons

[a] Seitz (1952).

counterparts, i.e., radicals and excited molecules, provides a subject for considerable debate. Within the framework and language of Table I, radiation damage represents an increase in the concentration of resident defects as a result of the action of the transient imperfections.

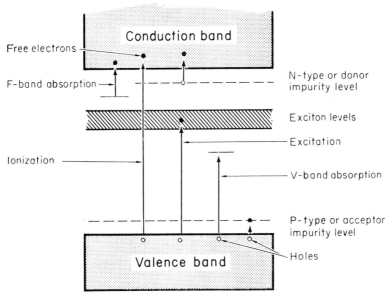

FIG. 1. Conventional energy-level scheme of an alkali halide crystal containing both donor and acceptor impurities. A hole may be generated by an electron jumping from the filled valence level to an impurity, exciton, or conduction level.

B. Ionization and Excitation

The transient defects of principal interest from the standpoint of damage are photons and neutrons which, because of their long range compared to charged particles, are more difficult to shield against. Photons interact through photoelectric, Compton, or pair-production processes to generate energetic primary electrons which move through the solid and account for the ionization and excitation. With heavy particles, nuclear reactions are possible but generally account for only a small percentage of the total effect in dielectrics susceptible to damage by ionization. Heavy particles such as deuterons and alpha particles transfer their entire energy in a very small volume; therefore, the concept of a thermal spike is frequently employed to describe their cataclysmic effect. Under these conditions, different yields and ratios of products may result. Neutrons can interact with solids either through a knock-on collision or through a nuclear reaction. The knock-on process predominates with fast neutrons, particularly in hydrogenous materials where protons are energized and are responsible for further interactions. In the average pile-type neutron spectrum, the knock-on process predominates and the thermal neutrons can be neglected, except in reactions with atoms having high capture cross sections, such as chlorine. In the terms of Table I, these reactions between transient defects, resident defects, and normal molecules can be described, for our purposes, as follows: (1) Excitation represents the formation of excitons, excited molecules, and/or phonons, depending on the energy transferred to the normal molecules or ions by the transient particle. (2) When sufficient energy is transferred, holes and free electrons are generated in ionic crystals and molecular ions, and free electrons are produced in the organic materials.

C. Reactions Involving Radiation Products

In a KBr crystal, ionization raises an electron from the filled band of electrons to the conducting state. The free electron moves in the conduction band until trapped by suitable defects, such as a hole or vacancy. Electron capture by a hole means annihilation for the hole, with the excess energy being liberated in the form of photons (luminescence) or phonons. When captured by a vacancy, the electron forms an F-center, a type of impurity. Excitation can also generate a hole, provided phonons supply the energy difference between the exciton level and the F-band. In the alkali halides the process of ionization or excitation can lead directly to a trapped atom, but this is not the case in organic compounds composed of neutral molecules. Here radical production follows as a second step. The initial process of energy absorption leads to an excited or ionized molecule which can dissipate the excess energy in a variety of competing reactions. The back reaction involving electron capture with the energy dissipated in photons or phonons is analogous to the hole capturing an electron in KBr; however, frequently the excited molecule is unable to cope with the excess energy and bond rupture ensues. This rupture may remove a hydrogen atom, cleave the main carbon chain, or break off a side chain. In the rigid framework of a solid, the motion of these fragments is limited by the surrounding molecules which, in effect, form a cage (Franck and Rabinowitch, 1934). This confinement by the "cage effect" encourages recombination or reactions leading to stable molecules. The important reactions can be classified in three groups: (1) back reactions, (2) cross reactions, and (3) molecular rearrangements. Back reactions re-form the original molecules and thereby erase the effects of irradiation. Cross reactions involve fragments or radicals from several molecules which frequently combine to form compounds of higher molecular weight. Cross-linking and polymerization are important cross reactions. The formation of stable molecules through rearrangement and double-bond production is the important reaction in chain scission and degradation where species of lower molecular weight are produced. The reaction is called a molecular process when the stable products are formed without benefit of a recognizable intermediate radical step. Although the cage effect favors back reactions and the formation of stable molecules, it is also important to the stabilization of radicals. If a fragment such as a hydrogen atom escapes from the cage, leaving a radical behind, the cage now becomes important in preserving the radical by interfering with cross reactions.

II. Characteristics of Radiation Damage

There is a growing tendency to bypass the expression "radiation damage" in favor of a more scientifically acceptable but insipid phrase "radiation effects," which avoids the judgment of good versus bad and acknowledges the fact that certain physical characteristics can be improved by suitable amounts of irradiation. Despite the merits of this trend an occasional "damage" has crept into this chapter when the situation called for a solid definitive term. The methods of studying trapped radicals covered in Chapters 6 through 8 depend on radiation-induced changes in the optical, magnetic, mechanical, and chemical properties of materials. The threshold for a measurable change varies widely as a function of the material and property considered. In engineering materials, a change in the mechanical properties of 25 % is frequently used as a threshold because the rate of change increases rapidly in numerous polymers beyond this point. In the simple compounds considered in this section, the optical and paramagnetic properties were altered long before physical and chemical properties. Numerous organic and inorganic compounds become colored by developing optical absorption bands in the visible and ultraviolet regions of the spectrum when exposed to ionizing radiation. The F-bands in the alkali halide crystals offer the classical example of coloring. Two other color centers, the U- and V-centers, which absorb in the ultraviolet portion of the spectrum and are produced by impurity atoms and holes, respectively, will be encountered later in discussions of the alkali halides. Since most inorganic crystals and glasses exhibit optical damage under irradiation, one of the problems associated with nuclear reactors and high-intensity radiation sources has been the limit imposed on the useful life of optical devices and windows by color-center formation. Although many organic compounds color under irradiation, they have not received the detailed study accorded the alkali halides. There appears to be some order in the coloring of a homologous series. For example, a number of the alcohols have been observed to develop a broad peak at about 5200 A as shown in Fig. 2 (Alger et al., 1959). The polyhydric alcohols, ethers, and paraffins also show some family traits in their coloring.

Considerable variation is observed in the effects of radiation on the mechanical properties of dielectrics (Sisman and Bopp, 1954). The effect on organic materials is several orders of magnitude larger than in the case of inorganic compounds. Under exposures of 10^7 to 10^{10} rad most organic solids become hard and brittle and ultimately rupture or crack. A few materials, such as Teflon and Lucite, continue to lose mechanical strength

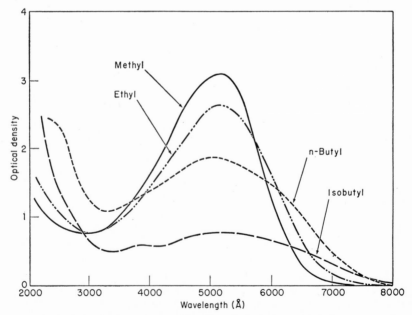

FIG. 2. Optical absorption spectra of irradiated alcohols at 77°K. (Alger et al., 1959.)

under irradiation until they are reduced to a powder or putty. Transient defects of the light-particle type have little effect on the mechanical properties of the alkali halides aside from a slight expansion of the lattice; however, heavy particles produce brittleness and a decrease in density. These effects can be removed by annealing at elevated temperatures.

The change in chemical properties induced by irradiation represents the largest distinction between organic and inorganic compounds. Aside from nuclear transformations, the reactions in the inorganic compounds are for the most part reversible by thermal or optical treatment. With organic compounds the reactions are generally irreversible. Both liquid and solid hydrocarbons evolve hydrogen and light hydrocarbon gases during irradiation. This escape of hydrogen changes the ratio of carbon to hydrogen in the solid and makes a return to the initial state impossible. In addition, molecules undergo degradation by other modes, and polymerize or cross-link to form new stable molecules.

The cost of radiation damage in terms of energy required to produce a particular color center, molecule, or degree of cross-linking is usually given in terms of G values, where G denotes the number of molecules of a particular species involved in reaction per 100 ev of energy expended. Table II lists typical G values for the formation of color centers and trapped radicals in several simple organic and inorganic species. The wide range of yields is related to such factors as molecular structure or crystal properties, temperature, and the presence of resident defects. The importance of temperature and the solid phase as prerequisites for radical stabilization have been emphasized in Chapters 4 and 10. At elevated temperatures the color centers and trapped radicals are destroyed through thermally activated reactions with other imperfections; therefore, their G values are reduced. Table II also shows evidence for a reduction in the G values at very low temperatures where the increased rigidity of the solid strengthens the cage effect and favors back reactions.

TABLE II

Yields of Radicals and Color Centers in Simple Compounds as a Result of Electron Bombardment[a]

Organic compounds	G_{rad}	Inorganic compounds	$G_{F\text{-center}}$
Methanol at 77°K	5.5	KCl F-center	0.8
Methanol at 4°K	2.3	KBr F-center	0.6
Ethanol at 77°K	8.3	NaCl F-center	0.9
Acetone at 77°K	1.6	KBr U → F	3

[a] Alger et al. (1959); Alger and Jordan (1955).

III. Radicals and Holes in Simple Compounds

A. Evidence of Radicals as Radiation Products

In assessing the role of free radicals in radiation damage to organic materials, it is first necessary to establish whether a sufficient number of excited molecules decay by free radical production to influence the over-all end products. The largest source of evidence is of chemical and spectroscopic nature. Unfortunately the chemical techniques (see Chapter 8) frequently involve the liquid phase, where the radicals produced are very short-lived. Although G values determined in the liquid state cannot be applied to solids because of the "cage effect," the results in the liquids show that the production of free radicals forms a substantial fraction of the decomposition process. A number of experiments have been performed using radical scavengers to separate radical and ionic processes from molecular reactions. For example, iodine has been observed to reduce the G_{H_2} yield in irradiated methanol by over one-half and in irradiated normal hexane by one-third (Meshitsuka and Burton, 1958; Dewhurst, 1958). Such experiments suggest that 30 to 50 % of the reaction processes may go by way of free radicals. Other evidence is found in the reaction kinetics of radiation-induced polymerization in materials such as styrene and methyl methacrylate. The kinetics are similar to those observed when polymerization is catalyzed by free radicals (Charlesby, 1958).

The results from electron paramagnetic (spin) resonance (ESR) spectroscopy offer some of the most convenient and convincing evidence for the presence of a substantial contribution from radicals to radiation effects. ESR has shown that trapped radicals are common products of irradiated organic solids. Paramagnetic resonance in over one hundred compounds, ranging from the simplest hydrocarbon, methane, to complex polymers and biological materials, has been observed after irradiation at suitably low temperatures (Smaller and Matheson, 1958; Abraham and Whiffen, 1958; Abraham et al., 1958; Gordy et al., 1955a, b; Alger et al., 1959). For example, the paraffins, alcohols and polyhydric alcohols, ethers, ketones, acids, esters, and many others, show the presence of unpaired electrons when irradiated as solids. Although the presence of radicals has been widely established, quantitative determinations of the proportion of radicals to other radiation products remain to be made. A few G_{rad} values listed in Table II suggest that in solids like methanol and ethanol radical production is an efficient process and must account for a substantial number of the reaction components. Although holes and color centers are the primary reaction products in the alkali halides, their G values fall within the range of those observed in organic plastics at room temperature.

In both the ionic crystals and the organic compounds the radicals are stable at suitably low temperatures, particularly when stored in the dark. Some of the reaction products, such as the color centers in the alkali halides and the alcohols, are photosensitive and can be bleached under appropriate visible or ultraviolet illumination. In the alkali halides, a photon liberates an electron from an F-center and if the process is repeated often enough all the electrons will return to holes, thereby eliminating the neutral halogen atoms. The bleaching of color centers in alcohols, and the associated changes in the ESR spectrum, are considered in a subsequent section. As an example of stability, samples of methanol, ethanol, and acetone stored for over two years in the dark at liquid nitrogen temperatures still retain color centers and spins (Alger *et al.*, 1959). From the preceding chapters, it is apparent that the term "suitably low temperature" actually covers a wide range of temperatures, depending on the molecules and crystals involved. At liquid nitrogen temperature the lifetime of many organic radicals is measured in years. In materials that are solid at room temperature, such as plastics, sugars, paraffins, and cellulose, the radicals have existed from hours to months. Of course, simple molecules like CH_4 require lower temperatures for stabilization, as discussed in the work at liquid helium temperatures.

B. IDENTITY AND PROPERTIES OF RADICALS FORMED BY RADIATION

The general methods of interpreting ESR hyperfine structure have been discussed in Chapter 7. Here we will apply the techniques to several examples of importance in interpreting the processes of radiation effects. In the one classic inorganic case of F-centers and holes, the nature of the reaction products and the coloring mechanism had been established with fair certainty by optical spectroscopy and chemical measurements prior to the advent of ESR. Now, ESR has added confirmation to the general model and clarified some of the finer details of the process. Two examples of radicals in alkali halide crystal will be considered.

(1) $h\nu$ + KCl crystal → hole + e

    e + vacancy → F-center

    hole = V_1 center

Initially, the V_1 centers were related to halogen atoms added to the crystal in a stoichiometric excess. The same ultraviolet absorption bands (V-bands) were later developed by ionizing radiation. These bands have been attributed to various combinations of the holes with other resident defects or ions. Recently ESR has identified one of these reaction products as Cl_2^- (Delbecq *et al.*, 1958).

(2) $h\nu$ + U-center → F-center + H atom

This second example is of particular interest because hydrogen atoms are evolved as one of the major reaction products. The U-center results from

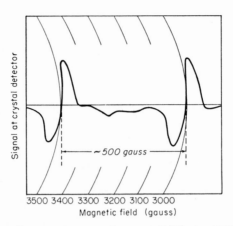

FIG. 3. Paramagnetic resonance spectrum at 80°K of a KCl-KH crystal after irradiation at 80°K. The derivative of the absorption bands is shown with F-center resonance in the center of the spectrum suppressed (Delbecq *et al.*, 1956, courtesy of the *Physical Review*).

the addition of KH to the crystal. Presumably the impurity ions (H^-) replace the halogen ions in the lattice, or in the nomenclature of Table I, the U-center is a combination of a foreign ion and a negative-ion vacancy. When the U-center is disrupted by energy from a transient imperfection, the electron is liberated to form an F-center and the hydrogen atom is left. The U $\rightarrow$ F reaction is reversible, and under illumination in the F-band the reaction can be reversed, re-forming U-centers. Although optical studies had established the behavior of the U $\rightleftarrows$ F reaction, the state and location of the hydrogen atoms remained undetermined prior to the ESR measurements. Figure 3 shows a typical ESR spectrum for irradiated KCl + KH, with the single central line due to the F-centers suppressed. The doublet has a splitting of 500 gauss, which agrees within the accuracy of measurement with the 506-gauss splitting normally associated with hydrogen atoms (Delbecq *et al.*, 1956). Additional confirmation of the ESR interpretation was obtained by substituting KD, which developed a triplet of 156 gauss, as is typical of deuterium atoms. Both the doublet and the triplet bleach rapidly above 108°K, indicating that the hydrogen atoms are stable only at low temperatures. At higher temperatures, the H atoms diffuse and pair spins, presumably by forming H_2. Finally, combined optical and ESR studies indicate that the H atoms are trapped at interstitial lattice positions, thereby confirming the idea that H atoms did the moving during the U $\rightarrow$ F reaction as originally suggested because of the lack of photoconductivity during the process. In both these examples, the radiation produced a charged center, the F-center, and a radical-type center.

The processes accompanying ionization and excitation of organic compounds are more involved; consequently, the analysis is generally not as complete as in the examples just cited. Two of the simpler compounds, solid methanol and CH_4, have received considerable attention and will serve as typical examples. When CH_4 is irradiated with Co^{60} gamma rays at about $4°K$ a paramagnetic species is developed which has an ESR hyperfine structure of four lines, with peak heights in ratios of $1:3:3:1$. According to the accepted methods of interpretation (Chapter 7) this spectrum indicates an electron interacting equally with three hydrogen nuclei, and so the radical is assumed to be CH_3. Furthermore, the 500-gauss doublet associated with hydrogen atoms was also present in concentrations about equal to the quartet (Smaller and Matheson, 1958). The experiment has been repeated with deuterated methane and the deuterium triplet develops (Wall et al., 1959a, b). In these cases the irradiation has produced free radicals apparently by the removal of hydrogen atoms. Since the hydrogen atoms are both very reactive and mobile they can be stabilized only near liquid helium temperature. The radicals responsible for the quartet were able to react at liquid nitrogen temperature, and the ESR signal disappeared rapidly. Such thermal bleaching of spins does not reverse the initial reaction to re-form CH_4.

The radicals in methanol have been studied after production by both low- and high-energy methods. When solid methanol is irradiated with X-rays, gamma rays, or energetic electrons, it becomes deeply colored (Fig. 2) and paramagnetic (Fig. 4) (Alger et al., 1959). The three lines in Fig. 4 are interpreted to mean that the unpaired electron is interacting with two hydrogen nuclei. Observations on CD_3OH and CH_3OD show that the hydrogens observed are attached to the carbon; however, the ESR spectrum, at this point, does not differentiate between CH_2^+ and CH_2OH. When the radicals are formed by ultraviolet photolysis of methanol containing a small percentage of hydrogen peroxide, the triplet also develops. In this case the excited peroxide apparently removes a hydrogen atom from the methanol, leaving a CH_2OH radical (Gibson et al., 1957). These radicals produced by ultraviolet excitation are colorless. The color centers produced directly by ionizing transient defects can be bleached with light in their absorption band. Such bleaching causes the ESR spectrum to change detectably, indicating that the color centers belong to another paramagnetic species which can be converted by optical excitation to the more stable radical responsible for the triplet (Zeldes and Livingston, 1959). The methanol ESR triplet is also sensitive to thermal and optical excitation; however, higher energies corresponding to ultraviolet photons are required. The second curve in Fig. 4 shows a doublet with a splitting of 129 gauss which develops when the triplet is destroyed with 2537-A illumination. A very small indication of the 129-gauss doublet also develops under electron

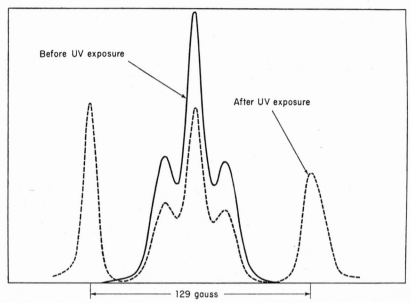

Fig. 4. Effect of 2537 A irradiation on the ESR hyperfine structure of irradiated methanol at 77°K. (Alger *et al.*, 1959.)

bombardment, showing that at least three paramagnetic species are produced during irradiation, but they are not completely independent, as demonstrated by the optical transformations. A large majority of the radicals are apparently formed by hydrogen removal, as in the case of methane; however, the hydrogen atom spectrum has not been reported in solid methanol. The uncertainty which prevails in identifying the radical responsible for the triplet in methanol stems from the fact that ESR does not necessarily see the entire radical, either because of spectrometer sensitivity or because the unpaired electron does not interact with all of the nuclei present in the radical. Obviously the uncertainty of interpretation tends to increase with the size and complexity of the molecule. This admonition should be kept in mind during the following discussion of family characteristics of the larger molecules belonging to a homologous series.

Upon irradiation chain-type molecules such as the paraffins, alcohols, and ethers develop some traits in their ESR spectrum characteristic of the series. The more obvious traits are splitting between lines and the number of lines. For example, in the paraffin series the splitting between lines is about 29 gauss and the number of lines progresses in an explainable fashion as the molecule is lengthened. The alcohols and ethers exhibit very similar spectra with a splitting of about 18 to 20 gauss, and the number of lines observed ranges from 3 to 7, always being odd (Alger *et al.*, 1959). In both these series the spectra are consistent with radicals formed through the loss of hydrogen atoms. The results in other groups are not always as straightforward. For example, the polyhydric alcohols, such as glycerin, sugar, starch, and cellulose, develop a basic doublet structure which is difficult to reconcile with the simple removal of a hydrogen atom. Also, the acids and esters tend to form even-lined spectra. In these last examples the splitting and general features of the spectrum of each particular series show family similarities although they are as yet unexplained.

C. Saturation Concentration of Stabilized Radicals

The presence and identity of a substantial number of trapped radicals have been established in some irradiated dielectrics. A connection between these radicals and changes in macroscopic properties of the materials recognizable as damage is the next concern. We approach the problem by considering the number and behavior of radicals as a function of irradiation dose. Obviously, from the difficulty encountered in measuring their presence, a few radicals, e.g., 10^{15} or 10^{16} per cubic cm, will have little effect on the materials unless they can start a chain reaction. There is also the possibility that the radicals are sufficiently like the normal molecules to constitute little damage other than optical absorption as long as they remain trapped. For example, acrylamide and related monomers when irradiated with Co^{60} gamma rays form trapped radicals at $-18°C$ (Mesrobian et al., 1954). As long as the sample remains at this temperature, there is no polymerization, but upon warming to room temperature the reaction proceeds rapidly until an explosion occurs. Although the basic mechanism has not been described, the more violent aspects of damage in this case were delayed until the radicals were thermally released and permitted to react.

There is a limit to the concentration of radicals that can be stabilized under certain conditions of sample material and temperature. The theoretical aspects of this problem have been considered in Chapter 10. Here we consider some experimental results in both the ionic and organic systems. Both cases show a trend toward saturation. The F-centers and holes in the alkali halides offer the simplest case and will be considered first. Figure 5 shows the relative number of F-centers stabilized in different materials as a function of energy absorbed from transient defects (Alger and Jordan, 1955). Although F-centers only were measured, in this case the number of holes produced should equal the number of F-centers. Initially, the number of defects is linearly related to exposure; however, above about 10^{16} centers per cubic cm saturation begins. The maximum concentration achieved is about 10^{17} F-centers or holes per cubic cm. Several processes contribute to the saturation effect: (1) back reactions, and (2) cross reactions, which form new defects. As the radiation exposure continues, the number of holes

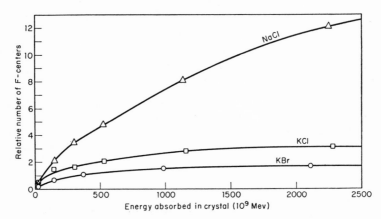

FIG. 5. Development of an equilibrium concentration of F-centers in several alkali halide crystals under 2-Mev electron bombardment. (Alger and Jordan, 1955.)

increases while the vacancies decrease owing to F-center formation. Consequently, the probability that a newly released electron will be captured by a hole increases with the total exposure. Capture by hole corresponds to a back reaction which re-forms the original ion or its equivalent. As the concentration of reaction products builds up with exposure, the probability increases for a cross reaction between these defects to form composite clusters such as F', V_2, V_3, M, R_1, R_2, and all the other known complex centers. In the case of the alkali halides, the number of cross reactions frequently can be controlled or favored by proper choice of material and temperature. In the examples of Fig. 5, cross-reaction products were not observed to an appreciable extent; therefore, the back reaction was assumed to be the principal limiting factor.

The results of a comparable experiment with irradiated methanol is illustrated in Fig. 6, which shows the concentration of radicals at liquid nitrogen temperature as a function of electron irradiation (Alger *et al.*, 1959). About 10^{20} radicals per cubic cm were formed at saturation, or the average spacing is about six molecules between radicals. Extending the exposure by a factor of 10 beyond the knee of the curve did not change the shape or area of the ESR hyperfine structure appreciably. The saturation concentration does depend on temperature. For example, when methanol samples irradiated to saturation at 77°K were cooled to 4°K, about 20 % more radicals could be added. It is reasonable to assume that at the onset of saturation normal molecules will continue to be excited and ionized, leading to the formation of radicals. At equilibrium radicals are destroyed as rapidly as they are created, and the reaction must result in the pairing of spins. The details of this reaction are, unfortunately, beyond the frontiers

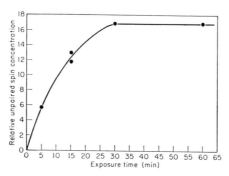

Fig. 6. Saturation of free radical concentration in irradiated methanol at 77°K. (Alger *et al.*, 1959.)

of information. It is known that radiation effects do not stop at this ESR saturation point. For example, the evolution of hydrogen gas has been measured in solid methanol undergoing irradiation and found to be proportional to the energy absorbed for exposures up to ten times the values used in the ESR experiments. Similar experiments on liquid methanol indicate that part of the gas comes by way of molecular reactions and part through radical processes (Meshitsuka and Burton, 1958). The information needed to bridge the gap between these end products and radicals in the solid remains to be obtained.

Another approach to the study of free radical yield and maximum concentration has utilized solid nitrogen, or nitrogen in a matrix of solid neon, argon, or xenon. Wall *et al.* (1959a, b) used ESR techniques to examine the radicals produced by irradiating nitrogen at 4.2°K with Co^{60} gamma rays. When the radical concentration and yield were measured as a function of radiation dose, the results differed in several respects from those in the two preceding examples. First, the yield increased with dose. For example the $G_{radical}$ averaged 0.4 but reached 0.8 at the maximum dose of 62 × 10^6 r. Secondly, when specimens were re-used after storage at room temperature, the yield over the entire exposure range was higher than for the virgin sample. Thirdly, the introduction of trace amounts of oxygen lowered the yield to the minimum values observed. It was suggested that this unexpected behavior could result if impurities initially present were removed from further participation by the initial irradiation.

Ultimately the concentration of radicals would be expected to reach saturation; however, the increasing yield indicates that saturation is well beyond the 0.06% concentration achieved (Wall *et al.*, 1959b). Various theoretical treatments provide estimates ranging from 0.3 to 10% for the concentration at saturation (see Chapter 10).

IV. Radicals in Polymers

Thus far the discussion has dealt primarily with optical and chemical forms of radiation damage in simple compounds. The remainder of this chapter will emphasize polymers, such as plastics and elastomers, where substantial mechanical changes are induced by transient defects. Unfortunately, the state of knowledge permits considerable divergence of opinion regarding the theory of damage. The procedure here will be to examine experimental results first and then compare theories and interpretations against this background.

In contrast to the simple compounds and molecules discussed thus far, the polymers exhibit a high sensitivity to radiation effects on their mechanical properties. This sensitivity arises from the relatively small number of cross links required to tie all the molecules together, or conversely the number of scissions required to rupture each molecule once. Charlesby (1958) has emphasized this feature by comparing a molecule with 20 bonds against a chain with 10^5 bonds. In either case one cross-link or bond change could double or halve the size of the molecule. However, this involves changing 5 % of the bonds in the smaller molecule compared to 0.001 % in the long chain.

A. Effects of Structure and State

Three general types of reactions have been observed and defined: (1) cross-linking—the joining of molecules to form a three-dimensional network; (2) degradation or chain cleavage, which divides the molecule into smaller fragments; and (3) unsaturation—rearrangement within the molecule to form double bonds. Of the several factors which influence the type of reaction that occurs, molecular composition or structure is probably the most important. Although all three reactions may occur simultaneously, certain polymers are predominantly degraded and others cross-linked, as indicated in Table III. Polyethylene and polyisobutylene are typical examples of cross-linking and chain cleavage, respectively. The effect of molecular composition is apparent in Teflon, which has the polyethylene-type structure but is degraded to a white powder under irradiation. A comparison of the molecular structure of the hydrocarbons in Table III indicates characteristics common to each type which seem to be generally related to the predominant reaction. Each carbon in the chain of cross-linking polymers has an attached hydrogen, whereas alternate carbons do not in the hydrocarbons which degrade.

Both the physical state, i.e., amorphous, crystalline, or glassy, and the molecular structure influence the yield from the radiation-induced reactions. As a result of a variety of mechanical tests on most of the commercially available polymers, Sisman and Bopp (1954) have divided the materials

TABLE III
PREDOMINANT REACTIONS OF TYPICAL IRRADIATED POLYMERS

Cross-linking		Degraded	
Polyethylene	$\begin{bmatrix} & H & H & \\ & \vert & \vert & \\ -& C-C & -& \\ & \vert & \vert & \\ & H & H & \end{bmatrix}_x$	Polyisobutylene	$\begin{bmatrix} & H & CH_3 & \\ & \vert & \vert & \\ -& C-C & -& \\ & \vert & \vert & \\ & H & CH_3 & \end{bmatrix}_x$
Polystyrene	$\begin{bmatrix} & H & H & \\ & \vert & \vert & \\ -& C-C & -& \\ & & \vert & \\ & \bigcirc\!\!-\!\!H & & \end{bmatrix}_x$	Polymethyl methacrylate (Lucite)	$\begin{bmatrix} & H & COOCH_3 & \\ & \vert & \vert & \\ -& C-C & -& \\ & \vert & \vert & \\ & H & CH_3 & \end{bmatrix}_x$
Methyl silicone	$\begin{bmatrix} & CH_3 & \\ & \vert & \\ -& Si-O & -\\ & \vert & \\ & CH_3 & \end{bmatrix}_x$	Tetrafluoroethylene (Teflon)	$\begin{bmatrix} & F & F & \\ & \vert & \vert & \\ -& C-C & -& \\ & \vert & \vert & \\ & F & F & \end{bmatrix}_x$
Nylon	$\begin{bmatrix} & O & \\ & \Vert & \\ -& C-N & -\\ & \vert & \\ & H & \end{bmatrix}_x$	Butyl rubber	$\begin{bmatrix} & H & R & \\ & \vert & \vert & \\ -& C-C & -& \\ & \vert & \vert & \\ & H & R & \end{bmatrix}_x$

roughly into three groups according to their radiation resistance. The first class consists mostly of elastomers which show a large decrease in strength for exposures of 10^8 to 10^9 rad. The second class is appreciably affected by 10^9 to 10^{10} rad, and includes nonrigid and some moderately rigid plastics. The third class, which shows good stability up to 10^{10} rad, comprises mostly highly rigid plastics and plastics containing suitable fillers. Although this classification is admittedly approximate, the tendency for rigid structures to be associated with radiation stability is apparent. This behavior has been observed in polyethylene when the rigidity was controlled by temperature. Plastics such as polyethylene are presumed on the basis of X-ray diffraction patterns to consist of crystalline regions surrounded by an amorphous sea. One molecule may extend readily into both regions and is presumed to be locked more rigidly in the crystalline portion, where a regular pattern of symmetry exists between molecules. At elevated temperatures the entire polymer becomes amorphous, and at low temperatures the amorphous area turns to a rigid glass. Lawton et al. (1958a, b) have observed that the rate of cross-linking reactions is highest in the amorphous

state. Both a high percentage of crystallinity and a rigid glassy state reduced the rate of cross-linking. As in the simple systems already discussed,
the temperature had a strong influence on the stability of the radicals and
the rate of reactions such as cross-linking and chain cleavage. Lowering
the temperature from about 300°K to 77°K reduced the yield of scissions
in polymethyl methacrylate by 60%, and the yield of cross-linkings in
polyethylene by 35% (Wall and Brown, 1957). Lest the consistency of
these results convey the impression that the experimental area of radiation
damage is in orderly control, we hasten to point out that both density and
temperature effects have been found to be zero in other experiments using
different supplies of polyethylene and different experimental techniques
(Epstein, 1957; Charlesby and Davison, 1957). This type of embarrassment
will overtake us more than once before we conclude.

The experimental evidence regarding the presence of trapped radicals in
irradiated polymers comes mostly from ESR measurements, whereas chemical and optical techniques supply the information on the types and rates
of the reactions. Frequently the degree of cross-linking is determined from
gel measurements in which the molecules that are not cross-linked are removed by a suitable solvent, leaving a residue in the form of an insoluble
gel. Similarly, when cross-linking is negligible, chain cleavage can be estimated from changes in viscosity of liquid samples prepared by dissolving
the irradiated polymer in a suitable amount of solvent. Infrared absorption
measurements have been used to follow the formation of double bonds and
to observe the reaction between trapped radicals and scavengers such as
oxygen. Other methods of observation include resonance vibration and
mechanical measurements of elasticity and hardness. ESR measurements
have shown that trapped radicals are a common radiation product in
polymers. In most cases the radicals are sufficiently stable to permit observation at room temperature. Table IV lists some of the common plastics,
with information about the ESR spectrum, ease of production or yield of
radicals in terms of G-values, and stability.

Polyethylene has been studied more extensively in regard to radiation
effects than have the other materials in Table IV. Nevertheless, the ESR
measurements have not converged on a single hyperfine structure pattern.
The six-line spectrum was obtained using a linear type of polyethylene,
Marlex-50, the same brand of highly oriented and crystallized material
which was observed by Lawton et al. (1958a) to exhibit a temperature and
phase influence on cross-linking. Presumably this material was devoid of
chain branches which were present and supposedly one of the contributors
to the extensive amorphous region in polyethylene formed by the high-
pressure process. The polyethylene seven-line spectrum was obtained with
a high-pressure material, which had numerous side chains. Several labora-

TABLE IV

ESR Spectra and Gas Yields in Some Irradiated Plastics

Material	Number of lines in ESR hf structure[a]	G_{rad}	Gas yield[c] (ml/gm-rad)	Stability of radicals from ESR Measurement[a]
Polyethylene	7, 6[b]	0.3	6.4×10^{-8}	Decayed to zero in few weeks
Polystyrene	3	0.2	2.5×10^{-10}	Unchanged over several months
Polymethyl methacrylate	2 sets 5 4	2.5	3×10^{-8}	Unchanged over several months
Polyethyl methacrylate	2 sets 5 4	3.0		
Nylon	4	1.7	2.0×10^{-8}	Changed to single line after week or two

[a] Abraham et al. (1958).
[b] Smaller and Matheson (1958).
[c] Bopp and Sisman (1953).

tories agree on the ESR hyperfine structure of irradiated Lucite and polystyrene. Although the G_{rad} values in Table IV are admittedly approximate, they were obtained on the same spectrometer and should permit a fair relative comparison. The superior radiation resistance of polystyrene is indicated by the comparatively low G_{rad}, which agrees with the evidence from the G_{H_2} also listed in Table IV. In terms of the Sisman and Bopp classification, polystyrene belongs in the third class, in comparison to Lucite, which is a factor of 10 less stable and comes in the second class. Table IV also contains qualitative remarks about stability of the trapped radicals at room temperature. Lifetimes of several months in polystyrene and Lucite indicate the presence of rigid cages or traps. Results from other experiments to be mentioned later will amplify these qualitative statements.

B. The Presence and Effects of Radicals

We now come to the evidence bearing on the relationship of free radicals to the three types of reactions responsible for radiation damage. Polyethylene is used as the main example because the preponderance of experimental evidence has been obtained on this polymer. Obviously as long as the radical remains trapped and visible by ESR it has not participated in a cross-linking or double-bond-forming reaction. In this sense trapped radicals contribute little to the change in mechanical properties; nevertheless, they represent a form of stored energy which ultimately can lead to a reaction. In studying polyethylene Lawton et al. (1958a) have considered both trapping and release of radicals. Locations offering sufficient stability to trap

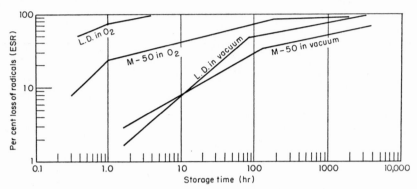

FIG. 7. Effect of storage conditions on trapped radicals in irradiated Marlex-50 (M-50) and low density (L.D.) polyethylene after 40 megaroentgen exposure at 25°C (Lawton *et al.*, 1958b, courtesy of *Journal of Polymer Science*, copyright 1958, Interscience, New York).

a radical are given as crystalline regions, low temperature glasses, and the matrix of a highly cross-linked material. The radicals were short-lived in amorphous regions and could be entirely dissipated by heating the materials above the temperature at which the crystals melt. Figure 7 shows the stability of trapped radicals in two types of irradiated polyethylenes stored at room temperature and measured by ESR. The influence of scavenger gas (oxygen) and the degree of sample crystallinity is also apparent. Presumably oxygen migrates more readily in the amorphous areas than in the crystalline regions, reacting with the radicals and eliminating the unpaired spins. The lower curves in Fig. 7 for samples stored in vacuum show that approximately 50 % of the radicals are lost in 100 days at room temperature. When similar Marlex-50 samples were stored in liquid nitrogen no radical loss was detected in 46 days. Similar evidence regarding stability is available from other polymers; e.g., when polymethyl methacrylate was heated to 100° the radicals disappeared at once, as compared to a life of several months at room temperature (Melville, 1957).

C. POLYETHYLENE

The facts are well established that polyethylene cross links when exposed to ionizing radiation and that radicals are present; however, the connection between these two observations is not settled. In the radical-escape experiments just described, the degree of cross-linking was found to increase as the radicals disappeared. Rapid thermal release of the trapped radicals always increased the number of cross links. Under these circumstances the cross links apparently resulted from radical reactions. One point of view has treated cross-linking as an extension of the polymerization process which has already been described as following the kinetics of radical-induced polymerization.

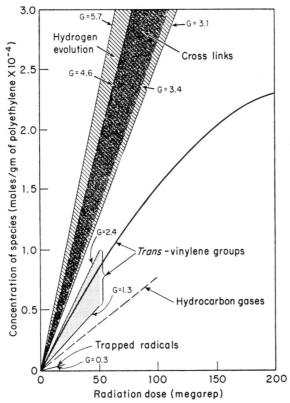

FIG. 8. Summary of radiation-induced reactions in polyethylene (after Pearson, 1957a, courtesy of *Journal of Polymer Science*, copyright 1957, Interscience, New York).

An attempt to summarize the experimental evidence dealing with the radiation products from polyethylene has been made in Fig. 8, which gives a composite presentation of the various yields as a function of radiation exposure. Substantially all of the reported yields fall within the appropriate shaded areas, and the G-values of the extremes are listed. In view of the variations between experiments, i.e., sample material, temperature during radiation, radiation sources, and dosimetry systems, the spread is not surprising. Analysis of the off-gases shows the composition to be 85 to 90 % hydrogen, with the remainder composed of hydrocarbons condensable at liquid nitrogen temperatures. At some time during the irradiation process the removal of a hydrogen is essential to the formation of a cross link or double bond; therefore, the yield of hydrogen molecules should account for both these reactions. Cross-linking is the predominant reaction, and since two chains are generally involved instead of one, as in the case of double-bond formation, the number of chains involved in cross-linking is larger than that indicated by the yield data. The line for trapped radicals is based on the G_{rad} value from Table IV. These trapped radicals represent a small component of the products, being an order of magnitude less than the number of cross links. It should be remembered that at this temperature the radicals are not completely stable, and presumably there is a maximum concentration of radicals that can be stabilized, as in the case of simpler compounds.

Some additional information is available from the use of scavengers or protective agents on elastomers. Because of relatively low resistance to radiation damage, the elastomers have been attractive systems for this type of investigation. Additives such as hydrazobenzene, 2,4-dinitrophenol, and 1,2-naphthoquinone have been found to reduce the cross-linking by as much as 60% (Turner, 1958). As will be seen in the subsequent discussion, several interpretations of the scavenger effect are possible.

Studies of the polymers wherein main-chain rupture predominates indicate that rupture is a random process, independent of the molecule length, and the number of scissions is linearly related to the irradiation dose, at least over several decades of exposures. Less energy is required to degrade the polymers at elevated temperatures. Trapped radicals are present in these materials, as indicated by ESR spectra mentioned in Table IV.

V. Damage Theories

Several assumptions are common to all the mechanisms discussed here. First, ionization and excitation are assumed to occur at random, independently of length or position in the chain molecules. Second, the structure of the polymer is considered to provide relatively few sites where the molecules are within cross-linking distance, compared to positions available for ionization, excitation, and chain rupture. Materials such as Lucite and polystyrene are in a glassy state, with random arrangements between molecules. Polyethylene, as previously mentioned, contains both amorphous and crystalline areas.

A suitable mechanism must explain several problems. In addition to predicting the proper reaction products, the mechanism should explain how the random position of ionization and excitation gets together with a suitable site for cross-linking, and what causes the materials on the left of Table IV to cross-link while those on the right are degraded. The proposed theories can be divided into two categories: free radical mechanisms and molecular-ion processes.

A. FREE RADICAL

Among the free radical processes there are two general approaches: (a) the stationary radical and (b) the migrating radical (Okamoto and Ishihara, 1956). There are several variations on each of these main themes. In the stationary-radical model, the essential net result of the passage of a transient imperfection is the production of a free hydrogen atom (H) and a radical site (R·). The hydrogen atom can abstract another hydrogen from an adjoining molecule (M) according to reaction (1), or from the parent chain reaction (2) to produce two radical sites or a double bond and a molecule of gas.

$$H + M \rightarrow H_2 + R·$$ (1)

$$H + R· \rightarrow H_2 + \text{double bond}$$ (2)

$$R· + R· \rightarrow \text{cross link}$$ (3)

The radicals then interact to form a cross link (3). A variation which yields comparable end results has the radical (R·) attack an adjacent molecule or itself to produce a cross link or double bond plus hydrogen. This stationary-radical mechanism leaves several of the problems unanswered. It provides no criterion for degradation versus cross-linking, and the ratio of cross links to double bonds is difficult to explain since the cross-linking reaction involves a molecule that is much farther away than the next carbon in the chain required to form a double bond.

The mobile-radical mechanism also commences with the evolution of a hydrogen atom, leaving an unpaired electron or radical site on the carbon chain. These radical sites are permitted to migrate along the chain by transferring a hydrogen atom from one carbon to another. Both transfer along the chain and transfer from chain to chain have been invoked. The variations on this mechanism are concerned with the reactions between the traveling radical sites. In the simplest case the two sites react to form a cross link (Lawton et al., 1958a). In another arrangement a double bond results from the encounter, and a cross link develops when a third traveling radical site meets the double bond (Pearson, 1957a). By this mechanism any obstacle to mobility of the radical site would reduce the opportunities for cross-linking in favor of main-chain cleavage. A carbon with no attached hydrogens could serve as such an obstacle if located in the main chain. Although this mobile-radical mechanism accounts for all phases of the problem, there is considerable divergence of opinion regarding the mobility of the radical site.

B. MOLECULAR ION

One of the reasons given for introducing the band picture of solids and for following the behavior of holes in alkali halide crystals containing imperfections was the extension of certain of these concepts to reactions induced in polymers by ionizing radiations.

In the molecular-ion process, or electron-hole mechanism, the band picture of solids is applied in various degrees. Energy transfer is assumed to occur while molecular ions (holes in the filled electron bands) and free electrons are present (Burton, 1958; Weiss, 1958). The site of ionization (hole) is permitted to move along the chain until a suitable location for cross-linking (4) or for double-bond formation is encountered.

$$e + M^+ + M \rightarrow RR + H_2 \tag{4}$$

The electron is assumed to be nonlocalized, as in the conduction band, and can return to neutralize the cross-linked molecule (RR) or the H_2^+, whichever is required. Double bonds are produced when the hole reacts with the adjacent carbon atom in the same chain. If the hole is localized (trapped) because of molecular structure, capture of an electron could produce chain rupture. This possibility is suggested to account for the degradation in polymers such as Lucite and polyisobutylene. The influence of scavengers and protective agents is explained on the basis of impurity levels which can trap electrons or fill holes (Fig. 1), thereby interfering with the radiation reaction. The exponents of the electron-hole mechanism point out that the existence of trapped radicals is not incompatible with this mechanism. Also, the presence of ESR alone is not sufficient proof of the existence of radicals, since the holes are also paramagnetic. Identification of the hyperfine structure is necessary before radicals can be separated from holes.

Obviously until more information is forthcoming on the behavior of free radicals and holes in these polymeric solids "the various theories must be allowed a state of peaceful, if competitive, coexistence" (Pearson, 1957b).

VI. Control of Radiation Damage

In terms of the nomenclature and reactions discussed in connection with Table I, there are three alternate paths for energy absorbed from transient defects to follow: (1) conversion to phonons, (2) conversion to photons of lower energy, (3) production of resident defects through ionization or excitation interactions. If radiation effects are to be avoided, paths (1) and (2) should be preferred and (3) should be limited to the production of harmless imperfections. Materials which inherently degrade the energy along paths (1) and (2) show superior stability under irradiation. In other materials where these characteristics are lacking, an improvement in stability can sometimes be achieved through the addition of so-called "protective agents" or scavengers. The various types of scavengers act to eliminate the excess energy by one or more of the three paths.

Certain molecular structures can absorb appreciable amounts of energy without rupturing a bond, and subsequently transfer this energy as heat to the solid. The benzene ring is the classic example of this type of stable molecule. Compounds composed of benzene structures such as anthracene, naphthalene, etc., or the phenyls, e.g., biphenyl and terphenyl, are some of the organic materials stable under irradiation. Other compounds containing a judiciously located benzene ring such as polystyrene also exhibit a high stability which is frequently ten to one hundred times that of other molecules. The addition of a protective agent may produce some increase in stability by the phonon path, but generally the increase is less than 50 %. For example, Dewhurst (1958) reduced the hydrogen evolution from n-hexane by the addition of benzene.

The dissipation of energy by photons is particularly attractive because these transient defects can leave the system; unfortunately the efficiency of this process is low. Even in the best organic scintillators, only 10 % or less of the absorbed energy escapes as visible and ultraviolet light (Swank, 1954). When these scintillators are incorporated in organic liquids or plastics, energy is transferred from the solvent to the solute but the fluorescence efficiency is generally less than half that of the pure scintillator.

If the energy cannot be dissipated before molecular rupture occurs, a free-radical scavenger can be used to produce a harmless reaction. This is a "last ditch" type operation, particularly when the scavenger is used up in the reaction. If the main reaction leads to cross-linking, a scavenger such as those mentioned in the case of natural rubbers (Turner, 1958) may react with the radicals and terminate the structure before a cross link develops. Scission, or chain cleavage, calls for a scavenger that can react with the two fragments or hold them together.

Finally, the possibility of balancing reactions involving chain cleavage against cross-linking has been attempted in mixed plastics. The results have not been very successful.

REFERENCES

Abraham, R. J., and Whiffen, D. H. (1958). *Trans. Faraday Soc.* **54**, 1291. Electron spin resonance spectra of some gamma-irradiated polymers.

Abraham, R. J., Melville, H. N., Ovenall, D. W., and Whiffen, D. H. (1958). *Trans. Faraday Soc.* **54**, 1133. Electron spin resonance spectra of free radicals in irradiated polymethyl methacrylate and related compounds.

Alger, R. S., and Jordan, R. D. (1955). *Phys. Rev.* **97**, 277. F-centers in pure and hydride-containing alkali halide crystals.

Alger, R. S., Anderson, T. H., and Webb, L. A. (1959). *J. Chem. Phys.* **30**, 3. Irradiation effects in simple organic solids.

Bopp, C. D., and Sisman, O. (1953). *U. S. Atomic Energy Comm.* **ORNL-1373**. Radiation stability of plastics and elastomers.

Burton, M. (1958). *In* "The Effects of Radiation on Materials" (J. J. Harwood, H. H. Hausner, J. G. Morse, and W. G. Rauch, eds.), p. 243. Reinhold, New York.

Charlesby, A. (1958). *In* "The Effects of Radiation on Materials" (J. J. Harwood, H. H. Hausner, J. G. Morse, and W. G. Rauch, eds.), p. 261. Reinhold, New York.

Charlesby, A., and Davison, W. H. T. (1957). *Chem. & Ind.* (*London*) **8**, 232. Temperature effects in the irradiation of polymers. (*Chem. Abstr.* **51**, 9205, 1957.)

Delbecq, C. J., Smaller, B., and Yuster, P. H. (1956). *Phys. Rev.* **104**, 599. Paramagnetic resonance investigation of irradiated KCl crystals containing U-centers. (*Chem. Abstr.* **51**, 7155, 1957.)

Delbecq, C. J., Smaller, B., and Yuster, P. H. (1958). *Phys. Rev.* **111**, 1235. Optical absorption of Cl_2^- molecule-ions in irradiated potassium chloride. (*Chem. Abstr.* **53**, 871, 1959.)

Dewhurst, H. A. (1958). *J. Phys. Chem.* **62**, 15. Radiation chemistry of organic compounds. II. *n*-hexane. (*Chem. Abstr.* **52**, 7877, 1958.)

Epstein, L. M. (1957). *J. Polymer Sci.* **26**, 399. Electron cross-linking of polyethylenes of two different densities.

Franck, J., and Rabinowitch, E. (1934). *Trans. Faraday Soc.* **30**, 120. Free radicals and photochemistry of solutions.

Gibson, J. F., Ingram, D. J. E., Symons, M. C. R., and Townsend, M. G. (1957). *Trans. Faraday Soc.* **53**, 914. Electron resonance studies of different radical species formed in rigid solutions of hydrogen peroxide after u. v. irradiation. (*Chem. Abstr.* **52**, 6948, 1958.)

Gordy, W., Ard, W. B., and Shields, H. (1955a). *Proc. Natl. Acad. Sci. U. S.* **41**, 983. Microwave spectroscopy of biological substances. I. Paramagnetic resonance in X-irradiated amino acids and proteins. (*Chem. Abstr.* **50**, 11816, 1956.)

Gordy, W., Ard, W. B., and Shields, H. (1955b). *Proc. Natl. Acad. Sci. U. S.* **41**, 996. Microwave spectroscopy of biological substances. II. Paramagnetic resonance in X-irradiated carboxylic and hydroxy acids.

Lawton, E. J., Balwit, J. S., and Powell, R. S. (1958a). *J. Polymer Sci.* **32**, 257. Effects of physical state during the electron irradiation of hydrocarbon polymers. Part I. (*Chem. Abstr.* **53**, 1815, 1959.)

Lawton, E. J., Powell, R. S., and Balwit, J. S. (1958b). *J. Polymer Sci.* **32**, 277. Effect of physical state during the electron irradiation of hydrocarbon "polymers." Part II. (*Chem. Abstr.* **53**, 1815, 1959.)

Melville, H. N. (1957). *Chem. & Ind.* (*London*) p. 1632. The use of radioactive isotopes and high-energy radiation in polymer chemistry.

Meshitsuka, G., and Burton, M. (1958). *Radiation Research* **8**, 285. Radiolysis of liquid methanol by Co^{60} gamma radiation. (*Chem. Abstr.* **52**, 16898, 1958).

Mesrobian, R. B., and Co-workers (1954). *Nucleonics* **12**, 62. Technical advances—storing radiation effects.

Okamoto, H., and Ishihara, A. (1956). *J. Polymer Sci.* **20**, 115. Crosslinking of polyethylene-type polymers by high energy irradiation.

Pearson, R. W. (1957a). *J. Polymer Sci.* **25**, 189. Mechanism of the radiation cross-linking of polyethylene. (*Chem. Abstr.* **51**, 13457, 1957.)

Pearson, R. W. (1957b). *Chem. & Ind.* (*London*) p. 209. Radiation cross-linking mechanisms.

Seitz, F. (1952). *In* "Imperfections in Nearly Perfect Crystals" (W. Shockley, J. H. Holloman, R. Maurer, and F. Seitz, eds.), p. 14. Wiley, New York.

Sisman, O., and Bopp, C. D. (1954). *In* "ONR-NRL Symposium on Radiation Effects on Dielectric Materials" (Dec. 14–15) 1954.

Smaller, B., and Matheson, M. S. (1958). *J. Chem. Phys.* **28**, 1169. Paramagnetic species produced by gamma-irradiation of organic compounds. (*Chem. Abstr.* **52**, 16899, 1958.)

Swank, R. K. (1954). *Nucleonics* **12**, 14. Recent advances in theory of scintillation phosphors.

Turner, D. T. (1958). *J. Polymer Sci.* **27**, 503. Radiation cross-linking of rubber-effects of additives.

Wall, L. A., and Brown, D. W. (1957). *J. Chem. Phys.* **61,** 129. γ-Irradiation of poly-methyl methacrylate and polystyrene.

Wall, L. A., Brown, D. W., and Florin, R. E. (1959a). *J. Chem. Phys.* **30,** 602. Electron spin resonance spectra from gamma-irradiated solid nitrogen.

Wall, L. A., Brown, D. W., and Florin, R. E. (1959b). *J. Phys. Chem.* **63,** 1762. Atoms and free radicals by gamma-irradiation at 4.2°K.

Weiss, J. (1958). *J. Polymer Sci.* **29,** 425. Chemical effects in the irradiation of poly-mers in the solid state.

Zeldes, H., and Livingston, R. (1959). *J. Chem. Phys.* **30,** 40. Paramagnetic resonance study of irradiated glasses of methanol and ethanol.

15. Trapped Radicals in Biological Processes

C. H. BAMFORD and A. D. JENKINS

Courtaulds Ltd., Maidenhead, Berks., England

I. Introduction

For nearly thirty years there have been speculations that some biological reactions may involve free radicals as intermediates. Whereas formerly the participation of radicals could be inferred only from indirect evidence, comparatively recent developments, particularly that of the electron spin resonance technique, allow the presence of these entities, or, strictly, unpaired electrons, to be established directly. The present is therefore a convenient time to review the evidence for the participation and trapping of radicals in biological processes.

Radicals are most likely to arise in biological systems from oxidation reactions and irradiation. We discuss in the following pages the general features of biological oxidation in order to provide a basis for understanding the chemical processes by which radicals can be produced and trapped, and for assessing the evidence for their participation. Subsequently, we consider the evidence for the presence and nature of radicals trapped in natural and irradiated biological materials, without specific reference to the chemistry of the processes by which they were formed. Finally, we give a brief discussion of hypotheses which have been advanced to explain carcinogenesis as a radical process.

II. The Mechanism of Biological Oxidation

The chemical processes which occur in living cells are almost without exception the result of catalysis by enzymes. Such reactions may be of diverse types, but the oxidation of cell metabolites is of special interest for present purposes in view of suggestions that free radical intermediates may be involved. We begin this section with a brief account of the main features of enzyme-catalyzed oxidation, and continue with a survey of the evidence on which the free radical hypothesis has been based.

A. Prosthetic Groups and Coenzymes

1 . Chemical Nature

An enzyme has been defined as "a protein with catalytic properties due to its power of specific activation" (Dixon and Webb, 1958a). The high specificity of enzyme action is a remarkable and characteristic phenomenon and is doubtless associated with the protein components of the enzymes. On the other hand, the catalytic activity in oxidation-reduction reactions involves the "prosthetic groups" or "coenzymes" which are more or less firmly bound to the protein constituents (Sect. II,A,2). The structural units of which these groups are composed are comparatively few in number; among the more important we may mention pyrrole, pyridine derivatives such as isoalloxazine (flavine), adenine (6-aminopurine), and D-ribose. Compounds of the heterocyclic derivatives and the pentose are called *nucleosides*, whereas the *nucleotides* contain combined phosphoric acid in addition. We shall now describe briefly the structure of some commonly encountered prosthetic groups and coenzymes.

The *flavoprotein enzymes* have prosthetic groups containing the flavine nucleus. In several cases the prosthetic group is flavine mononucleotide (FM, or FMN) (I) which is the 5′ phosphoric ester of riboflavine. Generally, however, these enzymes contain flavine adenine dinucleotide (FAD) (II). The structure of FAD has been confirmed by synthesis by Christie *et al.* (1954); this work constitutes the first synthesis of a dinucleotide. The flavoprotein enzymes are widely distributed, occurring, for example, in yeast [the first flavoprotein enzyme to be discovered was extracted from yeast by Warburg and Christian (1932, 1933)], liver, kidney, heart, snake venom, molds, and bacteria. They function biologically by catalyzing a wide range of oxidation-reduction reactions, e.g., the oxidation of amino acids, lactates, pyruvates, glycollates, oxalates, glucose, and the reduced form of coenzymes I and II by cytochrome c (see below).

Frequently enzymes contain metals or require the presence of metal atoms before they exhibit their characteristic activity. In a *"metalloenzyme"* the metal is combined either with groups in the protein, or as part

(I) Flavine mononucleotide (FMN)

of the prosthetic group. The latter is the case with the catalases and peroxidases, which bring about the oxidation of a variety of substances by hydrogen peroxide. The substrate for the catalases is often hydrogen peroxide itself, or ethanol, whereas the peroxidases are not active toward these

(II) Flavine adenine dinucleotide (FAD)

compounds, but bring about the oxidation of phenols, amines, etc. These enzymes belong to the class of *hemoproteins*, which contain prosthetic groups (hemes) similar to the hemin of hemoglobin (III). The heme molecules contain a central metal atom (often iron) surrounded by four linked pyrrole rings; the metal-free molecules are called porphyrins. The *cytochromes*, which Keilin (1925) showed to be of fundamental importance in cell respiration and of which over twenty are known, are also hemoproteins.

Some of them are enzymes, and others, such as cytochrome c, are oxidation-reduction carriers and are not true enzymes since they do not show catalytic activity in the absence of enzymes. The porphyrin of cytochrome c is

(III) Hemin

shown in (IV). The electronic structures of metal-porphyrin and related complexes are being studied actively at present; for information on this subject the reader may consult Gibson *et al.* (1958) and Griffith (1958) and earlier work referred to in these papers.

In other cases the presence of a metal ion is necessary for the enzyme to develop its catalytic activity; among such ions are Na^+, K^+, Rb^+, Cs^+,

(IV) The porphyrin of cytochrome c

Mg^{++}, Ca^{++}, Zn^{++}, Cd^{++}, Cr^{+++}, Cu^{++}, Mn^{++}, Fe^{++}, Co^{++}, Ni^{++}, Al^{+++}. Although the precise mechanism of activation by these cations has been established in only a few cases, it is realized that there are a number of ways in which the ions may cause activation, e.g., by acting as a binding link between enzyme and substrate or by changing the surface charge on the protein.

Harden and Young (1906a,b), in the course of their investigation of the fermentation of sugar by yeast, introduced the term *coenzyme* to denote a group which, like a prosthetic group, forms part of the catalytic system of an enzyme, but which is not bound firmly to the protein. These workers found that yeast juice loses its power of fermenting glucose on dialysis; the dialyzate was found to contain a substance of low molecular weight ("cozymase") which restored the full catalytic powers when added to the enzyme protein. Cozymase was obtained pure by von Euler *et al.* (1936) and by Warburg and Christian (1931) and is now identified with coenzyme I (V below). Coenzyme II, which has a similar structure (see V), was obtained from red blood corpuscles by Warburg and Christian (1936) and is the coenzyme of phosphoglucose dehydrogenase, which catalyzes the oxidation of D-glucose-6-phosphate to D-δ-gluconolactone-6-phosphate, and other animal dehydrogenases.

(V) R = H, coenzyme I; R = OP (OH)$_2$, coenzyme II

The distinction between prosthetic groups and coenzymes is by no means rigid; as mentioned above the original distinction was based upon the firmness of combination of the group with the enzyme protein. As pointed out by Dixon and Webb (1958d), however, there exists a continuous series of cases from true prosthetic groups to true coenzymes, and it is difficult to draw a dividing line at any point. For present purposes the distinction is not important.

2. Oxidation-Reduction Reactions of Prosthetic Groups and Coenzymes

The essential common feature of all prosthetic groups and coenzymes is that they can take part in reversible oxidation-reduction reactions. This will now be illustrated for the groups already considered. The flavine nucleus in the flavoprotein enzymes can undergo the following changes:

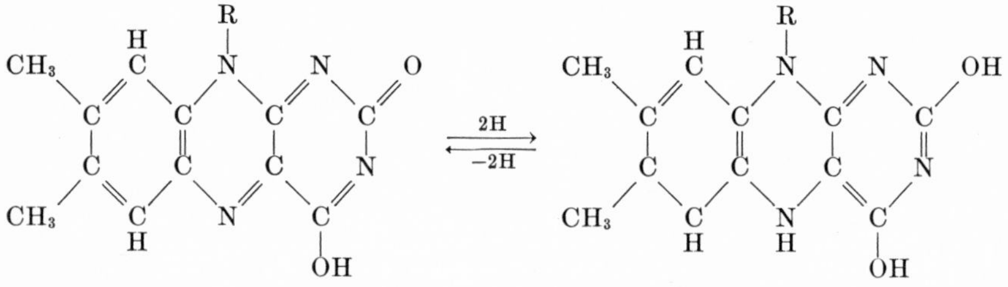

The leucoflavine formed on reduction can autoxidize with regeneration of the flavine. Thus, for example, the oxidation of a D-amino acid by D-amino acid oxidase (which contains FAD) proceeds according to Eq. (1),

$$CH_3CH(NH_2)COOH + FAD + H_2O \rightarrow CH_3COCOOH + NH_3 + FADH_2$$

$$FADH_2 + O_2 \rightarrow FAD + H_2O_2 \tag{1}$$

$FADH_2$ representing the leuco form of FAD. In general we may write

flavoprotein + substrate → leucoflavoprotein + oxidized substrate

leucoflavoprotein + O_2 → flavoprotein + H_2O_2 .

It was originally thought that in oxidation-reduction reactions undergone by the cytochromes only the valency of the central metal atoms is changed. The oxidation of succinic to fumaric acid would thus be represented as in Eq. (2).

$$\begin{array}{l} CH_2COOH \\ | \\ CH_2COOH \end{array} + 2 \text{ cytochrome (Fe}^{+++}) \rightarrow$$

$$\begin{array}{l} HCCOOH \\ \| \\ HOOCCH \end{array} + 2 \text{ cytochrome (Fe}^{++}) \tag{2}$$

However, the reduction is accompanied by changes in the absorption spectrum which show that the conjugated system has been modified, and there is no doubt that the process is more complicated than might be inferred from Eq. (2). This matter is further discussed in Section II, C, 3. The oxidized form of cytochrome c may be reduced chemically, e.g., by dithionite ($Na_2S_2O_4$) or polyphenols, and the reduced form is oxidized by hydrogen peroxide and ferricyanides, and also by oxygen in the presence of the enzyme cytochrome oxidase.

The reversible oxidation-reduction reactions of coenzymes I and II were shown by Warburg and Christian to involve the nicotinamide ring. It has subsequently been established (Pullman *et al.*, 1954; Loewus *et al.*, 1955) that the 1,4 positions in the ring are affected (Eq. 3).

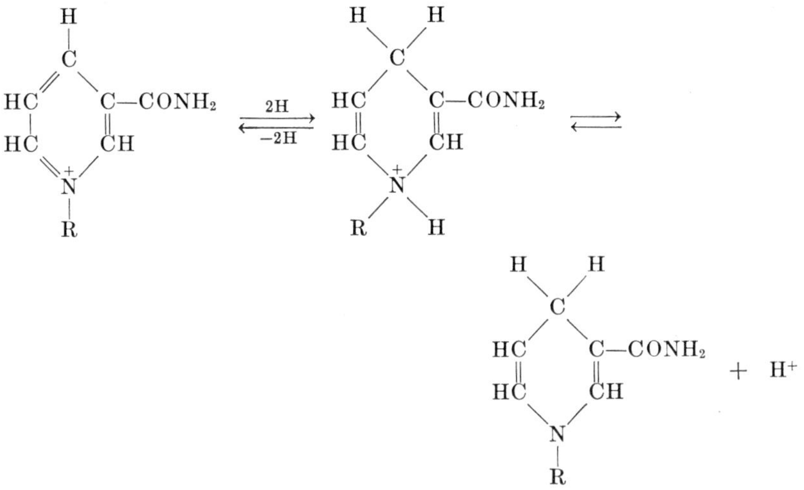

$$\tag{3}$$

The reduction can readily be carried out by chemical reducing agents (e.g., $Na_2S_2O_4$), or enzymatically. For example, in the oxidation of alcohol by alcohol dehydrogenases (such as are present in yeast) the following reaction occurs

$$CH_3CH_2OH + CoI \rightleftarrows CH_3CHO + CoIH_2 \tag{4}$$

CoI, CoIH$_2$ representing coenzyme I and the reduced form of the coenzyme, respectively. The reduced form can be oxidized chemically, e.g., by ferricyanides, or enzymatically, e.g., by cytochrome c in the presence of cytochrome c reductase.

B. The Mechanism of Oxidation-Reduction Reactions

1. Semiquinones

The reactions we have considered above illustrate the application to biological processes of Wieland's "dehydrogenation theory" (Wieland, 1921, 1931) according to which most oxidations, both *in vitro* and *in vivo*, are hydrogen abstractions rather than oxygen or hydroxyl additions. Many biological oxidations require the removal from each molecule of the substrate of two hydrogen atoms, which are taken up by the prosthetic group or coenzyme, e.g.,

$$\ce{>CHOH} \rightarrow \ce{>C=O} + 2H \tag{5}$$

$$\underset{H}{\overset{COOH}{\underset{NH_2}{\diagdown C\diagup}}} + H_2O \rightarrow \underset{O}{\overset{COOH}{\diagdown C\diagup}} + NH_3 + 2H \qquad (6)$$

The exact mechanism of hydrogen transfer has been the subject of much discussion. We are at present particularly concerned with the possibility that the process might proceed in two distinct stages, corresponding to the separate removal of the two hydrogen atoms; this would entail the formation of radicals from the substrate and the prosthetic group or coenzymes. (It is perhaps worth while to mention here that many workers consider oxidation-reduction reactions in terms of removal or addition of electrons, the removal of an electron together with a proton being formally equivalent to elimination of a hydrogen atom. We do not consider objections to this terminology (Dixon and Webb, 1958b) significant and we shall use it when convenient.)

Westheimer (1954a) has pointed out that "the distinction between one-electron and two-electron processes is a practical dichotomy, not a rigid distinction." Thus, from the chemical point of view, two successive single-electron transfers are indistinguishable from one two-electron transfer if the intermediate species has a half-life less than, say, 10^{-11} seconds, since a radical requires this period to break out of its solvent cage.

Michaelis and his colleagues (see Michaelis, 1951) first established that many oxidations in which the final product is derived from the substrate by the loss of two hydrogen atoms, or two electrons, per molecule proceed through the intermediate formation of ion-radical species with an odd number of electrons in the molecule. For example, the oxidation of hydroquinone in alkaline solution follows the path:

$$\qquad (7)$$

(VI)

The intermediate species (VI) is a (mesomeric) "semiquinone" radical, and its formation and existence in free radical form in solution were proved by potentiometric titration and by magnetic susceptibility measurements (the radicals being paramagnetic); its production may easily be demonstrated by carrying out the reduction polarographically, when two waves in the current-electromotive force diagram are obtained. The semiquinones are naturally more stable under alkaline conditions on account of the increased symmetry of the ion (VI) compared to the neutral radical. In some cases the semiquinone is stable enough to form definite salts; this appears to be so with durosemiquinone. The reactions shown above are reversible; the semiquinone may conveniently be obtained by reduction of the quinone by glucose in pyridine solution. Semiquinone radicals with intense colors are generally formed when indophenol and indamine dyes (e.g., Binschedler's green) are partially reduced. Semiquinones are also intermediates in the reduction of heterocyclic compounds in the quinoline (Breyer *et al.*, 1944) and acridine (Stock, 1944) series. The highly colored Würster salts, obtained when aromatic diamines are oxidized with bromine are similar in type to semiquinones, being formed from the parent amines by loss of one electron per molecule, and having the mesomeric structure (VII)

$$\left[\begin{array}{c} R^1 \\ \diagdown \\ \diagup \\ R^2 \end{array} N = = \hspace{-0.3em}\left\langle\hspace{-0.3em}\diagup\hspace{-0.3em}\diagdown\hspace{-0.3em}\right\rangle\hspace{-0.3em}= = N \begin{array}{c} R^3 \\ \diagup \\ \diagdown \\ R^4 \end{array} \right]^+$$

(VII)

Many semiquinones have been studied by the electron spin resonance technique; a summary of the results has been given by Ingram (1958a). In the main they are consistent with the previously accepted structures. Thus benzosemiquinone (VI) contains four equivalent protons in the molecule situated on equivalent carbon atoms, and would therefore be expected to give a hyperfine pattern of five lines with intensities in the ratios $1:4:6:4:1$. The experimental observations of Venkataraman and Fraenkel (1955) are completely consistent with these ideas.

C. Biological Implications

Michaelis' work on oxidation-reduction reactions provided the basis for the principle of compulsory univalent oxidation, which states that "any oxidation (or reduction) has to proceed in successive univalent steps" (Michaelis, 1939). We have seen above that many single-electron oxidation-reduction reactions are known; there is, however, little doubt that two-electron transfer reactions exist. Examples of these have been given by Westheimer (1954a). The point at issue here is not the general validity of the Michaelis principle, but whether it applies to the oxidation-reduction reactions of coenzymes and prosthetic groups in enzymes. We shall now consider the evidence available for the most important cases.

1. Flavoproteins

There is no doubt that the reduction of these groups can, in certain circumstances, be carried out in successive stages. Kuhn and co-workers (1937,

1944), Michaelis and co-workers (see Michaelis, 1951), and also Stern (1934) and Stare (1935) showed that the reduction of riboflavine in acid solution proceeds through a neutral free radical of the semiquinone type (VIII).

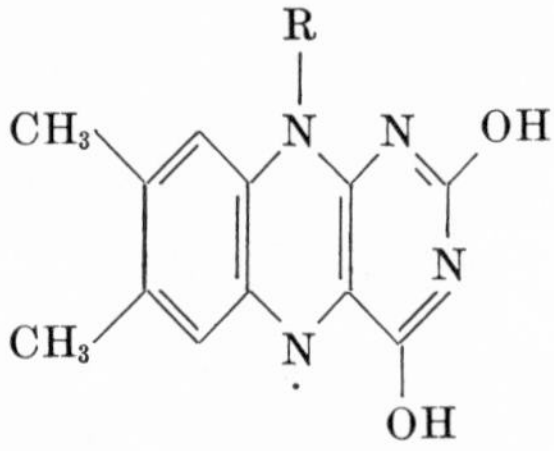

(VIII)

Their conclusion was based on evidence from potentiometric titrations and from magnetic and spectral observations. More recently a comprehensive examination of the spectral characteristics of flavines (FMN, FAD) at oxidation levels ranging from full oxidation to full reduction has been made by Beinert (1956). This worker observed absorption bands which are maximally developed at approximately 50% oxidation, and assigned them to semiquinone intermediates in their monomeric and dimeric forms. Spectra were recorded over a range of pH extending from 0 to 12.

Swallow (1955a) has demonstrated that reduction of riboflavine to the radical form (VIII) may be effected by free radicals ($CH_3\dot{C}HOH$) produced when aqueous sulfuric acid (2N) containing ethanol (0.5 M) is irradiated with X- or gamma rays. On exposure to air the radical autoxidizes to the original flavine. Barrón (1957) has carried out similar experiments on FAD, which, it will be recalled, is a component of D-amino acid oxidase. When irradiated by X-rays in aqueous solution reduction in univalent steps occurs, as for riboflavine. In the presence of alanine a similar dose of radiation causes a greater extent of reduction, and pyruvic acid is formed. Barrón suggests the following reactions, initiated by $H\cdot$ and $\cdot OH$ radicals produced from water by the radiation.

$$\underset{\underset{NH_2}{|}}{CH_3CHCOO^-} + \dot{O}H \rightarrow \underset{\underset{NH_2}{|}}{CH_3\dot{C}COO^-} + H_2O$$

$$FAD + H\cdot \rightarrow FADH\cdot$$

$$\underset{\underset{NH_2}{|}}{CH_3\dot{C}COO^-} + FADH\cdot \rightarrow \underset{\underset{NH}{\|}}{CH_3CCOO^-} + FADH_2 \qquad (8)$$

$$\downarrow H_2O$$

$$CH_3COCOO^- + NH_3$$

Barrón remarks that "this oxidation-reduction, carried out at neutral reaction and at 10°C is similar to that performed enzymatically." Although this is true with respect to the nature of the products, none of these experiments proves that the enzymatic reactions in which riboflavine participates involve free radicals.

Beinert and his colleagues (see Beinert, 1957) have made interesting observations on three fatty acyl coenzyme-dehydrogenases obtained from pig liver. These enzymes are flavoproteins, and a major part of the work has been done with the dehydrogenase "Y" which is most active with substrates from C_8 to C_{12}. A rapid-scanning spectrophotometer operating a cathode ray tube was used, the screen of the latter being photographed with a movie camera so that spectra could be recorded at very short intervals of time. When Y is reduced by dithionite a transient absorption band appears between 5000 and 6500 A which reappears on subsequent partial oxidation by oxygen. This band is characteristic of the semiquinone form of the flavine at pH 7 (Beinert, 1956). A similar absorption band is developed when the enzyme is reduced by its substrate, but in this case the semiquinone appears to be stable for some hours. Further, the semiquinone formed under these conditions is not easily oxidized by agents such as ferricyanide which oxidize the semiquinone instantly in the absence of substrate (Beinert and Page, 1957). Evidently the semiquinone is protected by a strong screening action of the substrate, and in these circumstances may be described as a trapped radical. [We may note that Michaelis (1951) considered that in enzymatic processes the semiquinone radicals exist within the complex, and not in the freely dissolved state.] A specific flavoprotein—the electron-transferring flavoprotein (ETF)—can, however, oxidize the semiquinone. Beinert and Page (1957) represent the equilibria which are set up when the substrate SH_2 (a fatty acyl coenzyme A) is added to Y as shown in Eq. (9).

$$Y + SH_2 \; \rightleftarrows \; Y \cdots SH_2 \; \rightleftarrows \; \overset{.}{Y}H \cdots \overset{.}{S}H \; \rightleftarrows \; YH_2 \cdots S \; \rightleftarrows \; YH_2 + S$$
$$\text{I} \qquad\qquad \text{II} \qquad\qquad \text{III} \underset{-e}{\searrow} \text{ETF} \underset{-e}{\swarrow} \text{IV} \qquad\qquad \text{V}$$
$$Y + S + ETFH_2 \tag{9}$$

(The full scheme contains additional complexes which are not important in the present connection.) Complex II is the Michaelis-Menten complex, and III a complex containing stabilized semiquinone and substrate radicals; IV is a complex between reduced enzyme and product. At equilibrium only small quantities of I and V are present. The action of ETF is represented as oxidation of complexes III and IV. Beinert (1957) has pointed out that if the semiquinone is an intermediate in the over-all reaction ($SH_2 \rightarrow S$) catalyzed by Y and ETF, its rate of formation must be at least equal to the rate of the over-all process. He has concluded that his observations are consistent with this.

Rather similar observations were made with L-amino acid oxidase and old yellow enzyme. During reduction of these enzymes with L-leucine and the reduced form of coenzyme II the characteristic absorption band between 5000 and 6500 A appears. The reactions of the enzymes with dithionite were found to be too rapid to follow. Beinert (1957) has postulated that it should be possible to obtain the semiquinone band with all flavoproteins under suitable conditions and has drawn attention to some data in the literature which seem to substantiate this idea. It seems that the fatty acyl dehydrogenases are particularly suitable for these studies.

Beinert (1957) has also recorded that electron spin resonance observations show that radicals are indeed formed when substrate is added to the acyl dehydrogenases, but no details are given.

Bray *et al.* (1959) have recently reported an electron spin resonance study of xanthine oxidase solutions. This enzyme is a flavoprotein which is activated by iron and molybdenum. Observations were made on the resting enzyme and also after treatment with xanthine or $Na_2S_2O_4$. It was concluded that the iron is always in the ferrous state and that in the presence of the substrate molybdenum is reduced and $FADH\cdot$ radicals (cf. VIII) are formed. The authors point out that although these changes have not been proved to be essential for the occurrence of the enzymatic reaction, they "furnish the strongest presumptive evidence which is yet available that molybdenum and also free radicals participate in the oxidation-reduction sequences of xanthine oxidase action." Unfortunately no further details of this work are yet available.

The experiments described clearly provide cogent reasons for believing that hydrogen transfer reactions catalyzed by flavoprotein enzymes proceed through the intermediate formation of the semiquinone flavine radicals.

2. Coenzyme I

In recent years Vennesland and co-workers (see Vennesland, 1955) have obtained important information about the mechanism of enzyme-catalyzed oxidations from isotope studies using deuterium. A number of different enzymes containing CoI were examined: alcohol dehydrogenase from yeast, lactic dehydrogenase from muscle, malic dehydrogenase from wheat germ, a bacterial testosterone dehydrogenase from *Pseudomonas*, and a bacterial transhydrogenase from *Pseudomonas*. In all cases two atoms of hydrogen are removed per molecule of substrate and reduce the coenzyme according to Eq. (3).

The results obtained with the alcohol dehydrogenase will be described for purposes of illustration. When labeled ethanol (CH_3CD_2OH) was used as substrate, it was found that the isolated reduced coenzyme and the acet-aldehyde formed both contain one D atom per molecule. The reaction is represented by Eq. (10).

$$\text{CH}_3\text{CD} \quad + \quad \text{(ring structure)} \quad \longrightarrow$$

$$\text{CH}_3\text{CDO} \quad + \quad \text{(CoIDH)} \quad \longrightarrow \qquad (10)$$

$$\text{CH}_3\text{CDO} \quad + \quad \text{(ring structure)} \quad + \quad \text{H}^+$$

The reduced coenzyme (CoIDH), on reoxidation by the alcohol dehydrogenase together with acetaldehyde, lost all its deuterium; this also occurred on reoxidation by the lactic dehydrogenase with pyruvate, in which case the lactate formed contained one D atom per molecule.

The reaction is clearly a direct transfer of hydrogen from a substrate C atom to the nicotinamide ring; at no stage is the atom being transferred in a position to undergo exchange reactions with the aqueous solvent. The technique has allowed some stereochemical features of the process to be elucidated. Thus it appears that the hydrogen atom attached to carbon is always transferred to C-4 of the nicotinamide ring (if it were transferred to N it would be able to exchange readily), and further that the reactions must involve transfers to and from one side of the ring only. By contrast, chemical reduction of the coenzyme (e.g., by reaction with $Na_2S_2O_4$ followed by hydrolysis—see page 453) introduces hydrogen atoms on both sides of the ring. The first three enzymes mentioned at the beginning of this section use the same side of the ring, whereas the last two use the opposite side.

It has been suggested (Dixon and Webb, 1958c) that in the transition state the substrate and coenzyme, combined with the protein, are brought into the configuration shown diagrammatically in (IX), the nicotinamide ring being seen partly edge on. The electronic changes

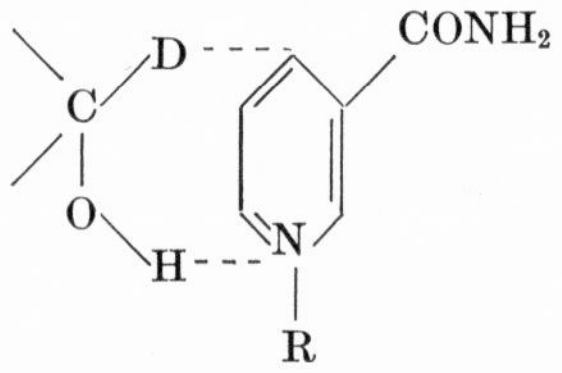

(IX)

corresponding to the forward and backward reactions shown in (X) resemble those of the concerted displacement mechanism postulated by Swain and Brown (1952) and result in the simultaneous transfer of two hydrogen

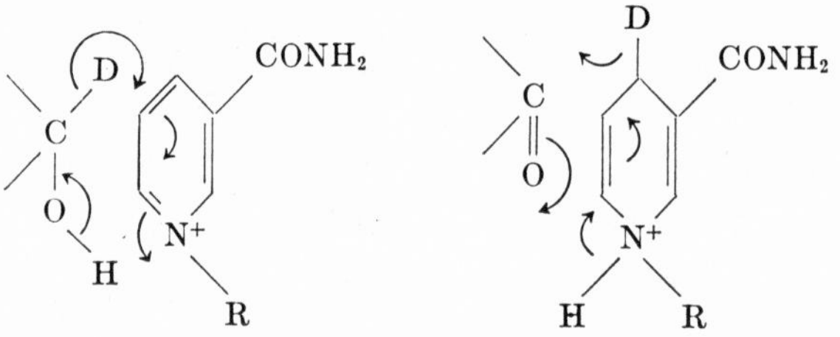

(X)

atoms. Dixon and Webb (1958c) consider that "this mechanism accounts simply and satisfactorily for the observed phenomena of dehydrogenase reactions."

Mackinnon and Waters (1953), on the other hand, suppose that the oxidation of the coenzyme occurs in two stages [Eq. (11)].

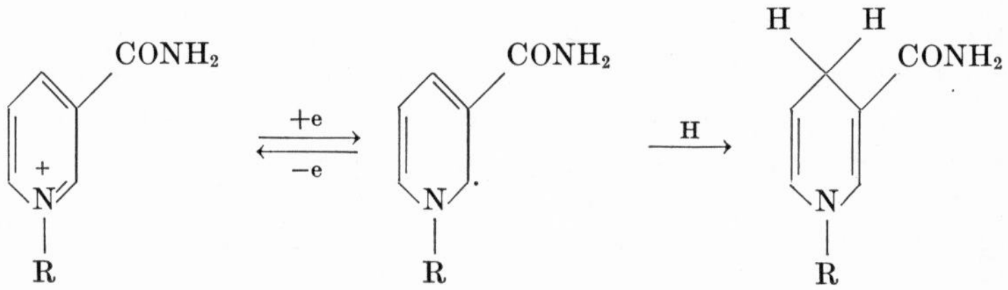

(11)

(Actually they represent the reduction as occurring in the 1,2-position, but this does not affect the argument.) The first is a reaction between the nucleotide and the half-oxidized substrate (radical) $\dot{S}H$, completing the oxidation of the substrate and forming a mesomeric semiquinone radical. The latter may then react with another substrate radical, the reduced form of another coenzyme, or a metal ion which can be oxidized to a higher valency state. In order that the process should be a cyclic one, as postulated by Mackinnon and Waters, an initiation reaction with some substance X must occur leading to the formation of $\dot{S}H$ from SH_2 ; $\dot{S}H$ then reacts as described above, and finally the reoxidation of the resultant semiquinone must take place with eventual regeneration of X. Mackinnon and Waters provide analogies for the various steps, some of them involving ferric or ferrous complexes; indeed the mechanism is mainly based on such analogies. They point out, however, that their scheme is not necessarily limited to systems in which a transition metal is present. Further, they state that the observation of direct hydrogen transfer (Vennesland, 1955) "strongly supports the view that the oxidation occurs by way of a cyclic electron transference within an enzyme-substrate complex from which even a proton cannot escape." Thus, although organic radicals are concerned, they are not *free* but trapped.

There does not appear to be any direct evidence that reduction of the nicotinamide ring can proceed in two chemically separate stages. A yellow compound is formed when coenzyme I is reduced by dithionite in alkaline solution, and it was assumed that this material is the free radical $CoIH\cdot$ (Adler *et al.*, 1936). On lowering the pH to about 7 the fully reduced coenzyme is obtained, a result attributed to the instability of the radical in neutral solution and consequent occurrence of the disproportionation

$$2CoIH\cdot \rightarrow CoI + CoIH_2 \tag{12}$$

It has been proved, however, that the yellow compound is not a free radical, but an addition compound, coenzyme sulfinate, formed by the reactions shown in (13) (Yarmolinsky and Colowick, 1954, 1956; Swallow, 1955b; Westheimer, 1954b.)

$$S_2O_4^{--} + H_2O \rightarrow SO_3^{--} + H_2SO_2$$

$$\tag{13}$$

This is stable in alkaline solution, but hydrolyzes at lower pH values (Eq. 14).

$$\tag{14}$$

3. Hemoprotein Enzymes

The oxidation-reduction reactions of the prosthetic groups of these enzymes are complex and have not been fully elucidated. Initially it was believed that only a valency change of the central metal atom is involved, but this now seems very unlikely, partly for reasons given in Section II, A, 2 and also because the electrons of the central atom are spread to some extent over neighboring atoms. The phthalocyanines, which resemble the porphyrins in structure, have been studied by electron spin resonance. During oxidation there is an intermediate state which gives a narrow resonance line with a g-value close to 2.0 (that of a free spin), showing that during oxidation electronic changes in the conjugated ring system occur, with appearance of a free radical (George et al., 1957). The oxidation of methemoglobin and metmyoglobin by hydrogen peroxide or periodate has also been shown to give rise to a free radical intermediate, but it has been suggested that this is derived from oxidation of part of the globin molecule (Ingram, 1958b).

The kinetic behavior of the hemoprotein enzymes catalase and peroxidase with hydrogen peroxide has received intensive study. Peroxidase forms four distinct complexes with hydrogen peroxide, of which only complexes I and II enter into the catalytic mechanism. Catalase forms three complexes, of which only I is active (Chance and Ferguson, 1954). The kinetics of the oxidation by hydrogen peroxide of a two-equivalent hydrogen donor AH_2 catalyzed by peroxidase seem to require the participation of a radical intermediate. The scheme given by Chance (1952) is shown in Eq. (15, a–d),

$$E + H_2O_2 \rightleftarrows ES_I \tag{15a}$$

$$ES_I + AH_2 \rightarrow ES_{II} + \dot{A}H \tag{15b}$$

$$ES_{II} + AH_2 \rightarrow E + \dot{A}H \tag{15c}$$

$$2\dot{A}H \rightarrow A + AH_2 \tag{15d}$$

in which ES_I, ES_{II} represent complexes I, II, respectively. George (1953) has proved by titration with ferrocyanide that ES_I possesses two, and ES_{II} only one, oxidizing equivalents above the enzyme. Thus, since the changes $ES_I \rightarrow ES_{II}$ and $ES_{II} \rightarrow E$ are one-equivalent processes, it is necessary to postulate the formation of the radical ȦH.

Chance and Ferguson (1954) record the examination by the electron spin resonance technique of the peroxidase complexes I and II. In solutions with concentrations in the range 0.5 to $1 \times 10^{-3} M$ no evidence for the presence of a free radical with a g-value between 1.7 and 2.3 could be obtained. In order to increase the sensitivity further, freeze-dried samples of the complexes were prepared. With an enzyme content of the sample of 4×10^{-7} moles, the maximum radical content appeared to be approximately 2×10^{-11} moles, i.e., the complexes contain less than one part in 20,000 of a free radical. These observations do not necessarily conflict with the kinetic scheme (15) given above, since the measured radical concentration, corresponding to a stationary value determined by the relative rates of formation and disappearance of ȦH, may well be low on account of the high rate of disproportionation of the radicals (15d). It is to be hoped that further work with high sensitivity electron spin resonance equipment will enable a decision to be made about the part played by radicals in these reactions.

4. Enzymes Containing Copper

The diphenol oxidases, which are widely distributed in plants and fungi, are copper proteins, which catalyze the oxidation by molecular oxygen of diphenols to quinones, and monophenols to o-diphenols. The catalytic activity is dependent on the presence of copper which acts as an oxidation-reduction carrier; after removal of the copper the enzyme may be specifically reactivated by cupric ions.

Forsyth and Quesnel (1957) and Bouchilloux and Lissitzky (1958) have shown that the products of the oxidation of catechol catalyzed by mushroom polyphenol oxidase are most readily accounted for by the intermediate formation of catechol semiquinone. Recently Le Clerc et al. (1959) have examined the absorption spectrum of catechol during photo-oxidation induced by brief exposure to intense ultraviolet irradiation in a rigid medium at 100°K and have identified an absorption band arising from the semiquinone. A similar band was observed when catechol in ethanol solution was treated with the pure enzyme at the same temperature, and Le Clerc et al. conclude that both the photo- and enzymatic oxidations proceed through the semiquinone.

D. CHAIN MECHANISMS IN ENZYMATIC REACTIONS

The first suggestion that chain processes may be involved in enzyme-catalyzed reactions came from Haber and Willstätter (1931) and arose from analogy with the mechanism of the autoxidation of sulfites. Haber and Willstätter pointed out that many enzymes contain metals such as iron and postulated that these might interact with a substrate forming free radicals which could then initiate chains. Subsequently Waters (1943, 1948) modified Haber and Willstätter's mechanism to take into account later developments and to include the functioning of prosthetic groups and coenzymes. The active form of the enzyme was supposed to be a radical which could initiate a chain by abstracting a hydrogen atom from a substrate SH_2. The resulting substrate radical $SH\cdot$ could then enter into repeating sequences, one version of which is shown below. Equation (16) represents 1,2 reduction of the nicotinamide ring, but similar equations could equally well be written for 1,4 reduction.

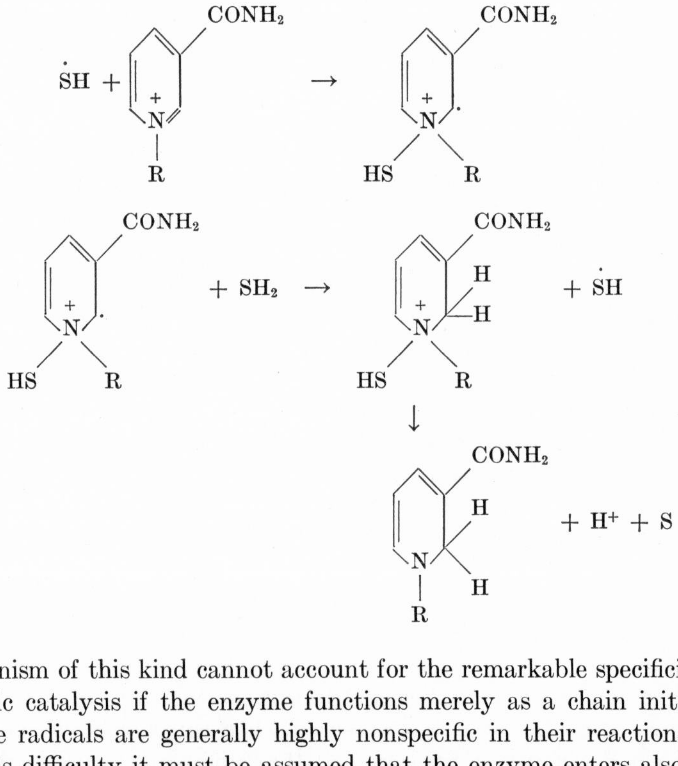

$$\tag{16}$$

A mechanism of this kind cannot account for the remarkable specificity of enzymatic catalysis if the enzyme functions merely as a chain initiator, since free radicals are generally highly nonspecific in their reactions. To avoid this difficulty it must be assumed that the enzyme enters also into the chain propagation reactions (16), as would be the case if the processes occurred within the protein-coenzyme-substrate complex. The reaction scheme suggested by Mackinnon and Waters (1953) and described earlier is of this form.

It is, of course, possible for an enzymatic reaction to proceed through radical intermediates without being a chain process; this is so, for example, with the fatty acyl coenzyme dehydrogenases according to Beinert and Page (1957) [Eq. (9)]. Thus the presence of radicals cannot be considered to be proof of a chain mechanism. It has been suggested that the kinetics of enzyme-catalyzed reactions provide evidence of their chain character; we shall now consider this matter.

The rates of many chain reactions do not show a simple dependence on the concentrations of the reactants. Moelwyn-Hughes (1937) drew attention to the similarity between chain and enzymatic reactions in this respect. At low concentrations enzymatic reactions are often of first order in substrate concentration, and tend to become of zero order at high concentrations. This behavior can, however, be readily explained on the basis of the well-established Michaelis-Menten scheme, without recourse to chain theories.

During the early stages of a chain reaction the rate increases with time, since the concentration of chain carriers is zero initially and requires a finite time to build up to its stationary value. The length of the period of acceleration depends on the nature of the system; in radical chain reactions in homogeneous solution the periods are generally quite short [e.g., a few seconds in some polymerization reactions (Bamford et al., 1958a)]. Acceleration periods are also found in enzyme-catalyzed reactions; they are usually short and can be followed only by refined techniques such as those developed by Chance. These observations do not provide evidence that the enzymatic reactions are chain processes, since any reaction occurring in more than one step can show this kind of kinetic behavior. The pre-steady state kinetics corresponding to the Michaelis-Menten scheme have been investigated by Gutfreund (1954, 1955), who has made measurements under nonstationary conditions and used them in deducing information about the absolute magnitudes of some of the velocity coefficients in the reaction scheme.

Both chain and enzymatic reactions are subject to inhibition. In the former the inhibitors destroy the chain carriers, or the initiating species which produce the chain carriers, and it has been suggested that a similar mechanism may apply to enzyme-catalyzed processes. Thus Leach (1954) has said that enzymatic oxidations are "highly sensitive to traces of the inhibitors which are known to inhibit radical chains, e.g. malonic acid, organic iodides, sulphites, cyanides and carbon monoxide." Schwab et al. (1933) claimed that the inhibiting powers of several compounds (ethanol, benzyl alcohol, resorcinol, hydroquinone) are parallel in the catalase-induced and photodecompositions of hydrogen peroxide. This conclusion was not confirmed by the work of Alyea and Pace (1933). The majority of enzymatic reactions are not affected by the compounds mentioned by Leach, and in general, inhibition of enzymatic processes can be satisfactorily accounted for by combination between the inhibitor and the enzyme, substrate, coenzyme, or metal activator.

It is difficult to avoid the conclusion that none of the arguments put forward provides compelling evidence that enzymatic reactions are chain processes. A further point to be considered is that, as already mentioned, if such a chain process is to be compatible with the stereospecificity of enzymatic reactions the various steps must occur within the enzyme-substrate complex, probably under conditions such that access of the reactant is partially restricted. The rates of these reactions would then become diffusion controlled, and the kinetic behavior would differ significantly from that encountered in a simple chain process. This situation is well-known in polymerization kinetics (Bamford *et al.*, 1958b).

III. Unpaired Electrons in Natural and Irradiated Biological Materials

A. NATURAL BIOLOGICAL MATERIALS

It is important to note that, except with apparatus of unusually high sensitivity, the presence of water in biological materials is a very considerable disadvantage in electron spin resonance studies since it causes a heavy damping of the signal. This can be overcome either by performing the experiments at very low (liquid nitrogen) temperatures or by freeze-drying the specimen prior to scanning. In both cases some modification of the material may result from the treatment, and to this extent some uncertainty must exist as to the significance of the measurements with regard to the natural material. In the first experiments on biological materials, made by Commoner *et al.* (1954), the materials were freeze-dried.

Table I gives estimates of unpaired electron contents for a variety of materials, animal and plant, the concentrations being based on a calibration with $\alpha\alpha$-diphenyl β-picryl hydrazyl. This table demonstrates (subject to the reservation implied above) that a remarkably high unpaired electron content exists in many biological materials, particularly in the *Coleus* leaf and unfertilized frog's eggs. In fact the average unpaired electron content of the materials cited above is higher than that of typical samples of polyacrylonitrile containing trapped radicals (Bamford *et al.*, 1958b). As a result of fractionation experiments it was found that the unpaired electron content of the tissues is associated with the protein components, denaturation of the latter resulting in destruction of the electron spin resonance absorption. Under refrigeration the absorption persists for several months.

An interesting correlation was found by Commoner *et al.* (1954) between

TABLE I

ESTIMATED FREE RADICAL CONTENT OF VARIOUS PLANT AND ANIMAL MATERIALS

Material	Estimated free radical content (moles × 10^8/gm dry weight)	Material	Estimated free radical content (moles × 10^8/gm. dry weight)
Nicotiana tabacum, leaf	65	Rabbit, muscle	20
		Rabbit, brain	25
Nicotiana tabacum, roots	10	Rabbit, liver	60
		Rabbit, lung	30
Coleus, leaf	180	Rabbit, heart	35
Barley, leaf	25	Rabbit, kidney	55
Snapdragon, petals	8	Mouse, liver	30
Digitalis, germinating seeds	10	Mouse, liver hepatoma	20
Carrot, root	8	Guppies, entire	2
Beet, root	6	Frog, unfertilized eggs	200
Bracket fungus	35	*Drosophila*, entire	4
Rabbit, whole blood	25	Firefly, lanterns	2

the concentration of unpaired electrons in biological material and the level of metabolic activity. Thus, etiolated barley leaves possess only about one-fifth of the concentration in the fully green leaves. Further, illumination with fluorescent lights progressively raises the unpaired electron content over a period of 24 hours to the normal value and during this time the leaves recover their normal green coloration. This correlation is consistent with Michaelis' suggestion (Michaelis, 1951) that metabolic electron transfer is associated with a stationary concentration of free radicals. The changes in unpaired electron content for different periods of illumination are illustrated in Fig. 1.

Commoner *et al.* (1954) also examined a number of polymers of biological origin including the following: melanin, gum guaiacum, liquorice, plant resins, humin, caramelized glucose, charcoals of organic origin, and various types of coals and tars. All were found to contain unpaired electrons which are stable at room temperature, and which, in some cases (e.g., melanin) survive refluxing for 24 hours in hydrochloric acid. All the natural products examined contain polymeric condensed ring structures; it seems possible therefore that, by analogy with the known presence of trapped radicals in coal extracts and chars, the unpaired electrons present are associated with the polymer lattice which both stabilizes them by resonance and shields them from attack by chemical reagents.

The existence of unpaired electrons in melanin constitutes indirect evidence that these entities can arise from irradiation of biological systems,

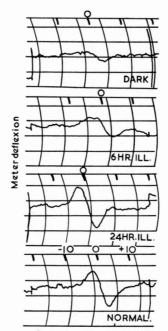

Relative increment in magnetic field (gauss)

FIG. 1. Electron spin resonance absorption (shown as first derivative) of leaves from barley seedlings. Uppermost curve: etiolated. Lowest curve: leaves from comparable seedlings grown in the light. Center curves: etiolated leaves after 6 and 24 hours' illumination. After Commoner *et al.* (1954).

for it has previously been shown that melanin formation in animals can be induced by ultraviolet or high energy radiation (McIlwain, 1946). The induction of electron spin resonance absorption by irradiation of etiolated barley leaves is direct evidence for the same phenomenon. Commoner *et al.* (1954) have discussed the possible relevance of the unpaired electrons in melanin to the functions of the latter in cell metabolism.

It seems likely that the unpaired electrons in the systems described are present as trapped radicals. The observation that in some cases denaturation of the protein components of the tissues leads to destruction of radicals suggests that the mechanism of trapping may be rather different from that encountered in vinyl polymers, in that more specific structures may be responsible for stabilization in the former case. For present purposes the important feature of free radical reactions is their high velocity coefficient; in this respect, however, they are not a unique class, but are qualitatively similar to proton exchanges between oxygen, nitrogen, sulfur, for example. Recently it has become clear that although these exchange reactions are extremely fast in simple systems, they occur only very slowly, if at all, in some native proteins (Lenormant and Blout, 1953; Hvidt and Linderstrøm-Lang, 1954, 1955a,b). These conclusions have been confirmed by experiments on synthetic polypeptides (Elliott and Hanby, 1958); thus the imino hydrogens in the α form of poly-γ-benzyl-L-glutamate do not appear to be readily exchangeable with deuterium, whereas those in the β form exchange rapidly. Obviously the steric features of the α-helix prevent the attainment of the transition state for the exchange reaction. It does not seem impossible that such a state of affairs should also be encountered with radicals. The essential point is that a radical might be stabilized by a relatively simple structure of a suitable type and that a gross entanglement of polymer chains may not be necessary. Obviously such stabilization could extend to radicals in solution, and it may in fact occur in the enzyme systems already mentioned. The stabilizing structure need not be the α-helix, but to be consistent with the observations of Commoner et al. (1954) it would have to be destroyed on denaturation of the protein.

B. Irradiated Biological Materials

1. In Photosynthesis

The work of Commoner et al. (1954), which indicated a close connection between the absorption of light by plant pigments and the presence of unpaired electrons in the plant tissue, led to a number of studies designed to investigate the possibility that free radicals participate in the process of photosynthesis. These studies have generally been made on isolated chloroplasts, which are clearly defined structures existing within the stroma of the cytoplasm and to which the photosynthetic pigments are confined. Chloroplasts are generally isolated by a procedure such as the following (Commoner et al., 1956). Gentle manual maceration of the leaf tissue in a buffer solution cooled in an ice bath is followed by filtration and low speed centrifugation (100 g) to remove the debris; the chloroplasts are then precipitated by centrifugation at a higher speed (2500 g). Finally, purification is effected by a further centrifugation in sucrose solution at a speed corresponding to 40,000 g followed by washing in a sucrose buffer solution in which the chloroplast is stored in the cold.

Using an apparatus of very high sensitivity, Commoner et al. (1956) were able to obviate the necessity for freeze-drying or low-temperature measurement, and under these conditions they examined the unpaired electron contents of chloroplast preparations from *Nicotiana tabacum* in the aqueous medium at ambient temperatures. The apparatus was so constructed that the chloroplast specimen in the cavity of the electron spin resonance spec-

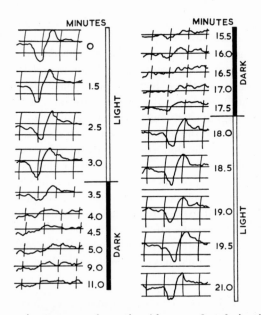

Fig. 2. Electron spin resonance absorption (shown as first derivative) of a chloroplast suspension at intervals during successive light, dark and light periods. The times noted represent minutes elapsed after the first curve (upper left corner) was obtained and refer to the center point of each absorption curve. After Commoner *et al.* (1956).

trometer could be irradiated by the light from a 50-candlepower automobile headlamp.

In the dark the chloroplast preparations give small signals which vary according to the age and history of the sample. Upon illumination with the headlamp a three- to sixfold increase in signal strength rapidly occurs; the resulting absorption is identical with that obtained from the great majority of organic free radicals. The response of the chloroplast preparation to illumination is completely reversible and reproducible. Figure 2 shows a typical run in which the absorption of a chloroplast sample was determined at approximately 30-second intervals during successive light, dark and light periods. It is clear that the induced absorption reaches its maximum in 1 minute or less and decays to the dark value with a half-life of the order of half a minute. More detailed studies showed that both build-up and decay follow exponential laws with half-lives of approximately 8 and 32 seconds, respectively; it is clear therefore that the absorption arises from a stationary concentration of unpaired electrons or reactive radicals rather than from stabilized radicals of long life.

The steady state concentration of the unpaired electrons present in these illuminated chloroplast suspensions increases with light intensity initially, but at high intensities approaches a maximum value asymptotically. This type of behavior is also characteristic of the photosynthetic activity of chloroplasts and suggests either that the unpaired electrons play a part in the process of photosynthesis or at least that both phenomena involve a common species. This conclusion is reinforced by the observation that both exhibit the same dependence on the wavelength of the incident light. A more detailed discussion of the connection between photosynthesis and unpaired electron formation is given by Commoner et al. (1956).

Observations of a similar nature by Sogo et al. (1957) confirmed the results just described and furthermore showed that within experimental error the signal growth time observed when wet chloroplast fragments are illuminated is the same at $-140°C$ and at $25°C$. This lack of temperature dependence appears to exclude the possibility that the signal arises as a result of enzymatic oxidation-reduction reactions. Further implications of these and other measurements have been discussed by Tollin and Calvin (1957), who were able to show by measurements of light emission that the decay of the luminescence of previously irradiated spinach chloroplasts is the resultant of at least two processes with different half-lives, the relative importance of the various processes depending on the temperature of the specimen. A comparison of the luminescence and electron spin resonance results has been shown to reveal the following similarities.

(1) Both the luminescence and the electron spin resonance absorption are excited by the same band of wavelengths and both are due to absorption of light by chlorophyll.

(2) The room temperature exponential decay constants are of the order of seconds for both processes.

(3) At $-140°C$ the electron spin resonance signals have decay constants of the order of hours, and the luminescence is undetectable, as would be expected if the decay constants for this process are of the same order of magnitude.

(4) At $60°C$ the electron spin resonance has a decay constant of the order of seconds, and at the same temperature a peak is observed in the luminescence.

It is reasonable to deduce that the light emission observed in the 7000 to 9000 A region is, at least partly, the result of the decay in the concentration of the light-induced unpaired electrons observed in the spin resonance studies.

Tollin and Calvin (1957) considered four possible mechanisms to account for their observations on electron spin resonance and luminescence in irradiated chloroplasts.

(1) The direct photodissociation of a chemical bond followed by interaction (with light emission) of the radicals produced.

(2) The excitation and decay of a triplet state.

(3) The reversible photosensitization of chemical or enzymatic reactions leading to the production and subsequent decay of radicals.

(4) The production of electrons trapped in a quasi-crystalline lattice. The thermal depopulation of these traps with subsequent radiative recombination of the electron and hole would thus result in light emission. It was concluded that only process (4) is consistent with all the observations, and a mechanism was suggested for the absorption and transformation of the incident quanta.

According to this suggestion the chloroplast behaves as a semiconductor. This concept is not new; Szent-Györgi (1941) suggested that the protein in the grana (disklike bodies present in the chloroplast) might act as a semiconductor so that the energy used in photosynthesis was carried by electrons in the conduction band. The idea was developed by Katz (1949), Bassham and Calvin (1955), and Arnold and Meek (1956) and recently has found other experimental support in the studies of Arnold and Sherwood (1957) on the glow curves and electrical resistance of chloroplasts. For example, the changes in electrical resistance on heating dried chloroplasts are consistent with the liberation of trapped electrons, and it was concluded that the nature of the primary photosynthetic act may have "at least as much in common with solid state physics as with the chemistry of solutions."

Shields et al. (1956b) have measured the electron spin resonance absorption of various plant materials and in some cases have found evidence for a copper derivative of chlorophyll, the spectrum obtained having a g-factor of 2.06, identical with that obtained by Ingram and Bennett (1955) for chlorophyll in which the magnesium had been replaced by copper. This result indicates that some substitution of copper for magnesium in chlorophyll occurs naturally in certain plants.

2. X- and Gamma-irradiated Materials

Gordy et al. (1955), using the electron spin resonance technique, have examined a considerable number of X-irradiated materials of biological origin or interest. These include simple amino acids such as glycine, alanine, valine, cystine and proteins such as hair, toenail, feather, and silk. In all cases evidence of the presence of unpaired electrons after exposure to 50-kv X-irradiation was clearly obtained.

Of the simple amino acids, glycine gives predominantly a triplet, indicating two protons equally coupled with the unpaired electron and possibly arising from the radical ions $\cdot CH_2^+$ or $\overset{+}{N}H_3\overset{\cdot}{C}H_2$ resulting from decomposition of the glycine molecule following removal of an electron by the X-irradiation. Alanine gives a quintet provisionally ascribed to the radical ion $(CH_2CH_2)^+$ and valine, a composite spectrum consisting of a septet and a quintet which may be due to the radical ions $(CH_3CH_3)^+$ and $(CH_2CH_2)^+$. Leucine also gives a septet which was ascribed to $(CH_3CH_3)^+$.

More recent experiments by Gordy and Shields (1956b) show that at liquid nitrogen temperatures the electron spin resonance spectra of X-irradiated glycine, valine, and leucine are more complex and less resolvable than their room temperature counterparts. This may result from quenching of certain types of motions which, at room temperature, tend to reduce anisotropic interactions, or from structural changes in the radicals themselves or their immediate environments. With alanine the only change at low temperatures is in the relative magnitudes of the five lines, which become almost equal.

Samples of X-irradiated alanine and glycine showed no diminution of their electron spin resonance even when heated to temperatures near their melting points (Gordy and Shields, 1956b), which demonstrates that the radicals present in these systems are virtually permanently trapped.

Different spectra were obtained for X-irradiated DL-leucine, DL-isoleucine, and L(−)-leucine; those for D- and L-alanine are similar (Gordy and Shields, 1956b). With single crystals of D-alanine the orientation dependence of the fine structure in the electron spin resonance spectrum indicates that the motions of the radical are highly restricted (van Roggen *et al.*, 1956), as might be expected from the evident high degree of occlusion. The electron spin resonance spectra of X-irradiated hair, toenail and feather are all closely similar to that of cystine (Gordy *et al.*, 1955) (see Fig. 3). This spectrum was considered to indicate that the unpaired electron is shared principally between the two sulfur atoms, contributing to a three-electron bond and *increasing* the strength of the —S—S— linkage. The suggested structure of the radical is shown in (XI).

$$[HOOCCH(NH_2)CH_2S\dot{\;-\;}SCH_2CH(NH_2)COOH]^+$$

(XI)

The similarity of the electron spin resonance spectra of the fibrous proteins hair, nail, and feather suggests strongly that they arise from the cystine in the proteins, although this is only a small fraction of the whole material. If this is the case the presence of cystine would appear to reduce greatly the irradiation damage otherwise suffered by the protein molecules and, if the structure shown above for X-irradiated cystine is correct, the cross-bonding through cystine links should, according to Gordy *et al.* (1955), be strengthened by irradiation.

The spectrum of irradiated cysteine is very similar to that of cystine (Gordy and Shields, 1956a). This could mean that the —S—S— link is

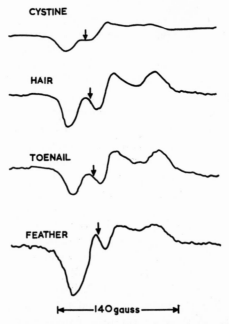

FIG. 3. Electron spin resonance spectra (first derivatives) of X-irradiated cystine and some fibrous proteins. Arrows indicate position for g of the free electron spin. After Gordy *et al.* (1955).

severed by the radiation, but it is more likely that hydrogen is eliminated from the cysteine.

Raw silk, cattle hide, fish scale, and other hydrogen-bridged structures give spectra of a different type (see Fig. 4). In this case the spectra are considered to arise from structures such as (XII).

(XII)

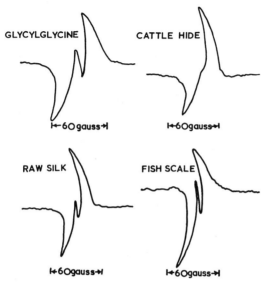

FIG. 4. Electron spin resonance spectra (first derivatives) of X-irradiated glycylglycine and some proteins. After Gordy *et al.* (1955).

Again the postulated structures indicate that the cross-bonding may be strengthened as a result of the formation of unpaired electrons within the proteins.

The electron spin resonance spectrum of an irradiated protein was found to decay when the specimen was preserved in air, but not when it was kept in vacuum. It thus appears that oxygen functions destructively by interaction with trapped radicals, probably to give mobile radicals which can subsequently undergo mutual destruction. A similar mechanism of radical destruction by oxygen was proposed earlier to account for the influence of oxygen on trapped polyacrylonitrile radicals (Bamford *et al.*, 1955, 1959). In the case of irradiated skull bone the spectrum is altered, but not destroyed, by oxygen.

Spectra obtained from other irradiated proteins (Gordy and Shields, 1956a), including histone, insulin, hemoglobin, and albumin, support the conclusions given above for cystine and raw silk. Results have also been obtained for some irradiated peptides (McCormick and Gordy, 1956), and hormones and vitamins (Rexroad and Gordy, 1956). With carbohydrate materials (Shields *et al.*, 1956a) various types of spectra were observed; some sugars, including ribose, give only a single sharp line, whereas lactose gives a quartet which may arise from $\dot{C}H_3$. Doublets from irradiated wood and plant cell tissue are ascribed to unpaired electrons, localized on oxygen atoms involved in hydrogen bonds, interacting with the bridging proton.

Some observations have also been made on the electron spin resonance spectra of gamma-irradiated amino acids. Combrisson and Uebersfeld (1954) found that several of the simple amino acids give a three-line spectrum, the only similarity with the results of Gordy *et al.* (1955) occurring with glycine. Ghosh and Whiffen (1958) in a detailed study of the electron reson-ance of single crystals of glycine after γ-irradiation identified $\overset{+}{N}H_3\overset{\cdot}{C}HCOO^-$ radicals. Later (1959) they suggest that this species may not be a primary product, and report the presence of other radicals.

IV. Carcinogenesis

Many attempts have been made to explain the carcinogenic activity of a wide variety of compounds in terms of a single common property, and we are concerned here with the suggestion that all such compounds are able to participate in reactions leading to the production of free radicals (Brues and Barrón, 1951). There is a clear link here with the carcinogenic effects of high energy radiations, which are known to generate radicals on interaction with water or almost any organic material.

It was shown in 1946 that the polynuclear aromatic carcinogens contain regions of high electron density (the so-called K region) corresponding to potential radical forms in these regions (Daudel, 1946; Pullman and Pull-man, 1946; Daudel and Daudel, 1949), and it has also been found that such compounds react readily with biochemically important sulfhydryl groups to give substituted hydrocarbons, e.g., (XIII) (Wood and Fieser, 1940).

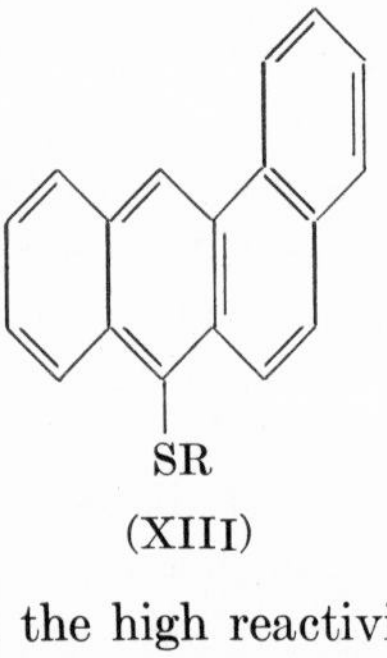

SR

(XIII)

This may be compared with the high reactivity of sulfhydryl compounds toward free radicals, e.g., in polymerization reactions or in aqueous systems during irradiation. According to Barrón and co-workers (Barrón, 1946; Barrón and Dickman, 1949; Barrón *et al.*, 1949), the reactions in the latter case are initiated by ȮH radicals and follow the stoichiometric equation

$$2RSH + 2\overset{\cdot}{O}H \rightarrow RS\!-\!SR + 2H_2O \qquad (17)$$

Butler (1950) has suggested that the nitrogen mustards function as carcinogenic agents by reason of a biradical form (XIV) of the derived cation.

$$R—\overset{+}{N}—CH_2\overset{.}{C}H_2 \leftrightarrow R—\overset{\overset{..}{+}}{N}—\overset{.}{C}H_2CH_2 \leftrightarrow R—\overset{++}{N}—CH_2\overset{-}{C}H_2$$
$$\quad\;\; | \qquad\qquad\qquad\quad | \qquad\qquad\qquad\quad |$$
$$\quad\;\, R' \qquad\qquad\qquad\quad R' \qquad\qquad\qquad\quad R'$$

(XIV)

Similar structures can be advanced to account for the activity of several other carcinogens such as the sulfur mustards, azo dyes, epoxides, stilbenes, and ethylene imine derivatives. It must be admitted, however, that the importance of biradical forms such as (XIV) is debatable and independent evidence is nonexistent.

Some analogies between carcinogenesis and radical chain reactions were pointed out by Park (1950). In particular, the latter reactions are both initiated and terminated by free radicals; it is also true that the same substances which can initiate cancer production may in addition alleviate the condition (Boyland, 1948).

Lipkin *et al.* (1953) have sought to explain the carcinogenic nature of hydrocarbons such as benzanthrene by reason of their ability to form negative hydrocarbon radical ions by interaction with mild reducing agents. This type of reaction is well known and is typified by that between sodium and naphthalene [Eq. (18)].

$$\text{(18)}$$

However, naphthalene and anthracene are noncarcinogenic since strong reducing agents, not present in the animal body, are essential for this type of reaction to occur. With more readily reducible hydrocarbons it is postulated that such reactions may proceed *in vivo*.

The idea that a radical can be carcinogenic only if it can exist as a charged species was put forward tentatively by Oppenheimer *et al.* (1953b) and is obviously in agreement with the mechanism suggested by Butler (1950) and Lipkin *et al.* (1953) in certain specific cases. The postulate of a radical ion as an intermediate in carcinogenesis has more recently been made by Nash (1957). This author supposes that an electron is accepted from its environment by the carcinogen to give a radical ion in which the portion of the molecule bearing the added electron undergoes a structure change to become optically active. It is thought that this species can undergo further reactions such as oxidative fission and linkage to protein materials which result in carcinogenesis. Nash also briefly discussed the relationship of this type of scheme to the carcinogenic activity of polar compounds of low molecular weight.

Many polymers have been found to be carcinogenic when present in the human or animal body (Oppenheimer *et al.*, 1953a), and it has been suggested that addition polymers generate or contain free radicals which arise from residual initiator or as a result of the mechanical treatment of the polymer (Fitzhugh, 1953). However, many other polymers, including regenerated cellulose and nylon, also have carcinogenic properties (Oppenheimer *et al.*, 1953a), and it seems difficult to account for the activity of all these polymers in the same way (Oppenheimer *et al.*, 1953b). A further difficulty arises in accounting for the transference of the supposed radical from the polymer into the neighboring cells (or for the diffusion of tumor inhibitors into the plastic).

In recent years much interest has been aroused by the suggestion that the occurrence of lung cancer is related to the habit of cigarette smoking. Carcinogenic hydrocarbons have indeed been isolated from mechanically inhaled cigarette smoke (Cooper and Lindsay, 1953), and Lyons *et al.* (1958) have shown, using the electron spin resonance technique, that the rapidly condensed smoke contains a high concentration of unpaired electrons some of which are short-lived but some of which are of indefinite life. Thus both active and stabilized radicals exist in cigarette smoke, and either or both may act as carcinogens. However, it has not, as yet, been found possible to relate the radicals directly to the carcinogenic hydrocarbons isolable from the same condensate.

REFERENCES

Adler, E., Hellström, V., and von Euler, H., (1936). *Z. physiol. Chem., Hoppe-Seyler's* **242**, 225. The reduction of cozymase.

Alyea, H. N., and Pace, J. (1933). *J. Am. Chem. Soc.* **55**, 4801. Inhibitors in the decomposition of hydrogen peroxide by catalase.

Arnold, W., and Meek, E. S. (1956). *Arch. Biochem. Biophys.* **60**, 82. Polarization of fluorescence and energy transfer in grana.

Arnold, W., and Sherwood, H. K. (1957). *Proc. Natl. Acad. Sci. U. S.* **43**, 105. Are chloroplasts semiconductors?

Bamford, C. H., Ingram, D. J. E., Jenkins, A. D., and Symons, M. C. R. (1955). *Nature* **175**, 894. Detection of free radicals in polyacrylonitrile by paramagnetic resonance.

Bamford, C. H., Barb, W. G., Jenkins, A. D., and Onyon, P. F. (1958a). "The Kinetics of Vinyl Polymerization by Radical Mechanisms," Chapter 2. Academic Press, New York.

Bamford, C. H., Barb, W. G., Jenkins, A. D., and Onyon, P. F. (1958b). "The Kinetics of Vinyl Polymerization by Radical Mechanisms," Chapter 4. Academic Press, New York.

Bamford, C. H., Jenkins, A. D., Symons, M. C. R., and Townsend, M. G. (1959).

J. Polymer Sci. **34,** 181. Trapped radicals in heterogeneous vinyl polymerization.

Barrón, E. S. G. (1946). *Manhattan District Declassified Document* **484.**

Barrón, E. S. G. (1957). *Ann. N. Y. Acad. Sci.* **67,** 648. The role of free radicals in biological oxidations.

Barrón, E. S. G., and Dickman, S. (1949). *J. Gen. Physiol.* **32,** 595. Studies on the mechanism of action of ionizing radiations. II. Inhibition of sulfhydryl enzymes by α-, β- and γ-rays.

Barrón, E. S. G., Dickman, S., Muntz, J. A., and Singer, T. P. (1949). *J. Gen. Physiol.* **32,** 537. Studies on the mechanism of action of ionizing radiations. I. Inhibition of enzymes by X-rays.

Bassham, J. A., and Calvin, M. (1955). *U. S. Atomic Energy Comm. Unclassified rept.* **UCRL-2853.**

Beinert, H. (1956). *J. Am. Chem. Soc.* **78,** 5323. Special characteristics of flavins at the semiquinoid oxidation level.

Beinert, H. (1957). *J. Biol. Chem.* **225,** 465. (This paper gives references to earlier work.) Evidence for an intermediate in the oxidation-reduction of flavoproteins.

Beinert, H., and Page, E. (1957). *J. Biol. Chem.* **225,** 479. On the mechanism of dehydrogenation of fatty acyl derivatives of coenzyme A. V. Oxidation-reduction reactions of the flavoproteins.

Bouchilloux, S., and Lissitzky, S. (1958). *Bull. soc. chim. biol.* **40,** 833. Sur la mécanisme de l'oxidation enzymatique du catéchol.

Boyland, E. (1948). *Yale J. Biol. and Med.* **20,** 321.

Bray, R. C., Malmström, P. G., and Vänngård, T. (1959). *Biochem. J.* **71,** 24P. Electron-spin resonance of xanthine oxidase solutions.

Breyer, B., Buchanan, G. S., and Duewell, H. (1944). *J. Chem. Soc.* p. 360. The reduction potentials of acridines with reference to their antiseptic activity.

Brues, A. M., and Barrón, E. S. G. (1951). *Ann. Rev. Biochem.* **20,** 350. Biochemistry of cancer.

Butler, J. A. V. (1950). *Nature* **166,** 18. Nature of nucleotoxic substances.

Chance, B. (1952). *Arch. Biochem. Biophys.* **41,** 416. The kinetics and stoichiometry of the transition from primary to the secondary peroxidase peroxide complexes.

Chance, B., and Ferguson, R. R. (1954). *In* "The Mechanism of Enzyme Action" (W. D. McElroy and B. Glass, eds.), p. 389. Johns Hopkins Press, Baltimore, Maryland. One and two electron reactions in catalases and peroxidases.

Christie, S. M. H., Kenner, G. W., and Todd, A. R. (1954). *J. Chem. Soc.* p. 46. A synthesis of flavin-adenine dinucleotide.

Combrisson, J., and Uebersfeld, J. (1954). *Compt. rend.* **238,** 1397. Détection de la résonance paramagnétique dans certaines substances organiques irradiées.

Commoner, B., Townsend, J., and Pake, G. E. (1954). *Nature* **174,** 689. Free radicals in biological materials.

Commoner, B., Heise, J. J., and Townsend, J. (1956). *Proc. Natl. Acad. Sci. U. S.* **42,** 710. Light-induced paramagnetism in chloroplasts.

Cooper, R. L., and Lindsay, A. J. (1953). *Chem. & Ind. (London)*, p. 1205. Presence of polynuclear hydrocarbons in ciagarette smoke.

Daudel, P., and Daudel, R. (1949). *Bull. soc. chim. biol.* **31,** 353. Possibility of formation of a complex between carcinogenic compounds and the tissues subjected to their action.

Daudel, R. (1946). *Rev. sci.* **84,** 37.

Dixon, M., and Webb, E. C. (1958a). "Enzymes," p. 5. Academic Press, New York.

Dixon, M., and Webb, E. C. (1958b). "Enzymes," p. 344. Academic Press, New York.

Dixon, M., and Webb, E. C. (1958c). "Enzymes," p. 351. Academic Press, New York.
Dixon, M., and Webb, E. C. (1958d). "Enzymes," p. 476. Academic Press, New York.
Elliott, A., and Hanby, W. E. (1958). *Nature* **182**, 654. Deuterium exchange in poly-peptides.
Fitzhugh, A. (1953). *Science* **118**, 783. Malignant tumours and high polymers.
Forsyth, W. G. C., and Quesnel, V. C. (1957). *Biochim. et Biophys. Acta* **25**, 155. Inter-mediates in the enzymic oxidation of catechol.
George, P. (1953). *Biochem. J.* **54**, 267. The chemical nature of the second hydrogen peroxide compound formed by cytochrome c peroxidase and horseradish peroxi-dase. 1. Titration with reducing agents.
George, P., Ingram, D. J. E., and Bennett, J. E. (1957). *J. Am. Chem. Soc.* **79**, 1870. One-equivalent intermediates in phthalocyanine and porphin oxidations investi-gated by paramagnetic resonance.
Ghosh, D. K., and Whiffen, D. H. (1958). *Faraday Soc. Symposium*, Sheffield, Eng-land. Electron resonance study of the free radicals trapped in a single crystal of glycine on γ-irradiation.
Ghosh, D. K., and Whiffen, D. H. (1959). *Mol. Phys.* **2**, 285. The electron spin reso-nance spectrum of a γ-irradiated single crystal of glycine.
Gibson, J. F., Ingram, D. J. E., and Schonland, D. (1958). *Discussions Faraday Soc.* **26**, 72. Magnetic resonance of different ferric complexes.
Gordy, W., and Shields, H. (1956a). *Bull. Am. Phys. Soc.* **1**, 199. Electron-spin res-onance in X-irradiated proteins.
Gordy, W., and Shields, H. (1956b). *Bull. Am. Phys. Soc.* **1**, 267. Effects of tempera-ture and isomeric structure on the electron-spin resonance of X-irradiated amino acids.
Gordy, W., Ard, W. B., and Shields, H. (1955). *Proc. Natl. Acad. Sci. U. S.* **41**, 983. Microwave spectroscopy of biological substances. I. Paramagnetic resonance in X-irradiated amino acids and proteins.
Griffith, J. S. (1958). *Discussions Faraday Soc.* **26**, 81. The electronic structures of some first transition series metal porphyrins and phthalocyanines.
Gutfreund, H. (1954). *Discussions Faraday Soc.* **17**, 220.
Gutfreund, H. (1955). *Discussions Faraday Soc.* **20**, 167. Steps in the formation and decomposition of some enzyme-substrate complexes.
Haber, F., Willstätter, R. (1931). *Ber.* **64**, 2844. Unpaarigkeit und Radikalketten im Reaktionsmechanismus organischer und enzymatischer Vorgänge.
Harden, A., and Young, W. J. (1906a). *Proc. Roy. Soc.* **B77**, 405. Alcoholic ferment of yeast juice.
Harden, A., and Young, W. J. (1906b). *Proc. Roy. Soc.* **B78**, 369. Co-ferment of yeast juice.
Hvidt, Å., and Linderstrøm-Lang, K. (1954). *Biochim. et Biophys. Acta* **14**, 574. Ex-change of hydrogen atoms in insulin with deterium atoms in aqueous solutions.
Hvidt, Å., and Linderstrøm-Lang, K. (1955a). *Biochim. et Biophys. Acta* **16**, 168. The kinetics of the deuterium exchange of insulin with D_2O. An amendment.
Hvidt, Å., and Linderstrøm-Lang, K. (1955b). *Compt. rend. trav. lab. Carlsberg. Ser. chim.* **24**, 385.
Ingram, D. J. E. (1958a). "Free Radicals as Studied by Electron Spin Resonance," p. 151 and ff. Butterworths, London.
Ingram, D. J. E. (1958b). "Free Radicals as Studied by Electron Spin Resonance," Chapter 9. Butterworths, London.
Ingram, D. J. E., and Bennett, J. E. (1955). *Discussions Faraday Soc.* **19**, 140. Para-magnetic resonance in phthalocyanine, haemoglobin and other organic deriva-tives.

Katz, E. (1949). "Photosynthesis in Plants," p. 291. Iowa State College Press, Ames, Iowa.

Keilin, D. (1925). *Proc. Roy. Soc.* **B98**, 312. (This is the first of Keilin's classic papers on the cytochromes.) Cytochrome, a respiratory pigment common to animals. yeast and higher plants.

Kuhn, R., and Ströbele, R. (1937). *Ber.* **70**, 753. Über den mechanimus der Oxydationsvorgänge.

Kuhn, R., and Wagner-Jauregg, T. (1944). *Ber.* **67**, 361. Über das Reduktions-Oxydations-Verhalten und eine Farbreaktion des Lactoflavins (Vitamin B_2).

Leach, S. J. (1954). *Advances in Enzymol.* **15**, 1. The mechanism of oxido-reduction.

Le Clerc, A.-M., Moudy, J., Douzou, P., and Lissitzky, S. (1959). *Biochim. et Biophys. Acta* **32**, 499. Mise en evidence pour oxydation photochimique ou enzymatique du radical libre du catéchol.

Lenormant, H., and Blout, E. R. (1953). *Nature* **172**, 770. Origin of the absorption band at 1550 cm.$^{-1}$ in proteins.

Lipkin, D., Paul, D. E., Townsend, J., and Weissman, S. F. (1953). *Science* **117**, 534. Observations on a class of free radicals derived from aromatic compounds.

Loewus, F. A., Vennesland, B., and Harris, D. L. (1955). *J. Am. Chem. Soc.* **77**, 3391. The site of enzymic hydrogen transfer in diphosphopyridine nucleotide.

Lyons, M. J., Gibson, J. F., and Ingram, D. J. E. (1958). *Nature* **181**, 1003. Free radicals produced in cigarette smoke.

McCormick, G., and Gordy, W. (1956). *Bull. Am. Phys. Soc.* **1**, 200. Electron-spin resonance in X-irradiated peptides.

McIlwain, H. (1946). *Nature* **158**, 898. The magnitude of microbial reactions involving vitamin-like compounds.

Mackinnon, D. J., and Waters, W. A. (1953). *J. Chem. Soc.* p. 323. The possible significance of Fenton's reaction in relation to oxidations effected by enzyme systems.

Michaelis, L. (1939). *Cold Spring Harbor Symposia Quant. Biol.* **7**, 33.

Michaelis, L. (1951). *In* "The Enzymes" (J. B. Sumner and K. Myrbäck, eds.), Vol. 2, Part 1, p. 1. Academic Press, New York. Theory of oxidation-reduction.

Moelwyn-Hughes, E. A. (1937). *Ergeb. Enzymforsch.* **6**, 23. The kinetics of enzyme reactions.

Nash, T. (1957). *Nature* **179**, 868. Chemical carcinogenesis.

Oppenheimer, B. S., Oppenheimer, E. T., Stout, A. P., and Danishefsky, I. (1953a). *Science* **118**, 305. Malignant tumors resulting from embedding plastics in rodents.

Oppenheimer, B. S., Oppenheimer, E. T., Stout, A. P., Danishefsky, I., and Eirich, F. R. (1953b). *Science* **118**, 783. Malignant tumors and high polymers.

Park, H. F. (1950). *J. Phys. & Colloid Chem.* **54**, 1383. The role of the unpaired electron in carcinogenesis.

Pullman, A., and Pullman, B. (1946). *Rev. sci.* **84**, 145.

Pullman, M. E., San Pietro, A., and Colowick, S. P. (1954). *J. Biol. Chem.* **206**, 129. On the structure of reduced diphosphopyridine nucleotide.

Rexroad, H., and Gordy, W. (1956). *Bull. Am. Phys. Soc.* **1**, 200. Electron-spin resonance in X-irradiated hormones and vitamins.

Schwab, G.-M., Rosenfeld, B., and Rudolph, L. (1933). *Ber.* **66**, 661. Zur Frage des Kettencharakters der Katalasewirkung.

Shields, H., Ard, W. B., and Gordy, W. (1956a). *Bull. Am. Phys. Soc.* **1**, 200. Electron-spin resonance in X-irradiated sugars and cellulose fibers.

Shields, H., Ard, W. B., and Gordy, W. (1956b). *Nature* **177**, 984. Microwave detection of metallic ions and organic radicals in plant materials.

Sogo, P. B., Pon, N. G., and Calvin, M. (1957). *Proc. Natl. Acad. Sci. U. S.* **43**, 387. Photo-spin resonance in chlorophyll-containing plant material.

Stare, F. J. (1935). *J. Biol. Chem.* **112**, 223. A potentiometric study of hepatoflavin.

Stern, K. G. (1934). *Biochem. J.* **28**, 949. Potentiometric study of photoflavin.

Stock, J. T. (1944). *J. Chem. Soc.* p. 427. The reduction of quinaldinic acid at the dropping-mercury cathode.

Swain, C. G., and Brown, J. F. (1952). *J. Am. Chem. Soc.* **74**, 2538. Concerted displacement reactions. VIII. Polyfunctional catalysis.

Swallow, A. J. (1955a). *Nature* **176**, 793. Reduction of riboflavin to a stable free radical using X-rays.

Swallow, A. J. (1955b). *Biochem. J.* **60**, 443. The reduction of diphosphopyridine nucleotide by sodium dithionite.

Szent-Györgi, A. (1941). *Science* **93**, 609. Towards a new biochemistry?

Tollin, G., and Calvin, M. (1957). *Proc. Natl. Acad. Sci. U. S.* **43**, 895. The luminescence of chlorophyll-containing plant material.

van Roggen, A., van Roggen, L., and Gordy, W. (1956). *Bull. Am. Phys. Soc.* **1**, 266. Electron-spin resonance in X-irradiated single crystals of amino acids.

Venkataraman, B., and Fraenkel, G. K. (1955). *J. Am. Chem. Soc.* **77**, 2707. Proton hyperfine interactions in paramagnetic resonance of semiquinones.

Vennesland, B. (1955). *Discussions Faraday Soc.* **20**, 240. Some applications of deuterium to the study of enzyme mechanisms.

von Euler, H., Albers, H., and Schlenk, F. (1936). *Z. physiol. Chem., Hoppe-Seyler's* **240**, 113. Chemical investigations on highly purified cozymase.

Warburg, O., and Christian, W. (1931). *Biochem. Z.* **242**, 206. Activation of the Robinson hexosemonophosphoric acid ester in the red blood cells and the method for preparation of activating enzyme solutions.

Warburg, O., and Christian, W. (1932). *Biochem. Z.* **254**, 438. A new oxidation enzyme and its absorption spectrum.

Warburg, O., and Christian, W. (1933). *Biochem. Z.* **266**, 377. The yellow enzyme and its functions.

Warburg, O., and Christian, W. (1936). *Biochem. Z.* **287**, 291. Pyridine, the hydrogen transferring component of the fermentation enzymes.

Waters, W. A. (1943). *Trans. Faraday Soc.* **39**, 140. A chemical interpretation of the mechanism of oxidation by dehydrogenase enzymes.

Waters, W. A. (1948). "The Chemistry of Free Radicals," 2nd ed., Chapter 12. Oxford Univ. Press, London and New York. Some possible mechanisms for biological processes.

Westheimer, F. H. (1954a). *In* "The Mechanism of Enzyme Action" (W. D. McElroy and B. Glass, eds.), p. 321. Johns Hopkins Press, Baltimore, Maryland. "One-electron" and "two-electron" oxidation-reduction reactions in inorganic and organic chemistry.

Westheimer, F. H. (1954b). *In* "The Mechanism of Enzyme Action" (W. D. McElroy and B. Glass, eds.), p. 356. Johns Hopkins Press, Baltimore, Maryland.

Wieland, H. (1921). *Ber.* **54**, 2353. Mechanism of oxidation processes.

Wieland, H. (1931). *J. Chem. Soc.* p. 1055. Recent researches on biological oxidation.

Wood, J. L., and Fieser, L. F. (1940). *J. Am. Chem. Soc.* **62**, 2674. Sulfhydryl and cysteine derivatives of 1,2 benzanthracene, 10-methyl-1,2-benzanthracene and 3,4-benzpyrene.

Yarmolinsky, M. B., and Colowick, S. P. (1954). *Federation Proc.* **13**, 327. Mechanism of DPN reduction by dithionite (hydrosulfite).

Yarmolinsky, M. B., and Colowick, S. P. (1956). *Biochim. et Biophys. Acta* **20**, 177. On the mechanism of pyridine nucleotide reduction by dithionite.

16. Chemical Utilization of Trapped Radicals

D. E. CARR

Phillips Petroleum Company,
Bartlesville, Oklahoma

I. Introduction

This is a difficult chapter to write. The reasons will be obvious to all my brother applied scientists to whose lot such an assignment might fall.

In the first place there is literally nothing in the literature which specifically covers the subject of utilization of trapped free radicals in the sense in which we are considering free radicals: namely, inherently active fragments, possessing unpaired electrons and of low molecular or atomic weight. Learning how to generate such species, stabilizing them in trapped form in some way or other, identifying them, and (hopefully) assaying their concentrations or proportions in at least a semiquantitative manner, has represented a colossal job. Certainly it has, even for the more amenable and docile species, not been completed; it has, in fact, been barely started. It is a job which, to a large degree, has been carried out, and justifiably so, by physicists. Although the chemists stand hungrily by and await their turn, they are understandably shy about manipulating extremely active pieces of matter whose compositions have not yet been definitely established, handed over to them in inconvenient and complicated combinations of very cold hardware. A chemist is somewhat less than enthusiastic when notified by the physicist that, for his edification and pleasure, he can play about with a bit of glasslike stuff (deposited on a rod held at 4.2°K in the depths of a triply Dewared cryostatic stronghold), believed but not certified to contain from 0.5 % to 1.7 % N atoms (but maybe also some NH, NH_2, N_3 and possibly a trace of H).

Of course, this is not to say that chemists have not been cognizant of, or working with, such species for many years. However, their work has been once or twice removed from the manipulation of trapped or stabilized free radicals of the kind we are discussing. Chemists produce such species *in situ* in various reaction sequences and follow their behavior vis-à-vis molecules or other free radicals. There is an enormous literature on this kind of work, including excellent books of broad scope by Steacie (1954), Trotman-Dickenson (1955), Walling (1957), etc.

In this chapter I shall use such references only in a special way. What I shall try to do is to speculate on what experiments an applied chemist would *like* to perform, *if* he had the small fragments in usable and manipulatable condition, well-tagged and assayed, available, so to speak, from a shelf in the icebox. To carry out such a series of speculations I shall attempt to select (1) chemical areas which are of huge practical importance and where theoretical elucidation is so far behind engineering application that more precise study of the free radical conversions known to be involved would aid in better understanding and better engineering control (combustion is an example) and (2) areas where "new chemistry" might evolve from well-planned exploitation of the available stabilized species (a particularly inviting area here is the chemistry of "hot" molecules, resulting from free radical recombinations).

When one enters the arena of speculation, my colleagues in applied chemistry, especially those who work, as I do, for profit-making organizations, will immediately recognize a second dilemma in composing such a chapter as this. Are one's speculations, presuming them to be reasonable and ultimately reducible to practice, "proprietary"? I have not dodged this issue, but I have included as ideas of my own only the more obvious and general extrapolations of work already performed in laboratories. The personal ideas propounded are syntheses from many areas and are invitations to research rather than implied inventions.

The techniques of generating and trapping free radicals have been ably handled in other chapters of this book. In this section I shall content myself with only the following comments which apply particularly to the utilization aspects.

II. The Generation of Free Radicals in Trappable Form

A. PHOTOLYSIS

Photochemical production of free radicals is an attractive route for industrial research workers, since it has recently become a fairly well understood technique and has been engineered to the point of commercial application. It is "clean"; the yields on the basis of quanta absorbed are predictable; high temperatures or pressure are not generally involved, nor is elaborate electronic circuitry required.

B. IONIZING RADIATION

One must distinguish a notable difference in the mechanism of free radical generation by powerful radiation such as gammas, betas, or neutrons, and by light in the visible or ultraviolet. Although the over-all results are often equivalent in terms of chemical conversion, the complicative variable of preliminary ionization enters. This is not present in photochemistry except in the case of materials of very low ionization potential, such as the alkali metals.

Thus in free radical-producing and -trapping programs we are, in the case of high energy radiation, dealing with "dirtier," more complex, systems and must familiarize ourselves with the rapidly accelerating, relatively new young giant "radiation chemistry." Here again the literature is becoming prodigious. The recent increased availability of radioactive materials such as cobalt-60 has made everyone with the inclination a practicing radiation chemist with a paper in the making.

Ionic reactions are extremely fast. Stevenson (1951) has observed that at reasonable radiation rates such ions could undergo more than 10^8 reactions with nonionized molecules or radicals before being neutralized by electrons. The possibility must be therefore borne in mind, even at pressures much higher than those used in mass spectrometers, that the transient appearance of ionic species may exert a perturbing effect on the course of free radical generation and trapping.

The method of alkyl radical preparation by irradiation of metal alkyls has a good deal of attraction because of its simplicity. It is disappointing that so little parallel work has been done with H-atom generation from metal hydrides, which obviously should be irradiated, however, at much lower temperatures.

An important source of information on preparation of free radicals by irradiation in the solid state is represented by industrial research teams working on individual government contracts. Aerojet-General Corporation, for example, has used a 1000-curie cobalt-60 facility (Baum *et al.*, 1956) to study free radical generation in conjunction with a contract with the Office of Scientific Research of the A.R.D.C.[1] Phillips Petroleum Company under contract with the same agency has utilized strong gamma irradiation from spent fuel elements at the Atomic Energy Commission's Materials Test Reactor, Idaho Falls, Idaho.

Although these and other contractual projects, originally set up as studies to explore the feasibility of preparing ultrahigh energy fuels for aircraft propulsion, have failed in their *ad hoc* objectives, interesting techniques and findings have come out of the activity. For example, the Phillips work showed that, in the case of crystalline ammonia at 4°K, the fraction of the absorbed energy which is stored appears to be independent of the gamma field intensity and of the total amount absorbed over the narrow ranges explored (range of intensities of 0.02 to 0.08 milliwatt per gram and total energy from 1.5 to 3 joules per gram). For this molecular system the stored energy is about 4% of that absorbed and is released at quite low temperatures. The energy release starts a few degrees above the boiling point of helium and is complete at 50°K.

Qualitatively, glassy solids appear to be able to store energy better than crystalline solids, at least as far as temperature of storage is concerned. Thus a glass with ammonia as the principal component will trap and store radicals at the temperature of liquid nitrogen whereas crystalline ammonia will not.

From the standpoint of practical future engineering programs on the utilization of trapped radical species, the radiation approach is a likely one. It shares with photochemistry the advantage of being wafted along by a rapidly developing technological know-how. It is probable that we will see many intrepid variations of the basic process, such as a combination of ultraviolet and gamma irradiation, the use perhaps of sonic or ultrasonic energy fields in conjunction with ionizing radiation (especially where solids or viscous liquids are exposed), and the application of effective sensitizers or energy-transfer agents to achieve certain specific free radical generation effects.

C. ELECTRICAL DISCHARGE

The interest of industrial research laboratories in applying electrical discharge methods is well exemplified by the very sophisticated laboratory

[1] Air Research and Development Command.

technique used by the Union Carbide Research Institute (1958). A curious and perhaps very significant fact to come out of this work was that atomic hydrogen was not produced as readily from highly purified molecular hydrogen as from ordinary cylinder hydrogen. The addition of less than 0.1 % N_2 to the highly purified gas also increased the atomic hydrogen yield. It is evident that in discharge techniques as in photochemical and radiation chemical methods, we may have to consider some kind of energy transfer or "sensitizing" reactions of a type not hitherto recognized and certainly not understood.

D. THERMAL OR CATALYTIC DECOMPOSITION

From an applied standpoint pyrolysis is still very much in the running, especially if appropriately simple and available molecular feed stocks can be used, followed by quenching and trapping devices that make economic sense. Although it is interesting from a scientific and a laboratory standpoint to decompose metal alkyls, for example, to obtain alkyl radicals, it is not going to make anybody much money—except insofar as such investigations help in the design of better metal alkyls for specific industrial purposes, or better to understand the behavior of metal alkyls for such purposes.

The same considerations apply to such interesting sources of free radicals as hydrazine. Rather than decompose N_2H_4 to various radicals, the industrial chemist would prefer to find out how to recombine active species from ammonia to form hydrazine. The pyrolysis of materials such as hydrogen, ammonia, nitrogen, CO_2, H_2O, and gaseous hydrocarbons to prepare active species represents a practical challenge which certain industrial research organizations are known to have heeded. However, a good deal of the pertinent information recently obtained in this field is undoubtedly regarded as proprietary, and its public release can be expected to await the establishment of patent positions.

E. FREE RADICAL RECOMBINATION

In reviewing free radical recombination reactions, a key study is that of Fontana (1958), who found that the nitrogen atom content of a deposit in an N_2 matrix may bear no relation, or possibly a very unhappy *reverse* relation to the concentration of N atoms in the gas as it approaches the condensing device.

In other words, at appreciable free radical concentrations in the gas phase, the deposition process itself results in accelerated recombination. There thus appears to be a serious need for much more work on deposition dynamics. If the relationships which Fontana observed prove to be general for most free radicals and matrix systems of interest, the gas phase generation of free radicals may prove technically unprofitable as a means of obtaining stable trapped species in reasonable concentrations. One might then wish to investigate more intensively the techniques involving the irradiation of solidified molecules. Most likely, however, in view of the serious experimental complexities of the latter approach, chemists would be content to study continuous flow systems in which a gas stream containing free radicals is placed directly in appropriate contact with another reactant. In effect, this would mean the abandonment of the subject of this book as a serious factor in practical chemical processing. In view of this criticality, it is obvious that all conceivable modes of attacks on the deposition-trapping mechanism should be examined to be certain no fruitful route has been overlooked. Unquestionably this is one of the most vital areas of free radical research and deserves more attention than it now appears to be receiving.

Assuming that this evidently grave problem can be solved, one would like to know more about recombination kinetics in the matrix mixture. Is there any temperature stage, for example, in the system where the three body mechanisms, common to gas phase recombination of free radicals, come into play? As Hinshelwood (1956) has pointed out, the efficiency with which the third body performs its function is not well understood, even in gaseous reactions. Pickup and Trapnell (1956) have concluded that in H atom recombination, where electron factors are operative, the rate-determining step in metal-containing systems must involve chemisorption of H on a metal surface. Is there anything in the matrix-free radical system analogous to chemisorptive bonding? Certainly the forces involved deserve very careful scrutiny.

III. Utilization of Trapped Free Radicals in the Study of Catalytic and Noncatalytic Conversions

A. General

In surveying the utility of active free radical species for reaction studies the industrial chemist will naturally think along two lines. If these materials are available from the physicist's cold storage, the chemist obviously will want to use them to test the validity of reaction patterns which have been constructed theoretically with such species as intermediates or as initiators. This I fear will not be his first desire, however. He will doubtless want to "play around" with the radicals and maybe come up with a new product or a new synthesis. He will have in mind the possibility that the free radicals may function as trigger mechanisms to induce chain propagations of a favorable kind. In radiation chemistry parlance, he will look for a big "G value" from these precious fragments.

It is therefore now appropriate to focus attention on the possibility of reaction families in which trigger effects are likely or are most in demand.

Before setting sail, however, on a sea of speculation or at least of intense extrapolation, it is well to review some of the reactions already studied with trapped radicals under difficult conditions. Without any particular regard to relative importance or chronology, the results which are mentioned in the following paragraphs may establish certain generalizations, and also may serve as starting points for further experiments of a more hypothetical nature.

The chemical reactions of O, H, OH, and N in frozen form were studied by Ruehrwein (1958). Reactions occurred below about 30°K. A 59 % conversion of oxygen to ozone was obtained in the O_2-O system, which compares very favorably with a standard laboratory ozonizer. Using other techniques, conversions of nearly 100 % were obtained. NO_2 gave 64 % conversion to ozone. An important point is that *in no experiment was there found a product of a reaction between a radical and a stable diagmagnetic molecule.*

H atoms react with solid oxygen at 20°K (Klein and Scheer 1959), giving on warm-up water and H_2O_2. The primary process is believed to be

$$H + O_2(S) \rightarrow HO_2.$$

It is interesting in this connection to compare the room temperature reactions of a beam of atomic H with molecular oxygen (Foner and Hudson 1955). Substantial quantities of OH radicals were formed, but no HO_2 was detected and no H_2O_2 resulted. The OH generation declined to zero when the temperature of the reactor was lowered to 77°K.

Although the color change to blue of MoO_3 is a useful detector for H atoms and for free atoms in general at temperatures close to 0°C, no reaction with H was found by Windsor (1959) at dry-ice temperatures.

The reaction of H atoms with solid olefins has been studied in a very constructive way by Klein and Scheer (1958). This gives promise of an elegant and precise method of preparing stabilized alkyl radicals in sizable concentrations.

Significant early work (Geib and Harteck, 1933, 1934) on the addition of H and O atoms at −190°C and −183°C, respectively, to a variety of molecular species should be reviewed in the light of present theory and practice. These experiments apparently established the fact that H does not react with HCN, N_2O, or H_2, although some incompletely defined reactions are indicated with SO_2, C_2H_2, C_2H_4, and C_6H_6. Of particular interest are the reactions of H with O_2 to form H_2O and OH and of H with O_2 and CO to give CO_2 and OH. HO_2 can also be formed.

B. HYDROGENATION

In petroleum refining as in coal liquefaction, a major problem is that of hydrogenative up-grading. Under this very broad term are included a number of processes differentiated mainly by the severity of pressure and temperature conditions and by the amount of hydrogen actually introduced into the molecular structures of the raw material. In practically all such processes catalysts are used. Hydrogen can be employed to modify profoundly the over-all structure by a combination of thermal cracking and saturation of the fragments ("hydrocracking"). It can be used at lower consumption levels to remove sulfur and nitrogen compounds and undesirable unsaturated hydrocarbons ("hydrotreating"); it can even be used with great benefit in various isomerization or even dehydrogenation reactions where hydrogen is not consumed ("hydroforming," "platforming," etc.).

It seems to be accepted that practically all catalytic reactions in which hydrogen is used involve H atoms in some conversion sequence connected with the functioning of the catalyst. This is probably true even in the so-called "hydrogen transfer" reactions, where a hydrogen donor, such as tetralin, may give up hydrogen under appropriate conditions to saturate an undesirable fraction of the feed stock, without the intercession of externally introduced hydrogen.

In all such catalytic conversions, hydrogen plays one particularly vital role which is not very clearly understood. This is to reduce the rate of formation of coke or carbon, which tends rapidly to deactivate the catalyst. If chemists could better understand this process, it would doubtless be possible to make more efficient catalysts or to make catalysts last much longer. Here, it would seem, is an area in which trapped H atoms could be profitably applied by the investigator to explore the mechanism of coke or carbon scavenging.

At some very critical stage in the pyrolysis of a hydrocarbon, active fragments are evidently formed, possibly C_2 or other hydrogen-poor radicals, which are on their way to lay down macroscopic cokes. Unquestionably the intervention of hydrogen in the form of H atoms, probably chemisorbed on the catalyst surface, interrupts the carbonization chain.

If one could, under cryogenic conditions, transfer trapped H atoms from a stable matrix to a catalyst surface without excessive recombination, it might be possible to study the coke-forming process and the coke-inhibiting process in "slow motion," so to speak, at low temperatures. Various hydrocarbon skeleton fragments, including CH, C_2, possibly C_3, or perhaps active and carbon-forming molecules such as C_2H_2 could be introduced into the system and studied at various levels of catalyst H atom concentration. Perhaps the catalyst itself could be used as a trapping matrix.

Indeed it is not inconceivable that catalysts which at low temperatures chemisorb H atoms, formed externally rather than by extremely rapid high temperature events, would find themselves permanently modified and improved, when put to work at higher temperature. Perhaps H atoms diffuse through the pore structure more thoroughly under these conditions. So little basic information is in hand concerning the carbonization of catalysts or the role of hydrogen in conjunction with the catalyst surface chemistry that anything we could learn in this dark chamber of technical ignorance would be expected to help us.

The fact that molybdenum trioxide is affected above a critical temperature level by H atoms was mentioned above. It is significant that molybdenum-containing catalysts are among the most widely used in hydrogen conversion processes.

To my knowledge there has been, so far, very little indication of an approach in modern free radical chemistry to these massive industrial chemical problems. The above-mentioned excellent work of Klein and Scheer on H atom saturation of olefins is virtually a pioneering expedition.

Another research situation where the availability of trapped free radicals, in this case CH_2 as well as H, might have been exploited with profit is the hydrocracking investigation on 5- and 6-membered cyclanes conducted by Shuikin et al. (1953). A melange of compounds from methyl cyclopentane, including cyclopentane, pentane, hexane, isohexanes, benzenes, etc., were obtained. The intermediate formation of CH_2 was regarded as a crucial step in the sequence of reactions. The availability of trapped CH_2 along with H would undoubtedly go a long way to clarify this and other hydrocracking mechanisms.

Norrish and Purnell (1958) in their study of the reaction of H atoms with n-hexane under mercury-sensitized photolytic conditions found that the reaction $C_6H_{14} + H \rightarrow C_6H_{13} + H_2$ has a low activation energy (6 kcal). The hexyl radical may polymerize or may crack to lower hydrocarbons. If trapped H atoms had been applied, without the complication of the presence of mercury and of possible photocatalyzed side reactions, it could be argued that a different kind of reaction sequence might have been obtained, possibly isomerization and even cyclization. Such possibilities greatly intrigue the industrial research chemist.

C. DEHYDROGENATION

The abstraction of H atoms from hydrocarbons by other free radicals represents a kind of reaction in which study by means of trapped species could bear important fruit in hydrocarbon conversion technology. The dehydrogenation of butane to butenes and to butadiene is one of the most crucial commercial processes in world chemistry. So also is the dehydrogenation of cycloparaffins to aromatics.

The work of Norrish and Purnell on abstraction of H by external H itself has been noted above. Another recent investigation (James and Steacie, 1958) includes the use of ethyl radicals alternatively to abstract hydrogen from olefins, forming an intermediate allyl free radical, followed by addition of ethyl to form a branched-chain olefin or by the alternative route of direct addition of the ethyl radical to the double bond. Here again the situation is complicated by the fact that ethyl radicals are prepared *in situ* by the photolysis of diethyl ketone. Availability of trapped ethyls would presumably have provided a more clean-cut and firmer insight into this important series of reactions.

One is not sure actually that the availability of externally produced and trapped free radicals would not in such cases introduce propagating chains of considerable length so that the trapped ethyl would function in a true trigger sense. The same might be true of hydrogenation. Although there is no evidence, except in the fields of polymerization, sulfoxidation, and combustion that long-chain or branched-chain propagations can be started by the introduction of free radicals, there is certainly no conclusive evidence on the other hand that they cannot be. This is a highly exploratory area where there has simply not been enough work done. If by a few key experiments, the use of trapped free radicals could be shown to start such chain reactions in hydrocarbon conversions of technical interest—chains not producible by the generation of free radicals *in situ*—there would be an immediate and eager invasion of the field by industrial investigators. It is hoped that the possibility of such initiating processes will be exhaustively examined.

D. CRACKING

The hydrocarbon cracking process can be said to have twice revolutionized the petroleum industry, first in the form of thermal decomposition and second in the application of catalytic equipment. Pyrolytic processes will always be of huge interest to industrial chemists and engineers. It is thus useful to review briefly in what way they could be affected by the application of free radical techniques.

Decompositions involving the intercession of free radical species are unquestionably encountered in many technical areas not directly associated with refining or petrochemical technology. This is particularly true in combustion and will be discussed in a separate section concerned with that monstrous complex of problems.

Next to the decomposition of hydrogen itself to form H atoms, supposedly the simplest cracking process would be the decomposition of methane. That this is not so simple as might be hoped is indicated by conflicting results from various investigators on the effect of hydrogen on pyrolysis. Hill (1956) has recently reviewed this confused picture and believes that the initiating steps arise from the generation of H atoms and subsequent abstraction by them of hydrogen atoms from the methane. Since at various stages of methane decomposition such radicals as CH, CH_2, CH_3, C_2 and possibly C_3 undoubtedly also come into the picture, it would be instructive to attempt to study "cold cracking" of CH_4 by exposing methane to such species, including H, in trapped form. From other than the heuristic standpoint, the interesting possibility that such industrially significant processes as conversion in high yield of methane to acetylene might be streamlined and rendered more efficient cannot be ignored.

It is technically interesting to note that the pregeneration of H atoms by exposing molecular hydrogen to an electric arc and impinging the stream on lower hydrocarbons such as propanes, has been patented by the European firm of Knapsack-Griesheim (1957). As high as 85 % acetylene yields are claimed. [This technique should not be confused with the Schoch (1950) process, in which the hydrocarbon feed itself is exposed to an arc. An excellent interpretation of the Schoch-arc reactions has been presented by Burton and Magee (1955) in which the concept of successive excitation by low energy electrons is introduced.]

One of the most complete theoretical treatments of the cracking of the lower paraffins is that of Stepukhovich (1954). By using azomethane as a cracking initiator, he has studied the role of the $C_2H_5 \rightarrow C_2H_4 + H$ equilibrium in chain propagation. The cracking of propane and i-butane appears, from his work (1955 and 1956) indubitably to be initiated by H atoms. The addition of a simple diene, such as allene, strongly inhibits such initiation.

Of salient importance is the fact, however, that the presence of dienes does not suppress the cracking of n-butane, which in Stepukhovich's opinion is initiated by the CH_3 free radical. In an ingenious application of deuterated methyl (CD_3) produced from the photolysis of acetone-d_6, Gordon and McNesby (1957) have shown that although CD_3 may add to the double bond of butenes or abstract alpha H atom, it also initiates cracking reactions, leading to CH_3 evolutions.

These apparent specificities are of the highest degree of interest, and study of them could undoubtedly be benefited by application of free radical trapping techniques. In particular, the use of deuterated radicals under cryogenic conditions may be helpful in unraveling the more complicated reaction patterns concerned.

E. ISOMERIZATION

Very little work relating hydrocarbon isomerization to free radical chemistry can be cited, at least in a manner that suggests constructive invitations to research with the use of trapped radicals. In this, as in the field of hydrocarbon alkylation, the "carbonium ion" theory has captured the fancy of hydrocarbon chemists.

However, there are signs that H atoms may play a crucial role in alkane isomerization if dehydrogenation and hydrogenation sequences are involved such as have been recently suggested by Swegler and Weisz (1957). They picture the isomerization of *n*-hexane to isohexanes over a platinum-type catalyst to proceed through the following mechanism: (1) dehydrogenation of *n*-hexane to *n*-hexene over individual platinum sites, (2) isomerization of *n*-hexene to *i*-hexene over acidic sites on the catalyst support, and (3) hydrogenation of the *i*-hexene to *i*-hexane, again over the platinum sites. The intercession of dehydrogenation-hydrogenation coupling again suggests a "slow motion picture" study by the use of trapped H atoms, using preferably the catalyst itself as a cold matrix. (It should be emphasized that such isomerizations commercially are carried out in the presence of hydrogen.)

Entropy changes in isomerization processes are especially instructive and have been reviewed recently by Gascoigne (1958). In view of the possible transient existence of small ring hydrocarbon molecules and free radicals in certain isomerization reactions, the excellent work of Roberts and his associates (1952) in this very difficult field deserves a good deal of attention. The use of cryogenic techniques for stabilizing such interesting intermediates will doubtless become standard practice, as familiarity with this technique grows.

F. ALKYLATION AND OTHER ADDITION REACTIONS

In considering alkylation-type hydrocarbon reactions we enter a field which can best be broadened to include addition reactions of various kinds, insofar as they have to do with free radical mechanisms. This is a tremendous area of study, and as before the literature must be selectively examined for references of the likeliest applicability to the use of trapped radical manipulations.

In references already cited it has been reported that there exist alternative routes for the reaction of free radicals and of atoms with various molecules. In the important work of Scheer and Klein the cryogenic reaction of H with olefins follows the stepwise addition pattern. At higher temperatures H atoms may act as dehydrogenating agents. Alkyl radicals may similarly play bifunctional roles, that of alkylation or of hydrogen abstraction or even cracking initiators. Let us look at the addition reactions.

In a notable series of experiments Szwarc and his co-workers (1957) have studied the addition of methyl radicals to *cis* and *trans* isomers of olefins. The two reactions proceed evidently through different transition states and the differences in reactivities for the *cis* and *trans* isomers are accounted for on the basis of differences in resonance energy in the respective transition states. Szwarc has even attempted to decide from which direction the radical approaches a double bond and has tentatively suggested that the most probable approach is along the double-bond axis. In view of the work of Klein and Scheer, it would appear obvious that stereospecific alkylations of this type could more precisely be studied cryogenically, using either the olefin or the alkyl radical (or both) in a stabilized matrix.

Such experiments are of diagnostic value, but in view of the essential lack of triggering effects (unless polymerization of the olefin is initiated) they offer little hope for commercializing a process based on trapped radicals. The scientific value, however, is so considerable for obtaining a better insight into addition mechanisms that enlightened investigators will wish to pursue the subject further.

The reactions of the methylene radical, CH_2, have been studied in recent years by a number of very capable investigators. Methylene is a popular radical since it can conveniently be generated from diazomethane or by the photolysis of ketene. The "carbene" groups, as Urry (1959) generically designates methylene and its free radical derivatives, have been shown by him and Buttery et al. (1956) to react additively with n-pentane to give a statistical distribution of hexane and isohexanes at 80 to 70°C:

$$C-C-C-C-C + CH_2 \rightarrow C-C-C-C-C-C$$
$$\begin{matrix} & & & & & C & & & & C \\ & & & & & | & & & & | \\ & & & & + C-C-C-C-C & + & C-C-C-C-C \end{matrix}$$
$$(49\%) \qquad\qquad (34\%) \qquad\qquad (17\%)$$

Knox et al. (1958), studying the reactions of methylene (from ketene photolysis) with isobutene, found that addition of CH_2 to the double bond is several times more rapid than insertion into the C—H bonds. It is remarkable that the activation energy for the olefin reaction is greater than for the reaction of methylene with a saturated hydrocarbon. The absolute values involved are quite low, and Knox believes them to be determined by steric requirements rather than by the strengths of the bonds broken or formed. The high reactivity of the double bond appears to be due to a high entropy of activation, not to a low energy of activation. This is a crucial point and suggests that at cryogenic temperatures, using trapped systems, repetition of such methylene reactions might throw much needed light on the badly confused criteria of reactivity. It is becoming all too easy to blame reversals of reaction tendency vis-à-vis activation energy on the "pre-exponential factor" without defining exactly what this signifies and how it may help us in some way to better predict reactivity.

A fascinating phase of the CH_2-isobutene reaction mentioned is the presence in the product (and presumably functioning as an intermediate to the "alkylates") of the energetic molecule dimethylcyclopropane. The existence of a transitory cyclopropane molecule has been noticed by Frey and Kistiakowsky (1957) in somewhat similar CH_2 experiments with ethylene, in which the stable end product is propylene. In the reaction of CH_2 with ethane to yield propane, Knox and Trotman-Dickenson (1957) see no evidence of intermediate species formation.

As noted previously in regard to the studies of Roberts *et al.*, such ephemeral cyclics might be prepared and stabilized more effectively by using the trapped radical approach.

Although N as generated by electrical discharge prefers to form HCN as the main product with practically any molecule containing hydrogen, Aronovich and Mikhailov (1956) have found that addition reactions also occur to some extent, especially with aromatics or olefins. Benzene gives some pyridine, C_6H_5CN, and C_6H_5NC. Naphthalene yields a little quinoline and isoquinoline. Solid compounds of unknown structure containing as much as 16 % nitrogen are also formed. It would appear that much more clean-cut reactions, perhaps minimizing the HCN-producing reaction, might be obtained via the stabilized radical technique. It is not entirely chimerical to anticipate a practical process for synthesizing difficult nitrogen-containing compounds, perhaps pharmaceuticals, via a trapped N-atom method.

A notable series of studies was carried out by Kharasch and associates (1953) on the reactions of atoms and free radicals in solution. One especially intriguing reaction is the peroxide-generated free radical-induced addition of cyclohexanone to 1-octene. In view of the great technical importance of the organic peroxides and hydroperoxides as sources of free radicals, it would be interesting to expose these materials to photolysis or energetic radiation under solidified cryogenic conditions. A considerable wealth and variety of cryogenic radicals could be expected from such treatment.

G. PARTIAL OXIDATION

A good deal of the extremely voluminous literature on partial oxidation processes is concerned more or less directly with investigating what might be called the "prelude to combustion." It has been hoped by a gradual approach to the severity and the wildly branched chain mechanisms of flame chemistry to "sneak up" so to speak, on the secret of this great problem. However, in the application of free radical principles a goodly amount of by-product chemistry has also been developed.

A very recent and profitable example is the series of papers published by Coffman and his associates (1958), some of which deal with novel utilizations of the OH radical—a favorite in combustion science. In this work OH is generated by the reaction of ferrous sulfate and H_2O_2 (Fenton's reagent), $H_2O_2 + Fe^{++} \rightarrow OH + OH^- + Fe^{+++}$. It has been found that very efficient dimerization reactions can be produced of the type

$$RH + OH \rightarrow R + H_2O$$

$$2R \rightarrow R-R$$

This is found to be applicable to carboxylic acids, nitriles, amines, amides, alcohols, and ketones. The OH radical in the presence of 1,3-diene will initiate "additive dimerizations" to give novel long-chain compounds (Coffman and Jenner, 1958). Other unusual additive dimerizations can be more effectively caused by free radical sources such as organic peroxides.

In the last examples cited we are in a realm of trigger functions, where trapped free radicals begin to emerge from the academic grove and make future economic sense.

Cruising lightly through the overabundance of literature, one finds here and there other examples of possible approaches by the trapped radical route.

Mullen and Skirrow (1958) in the study of the gas phase partial oxidation of olefins find an inhibiting effect of excess olefin ascribed to relatively unreactive allyl radicals. Cryogenic isolation would be interesting.

In a study of the oxidation of methyl radical at room temperature, Calvert and Hanst (1959) are able to explain the main products only by postulating the reaction

$$CH_3OO + O_2 \rightarrow CH_3O + O_3$$

In this case CH_3OO is a "hot" vibrationally excited radical. This is a theoretical recourse of increasing popularity. The exploration of such hypothetical hot free radicals under trapped conditions would indeed be a fascinating study. They are extremely difficult to detect spectroscopically under normal conditions. (This kind of reaction, incidentally, is important to the Los Angeles smog problem.)

In partial oxidations, wall effects are likely to be very significant, since the surface can act as a free radical sink. As emphasized by Badriyan and Furman (1956) the homogeneous reaction sequence can be favored, as in the oxidation of propane to alcohols, by increasing pressure which retards alkoxy radical breakdown on the wall.

The negative temperature coefficients in some partial oxidations over critical ranges have long puzzled many investigators, especially near the "cool flame" region. This is attributed to peroxy-radical decomposition and to disproportionation of hydrocarbon free radicals (Orchin and Swarts, 1956).

Ethylene oxide as a source of very active free radicals, such as CH_2OCH_2, is a likely candidate for the cryogenic approach (Zimakov, 1955). Its rapid decomposition above 400°C has been used in gas generation.

Gray (1956) has established through a clever study of nitrites, nitrates, and peroxides, the enthalpies of formation of various oxygen-containing radicals, many of them of considerable technical significance. These vary from $+5$ kcal per mole for OH to -30 kcal per mole for tertiary AmO, all in the gas state at 25°C. Similar data have been obtained by Luft (1956).

The work of Cvetanovic (1956) on O-atom reactions with olefins is very suggestive because of the wide variety of products formed by the individual unsaturates. Some oxidations, performed with either O in a solid matrix or the hydrocarbons in solid form, would be expected to yield much additional information, and perhaps low temperature trigger effects would be observed.

IV. Combustion

The various fast chemical processes included under this term represent some of the most contentious scientific enigmas and, at the same time, as a whole probably add up technically and basically to the "number two" reaction complex of our planet. (Photosynthesis is probably the number one process and, at the rate the bioscientists are going, this may be reduced to understandable terms before combustion is thoroughly understood, this in spite of the enormous congregation of both scientific and engineering brains that have been focused on it.)

Two salient features characterize combustion phenomena: (1) it is an extremely fast process, and therefore eludes precise and leisurely analytical methods; and (2) from all we now know of modern concepts, it unquestionably involves branched free radical chains.

It is my personal conviction that the availability of techniques for trapping small and active radicals now offers combustion science its greatest opportunity since the reaction chain propagation theories earned for such brilliant pioneers as Semenov and Hinshelwood the formal accolade so justly deserved by them.

Let us examine, for instance, one facet only of flame theory; the interminable and perhaps essentially semantic tug of war between the proponents of the "heat transfer" and the "radical transfer" (Tanford, 1949) explanations of flame propagation. This is covered in all texts on combustion (for example, Lewis and von Elbe, 1951; Jost, 1946) and has resulted in the recruitment of innumerable legions of investigators eager to set up a critical experiment which would decide the issue. The issue is still in doubt. Perhaps there *is* no issue. Moreover (and more importantly), there is still no systematic dogma that would teach the unlearned to predict the flame speed, laminar or turbulent, of a given chemical compound from its structure, its bond energies, its spectroscopy, its viscosity, or its anything. Obviously this old and somewhat weary discipline needs a "shot-in-the-arm." Could not trapped radical technology provide this critical infusion?

How would one go about this? Assuming a good deal of favorable extrapolations in the trapping and disengagement methods for supposedly crucial radicals, such as H, OH, HO_2, etc., one would develop a technique for the introduction of such pedigreed radicals into selected zones of a combustion system and would follow as quantitatively as possible the effects of such introduction. However, the definition of a combustion system need not presuppose a vigorous premixed or diffusion-type flame. One would want to start with experiments designed to see if one could start a fire at cryogenic temperatures. Relationships could perhaps be established that would show a continuous extrapolation possible to more meaningful temperature levels. Level by level, as skill in the trapping and untrapping techniques accumulated, one might be able to pierce the secrets of flame, so to speak, from its soft (and cold) underbelly.

Discussions of combustion seem inevitably to precipitate one into metaphor or insoluble partial differential equations. It may be appropriate now to conduct another high light cruise through the unmanageably vast and discordant literature. Again the selections are private ones and reflect only personal impressions of what seems pertinent to our theme.

Recent work by Kaskan (1958) on lean hydrogen-air flames held on porous burners (a justifiably popular technique) has demonstrated the vital fact that the OH radical is always in excess of equilibrium ($2OH \rightleftarrows H_2O + O$) near the flame front but decays downstream. When H_2-rich flames are studied, the same OH excess is encountered but the very fast equilibrium $H_2 + OH \rightarrow H_2O + H$ enters in. Both radicals decay by three-body recombinations.

One of the outrageously unexplainable problems of combustion systems is the electron population of flames; it is too high under any reasonable hypothesis. It is quite possible that this indeed represents a crucial dilemma that is connected with our general incapacity to predict flame events. Perhaps in the elaborate area of free radical - ion interactions we may discover a touchstone. This also would seem a problem amenable to the trapping, untrapping approach. Knewstubb and Sugden (1956) have made crucial contributions here, as have Calcote and King (1955).

Reference has been made to free radical reactions in the operation of jet combustors. An equally powerful case could be made for studies in rocket combustion. I do not refer here to the notion of preparing a solid tank of high free radical concentration, the recombination behavior of which would lead to powerful heat release effects, but to the role of free radical reactions in "normal" rocket combustion systems, both of the liquid and solid propellant type. In liquid rocket systems, in which nitric acid is used as oxidizer, a very helpful review with respect to the free radical aspect emphasizes the importance of the reactions (Casaletto, 1957). Of indirect pertinence to such processes as $HNO_3 \rightarrow OH + NO_2$ and $NO_2 \rightarrow NO + O$ are the data reported by Brown and Pimentel (1958) on the photolysis of nitromethane and of methyl nitrite in an argon matrix, with identification of the nitroxyl radical HNO.

Radiation of combustion systems *in situ* has been carried out by Churchill and Weir (1955), using Au^{198} and Au^{199}. Rates of flame propagation were increased proportionately to an increase of emission of C_2 and OH, which appeared earlier in the flame than the CH maximum. This supports faith in the well-known hypothetical reaction for CO production by

$$C_2 + OH \rightarrow CH + CO.$$

Among many curious effects in the so-called "cool flame" phenomena of combustion, which warrant consideration for cryogenic research, is the fact that a little argon added to a mixture which normally exhibits a cool flame will change the cool combustion to hot burning (Myerson *et al.*, 1956). Neu (1956) has found that the products during the cool flame of n-butane are mostly olefins, whereas just before the appearance of cool flame they tend to be oxygenated materials. This is important from a technical petrochemical standpoint. I shall not discuss the cool flame or other "prereactions" as related to automobile engine "knock." Although it is quite obvious that knock reactions involve free radical effects, the application of trapping techniques to this phenomenon is too indirect to suggest usefully specific experiments. Knock is inherently a "scale" effect (very small fast engines do not experience it) and thus is connected to a large degree with parameters extraneous to cryogenic studies.

Surface effects in combustion appear to be a field in which resourceful planning of free radical trapping research should offer great promise. The phenomenon of "surface combustion" itself has in fact never been comprehensively examined since the Semenov-Hinshelwood revolution in chemical kinetics.

Among the work of a great many others, the recent treatment by Walsh (1958) of the effect of different categories of surfaces on the slow oxidation of methane and the second explosion limit of hydrogen is especially cogent. One category (silicic acid, boric acid, phosphoric acid) favors the reduction of the HO_2 radical to H_2O_2 by proton transfer; the other category (most metal oxides and salts) favors the decomposition of HO_2 to H_2O and O_2 by donation of an electron. Although Walsh extrapolates his considerations to an explanation of the effect of tetraethyl lead as an antiknock, we can be content for our purposes with pointing out the advantage of studying these different surface structures as cryogenic matrices for HO_2 and other important free radicals. Results similar to those of Walsh and his associates have been obtained by others (Dalmai *et al.*, 1957; Warren, 1957).

There appears to be a growing feeling among combustion physicists that at least a part of the surface combustion capability of platinum is due to its peculiar ability to adsorb free O atoms and to combine them catalytically. Fryburg (1956) has studied systems of platinum and moist activated oxygen. Eyraud *et al.* (1956) passed air or oxygen over a hot platinum filament, then mixed it immediately with methane, obtaining much lower methane oxidation rate than when methane and oxygen were simultaneously passed over the catalyst. Obviously such a problem recommends itself to the cryogenic methods.

V. Aerodynamic Chemistry

The illustrious and versatile Imperial College investigator, Ubbelohde (1958), has expressed his satisfaction that recent trends in the study of detonation systems are taking a chemical, rather than the usual formalistic fluid dynamics, approach.

One may see the beginnings of a similar shift in emphasis in a field of inquiry which promises in the decades ahead to achieve major status in the massive hierarchy of problems which face future applied science—the problem of supersonic air.

Although on first glance, it would seem thoroughly implausible that predominantly cryogenic experiments with free radicals and ions could even remotely have anything to offer toward the elucidation of very hot and extremely fast systems, such as the chemical reactions taking place around intense shock waves associated with an object possessing velocities in the range of Mach 7.0 to Mach 25, it is the personal opinion of this author that it may be only by the use of these "frozen" or "slow motion" techniques that we will be able to understand in a fundamental way the extraordinarily complicated chemistry involved. Fast spectroscopy, high response ionprobes, all the paraphernalia of microsecond systems are needed, of course, but in such regimes they have reached the ragged edge of useful precision. If some degree of relationship or the validation of vast time-temperature extrapolations can be established between the cryogenic and the hypersonic situations, we can look to very exciting progress indeed.

One aspect of the problem is the surface catalysis of the recombination of radicals and ionic species and how it is affected by the chemistry of the surface. A bit of literature orientation may be indicated at this point.

Hirschfelder (1956) in a useful review has pointed out the non-Fourier character of heat conduction at the surfaces of high-speed missiles because of the presence of metastable electronic or molecular species in the gas mixture. Heat fluxes are observed which depend to a very sensitive degree on the boundary conditions and cannot be expressed in terms of an effective coefficient of thermal conductivity. When it is realized that at Mach 15 about 60 % of the total energy of a missile may be consumed in dissociating air, the magnitude of the heat transfer situation may be realized.

In attempting the spectroscopic identification of free radicals formed in such systems, Cornell Aeronautical Laboratory by the use of single-pulse shock tube techniques, has gone about as far as present instrumentation permits (Wurster, 1958). The same facility has published useful tables (Logan, 1956) summarizing the known thermodynamic properties of air at temperatures from 1000°K to 20,000°K, assuming at higher temperatures the components: O_2, O, N, O^+, N^+, e, NO, NO^+, N_2, O^-, N_2^+, O_2^+, and A. Cornell investigators have emphasized that the specific heat of air at constant pressure will exceed the ideal gas value by factors of about 12 at 7000°K.

Dommett (1956) has emphasized the little-realized effect of ionization (at temperatures above 10,000°K) on the viscosity and on self-induced magnetoelectric fields and forces. Thus at very high speeds the overworked aerodynamicist must also qualify himself in the fresh and hideous complications of magnetohydrodynamics!

Of particular interest to our special theme, because it stresses catalytic recombinations of free radicals, is a paper by Lees (1956). Hansen (1957), Stalder (1957), and Eggers (1957) describe such typical shock layer events as direct recombination of radicals leading to an excited molecule and the airglow radiation and three-body collisions between two oxygen atoms and a third particle. Since at the altitudes associated with high-speed missile flight, there is considerable residence time in the E layer of ionization, O ions can form and react with O atoms to yield the molecular state and a free electron—a sort of chain process which needs much more study. Note also that at escape velocity, a molecule of nitrogen strikes with 18.2 ev energy, whereas the vaporization energy of iron is but 4.2 ev per atom. Thus in addition to the bulk surface effect, we may have catalytic amounts of vaporized or finely divided metal in the shock layer. This effect may be still more immensely complicated by use of the successful ablation method of protecting missile nose cones in which the evaporation and exothermic decomposition of sacrificial coatings are used to absorb heat. Such "clouds" of coating material evidently confer the additional benefit of holding at a distance the recombination of O atoms so that the usually large, although difficult-to-calculate variety of heat transfer increase due to the surface recombination effect is considerably reduced.

In the ion–free radical–molecule interaction field, which is a matter of immediate cruciality in missile aerodynamic chemistry, the development of the stabilized arc and of other processes for preparing hot, high velocity plasmas has broadened the possibilities of laboratory study. Drastic improvements in quick-quenching and cryogenic probe sample stabilization should offer another road to the large increments of knowledge needed. Perhaps the trapping technique, however, could best be first applied in the study of surface catalysis of O-atom recombination. This is a broad, attractive, and practically a virgin pasture, although it does have something in common with the surface combustion phenomena mentioned above briefly and the problem of solid catalyst fundamentals to be discussed later.

VI. Solid Catalysts

Although a good deal of brilliant work has been devoted to the basic investigation of solid catalyst mechanisms, it cannot be said that chemists have a sufficiently productive theory to enable them to proceed with solid catalyst development on other than an Edisonian plan. This has become a source of annoyance and even of intellectual scandal. With all the new physical tools available (nuclear magnetic resonance, electron spin resonance, high-resolution surface infrared spectroscopy, etc.), why do we know so little? Why are we able to predict so little?

Is it not possible that we need an entirely new approach and that the frozen free radical technology now in its early adolescence could provide such an approach? The chances appear to be favorable that it could. In line with the defect-structure studies of the solid state, cryogenically studied matrices in which small free radicals find a protective place to hide, would seem to offer once again a slowed-down system susceptible to more precise measurements than the usual hot investigations.

Again a skeletonized résumé of recent and pertinent literature is in order.

What Tolkachov and Trofimov (1955) refer to as "half-reduced" atoms in the technically very important difficultly reduced oxide catalysts such as Cr_2O_3, V_2O_5, and ZnO, are the active catalytic centers, according to their conclusions, which are based on luminescence data. The crystalline phase itself is catalytically inactive and functions only as a sort of matrix for the "atom" phase. Closely related to it are the considerations of Kobozev and Mal'tsev (1955) on platinum systems, in which single atom centers are interpreted as forming on carrier regions having high reflectance and luminescent power, such platinum atoms being associated with a high degree of paramagnetism. The active atomic sites are superior in catalytic power to the bulk metal. A notable contribution to luminescence and the theory of "exo-electrons" initiated from high energy spots in catalysts has been made by Nassenstein (1955). "Trap distribution" on a catalyst surface can be plotted by this means.

According to Block (1956) defect structure catalysts, whether p-type or n-type semiconductors, function only when electronic transfer to and from the catalyzed system is a rate-controlling process. That this may not always be the case and that internal molecular arrangements without electron transfer may become rate-determining is too easily overlooked in modern catalyst theory. Nevertheless the correlation between lattice imperfection and catalytic activity is very impressive, and indeed Krause (1956a) has emphasized that degree of lattice imperfection may be measured more sensitively by catalysis than by the conventional methods.

Interstitially built-in H atoms are among the "defects" considered by Ruchkina (1956) in the catalytically active palladium surface formed on platinum by electrolysis of $PtCl_2$. This work certainly deserves careful scrutiny and calls attention to the possibilities of close "tailoring" of defects by electrochemical means.

Accommodation mechanisms are, of course, importantly concerned in solid catalysis. The attention being given to accommodation coefficients in cryogenic free radical trapping shows possibilities of useful convergence in this area.

The technically important field of catalyst poisoning deserves some new method of research attack and would appear to be especially amenable to cryogenic techniques. Why does a trace of water, insufficient to form even a monomolecular layer, often result in complete deactivation of some systems? Such facts constitute obviously significant arguments for the active-center hypothesis. From the adsorption standpoint, means of distinguishing between physical and chemical adsorption may be particularly useful in the poisoning problem. The work of Kipling and Peakall (1957) on the adsorption of water vapor and alcohols in alumina, silica, and titania gels is enlightening and indicates chemisorption is possible only if oxide ions are present on the surface. Another relatively new concept is the occurrence of physical adsorption on the chemisorbed layer.

In an excellent review of work at Princeton University on catalyst fundamentals, Taylor (1956) has emphasized the rate-determining importance of adsorbed hydrogen atoms in various hydrogenation reactions. Note also that Harrison and McDowell (1953) find that solid free radicals markedly accelerate hydrogen-deuterium exchange reactions in the case of ZnO. Such observations could obviously be sharpened considerably in cryogenic experiments. Very little low temperature catalytic work has been done, but one should bear in mind that there has been some. Attention is called, for example, to the studies of Krause (1956b) in which it was found that ferric hydroxide at liquid air temperature showed a substantial increase in catalytic activity.

One possible aspect of catalytic mechanism that *cannot* be directly observed in cryogenic work should be mentioned. Although chemists have hitherto tended rather laxly to view the vaporization of metal oxides and hydroxides as a rather precarious process, probably involving at least partial decomposition, recent research by Atomic Energy Commission laboratories, which are naturally very much concerned with such phenomena, discloses that polymerization rather than dissociation may be the rule. For example, the vapor in equilibrium with powdered MoO_3 is predominantly composed of Mo_3O_9, Mo_4O_{12}, and Mo_5O_{15} (Berkowitz et al., 1957).

Finally there is one recently announced effect the elucidation of which appears ready-made for application of trapping methods by free radical chemical physicists. This is the evident enhancement of rate and extent of gas-solid chemisorption reactions by gaseous electrodeless discharge (Low and Taylor, 1957). This is especially pertinent to the practical application of catalysis in hydrogenation systems.

REFERENCES

Aronovich, P. M., and Mikhailov, B. M. (1956). *Izvest. Akad. Nauk S.S.S.R., Otdel. Khim. Nauk*, **1956**, 544. (*Chem. Abstr.* **51**, 1892, 1957.) Action of active nitrogen on organic compounds.

Badriyan, A. S., and Furman, M. S. (1956). *Doklady Akad. Nauk S.S.S.R.* **108**, 861. Pressure effects on the formation of intermediate propane-oxidation compounds.

Baum, L., Graff, H., Hormats, E. I., and Moe, G. (1956). *ASTIA Document No.* **AD 95 432**. Research on ultra-energy fuels for rocket propulsion.

Berkowitz, J., Inghram, M. G., and Chupka, W. A. (1957). *J. Chem. Phys.* **26**, 842. Polymeric gaseous species in the sublimation of molybdenum trioxide.

Block, J. (1956). *Werkstoffe u. Korrosion* **7**, 127. (*Chem. Abstr.* **50**, 17255, 1956.) Selection of material for contact catalysis. II. Behavior and properties of semiconducting catalysts.

Brown, H. W., and Pimentel, G. C. (1958). *J. Chem. Phys.* **29**, 883. Photolysis of nitromethane and of methyl nitrite in an argon matrix; infrared detection of nitroxyl, HNO.

Burton, M., and Magee, J. L. (1955). *J. Chem. Phys.* **23**, 2195. Successive excitation by low-energy electrons: application to electric discharge in methane.

Buttery, R. G., Doering, W. von E., Laughlin, R. G., and Chaudhuri, N. (1956). *J. Am. Chem. Soc.* **78**, 3224. Indiscriminate reaction of methylene with the carbon-hydrogen bond.

Calcote, H. F., and King, I. R. (1955). "Fifth Symposium on Combustion," p. 423. Reinhold, New York.

Calvert, J. G., and Hanst, P. L. (1959). *J. Phys. Chem.* **63**, 71. The oxidation of methyl radicals at room temperature.

Casaletto, G. (1957). *ASTIA Document No.* **AD 140 077**. Research on nitric acid-supported combustion.

Churchill, S. W., and Weir, A., Jr. (1955). *ASTIA Document No.* **AD 90 142**. Combustion studies with a 1200-Curie gold source.

Coffman, D. D., and Jenner, E. L. (1958). *J. Am. Chem. Soc.* **80**, 2872. Syntheses by free-radical reactions. II. Additive dimerizations effected by hydroxyl radicals.

Coffman, D. D., Jenner, E. L., and Lipscomb, R. D. (1958). *J. Am. Chem. Soc.* **80**, 2864. Syntheses by free-radical reactions. I. Oxidative coupling effected by hydroxyl radicals.

Cvetanovic, R. J. (1956). *J. Chem. Phys.* **25**, 376. Mechanism of the interaction of oxygen atoms with olefins.

Dalmai, G., Delbourgo, R., and Laffitte, P. (1957). *Compt. rend.* **244**, 897. Influence des parois sur les domains d'inflammabilité des mélanges de butane et d'air aux basses pressions.

Dommett, R. L. (1956). *ASTIA Document No.* **AD 115 386.** Thermodynamic properties of air at high temperatures.

Eggers, A. J., Jr. (1957). *Jet Propulsion* **27,** 1147. Performance of long range hypervelocity vehicles.

Eyraud, C., Domanski, B., and de Mourgues, L. (1956). *Bull. Soc. chim. France* **1956,** 808. Catalytic activation of platinum wire. Combustion of the methane-air mixture.

Foner, S. N., and Hudson, R. L., Jr. (1955). *J. Chem. Phys.* **23,** 1974. OH, HO_2, and H_2O_2 production in the reaction of atomic hydrogen with molecular oxygen.

Fontana, B. J. (1958). *J. Appl. Phys.* **29,** 1668. Thermometric study of the frozen products from the nitrogen microwave discharge.

Frey, H. M., and Kistiakowsky, G. B. (1957). *J. Am. Chem. Soc.* **79,** 6376. Reactions of methylene. I. Ethylene, propane, cyclopropane and n-butane.

Fryburg, G. C. (1956). *J. Chem. Phys.* **24,** 175. Enhanced oxidation of platinum in activated oxygen.

Gascoigne, R. M. (1958). *J. Chem. Soc.* p. 876. Entropy changes in isomerism.

Geib, K. H., and Harteck, F. (1933). *Ber.* **66B,** 1815. Anlagerungs-Reaktionen mit H- und O-Atomen bei tiefen Temperaturen.

Geib, K. H., and Harteck, F. (1934). *Z. physik. Chem. (Leipzig)* **A170,** 1. (*Chem. Abstr.* **28,** 7121, 1934.) Oxidation reactions with the aid of uncoupled hydrogen atoms.

Gordon, A. S., and McNesby, J. R. (1957). *J. Am. Chem. Soc.* **79,** 5902. Reactions of CD_3 radicals with the butenes.

Gray, P. (1956). *Trans. Faraday Soc.* **52,** 344. The chemistry of free radicals containing oxygen. Part 1. Thermochemistry of the alkoxy radicals RO and dissociation energies of oxygen bonds.

Hansen, C. F. (1957). *Jet Propulsion* **27,** 1151. Some characteristics of the upper atmosphere pertaining to hypervelocity flight.

Harrison, L. G., and McDowell, C. A. (1953). *Proc. Roy. Soc.* **A220,** 77. The catalysis of the *para*-hydrogen conversion by the solid free radical $\alpha\alpha$-diphenyl-β-picryl hydrazyl.

Harrison, L. G., and McDowell, C. A. (1955). *Proc. Roy. Soc.* **A228,** 66. The catalysis of the *para*-hydrogen conversion by zinc oxide.

Hill, G. R. (1956). *ASTIA Document No.* **AD 109 608.** Kinetics and thermodynamics of combustion.

Hinshelwood, Sir Cyril (1956). *Science* **124,** 708. Amedeo Avogadro.

Hirschfelder, J. (1956). *ASTIA Document No.* **AD 105 287.** Heat conductivity in polyatomic electronically excited or chemically reacting mixtures.

James, D. G. L., and Steacie, E. W. R. (1958). *Proc. Roy. Soc.* **A244,** 289. Reactions of the ethyl radical. I. Metathesis with unsaturated hydrocarbons.

Jost, W. (1946). "Explosion and Combustion in Gases." McGraw-Hill, New York.

Kaskan, W. E. (1958). *Combustion and Flame* **2,** 286. The concentrations of hydroxyl and of oxygen atoms in gases from lean hydrogen-air flames.

Kharasch, M. S., Kuderna, J., and Nudenberg, W. (1953). *J. Org. Chem.* **18,** 1225. Reactions of atoms and free radicals in solution. XXXIII. Photochemical and peroxide-induced addition of cyclohexanone to 1-octene.

Kipling, J. J., and Peakall, D. B. (1957). *J. Chem. Soc.* p. 834. Reversible and irreversible adsorption of vapours by solid oxides and hydrated oxides.

Klein, R., and Scheer, M. D. (1958). *J. Phys. Chem.* **62**, 1011. The reaction of hydrogen atoms with solid olefins at −195°.

Klein, R., and Scheer, M. D. (1959). *J. Chem. Phys.* **31**, 278(L). Reaction of hydrogen atoms with solid oxygen at 20°K.

Knapsack-Griesheim, A.-G. (1957). *Chem. Eng. News* **35**, (24) 80. More European acetylene processes.

Knewstubb, P. F., and Sugden, T. M. (1956). *Research Correspondence, "Research"* **9**(8), S32. Ionization produced by compounds of lead in flames.

Knox, J. H., and Trotman-Dickenson, A. F. (1957). *Chem. & Ind. (London)* p. 268. The reactions of methylene.

Knox, J. H., Trotman-Dickenson, A. F., and Wells, C. H. J. (1958). *J. Chem. Soc.* p. 2897. Reactions of methylene with isobutene.

Kobozev, N. I., and Mal'tsev, A. N. (1955). *Zhur. Fiz. Khim.* **29**, 291. (*Chem. Abstr.* **50**, 16316, 1955.) Active centers in hydrogenation of ethylene. II. Monoatomic hydrogenation centers.

Krause, A. (1956a). *Roczniki Chem.* **30**, 1047. Latent lattice imperfections and their importance in the investigations of contact catalysts.

Krause, A. (1956b). *Roczniki Chem.* **30**, 1095. The effect of low temperatures on structure and catalytic activity of certain ferric hydroxides.

Lees, L. (1956). *Jet Propulsion* **26**, 259. Laminar heat transfer over blunt-nosed bodies at hypersonic flight speeds.

Lewis, B., and von Elbe, G. (1951). "Combustion, Flames, and Explosions of Gases." Academic Press, New York.

Logan, J. G., Jr. (1956). *ASTIA Document No.* **AD 95 200**; The calculation of the thermodynamic properties of air at high temperature.

Low, M. J. D., and Taylor, H. A. (1957). *J. Electrochem. Soc.* **104**, 439. Enhanced surface reactions.

Luft, N. W. (1956). *Z. Electrochem.* **60**, 94. Bildungswärmen der Alkoxy-Radikals und OO-Dissoziationsenergien.

Mullen, J. D., and Skirrow, G. (1958). *Proc. Roy. Soc.* **A244**, 312. Gas-phase oxidation of propylene.

Myerson, A. L., Taylor, F. R., and Faunce, B. G. (1956). *ASTIA Document No.* **AD 110 509**. Studies of the ignition limits and multistage flames of the system propane-nitrogen dioxide-nitric oxide.

Nassenstein, H. (1955). *Z. Naturforsch.* **10a**, 944. Electron emission from surfaces of solids after mechanical treatment and irradiation.

Neu, J. (1956). *J. Phys. Chem.* **60**, 320. Infrared spectrographic studies of preflame reactions of n-butane.

Norrish, R. G. W., and Purnell, J. H. (1958). *Proc. Roy. Soc.* **A243**, 449. The decomposition of n-hexane. II. By reaction with atomic hyrdogen.

Orchin, M., and Swarts, D. E. (1956). *Natl. Advisory Comm. Aeronaut., Tech. Note* **3579**. Vapor-phase oxidation and spontaneous ignition. Correlation and effect of variables.

Pickup, K. G., and Trapnell, B. M. W. (1956). *J. Chem. Phys.* **25**, 182. Recombination of hydrogen atoms at metal surfaces.

Roberts, J. D., Streitwieser, A., Jr., and Regan, C. M. (1952). *J. Am. Chem. Soc.* **74**, 4579. Small-ring compounds. X. Molecular orbital calculations of properties of some small-ring hydrocarbons and free radicals.

Ruchkina, V. D. (1956). *Izvest. Akad. Nauk S.S.S.R., Ser. Fiz.* **20**, 761. X-ray investigation of the dependence of the structure of a palladium catalyst on the conditions of electrical formation and thermal treatment.

Ruehrwein, R. H., Hashman, J. S., and Edwards, J. W. *J. Phys. Chem.* (In press.) Chemical reactions of free radicals at low temperatures.

Schoch, E. P. (1950). *Texas, Univ. Publ. No.* **5011.**

Shuikin, N. I., Berdnikova, N. G., and Novikov, S. S. (1953). *Problemy Mekhanizma Org. Reaktsii, Akad. Nauk Ukr. S.S.R., Otdel. Fiz.-Mat. i Khim. Nauk* **1953,** 24. Contact-catalytic transformations of five- and six-membered cyclanes under conditions of elevated temperatures and pressure of hydrogen.

Stalder, J. R. (1957). *Jet Propulsion* **27,** 1178. A survey of heat transfer problems encountered by hypersonic aircraft.

Steacie, E. W. R. (1954). "Atomic and Free Radical Reactions." Reinhold, New York.

Stepukhovich, A. D. (1954). *Zhur. Fiz. Khim.* **28,** 2088. Equilibria in free radical reactions. (*Chem. Abstr.* **50,** 16294, 1955.)

Stepukhovich, A. D. (1956). *Zhur. Fiz. Khim.* **30,** 556. The kinetics and mechanism of cracking paraffin hydrocarbons. (*Chem. Abstr.* **50,** 13575, 1955.)

Stepukhovich, A. D., and Derevenskikh, L. V. (1955). *Zhur. Fiz. Khim.* **29,** 2129. Kinetics and mechanism of decomposition of hydrocarbons. VII. The kinetics and the mechanism of the gaseous decomposition of alkanes in the presence of propadiene. (*Chem. Abstr.* **50,** 13711, 1955.)

Stevenson, D. P. (1951). *Discussions Faraday Soc.* **10,** 35. Ionization and dissociation by electronic impact.

Swegler, E. W., and Weisz, P. B. (1957). *Science* **126,** 31. Stepwise reaction on separate catalytic centers: isomerization of saturated hydrocarbons.

Szwarc, M., Bader, A. R., Buckley, R. P., and Leavitt, F. (1957). *J. Am. Chem. Soc.* **79,** 5621. Addition of methyl radical to *cis* and *trans* isomers.

Tanford, C. (1949). "Third Symposium on Combustion and Flame and Explosion Phenomena," p. 140. Williams & Wilkins, Baltimore, Maryland.

Taylor, H. (1956). *ASTIA Document No.* **AD 112 551.** Solid state properties and catalytic activity.

Tolkachov, S. S., and Trofimov, A. K. (1955). *Doklady Akad. Nauk S.S.S.R.* **104,** 54. An investigation of the $\gamma \rightarrow \alpha Al_2O_3$ polymorphic transformation by the luminescence spectra. (*Chem. Abstr.* **50,** 7597, 1955.)

Trotman-Dickenson, A. F. (1955). "Gas Kinetics." Butterworths, London.

Ubbelohde, A. R. (1958). "Seventh Symposium on Combustion." Butterworths, London.

Union Carbide Research Institute (1958). Unpublished information.

Urry, W. H. (1959). "Chemistry of Carbenes." Am. Chem. Soc. Meeting, Southeast Texas Section.

Walling, C. (1957). "Free Radicals in Solution." Wiley, New York.

Walsh, A. D. (1958). "Seventh Symposium on Combustion." Butterworths, London.

Warren, D. R. (1957). *Trans. Faraday Soc.* **53,** 199. Surface effects in combustion reactions. Part 1. Effects of wall coating on the $H_2 + O_2$ reaction.

Windsor, M. (1959). (Unpublished results.) Temperature effect on molybdenum oxide as a detector for hydrogen atoms.

Wurster, W. H. (1958). *ASTIA Document No.* **AD 148,054.** Final report summarizing research in rate of high-speed reactions.

Zimakov, P. V. (1955) *Zhur. Fiz. Khim.* **29,** 76. Ethylene oxide behavior at higher temperatures.

AUTHOR INDEX

Numbers in italic indicate the pages on which the references are listed.

SUBJECT INDEX

A

Absorption spectra, *see* individual molecules

Acrylonitrile polymer,
trapped radicals in, 374

Active nitrogen,
atom concentrations, 260
characteristics, 259–260
electron concentration, 262
excited molecules, 263–269
mass spectrometry, 260
origin, 261, 268
reactions, 263

Active oxygen,
characteristics in solids, 285
mass spectrometric studies, 284

Addition reactions, 486–488

Afterglow,
see also Active nitrogen,
mechanism in solid nitrogen, 275

Alcohol radicals,
ESR spectra, 246

Alkyl radicals, 34

Alkylation, 486

Allyl radical,
ESR spectrum, 104
in solids, 104

Ammonia, *see* NH_3

Amorphous materials,
definition, 309
x-ray diffraction by, 310

Ar,
matrix, 17
x-ray diffraction, 318

Argon, *see* Ar

Aromatic radicals, 35
absorption spectra, 200
electron ejection, 201
photodissociation, 200
ring fission, 201

B

Benzyl radical,
production by photolysis, 85

B_2H_6 , x-ray diffraction, 324

Biological oxidation, 440

Bulk polymers,
trapped radicals in, 374–379

C

C_2 , 32
absorption spectrum, 192

C_3 , 32

Cage effect, 72, 95, 173
in radiation damage, 415

Carcinogenesis,
free radical effects, 468

Catalytic conversions, 480–490

Catalysts, solid, 495

CD_3 , ESR spectrum, 240

CD_4 , u.v. photolysis, 240

CF, 33

CF_2 , 33

CF_3 , 33

CH, 33
absorption spectrum, 192

CH_2 , 33
absorption spectrum, 196
production by photolysis, 90

CH_3 ,
absorption spectrum, 198
discovery, 5
ESR spectrum, 103, 239, 421
free rotation in matrix, 243
hindered rotation in matrix, 243
production by gamma rays, 421
in solid matrices, 100, 102

C_2H, 33

C_2H_5 ,
ESR spectrum, 103, 244
in solids, 102

C_3H_7 ,
ESR spectrum, 244

Chain mechanisms,
in enzymatic reactions, 456–458

Chain reaction, 337–339
critical concentrations, 339

$C_3H_5D_2I$,
u.v. photolysis, 244

in matrix, 16
in rapidly condensed solids, 75
Dislocations in crystals, 313
DNO,
  absorption spectrum, 197

E

Electric discharge,
  circuits, microwave, 56
  discharge parameters, microwave, 58
  discharge tubes, 50
  dissociation in, 48–50
  electrodeless, 53
  in hydrogen, 52, 61
  low frequency, 48, 51
  microwave, 55
  in nitrogen, 62
  in oxygen, 62
  radio-frequency, 52
  recombination in, 50
  thermal dissociation, 59
  water vapor, effect of, 61–63
Electron irradiation,
  of hydrogen, 230
Electron spin resonance,
  see ESR
Emulsion polymerization,
  trapped radicals in, 371
Energy of chemical reactions, 388
ESR,
  detection, 227
  double spin resonance, 228
  experimental techniques, 225
  field modulation, 227
  frequency stabilization, 226
  measurement techniques, 225
  microwave cavity, 226
  microwave field stabilization, 226
  microwave techniques, 225
  principles, 214–224
  sensitivity, 228
ESR hyperfine interaction,
  contact, 217
  dipolar, 219
  electronic spin, 221
  isotropic, 217
  magnetic, 217
  orbital moments, 221
ESR hyperfine spectrum,
  effect of crystal field, 222

intensities, 224
nuclear statistics, 224
ESR spectrum,
  coupling constant, 248
  g-factor, 248
  half-width, 249
  of irradiated biological materials, 461,
    464
  matrix effects on, 248
  of polymers, 368, 427
  in radiation damage, 420
  relaxation time, 249
Ethyl iodide, see C_2H_5I
Ethyl radical, see C_2H_5

F

Faults in crystals,
  stacking, 313
F-centers, 419
  in HN_3, 36
Flavoproteins,
  oxidation-reduction reaction, 448
Free radical propulsion,
  see Propulsion systems
f-values,
  of trapped radicals, 204

G

Gamma irradiation,
  of acids, 100, 228
  of $Ca(OH)_2$, 102
  of CH_4, 100, 103, 229, 239
  of C_2H_6, 103, 244
  of D_2, 100, 230
  of dimethylmercury, 103
  of formic acid, 102
  of glycine, 104
  of H_2, 100, 230
  of H_2O, 99, 102, 251
  of H_2SO_4, 100
  of N_2^{15}, 231
  of NH_3, 100
  of propylene, 104
Gas-handling equipment, 142
Gas thermometer, 152
Gel effect, 367
Graft polymerization,
  gel effect in, 380
  initiation by trapped radicals, 379
  irradiation effects, 379